Machine Tool
Technology

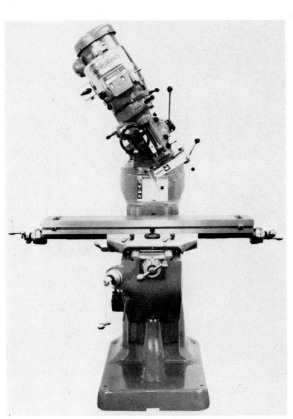

(Bridgeport Machines—A Textron Company)

(LeBlond Makino Machine Tool Company)

(Cincinnati Milacron)

Machine Tool
Tool
Technology

Victor E. Repp, Ed. D.
Professor of Manufacturing Technology
School of Technology
Bowling Green State University
Bowling Green, Ohio

Willard J. McCarthy
Associate Professor, Emeritus
College of Applied Science
and Technology
Illinois State University
Normal, Illinois

Glencoe McGraw-Hill

Glencoe/McGraw-Hill

A Division of The McGraw·Hill Companies

Printed in the United States of America

Send all inquires to:
Glencoe /McGraw-Hill
3008 W. Willow Knolls Drive
Peoria, IL 61614-1083

ISBN 0-02-671570-8 (Student Text)

16 17 18   026   05

Cover photos courtesy of **Valenite Division, The Valeron Corporation**

Front cover: A Valenite Val-U-Mil® Cutter performing a rough-milling operation on steel.

Back cover: (Lower photo) A semi-finishing turning operation on steel using a Valenite V-32-grade hot-pressed ceramic insert.

(Upper photo) Slot milling solid stock 1020 steel with a Valenite Centre-Dex® machining tool.

Preface to the Fifth Edition

This book provides the most comprehensive introduction to machine tool technology that is practical in a single volume. It provides content appropriate for a first course in machine tool processing, regardless of the level at which it is used. The content is generously illustrated, and the language used is simple and direct. While **Machine Tool Technology** has been prepared as a text, its comprehensive nature also makes it a valuable general reference.

Measurements throughout the book are given according to the SI Metric system of measurement, followed immediately by the U.S. customary equivalent in parentheses. These dual metric/customary measurements are intended to speed students' learning of the metric equivalents of familiar customary measurements. This feature also permits the continued use of the book throughout an extended period of conversion to the metric system.

Several content areas in this edition of **Machine Tool Technology** have been improved by selective revision. In Section 10, "Milling Machine Operations," vertical milling machine operations have been given more visibility. Section 10 also includes three new units, one of which provides instruction on drilling and boring holes in precise locations. The other two units introduce digital readouts and provide instruction on using a digital readout. In Section 11, "Shaping, Slotting, Planing, and Broaching Operations," the coverage of shaper operations has been reduced to its essentials, whereas the treatment of broaching operations has been expanded. Sections 15 and 16, "Nontraditional Machining Processes" and "Numerical Control Programming," have also been revised. For example, more detailed illustrations of EDM and ECM machines appear, and an introduction to computer numerical control capabilities has been added. Other changes have been made throughout the book as needed to bring the text up-to-date with recent technological developments.

General safety practices for working in machine shops are covered in Section 2. Additional safety practices and precautions concerning each basic machine tool or process are included in succeeding sections. Their presentation in bold color indicates their importance and attracts the attention of the reader.

Review questions are included at the end of each section. They are a valuable aid for study or review of the material. A new instructor's guide and revised study guides are also available for use with this new edition.

Acknowledgement and appreciation are expressed by the co-authors to Robert E. Smith, who wrote the first and second editions of this work.

Suggestions for improvements of future printings and editions of the book are welcomed.

Victor E. Repp
Willard J. McCarthy

Table of Contents

Section 1

Introduction to Machine Tool Technology

Section 2

Safety

Section 3

Measurement, Layout, Inspection, and Setup Tools

4 Section

Threads and Threading

5 Section

Tool and Utility Grinder Operation

6 Section

Stock Cutoff Machines

7 Section

Band Machining

Section 8

Drills and Drilling Machine Operations

Section 9

The Lathe and Lathe Operations

10 Section

Milling Machine Operations

11 Section

Shaping, Slotting, Planing, and Broaching Operations

Section 12

Abrasives and Grinding Wheels

Section 13

Abrasive Machining

Section 14

Cutting Action and Cutting Fluids

15 Section

Nontraditional Machining Processes

16 Section

Numerical Control Programming

17 Section

Metals and Alloys

Section 18

Basic Metallurgy and Heat Treatment of Steel

Section 19

Career Opportunities in Machine Tool Processing

Tables in Text

Tables in Appendix

Section

1

Introduction to Machine Tool Technology

Unit

1

Importance of Machine Tools

Why Machine Tools Are Important

Every product requires the use of metal cutting machine tools, either for its manufacture or for manufacture of the vehicles needed to get the product to market. Our modern civilization depends on machine tool technology. Hundreds of different types of machine tools have been developed to perform cutting operations. Tools range from simple hand operated machines to complex computer controlled machines capable of great precision, Fig. 1-1.

(LeBlond Machine Tool Company)

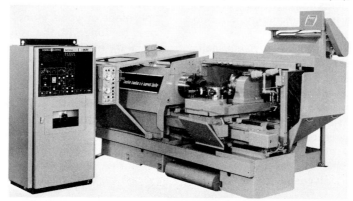

Fig. 1-1. Super precision—a lathe with a computer numerical control system.

Fig. 1-2. Machine tools produce machines and equipment for all major divisions of industry.

(General Motors Corporation)

Fig. 1-3. Specialized production tools are used on the auto production line.

Machine tools serve all types of industries, Fig. 1-2. Here is a partial listing of machine tool products:

1. Machinery and instruments for exploration, excavation, and raw material processing.
2. Transportation systems and materials handling vehicles.
3. Equipment for generation, distribution, and reception of power and communications.
4. Machinery and hand tools for manufacturing, assembling, and packaging.
5. Cylinders for rolling, pressing, embossing, and mixing.
6. Molds for forming metal, plastics, rubber, soap, food, and other molded objects.
7. Dies for stamping, forming, cutting, piercing, embossing, and coining.
8. Precision instruments and measuring devices.
9. Replacement parts for machinery, automobiles, and appliances.
10. Fasteners such as nuts, bolts, rivets, nails, screws, and mechanisms for cabinets, furniture, and doors.

If a country has a highly developed machine tool technology, its workers usually enjoy a high standard of living. Having precision machine tools that can produce interchangeable parts and mass produce those parts make that possible. Mass production machinery produces high quality products at low per unit cost, Fig. 1-3. This permits mass consumption. High productivity by machines has made high wages possible for workers. High income gives the workers more buying power than needed for food, clothing, and shelter. This is why America has one of the highest standards of living in the world.

Origin and Development of Machine Tools

Machine tools evolved from hand tools. The earliest known hand tools were made of wood, stone, animal bones, and animal teeth. They were operated entirely by human muscle power. Then, about 2400 B.C., when the Bronze Age began, people began to make hand tools of copper and bronze. Also, they learned to use animals to power some of the tools.

About 1000 B.C., hand tools were made of iron. This was the beginning of the Iron Age. These tools were of better quality than earlier tools made of wood, stone, or bronze. More work could be done with the iron tools and more products were made. Domesticated animals became commonplace and were more widely used to replace human power. Harnessing water power with the water wheel provided the **first great impetus** for the development of power driven machine tools and manufacturing machinery. In fact, the machine tools used to build the first steam engines were water powered, Fig. 1-4.

The first practical steam engine was invented by James Watt in 1776. This is generally ac-

(DoAll Company)

(South Bend Lathe, Inc.)

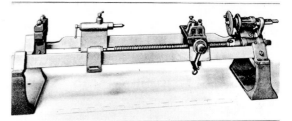

Fig. 1-5. Screw-cutting lathe invented by Henry Maudslay in 1797.

Fig. 1-4. A model of Wilkinson's boring machine. The original was used for machining the cylinders of James Watt's steam engine.

cepted as the beginning of the Industrial Revolution. For the first time in history, a source of power had been harnessed that could be located almost anywhere it was needed to do work. Thus, the invention and development of machine tools was given its **second great impetus.** The screw cutting lathe, invented by Henry Maudslay in 1797, became one of the first and most important of the power driven machine tools, Fig. 1-5.

Other inventions soon followed:
1. Roberts' metal cutting planer in 1817.
2. Whitney's milling machine in 1818.
3. Nasmyth's 'steel arm', the shaper, in 1836.
4. Nasmyth's power feed drill press in 1840.

5. The turret lathe by 1854.
6. The first automatic lathe by Spencer in 1893.

By 1910, all of the basic machine tools using mechanical cutting techniques had been developed.

Today's huge array of machine tools is far superior to those of the Industrial Revolution. Machine tools are built with greater power and durability. This makes possible **heavier cuts, greater speed,** and **accuracy.** Increasing numbers of machine tools are being controlled by built-in computers. The computers program the machines automatically to turn on and off, guide the tool path, change tools, and change speeds and feeds. In some cases, computerized machines adjust for tool wear, sense and replace broken tools, and even load and unload the workpiece. Figure 1-6 shows a machining center equipped with a tool changing system; machining and tool changing are computer controlled. Research and development will provide even better, more efficient machine tools.

(DeVlieg Machine Co.)

Fig. 1-6. Machining center with automatic tool-changing system and computer numerical control.

Unit 2

The Five Basic Arts of Machining Metal

Metal machining technology can be grouped into five basic machining techniques or arts:

1. Mechanical cutting, including use of metal or abrasive cutting tools.
2. Electrical erosion as in electrical discharge machining (EDM).
3. Electrochemical erosion as in electrochemical machining (ECM).
4. Chemical erosion as in chemical milling (CHM).
5. Controlled melting as in electron beam machining (EBM), laser beam machining (LBM), and plasma arc machining (PAM).

Specific machining operations such as drilling, reaming, turning, shaping, planing, milling, grinding, etc., can be classified under one or more of the above techniques.

Traditional mechanical machining operations can be performed on several different types of machine tools. However most of these operations are commonly performed on five basic machine tools:

1. The drill press.
2. The metalworking lathe.
3. The shaper or planer.
4. The milling machine.
5. The grinder.

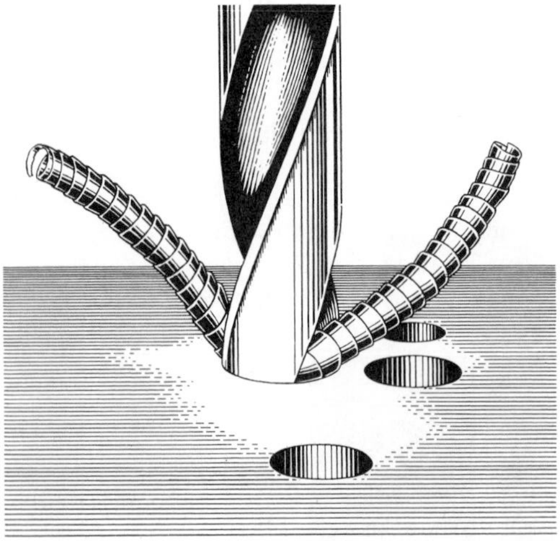

Fig. 2-1. Using a twist drill to machine a hole.

Fig. 2-2. Drilling with a floor model drill press.

(Clausing Corporation)

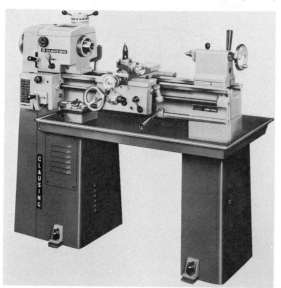

Fig. 2-3.   A modern toolroom lathe.

Each of the nontraditional machining techniques requires highly specialized equipment. Therefore nontraditional techniques can only be performed on machines designed specifically for that type of cutting.

The variety and number of combinations of machine tools in use today are almost unlimited. Some are so small they are mounted on a workbench. Others are as large as a three-story house. They range in cost from a few hundred dollars to hundreds of thousands of dollars. Some machine tools weigh several hundred tons and require a massive foundation for proper support. Large or small, most of the machine tools can be assigned to one of the **five basic techniques** of machining metal just listed.

Mechanical Cutting Techniques

Drilling

Drilling is performed with a rotating tool called a **drill.** Most drilling in metal is done with a twist drill, Fig. 2-1. The basic machine used for drilling is called a **drill press,** Fig. 2-2. Operations, such as **reaming** and **tapping,** are also classified as drilling. Reaming consists of removing a small amount of metal from a hole already drilled. The tool is called a **reamer.** Tapping is the process of

cutting a thread inside a hole so that a cap screw or bolt may be threaded into it. Drilling operations can also be performed on lathes, jig boring machines, milling machines, and on other special machines.

Turning and Boring

The basic machine used for turning is the **metalworking lathe,** Fig. 2-3. The lathe is commonly called the father of the entire machine tool family. For turning operations, the lathe uses a single-point cutting tool which removes metal as it travels past the revolving workpiece, Fig. 2-4. Turning operations are required to make many different cylindrical shapes, such as axles, gear blanks, pulleys, and threaded shafts. Turning operations are also performed on other specialized machines, such as turret lathes and automatic screw machines. Boring operations are performed to enlarge, finish, and accurately locate holes. A hole is bored by means of a single-point cutting tool which travels along inside the work, Fig. 2-5. When boring is done in boring and milling machines, the work is stationary while the boring tool revolves.

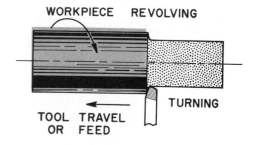

Fig. 2-4.   Turning on a lathe.

Milling

Milling removes metal with a revolving, multiple cutting edge tool called a **milling cutter,** Fig. 2-6. Milling cutters are made in many styles and sizes. Some have as few as two cutting edges and others have 30 or more. Milling can produce flat or angled surfaces, grooves, slots, gear teeth, and other profiles, depending on the shape of the cutters being used. The basic machine used for milling operations is called a **milling machine,** Fig. 2-7. There are many different kinds of milling machines, such as the horizontal, vertical, plain,

(South Bend Lathe, Inc.)

Fig. 2-5. Boring work mounted in a lathe chuck.

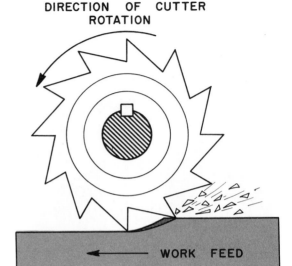

Fig. 2-6. Milling with a plain milling cutter.

and universal. Milling cutters are also used on jig boring machines, engraving machines, and many specialized production machines.

Shaping and Planing

Shaping and planing produce flat surfaces with a single-point cutting tool. In **shaping**, the cutting tool on a shaper reciprocates or moves back and forth while the work is fed automatically towards the tool. A small amount of material is removed with each stroke, Fig. 2-8. The shaper is commonly used for machining short, flat surfaces and for making grooves and keyways. The surfaces may be horizontal, vertical, angular, or irregular. See Fig. 2-9

In **planing**, the workpiece is attached to a worktable that reciprocates past the cutting tool. The cutting tool is automatically fed into the workpiece a small amount on each stroke, Fig. 2-10. The planer is used for machining large flat surfaces and long grooves. It can machine horizontal, vertical, angular, and irregular surfaces also. See Fig. 2-11.

Shaper and planer operations are similar in many ways. Both use single-point cutting tools, but the planer tools are usually larger so they can withstand the forces of heavy cuts. They both perform similar operations. The largest shapers, however, can only handle workpieces up to about one cubic meter (39-inch cube). On the other hand, the largest planers can accept workpieces larger than 6 × 6 × 12 meters (20 × 20 × 39 feet). Since planers are very large and expensive, they are not often found in schools.

(Jet Equipment and Tools)

Fig. 2-7. Manually operated toolroom horizontal milling machine.

Broaching

Broaching performs many of the operations once done by shaping. Broaching is more efficient, however, since it uses a multiple-tooth cutting tool which normally completes its cutting operation in one stroke. The teeth of a broaching tool are equally spaced. They are positioned so that as the tool advances into the workpiece, each tooth removes a predetermined amount of metal. See Fig. 2-12.

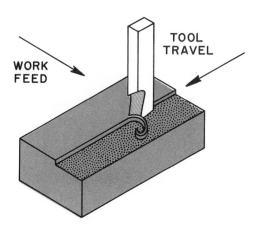

SHAPING

Fig. 2-8.   Tool in relation to the workpiece when shaping.

(Cincinnati Incorporated)

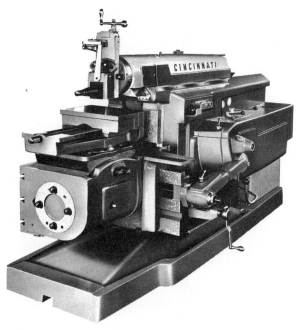

Fig. 2-9.   Medium-size, heavy-duty shaper of horizontal design.

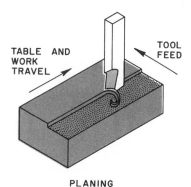

PLANING

Fig. 2-10.   Tool in relation to workpiece when planing.

(G.A. Gray Company)

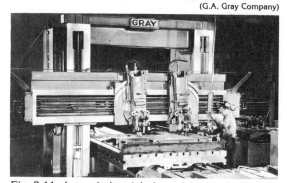

Fig. 2-11.   Large industrial planer with two toolheads.

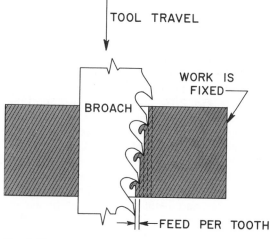

Fig. 2-12.   Action of a broaching tool.

Grinding

Grinding makes use of abrasive particles to do the cutting. These particles may be bonded together to form wheels or stones; bonded to paper or cloth backing to make sheets, discs, cylinders, cones, or belts; combined with paste-like materials to form compounds; or used as loose grains.

Grinding operations may be classified as precision or nonprecision, depending on the purpose. Precision grinding is concerned with grinding to close tolerances and very smooth finish. Nonprecision grinding involves the removal of metal where accuracy is not important. Precision grinding machines are made for many types of grinding. This includes grinding of flat surfaces, external and internal cylindrical grinding, tool and cutter grinding, and thread grinding.

Precision flat surfaces are ground on a **surface grinder.** The workpiece is mounted on the grinding wheel table which reciprocates under the grinding wheel, Fig. 2-13. When each stroke is completed, the work is fed farther under the wheel. This cycle is repeated until the entire surface is ground.

Cylindrical workpieces are ground on **cylindrical grinders.** See Figs. 2-14 and 2-15. Grinding is described in greater detail in Sections 12 and 13 of this book.

Ultrasonic Machining (USM)

Ultrasonic machining, also called **impact grinding,** bombards the workpiece with abrasive particles. The tool is made to vibrate a few hundredths of a millimeter (thousandths of an inch) at an ultrasonic frequency of about 20,000 Hz. Fine abrasive particles suspended in a fluid, usually water, are circulated between the tool and workpiece, Fig. 2-16. The pumping action of the tool causes the abrasive particles to strike the workpiece with high velocity, thus grinding the workpiece to the shape of the tool.

Since **abrasives** do the cutting, almost any material can be machined. Graphite, plastics, and glass are cut as readily as ceramic materials, metal oxides, and metal carbides. A particular advantage of USM is that it can **cut materials that do not conduct electricity.** Note the small USM machine, Fig. 2-17.

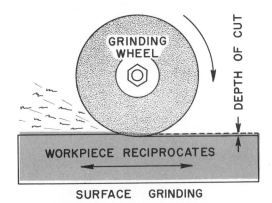

Fig. 2-13. Action of grinding wheel in surface grinding.

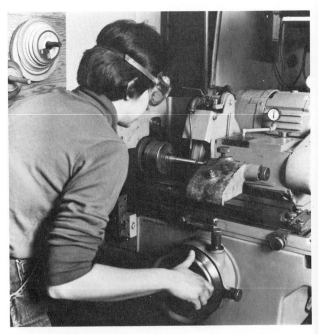

Fig. 2-14. Grinding a part in a plain cylindrical grinding machine.

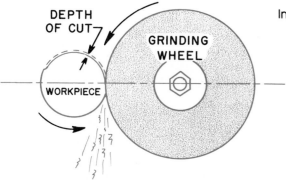

CYLINDRICAL GRINDING

Fig. 2-15. Relationship of grinding wheel and workpiece in cylindrical grinding.

(Speed Fam Corporation)

Fig. 2-18. A machine for lapping flat surfaces.

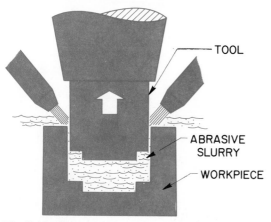

Fig. 2-16. Basic elements of an ultrasonic machining system.

Fig. 2-17. A small ultrasonic machine tool.

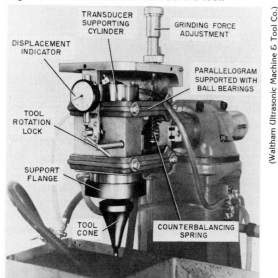

(Waltham Ultrasonic Machine & Tool Co.)

Lapping and Honing

Lapping is a grinding technique which involves the use of abrasive pastes or compounds to cut away very small amounts of metal. It produces precision flat or cylindrical surfaces which require a very smooth finish. Special tools are made for both hand and machine lapping, Fig. 2-18.

Honing employs abrasive sticks mounted in a rotating tool. It is used for extremely accurate sizing of holes that require a very smooth finish. Equipment for honing the external surfaces of cylinders and for honing gears is also available. Note the honing machine, Fig. 2-19.

Electrical Discharge Machining (EDM)

Electrical discharge machining removes metal by controlled electrical arcing between the tool and the workpiece while both are submerged in a dielectric (insulating) fluid, Fig. 2-20. Figure 2-21 shows a typical EDM machine. EDM is capable of **machining complex shapes in metals of any hardness.** Therefore, it is being used increasingly in the production of metal stamping dies and

(Chevrolet)

Fig. 2-19. Honing machine for finishing holes.

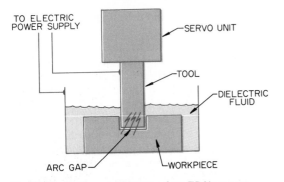

TO ELECTRIC POWER SUPPLY

SERVO UNIT

TOOL

DIELECTRIC FLUID

ARC GAP

WORKPIECE

Fig. 2-20. Basic components of an EDM system.

(Charmilles Corporation of America)

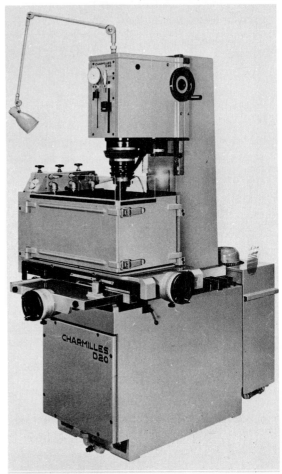

Fig. 2-21. An electrical discharge machine of a size typically found in tool and die shops.

molds for making parts of plastic, rubber, metal, and other materials. A disadvantage of the process is that **only materials which can conduct electricity can be machined.**

Electrochemical Machining (ECM)

The electrochemical machining process is based on the same principles as electroplating in which a workpiece is plated or coated with metal. With ECM, however, the electroplating process is reversed so that the metal is deplated or removed from the workpiece instead of being added to the workpiece. ECM can machine any metal regardless of hardness, so that like EDM, it is finding increasing use in tool, die, and mold making. See Figs. 2-22 and 2-23.

Electrochemical deburring (ECD) and electrochemical grinding (ECG) are adaptations of the ECM process. They are discussed in Section 15, Unit 119.

Chemical Milling (CHM)

The chemical milling process shapes metal by using strong acid or alkaline solutions to dissolve away unwanted metal. Parts to be chemically milled are first cleaned to remove dirt and grease. They are then coated with a material that resists chemical attack and can later be stripped off easily. Areas to be chemically milled are exposed by

cutting through and peeling away the masking. The parts are then submerged in the chemical milling solution until the unwanted metal is dissolved. Rinsing with water and demasking complete the process. See Fig. 2-24.

The aircraft and aerospace industries are using chemical milling to replace conventional milling with expensive machine tools. This has saved millions of dollars in the cost of producing airframe parts of complex shape.

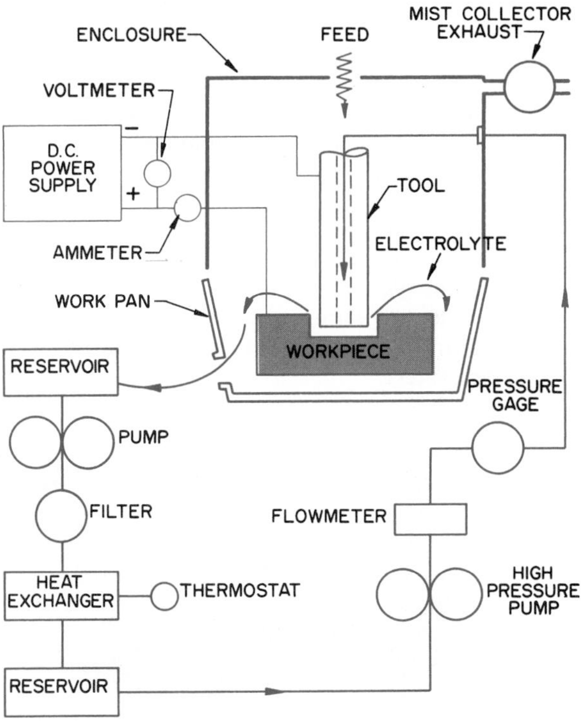

Machining by Controlled Melting

Electron Beam Machining (EBM)

Electron beam machining is accomplished by focusing a high-speed beam of electrons on the workpiece. The heat generated is sufficient to vaporize any known material, but the amount of material removal is very small. Best results are obtained when the work is carried out in a vacuum chamber. High equipment cost, slow cycle times, and other disadvantages make the process too costly to be competitive with most conventional machining techniques. Therefore,

Fig. 2-22. Basic components of an ECM system.

Fig. 2-23. Electrochemical machine of medium size.

(Chemform)

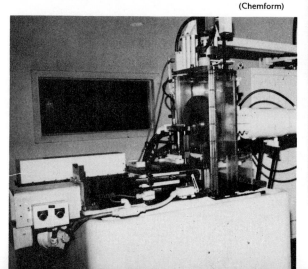

Fig. 2-24. An airline company worker examines an airframe section that has been chemically milled.

EBM is used mainly for machining extremely hard materials such as ceramic oxides, carbides, and diamond. See Figs. 2-25 and 2-26.

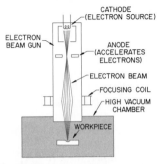

Fig. 2-25. Electron beam machining system.

(Leybold-Heraeus Vacuum Systems, Inc.)

Fig. 2-26.   An electron beam gun used in an electron beam machining system.

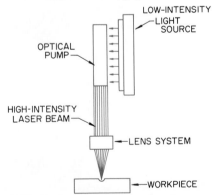

Fig. 2-27. Basic components of a laser machining system.

Laser Beam Machining (LBM)

The source of energy for LBM is a concentrated beam of light. (Laser is an acronym for light amplification by stimulated emission of radiation.) See Fig. 2-27. The intense heat of the laser beam enables it to vaporize any known material. However, the rate of metal removal is so small that most lasers are only used for jobs where small amounts of metal are to be removed. Lasers are useful for such jobs as drilling tiny holes in diamond or carbide wire drawing dies. It is also useful for removing minute amounts of metal to balance parts which spin at very high speed.

Newer computer controlled laser cutting systems use oxygen to boost the heat of the laser beam. This generates enough heat to make fast, clean cuts in plain carbon steel sheets up to 6.35 mm (1/4 in.) thick. Figure 2-28 shows a laser cutting system.

(Hughes Aircraft Company — Industrial Products Div.)

Fig. 2-28. Computer-controlled laser cutting system.

(Union Carbide Corporation — Linde Div.)

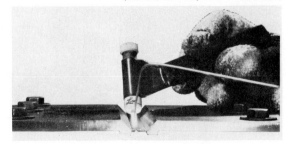

Fig. 2-29. Plasma arc cutting torch.

Plasma Arc Machining (PAM)

Plasma arc machining is done with a cutting torch that produces an extremely hot jet of ionized gas, called **plasma.** Temperatures of 11,100°-27,760° C (20,000°-50,000° F) are obtained by heating a gas with an electric arc inside a specially designed, water cooled torch, Fig. 2-29. The process is used mainly for profile cutting of stainless steels and aluminum alloys. These metals and others, such as titanium, magnesium, and copper/nickel alloys, cannot be readily cut with oxyacetylene torches. PAM can cut stainless steel as thick as 125 mm (5 in.), and aluminum as thick as 150 mm (6 in.). For straight cuts in production quantities, even low carbon steel up to 50 mm (2 in.) thick, which is readily cut by oxyacetylene torches, can be cut economically by PAM. Figure 2-30 shows equipment for PAM.

(W.A. Whitney Corp.)

Fig. 2-30. Plasma arc cutting system.

Test Your Knowledge of Section 1

Unit 1: Importance of Machine Tools

1. Explain briefly why metal cutting machine tools are important to modern civilization.
2. What were the principal sources of power used to drive machine tools at the beginning of the Industrial Revolution?
3. Why did the Industrial Revolution begin in 1776?
4. Which machine tool was one of the earliest and most important to be developed?
5. Most of the basic machine tools using mechanical cutting techniques had been invented by what year?
6. How are today's machine tools different from their Industrial Revolution ancestors?
7. How can we be sure that future machine tools will be even better than those we have today?

Unit 2: The Five Basic Arts of Machining Metal

1. List the five basic metal machining arts.
2. Explain what drilling is, and list several machines which do drilling operations.
3. Explain what turning is, and name the basic machine on which it is done.
4. Explain what is meant by boring, and list several machines on which boring may be done.
5. Explain what milling is, and tell which basic machine is used for milling.
6. Explain what shaping is, and tell what basic machine is used for shaping.
7. Explain what planing is, and tell what machine is used for planing.
8. Explain what broaching is, and tell how it is done.
9. Explain what grinding means.
10. List several types of precision grinding machines.
11. Explain how ultrasonic machining is done, and list advantages of the process.
12. What grinding materials are used for lapping, and why is it done?
13. Describe the honing process and tell why it is done.
14. Describe the EDM process and list major advantages and disadvantages.
15. Explain how the ECM process works.
16. How is chemical milling different from mechanical milling?
17. Tell how cutting is accomplished in EBM.
18. What is the energy source used for LBM? For what kind of work is it used?
19. Describe the PAM process and discuss its applications.

Section

2

Safety

Unit

3

Safety in the Machine Shop

Machine shop safety can be divided into two broad areas of concern:

1. Protection against personal injury, and
2. Prevention of damage to tools, machines, and equipment.

In sections of this book concerned with each family of machines, specific practices and precautions are recommended for safe and efficient use of each basic machine tool. This unit is about personal safety.

Personal Safety

Working with metals in a machine shop makes personal injury possible, Fig. 3-1. Metals are hard, unyielding materials. They have raggedly sharp edges. Hot, sharp metal chips produced in cutting operations can burn and cut the worker. Grinding wheels can throw abrasive particles into unprotected eyes. Rotating tools and workpieces can catch loose clothing and hair. A careless worker can be painfully injured. Workers who **think safety** and **work safely** can avoid these and many other hazards. They must dress properly, follow correct work procedures, and work harmoniously with fellow workers.

How to Dress Safely

1. For maximum eye protection, wear clean, properly fitted, approved safety glasses with side shields, goggles, or a face shield in the machine shop, Fig. 3-2. Laws in most states require everyone to wear eye protection in school shops, laboratories, and factories. Even if this were not so, **your eyes cannot be replaced. Protect them at all costs.**
2. Wear close fitting clothing made of a hard, smooth finished fabric. Such fabric will not catch easily on sharp edges or be wrapped around drills or other rotating tools. A fuzzy sweater is particularly bad. Wear it only under a hard finished garment like a shop coat.

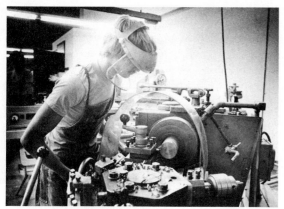

A. Protect yourself against hot, sharp chips produced by turning.

B. Long, loose hair can get caught in machinery. Confine it in a cap or tie it back securely.

Fig. 3-1.   Workers in a machine shop must be cautious.

Wear short sleeved shirts. If long sleeves are worn, they should be close fitting or rolled up past the elbow. A necktie should be removed or tucked into the shirt. Wear a close fitting apron or shop coat to protect street clothes from the usual grime of metalworking, Fig. 3-3.

3. Protect your feet against hot, sharp chips and sharp or heavy falling objects. Safety shoes offer the best protection, but ordinary leather shoes also provide considerable protection. **Canvas shoes and open toe sandals offer no**

protection and should not be worn in the machine shop.

4. Rings, wrist watches, bracelets, and necklaces can get caught on equipment and cause serious injury. **Always remove all jewelry before working with tools and equipment.**

5. **Long, loose hair is dangerous around machine tools.** Confine long hair under a close fitting cap or tie it back securely.

6. **Never wear gloves while operating machines.** They are easily caught in moving parts which can cause serious injury to the hand.

A. Safety glasses with side shields.

B. Goggles.

C. Face shield for full-face protection.

Fig. 3-2.   Choose the right equipment to protect your eyes and face.

Fig. 3-3. Machine operator dressed properly for working safely at metal lathe or other machine tools.

Safe Work Practices

1. Before starting a machine, be sure that all its safety devices are in place and working properly.
2. Always be sure that the workpiece and the cutting tool are mounted securely before starting the machine, Fig. 3-4.
3. Keep your hands away from moving machinery and tools.
4. Handle materials carefully to avoid getting cut, Fig. 3-5.
5. Avoid feeling the machined surface of the workpiece while the machine is running.
6. Never leave a machine while it is running or in motion, even if it is coasting with the power off. Someone else may not notice that it is still in motion and may be injured.
7. Always stop the machine to perform an operation where there is danger of the tool catching as, for example, when using an inside caliper.
8. Always stop machines to oil, clean, adjust, or repair them. If extensive repairs are required,

Fig. 3-4. Tools and workpieces must be rigidly supported for safety.

Fig. 3-5. Raggedly sharp edges call for care in materials handling.

disconnect the machine from its power source so that it cannot be accidentally turned on. The National Safety Council recommends that out-of-order equipment be identified with a blue tag or sign, Fig. 3-6.

9. Never use your hands to stop a machine or a moving part, such as a lathe or drill press chuck.
10. Use a brush, a piece of cardboard, or a thin strip of wood to remove metal chips, Fig. 3-7. **Do not use your hands!**

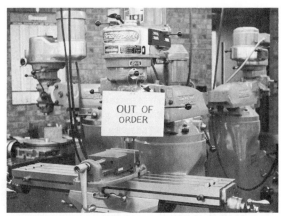

Fig. 3-6. Out-of-order equipment marked with blue tag or sign.

Fig. 3-7. Brushing is a safe way to remove chips.

11. Whenever you remove guards, change the normal position of a machine, or remove parts to perform a specific operation, be sure that all are properly replaced before leaving the machine.

12. When closing electric switches, always grasp the switch by the insulated handle. **Keep your hands away** from the metal parts of the switch or the switch box itself.

13. If you must change speeds on a cone pulley drive system, wait until the machine comes to a **complete stop** before shifting the belt.

14. Be very cautious when using equipment with projecting setscrews. Where possible, replace them with flush setscrews.

15. Do not use compressed air to blow metal chips from a machine or work station. Be sure that you and others in the area are wearing **safety goggles when working with compressed air for ANY purpose.**

16. Avoid walking through restricted areas designated by floor markings or provided with barriers.

17. Avoid touching metal you suspect is hot. If doubtful, test the piece by sprinkling a few drops of water on it.

18. Return long metal bars to the proper storage rack after cutting off the necessary stock.

19. Keep aisles free of stock, small metal remnants, metal chips, and other waste. These are hazards to safe travel.

20. Clean up oil, grease, or other liquid spilled on the floor. Otherwise, it may cause someone to slip and fall.

21. Ask for help in lifting and handling heavy stock or machine accessories such as milling machine vises. Remember to lift with your legs, not your back, Fig. 3-8. Also, ask for help in handling long pieces of stock so as to avoid injuring someone or damaging equipment.

Fig. 3-8. Learn to lift heavy objects by using the strength in your legs, not your back.

22. When working with another person on a machine. agree **beforehand** on who will operate the switches and controls.
23. Make it a habit to **stop, look, and think** in unfamiliar or possibly dangerous situations.
24. Always try to be **alert, patient, courteous, and willing to help.** This is especially necessary in school shops where facilities are often scarce, and you may have to lose precious time waiting for a tool or for your instructor's assistance.

Safety with Hand Tools

1. Always use the **right tool** for the job. Using makeshift tools increases the chance of accident.
2. Keep hands and tools wiped clean and free of dirt, oil, and grease. Dry tools are safer to use than slippery tools, Fig. 3-9.
3. Keep tools sharp and in good adjustment. Dull or poorly adjusted tools must be forced, causing accidents more easily.

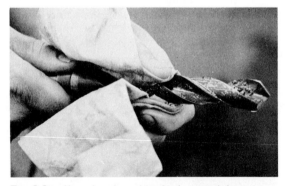

Fig. 3-9.   Keep hands and tools clean and dry.

Fig. 3-10. A scriber is needle-sharp and hard enough to cut steel. If you are giving it to someone else, offer it handle first.

4. Carry sharp-edged tools with the edges or points down. **Never carry them in your pockets.**
5. When handing a tool to another worker, be sure to offer it handle first, Fig. 3-10.
6. Check that heads of punches and cold chisels are properly dressed. When they start to mushroom or check, small pieces may break off and cause serious injury.
7. When using chisels, be careful that flying chips do not hit others in the shop.
8. Always use the right wrench for the job. Knuckles or hands are easily injured when a poor fitting wrench slips.
9. Check that handles are securely attached to files.
10. Damaged tools or ones in poor condition are dangerous to use. Always report tool damage to the instructor.

Fire Prevention

1. Learn the location of the nearest fire alarm in the shop or building as well as the nearest fire exit.

Fig. 3-11. Approved metal container for oily rags.

Fig. 3-12. Give cuts or bruises prompt attention.

2. Learn the location and proper use of fire protection equipment in the building. Fire extinguishers which use a dry chemical or carbon dioxide should be readily available at all times.
3. Place oily rags or waste in the proper metal containers, Fig. 3-11. This guards against possible fire from spontaneous combustion.
4. Always close containers of inflammable materials such as paints or oils after use. Return them to the proper metal storage containers.

First Aid

1. Always notify the instructor immediately when injured, regardless of how slight the injury may be.
2. Always get first aid treatment for cuts or bruises promptly, Fig. 3-12. It is good practice to allow slight and moderate cuts to bleed for a few moments before stopping the flow of blood. Free bleeding carries infectious particles out of the wound. **Severe cuts or bruises should receive the immediate attention of a physician.**
3. Always treat burns promptly, according to how severe they are. A first degree burn is one in which the skin is merely reddened. In a second degree burn, the skin is blistered. In a third degree burn, the flesh is seared or charred. Treat first degree burns with applica-

tions of cold water. Then apply a sterile dry bandage. **Second and third degree burns should receive a physician's attention immediately.**
4. If you are concerned about either injury or an illness, get professional help as soon as possible.

Test Your Knowledge of Section 2

Unit 3: Safety in the Machine Shop

1. How can you protect your eyes and face in the shop?
2. What type of clothing is considered safest to wear in the machine shop?
3. Next to safety shoes, what kind of shoes offer feet good protection for shop work?
4. What kinds of jewelry should be taken off when working in the shop? Why?
5. What danger is involved in having long, loose hair when working with machines?
6. Why are gloves potentially dangerous to wear when working with machines?
7. Describe how to shift a drive belt safely.
8. What danger is involved in blowing metal chips off a machine with compressed air?
9. Describe how to check whether it is safe to pick up a piece of metal you suspect of being hot.
10. Describe the recommended body posture to use when lifting heavy objects.
11. Why must the workpiece and the cutting tool be mounted securely before starting the machine?
12. What possible dangers are involved if you attempt to operate a machine before you have had instructions on how to use it?
13. List several safety precautions which should be taken when working with hand tools.
14. What precautions should be taken to help prevent fires in the shop?
15. Who should be notified immediately in case you are injured, even if the injury is slight?
16. What types of injuries should receive first aid treatment?
17. Describe the appearance of a first degree burn. What kind of first aid treatment should be given?
18. What type of burn injuries should always receive the immediate attention of a physician?

Section 3

Measurement, Layout, Inspection, and Setup Tools

Unit

4

Principles of Linear Measurement

Precision measurement is the key to producing interchangeable parts and mass producing consumer goods. Every part must be made accurately, within specified limits, to the size and shape specified by the designer. Inaccurately made parts will not assemble and fit properly with mating parts. Hence, the finished product may not operate properly, or may wear out sooner than it should.

All workers in the machine shop must be responsible for accurate work. Accurate workmanship depends primarily on accurate measurement and layout work. To insure accuracy, machinists must know the principles of measurement. They also must know how to use the common hand tools, measuring instruments, and gages used in the trade.

Machinists use many instruments and gages for making measurements. Some of these measurements need only be accurate to half a millimeter (or 1/64"). At other times, measurements must be made within two-hundredths of a millimeter (0.001") or less. In the manufacture of certain gage blocks and measuring instruments, measurement must be made to within 50 μmm (2-millionths of an inch). To make accurate measurements, the machinist must learn how to use tools such as rules, micrometers, vernier calipers, gage blocks, and special optical instruments. These tools are presented in the next unit, after a review of some of the principles of measurement.

Linear Measurement

The United States is the last industrialized nation to continue use of the English system of linear (straight line) measurement. But with legislation passed in 1975, conversion to the modern form of the metric system, The International System of Units (abbreviated SI), is expected to accelerate. Some large corporations have already converted wholly or partially to the SI metric system of measurement. Many others have begun to convert by dimensioning part drawings both in metric and in English (inch) units (dual dimensioning). Until a complete conversion to the SI metric system can be made, textbooks must provide information on both the metric and the inch system of measurement. That practice is used in this book.

Metric System

The standard SI metric unit of linear measurement is the **meter.** The meter is subdivided into the following parts:

1 meter = 10 decimeters (dm)
1 decimeter = 10 centimeters (cm)
1 centimeter = 10 millimeters (mm)

Hence, one **decimeter** is one-tenth meter, one **centimeter** is one-hundredth meter, and one **millimeter** is one-thousandth meter. Other subdivisions of the meter are also included in the metric system. One meter is equal to 39.37 inches. One inch is equal to 2.54 centimeters or 25.4 millimeters. Both dimensions are exact and are not rounded. (Further decimal positions would be zeros.) For measurements finer than one millimeter, steel rules are available with 1/2 mm graduations. For precision measurement, micrometers are available which measure as finely as 0.001 mm.

Occasionally the machinist must convert measurements from the inch system to the metric system or vice versa. The metric units of linear measure and equivalent inch units are included in Tables A-12 and A-13, Appendix.

Inch System

The inch is divided into common fractions such as 1/2", 1/4", 1/8", 1/16", 1/32" and 1/64". It also may be divided into decimal fractions such as 1/10", 1/100", and 1/1000". These fractions may be expressed with a numerator and a denominator as shown. However, in machine shop and on drawings and blueprints, they frequently are expressed in decimal form such as 0.1", 0.01", and 0.001". Decimal fractions are expressed in the following manner:

$$\text{One-tenth inch} = \frac{1}{10}'' = 0.1''$$

$$\text{One-hundredth inch} = \frac{1}{100}'' = 0.01''$$

$$\text{One-thousandth inch} = \frac{1}{1,000}'' = 0.001''$$

$$\text{One ten-thousandth inch} = \frac{1}{10,000}'' = 0.0001''$$

Common fractions, such as the following, also are expressed in decimal form:

1/2" = 0.500"	1/16" = 0.062 5"
1/4" = 0.250"	1/32" = 0.031 25"
1/8" = 0.125"	1/64" = 0.015 625"

Learning the decimal equivalents of common fractions of an inch is a practical necessity for working with the inch system of measurement. Steel rules can be used for measurements as fine as 1/64". For precision measurement, micrometers are available which measure as finely as 0.0001".

Standards for Measurement

Standards for linear measurement were established by the International Bureau of Weights and Measures. This bureau was created in 1875 and is located near Paris, France. It has representatives from most nations of the world. It keeps models or standards for units of metric measurement, including a standard for the meter.

The international standard for the meter is the length between two finely scribed lines on a platinum-iridium-alloy metal bar at a temperature of 0°C (32°F). This metal meter bar was declared the International Prototype Meter, and all member nations received exact duplicate copies of it. The United States received its copy in 1889. It is kept at the Bureau of Standards at Washington, D.C.

The U.S. Bureau of Standards adopted the metric system in 1893 as a **standard** for legally defining the pound and the yard. The length of the U.S. yard was defined as $\frac{3600}{3937}$ meter. One inch was defined as 2.54 centimeters, exactly. Thus the units of linear measurement in the inch system are defined in terms of equivalent metric units.

The International Bureau of Weights and Measures also defined the length of the meter in terms of a specific number and kind of light wavelengths. One meter equals 1,650,763.73 wavelengths of orange light emitted by Krypton-86 atoms in an electrical discharge. This is now the accepted standard. Since no standard of length is maintained for the inch system, the International Inch, as defined in terms of metric units (2.54 centimeters), can be stated in terms of wavelengths of krypton light as follows:

    1 inch = 0.025 4 meters × 1 650 763.73 wavelengths per meter
    1 inch = 41 929.398 7 wavelengths
    1 wavelength = 0.000 023 8 inch

Since light waves do not vary significantly with temperature and atmospheric conditions, this method of precise measurement may be duplicated in all parts of the world.

Without a standard for units of linear measurement, precision measurement, mass production methods, and interchangeability of parts would not be possible. Precision **gage blocks,** as shown in Fig. 11-1, are the practical standard of measurement used in machine shops the world over. Gage blocks are used to check the accuracy of various measuring tools and instruments. They are also used for making extremely accurate measurements, as required for special work. Precision gage blocks are available with tolerances of plus or minus 50 μmm, (2-millionths or 0.000 002 inch). In their manufacture, size is measured to light-wave accuracy with optical instruments. Gage blocks and their use are explained in detail in Unit 11.

Limits and Tolerances for Interchangeability

Generally, it is considered impossible to produce parts to absolute size, since some error or inaccuracy would always exist, even if a part were made to within 25 μmm (one-millionth of an inch). Therefore, in most cases, it is impractical, costly, and wasteful to machine parts to a greater degree of accuracy than that required and specified on the drawing.

Interchangeability

Modern mass production methods require that parts be machined to size limitations which provide for interchangeability in use. The size limita-

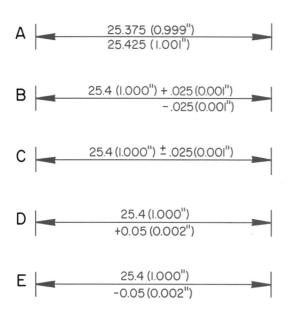

Fig. 4-1.   Dimension limits expressed in different ways. In each case, the tolerance is 0.05 mm (0.002″).

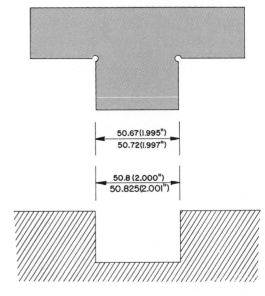

Fig. 4-2.   An example of limit dimensioning. Tolerance on tongue is 0.05 mm (0.002″). Tolerance on groove is 0.025 mm (0.001″). Allowance is 0.075 mm (0.003″) (positive) for clearance fit.

tions are indicated in notes or dimensions on working drawings.

The automobile is an example of the importance of parts being interchangeable. The parts are manufactured in different sections of the country and are shipped to an assembly plant where they are assembled. Other parts are sent to parts distributors where they are used later to replace worn-out parts. All of the parts, therefore, must be made within specific size limitations. These limitations allow the parts to be interchangeable on all automobiles of the make and model for which they were designed.

Terms Used

The terms used in dimensioning limits of size are so interrelated that they should be understood clearly for correct interpretation of dimensions. The following definitions are adapted from those by ANSI*:

Actual size is a measured size.

Basic size is the size from which limits of size are derived by the application of allowances and tolerances. The basic size for the dimensions in Fig. 4-1 is 25.4 mm (1.000″).

Limits of size are the maximum and minimum permissible sizes which apply. Several methods are used in dimensioning to indicate limits of size, as shown in Fig. 4-1.

Tolerance is the total permissible variation of a size. It is the difference between the maximum and minimum limits of size. The tolerance for each of the examples in Fig. 4-1 is 0.05 mm (0.002″).

Fit is a general term which is used to signify the range of tightness that exists between two mating parts as a result of a specific combination of tolerances and allowances.

Clearance fit is a fit that has limits of size which always provide a clearance between mating parts when assembled.

Interference fit is a fit that has limits of size which always provide for an interference between mating parts when assembled.

Allowance is an intentional difference between the maximum material size limits of mating parts. It is the minimum **clearance** (positive allowance) or maximum **interference** (negative allowance) between mating parts. Figure 4-2 shows two parts which are designed to

*American National Standards Institute, **Preferred Limits and Fits for Cylindrical Parts** (B4.1-1955), The American Society of Mechanical Engineers, 29 West 39th St., New York, New York. (Permission granted.)

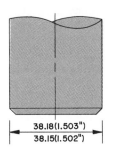

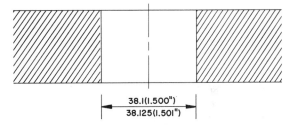

Fig. 4-3. Shaft and hole designed with allowance (negative) for interference fit. Loosest figure is 0.025 mm (0.001″) interference. Tightest figure is 0.075 mm (0.003″) interference.

mate together with a clearance for a free-sliding fit. In this example, the tongue at a maximum material size 50.72 mm (1.997″) and the groove at maximum material size 50.8 mm (2.000″) provide for a 0.075 mm (0.003″) **positive** allowance. The positive allowance provides for a free fit which is called a **clearance fit.**

The shaft and hole in Fig. 4-3 are designed with a **negative** allowance for a tight fit, which is called an **interference fit.** The loosest permissible fit has a 0.025 mm (0.001″) interference. The tightest permissible fit has 0.075 mm (0.003″) interference.

Reading Tolerances

Tolerances on a drawing may be **general, i.e.,** specified with a note; or they may be **specific, i.e.,** specified with a dimension. Whenever specific tolerances are not indicated on the drawing, general tolerances apply. When no tolerance is specified, the tolerance generally is assumed to be ± 1/2 mm (± 1/64″) for fractional dimensions and ± 1/2° for angular dimensions. When no tolerance is specified for decimal dimensions, the tolerance generally is assumed to be ± one figure, to the nearest significant figure. For example, with a two-place decimal specifying millimeters, the tolerance is ±0.01 mm; for a two-place decimal specifying inches, the tolerance is ±0.01″.

Specific limits and tolerances may be indicated in several ways with a dimension value as shown in Fig. 4-1. The tolerance at **A** is indicated by writing the two limit sizes above and below the lines. The tolerance at **B** and **C** is shown by indicating the basic size, followed by the tolerance plus and minus. When only one tolerance is indicated, as at **D** and **E**, the other is assumed to be zero.

Unit

5

Measuring and Layout Tools

In machine shop work, the term **laying out** means the marking of lines, centers, or circles on metal workpieces. Layout work shows size, shape, hole locations, or areas to be machined.

The machinist uses many common tools for measuring and laying out parts to be machined. This unit is concerned with the study of measurement and layout tools such as rules, squares, calipers, and other common measuring tools. Some of these tools also are used for inspection work.

Rules

The machinist uses the steel rule to make rough measurements, Fig. 5-1. Steel rules are available in lengths from 25 mm (1″) to 1 800 mm (72″). Commonly used lengths are 150 mm (6″), 225 mm (9″), and 300 mm (12″). The better ones are made of spring steel, hardened and tempered, and may have graduated measurements on one or both sides.

Metric steel rules have millimeter and half-millimeter graduations. On inch steel rules, the graduations on one side may be 8ths and 16ths and on the other 32nds and 64ths. Some rules also have graduations in 10ths, 50ths, and 100ths.

Caliper Rule

The caliper rule, Fig. 5-2, is used to make rapid measurements to scale dimensions. Outside measurements of various thicknesses are read at the **out** graduation line; inside measurements, at the **in** graduation line.

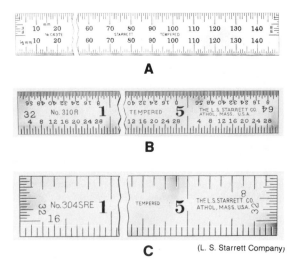

A

B

C

(L. S. Starrett Company)

Fig. 5-1.   Machinist's steel rules. A is a metric rule; B and C are inch rules.

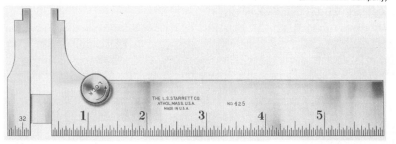

Fig. 5-2. Caliper rule. Metric rules are also available.

Fig. 5-3. Rule depth gage.

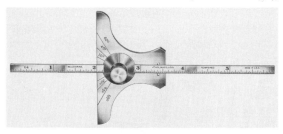

Fig. 5-4. Combination depth and angle gage. Metric gages are also available.

Rule Depth Gage

The rule depth gage, Fig. 5-3, is used to measure the depth of recesses accurately and quickly. The rule is adjusted to the depth of the recess being measured and is locked with a knurled nut and friction spring. Graduations are 1/2 mm (or 1/64").

A **combination depth and angle gage** is shown in Fig. 5-4. In addition to measuring the depth of recesses, this tool may be used to measure or lay out angles of 30°, 45°, and 60°. Angles are set by swinging the rule so that the line on the center turret is aligned with the desired angle line. The rule graduations are in 1/2 mm (or 1/64").

Squares

Squares are important tools which serve several purposes, including laying out lines on parts and checking two surfaces for right-angle squareness. Several types of squares are available.

Combination Set

A combination set consists of the steel rule or blade, a square head, a center head, and a protractor head, Fig. 5-5. The square head and the

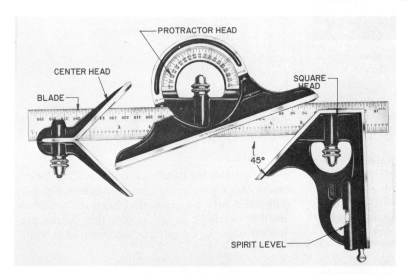

Fig. 5-5. Combination set.

Fig. 5-6.   Measuring height with a combination square.

Fig. 5-7.   Measuring an angle with a protractor head.

Fig. 5-8.   Locating centerline with a center head.

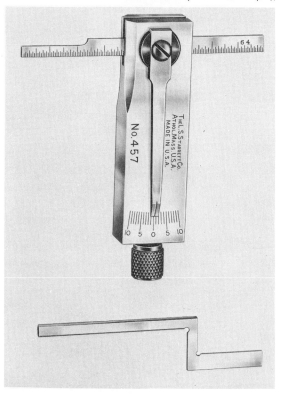

Fig. 5-9.   Diemaker's square (top) with offset blade (bottom). The tool is also available in metric measurement.

protractor head are furnished with a spirit level. Although the level is not a precision level, it is an aid in measuring angles in relation to the vertical or horizontal plane.

The blade and the square head together make up a **combination square**. It may be used to lay out or test 90° or 45° angles and to lay out lines parallel to an edge. It also may be used to measure the height of parts, Fig. 5-6, or the depth of slots or grooves.

The **protractor head** is used to test, measure, or lay out angles to within 1° accuracy, Fig. 5-7. The head may be graduated from 0° to 90° or from 0° to 180° in either direction. Some protractor heads, called the **nonreversible type,** have a shoulder extending from one side of the blade only. A second type, the **reversible type,** has a shoulder on both sides for measuring from either side.

The **center head** is used in locating and laying out the center of round bars or other round objects, Fig. 5-8. It also is used to locate the center of square bars or square objects.

Thus, the combination set may serve as a steel rule, height gage, bevel protractor, level, depth gage, marking gage, or plumb. The combination set is used for work which generally does not require extreme accuracy.

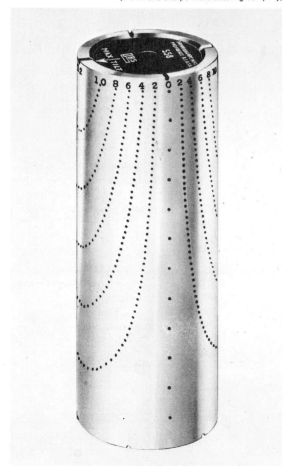

Fig. 5-10. Cylindrical square.

Precision Steel Square

The steel square is a precision tool which is used when extreme accuracy is required. It is used for laying out lines, Fig. 5-19, or for testing the squareness of two surfaces with each other. Since it has no movable parts, it is extremely accurate. Compared to other types of squares, it is expensive. It is available in several sizes. The steel square is widely used by toolmakers and machinists for checking work on both surface plates and machine tools. Although it is hardened and tempered, the steel square should be handled carefully. Dropping or severe abuse may spoil its accuracy.

Die Maker's Square

A die maker's square, Fig. 5-9, is used for measuring clearance on dies. It also is used to check draft angles on foundry patterns. The blade has 1/2 mm (or 1/64″) graduations and may be set at any angle up to 10°. The angle is indicated on the beam of the square by the line on the pointer. The offset blade may be used for measuring angles in places where the straight blade cannot be used.

Cylindrical Square

The cylindrical square is used to tell the **out-of-squareness** of work in units of 0.005 8 mm (0.000 2″) without the use of transfer tools. See Fig. 5-10. It is a true cylinder, with one end lapped perfectly square and the other end lapped at a fixed angle in relation to the sides.

To check a workpiece for squareness, the cylindrical square is placed on a surface plate with the angular end down. The base of the cylindrical square is placed against the work and rotated until light between the cylindrical square and the workpiece is shut out. The topmost dotted curve in contact with the part is read, and the number at the top of the cylindrical square indicates the out-of-squareness of the part in 2, 4, 6, 8, 10, or 12 ten-thousandths of an inch. The same reading may be obtained at two places on the circumference of the cylindrical square. Thus, the tool is self-checking. (Note: The manufacturer has no plans at present to produce a metric version of this tool.)

A   **B**   **C**   **D**

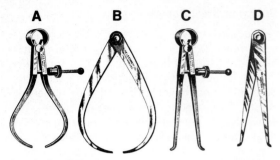

Fig. 5-11. Calipers.   (L. S. Starrett Company)

(L. S. Starrett Company)

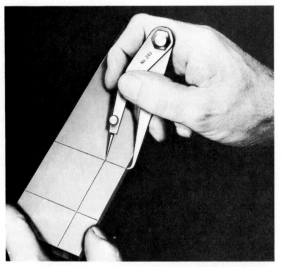

Fig. 5-12. Scribing lines parallel to a surface with hermaphrodite caliper.

Calipers

Calipers are used chiefly for determining diameters. **A** and **B** in Fig. 5-11 are outside calipers used for measuring outside diameters. **C** and **D** are inside calipers for measuring inside diameters. **B** and **D** are called firm joint calipers, while **A** and **C** are spring type, commonly called **spring calipers**. Calipers are available in sizes from 76 to 254 mm (3″ to 10″) or more. The procedures for setting and using calipers are outlined in Unit 6.

Hermaphrodite Calipers

The hermaphrodite caliper, Fig. 5-12, is used by machinists for scribing lines at a desired distance parallel to a flat or curved surface. It also is used for locating the center of circular objects, as described in Unit 6.

Center Gage

The center gage, Fig. 5-13, is used principally when grinding and setting thread-cutting tools.

Screw Pitch Gage

The screw pitch gage, Fig. 5-14, is used to determine the pitch of a thread or to compare the threads of different objects, as, for example, the thread on a bolt with that on a nut. The notches on each blade are cut to match the pitch of a standard thread. Each blade is stamped with the pitch of threads it represents.

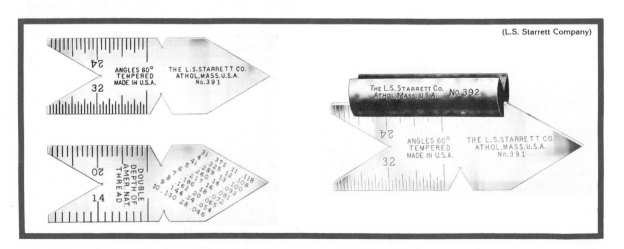

(L.S. Starrett Company)

Fig. 5-13. Center gages in inch measurement.

(L. S. Starrett Company)

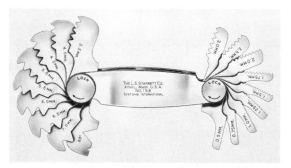

Fig. 5-14. Screw pitch gage.

(Brown and Sharpe Manufacturing Company;
L.S. Starrett Company)

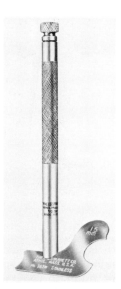

Fig. 5-15. Radius gages.

Radius Gage

The radius gage, Fig. 5-15, is used for determining the radius of fillets and rounds on machine parts. It also may be used for laying out fillets and rounds. Each blade is marked with its radius. Other styles of radius gages also are available in larger sizes.

Layout Tools

In layout work, the machinist scribes lines to locate centers for holes and circles and to define the outline of parts. Layout work is similar to drafting. However, it generally is performed to greater accuracy, usually to 1/2 mm (1/64") or closer tolerances.

Machinist's Vise

The machinist's vise, Fig. 5-16, is essential in the machine shop, toolroom, and maintenance shop. It is mounted on a work bench and holds work for various operations performed there.

Ball Peen Hammer

The hammer most widely used by the machinist is the ball peen hammer, Fig. 5-17. It is available in sizes which vary in weight from 57 grams (2 ounces) to 1.36 kilograms (48 ounces).

(Warren Group)

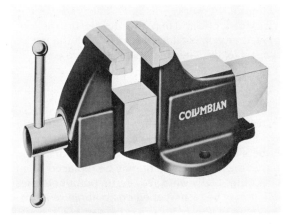

Fig. 5-16. Machinist's vise.

(Stanley Tools)

Fig. 5-17. Ball peen hammer.

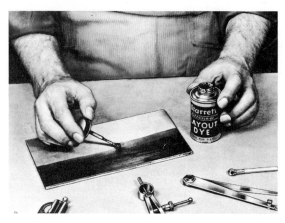

Fig. 5-18. Applying layout dye before laying out work.

Fig. 5-19. Scribing lines on dyed surface with scriber and precision steel square.

Layout Dye

Several kinds of layout dye in various colors are available to color metal workpieces before laying out the work, Fig. 5-18. The workpiece is first wiped clean. Then the dye is brushed or sprayed on and is allowed to dry. With the use of layout dye, the lines may be seen more easily.

Scriber

The scriber, Fig. 5-19, is used for marking layout lines on workpieces. Several types of scribers are available. They usually are designed so that the hardened point may be replaced when it becomes badly worn. When the point becomes dull, it may be resharpened on an abrasive stone. The point of some scribers is bent at a 90° angle for use in scribing the inside of cylindrical objects.

Dividers

Dividers, Fig. 5-20, are used chiefly for spacing, scribing circles, and laying out work. Distances may be transferred directly from a rule to the work. **A** in Fig. 5-20, has a solid nut, while **B** has a quick-adjusting, automatic-closing nut. Dividers are available in sizes from 51 to 305 mm (2″ to 12″).

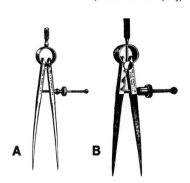

A   **B**

Fig. 5-20. Dividers.

Prick Punch

A prick punch is used to mark sharp, small points along layout lines, or to prepare for center punching before drilling. The prick punch is similar to a center punch, but has a sharper point. The included angle of the point on a prick punch usually is about 30°. See Fig. 5-21.

Center Punch

A center punch, Fig. 5-21 (bottom), is a hardened tool whose point usually is ground to an included angle of about 60° to 90°. This tool is used for **center punching** preparatory to drilling. The center punch mark aids in guiding the drill so that it will drill in the desired location without drifting to the side.

Surface Plate

A surface plate, Fig. 5-22, is a heavy plate of steel, cast iron, or granite, with a precision flat surface. The surface is machined, ground, and scraped for extreme flatness. It provides the necessary flat surface for making the accurate measurements required in precision layout and inspection work, Fig. 5-23.

Surface plates are available in sizes ranging from 305 mm (12″) square to 1 219 × 3 048 mm (4′ × 10′) or larger. They are expensive and should be used carefully. A surface plate never should be hammered or struck, since the smallest nick or dent will affect its accuracy. A thin film of oil should be applied to the surface of iron or steel plates to prevent rust when not used regularly.

Straightedge

The straightedge in Fig. 5-24 is made of cast iron. This tool is portable and is placed on machined surfaces to check them for flatness or straightness. The straightedge is a scraped surface which should be protected with a wooden cover when not in use. Straightedges are available in various lengths ranging from 457 mm (18″) to 4 572 mm (180″).

V-Blocks

Several types of V-blocks are used in doing layout work and in making machining setups. V-blocks may be used singly or in pairs. A single

(L. S. Starrett Company)

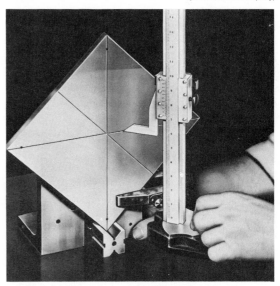

Fig. 5-23. Precision layout work on a surface plate.

(L. S. Starrett Company)

Fig. 5-21. Prick punch (upper) and center punch (lower).

(DoAll Company)

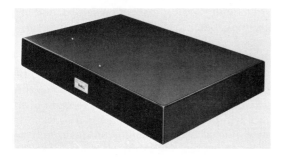

Fig. 5-22. Surface plate.

(Brown and Sharpe Manufacturing Company)

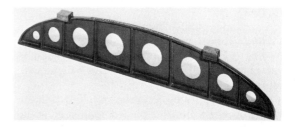

Fig. 5-24. Cast iron straightedge.

Fig. 5-25. V-block.

Fig. 5-27. Toolmaker's clamps.

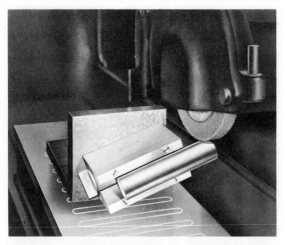

Fig. 5-26. Magnetic V-block held firmly on an angle plate resting on a magnetic chuck.

Fig. 5-28. Surface gage.

V-block is used in a setup for layout work in Fig. 5-23.

A second type of V-block, Fig. 5-25, is used in holding work for machining operations. This V-block is held to the worktable of the machine with clamps. The work is held in the V-block with adjustable screws.

A third type of V-block is shown in Fig. 5-26. In this setup, the V-block is placed against an **angle plate** where it is held magnetically when the control knob is turned to the **on** position.

Toolmaker's Clamps

The toolmaker's clamps, Fig. 5-27, are used for clamping workpieces and accessories together for machining setups, inspection setups, and layout setups. See Fig. 5-23.

Surface Gage

The surface gage, Fig. 5-28, is used chiefly for locating distances from a base and for locating points at a given height on opposite ends of an object lying on a flat surface. It is also used to draw a line at a given height on an irregularly shaped object, as, for example, a line on an irregularly shaped casting. It also may be used to indicate the accuracy or parallelism of a surface.

Telescoping Gages

Telescoping gages are used to gage inside diameters or distances, as shown in Fig. 5-29. The gage is equipped with a plunger which is under spring tension when retracted. When a part is gaged, the tool first is inserted with the plunger retracted. The plunger is released and the tool positioned for accurate gaging. The knurled nut on the handle is then tightened, and the gage is extracted from the part. The distance across the ends of the gage then is measured with a micrometer caliper, Fig. 7-1. Telescoping gages are available in sizes which measure distances from 8 mm (5/16") to 150 mm (6").

Small-Hole Gage

Small-hole gages, Fig. 5-30, are used for gaging the size of small holes and narrow slots. They are available in sizes which measure the diameters of holes or recesses from 3 mm (1/8") to 13 mm (1/2"). In using the gage, select a gage of the proper size, insert the gage in the hole or recess to be measured, and turn the knurled screw on the handle to expand the ball-shaped end to size. (A very slight pressure or drag will be felt when the screw tension is right.) Then use a micrometer caliper to measure the diameter of the ball end.

Adjustable Parallels

The edges on adjustable parallels are precision ground and parallel with each other. The parallels are made in various sizes and are adjustable in width within a range for each size. A screw is provided to lock the tool firmly at the adjusted width. Adjustable parallels may be used to gage the width of grooves or slots, Fig. 5-31. The size of the parallel then may be measured with a micrometer.

Adjustable parallels may be used as spacers for part location in accurate assembly work, or they may be set at a specific size to serve as gage blocks. They also may be used in machine vises for setting work at the proper height for drilling, shaping, grinding, or milling, although solid parallels are usually used for this purpose.

Fig. 5-30. Small-hole gage.

Fig. 5-29. Using a telescoping gage.

Fig. 5-31. Measuring width of slot with adjustable parallels and micrometer.

(L. S. Starrett Company)

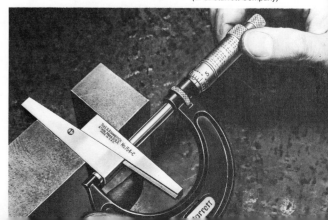

Master Planer and Shaper Gage

The master planer and shaper gage, Fig. 5-32, has many possible uses in making layouts, machining setups, and precision measurement and inspection work. Tools of this type originally were designed primarily for use in establishing the height of the cutting tool on shapers and planers. The height or width of the gage is measured to micrometer accuracy. The cylindrical extension may be screwed into the gage for added length, as shown in Fig. 5-32. The gage is fastened firmly at the desired size with the knurled nut.

The master planer and shaper gages can be used in several other ways: as an adjustable parallel; in conjunction with gage blocks in building up work on a surface plate; and with a dial indicator in transferring measurements.

The gage shown in Fig. 5-33 is a **universal precision gage.** It may be used for the same purposes for which planer and shaper gages are used, and for several other uses as well. It is equipped with a fine-adjustment thumbscrew for ease in making sensitive adjustments required when used with micrometers, vernier height gages, vernier calipers, or gage blocks.

The universal precision gage also is equipped with a scribing attachment for use in layout work or as a height gage. In addition, the screw which holds the beveled scriber may be used to hold a dial indicator.

(L. S. Starrett Company)

Fig. 5-32. Master planer and shaper gage used to establish height of cutting tools.

(L. S. Starrett Company)

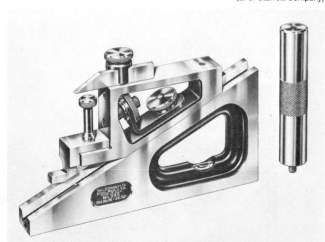

Fig. 5-33. Universal precision gage with extension bar.

Metal workers commonly use the outside, inside, and hermaphrodite calipers. Outside calipers are used for gauging outside diameters and exterior surfaces. Inside calipers are used for gauging inside diameters and width of slots. Hermaphrodite calipers are used mostly for locating centers of round stock. Both outside and inside calipers are made with a spring joint, a firm joint, or a lock joint. The latter has one loose leg and commonly is called a **transfer caliper,** Fig. 6-1. An inside transfer caliper is used for gauging recesses that are wider at the bottom than at the top. An outside transfer caliper is used for gauging the width of a small projection or rib located behind a wider part of the object and inaccessible to an ordinary outside caliper.

Calipers are used in one of two ways:

1. By adjusting the caliper until its legs just contact the object being measured, then measuring the distance across the caliper legs, or
2. By setting the caliper to a known dimension and using it as a gage to test the nearness of the object to the size desired.

Outside calipers are set to a known dimension as in Fig. 6-2. Inside calipers are set as in Fig. 6-3.

Procedure for Using a Caliper to Measure a Workpiece

.Accuracy in using calipers is dependent on developing a keen sense of touch. Always hold the caliper very lightly so as to be able to detect the lightest possible contact of the caliper with the workpiece.

(Brown and Sharpe Manufacturing Company)

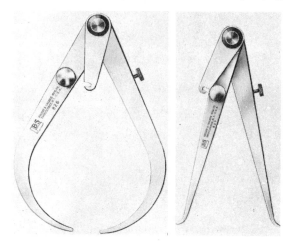

Fig. 6-1.   Loose leg or transfer calipers.

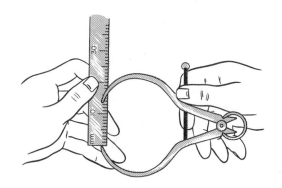

Fig. 6-2.   Setting the outside caliper.

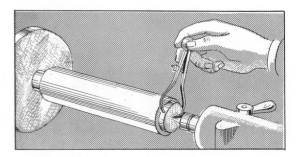

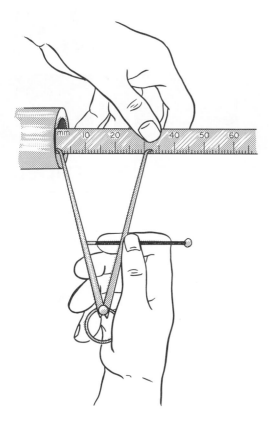

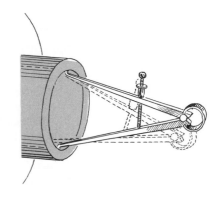

Fig. 6-4.   Using the outside caliper.

Fig. 6-5.   Using the inside caliper.

Fig. 6-3.   Setting the inside caliper.

Caution
The work to be calipered must not be moving!

1. For external work, open the caliper wide enough to admit the work freely, Fig. 6-4. For internal work, close the caliper until it will enter the hole loosely, Fig. 6-5.
2. Holding the caliper perpendicular to the workpiece axis, adjust the caliper slowly while moving it back and forth across the diameter of the workpiece. With the outside caliper, stop adjusting the instant that contact with both legs is felt. With the inside caliper, it is necessary first to work the caliper across the mouth of the hole to find the point where pressure on the caliper legs is lightest. Then, pivoting on one leg held firm, work the other caliper leg in and out of the hole, while ad-

justing the caliper to obtain the lightest contact possible.
3. With a steel rule, measure the distance across the caliper legs as in Fig. 6-2 for outside calipers, and as in Fig. 6-3 for inside calipers.
4. Make additional cuts as necessary, repeating the calipering after each cut. Each new cut should be allowed to proceed only a distance of about 6 mm (1/4″) before taking a new measurement. In this way, if a cut is too deep, it can be corrected in time to save the part.

Procedure for Locating Centers with the Hermaphrodite Caliper

1. Paint the stock end with layout fluid.
2. Set the caliper to about half the diameter of the stock, preferably slightly less than half.
3. Hold the caliper in a vertical position with the bent leg against the outside surface of the stock.

4. Describe four short arcs, as in Fig. 6-6.

When the caliper is set at exactly half the diameter of the stock, the arcs will have a form as at **A,** Fig. 6-6. When slightly less than half the diameter, the arcs will have a form as at **B**; when slightly greater, the arcs will appear as at **C**.

5. Make a punch mark in the center of the area formed by the intersection of the arcs, or at the exact point of intersection, as would be the case at **A**, Fig. 6-7.

6. Check the accuracy of the punch mark by placing the point of the straight leg of the caliper in the punch mark and rotating the caliper. Observe if the point of the bent leg remains in contact with the work throughout the course of rotation.

7. If the punch mark is not located at the center of the work, draw the punch mark in the direction desired; then test as in step 6.

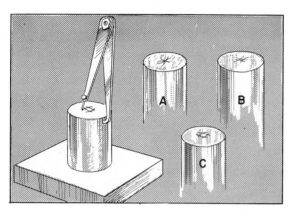

Fig. 6-6. Locating the center with the hermaphrodite caliper.

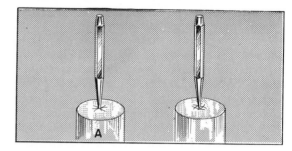

Fig. 6-7. Locating the center.

Unit

7

Micrometers

Micrometers are the precision measuring tools most commonly used by machinists. They are available in a variety of types and sizes, but the most common is the outside micrometer caliper. Figure 7-1 shows a 0-25 mm outside micrometer of traditional design. Several micrometers of more recent design include the following:

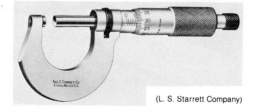

(L. S. Starrett Company)

Fig. 7-1. A 0-25 mm outside micrometer.

1. Indicating micrometer, Fig. 7-2, which can also be used as a comparator for quickly checking parts in quantity;
2. Direct reading micrometer, Fig. 7-3, which provides a numerical display of the micrometer reading;
3. Dual reading micrometer, Fig. 7-4, which reads both in metric and in English measurement; and
4. All electronic micrometer, Fig. 7-5, which has a motorized spindle and electronic digital readout.

In addition, there are many special purpose outside micrometers.

A micrometer often is called a "mike". Plain metric micrometers measure accurately to one-hundredth of a millimeter (0.01 mm). Plain inch micrometers measure accurately to one-thousandth of an inch (0.001″). Some metric micrometers are equipped with a vernier which makes it possible to measure accurately to one-thousandth of a millimeter (0.001 mm). Vernier equipped inch micrometers measure accurately to one ten-thousandth of an inch (0.0001″).

Principal Parts

The parts of a traditional micrometer are shown in Fig. 7-6. The principal parts are the **frame,** the **anvil,** the **spindle** with a precision

(L. S. Starrett Company)

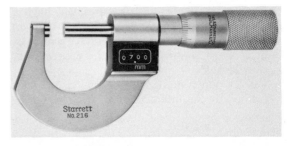

Fig. 7-3.   Direct reading micrometer.

(MTI Corporation)

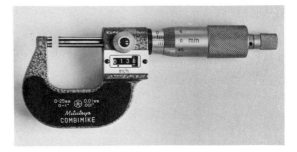

Fig. 7-4.   Dual reading micrometer.

(Quality Measurement Systems)

Fig. 7-5.   Electronic micrometer.

(Federal Products Corporation)

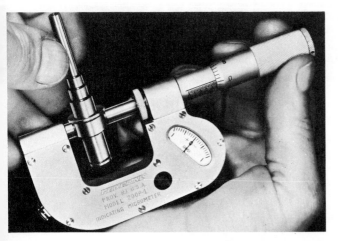

Fig. 7-2.   Dial indicating micrometer.

screw thread, the **sleeve** which is also called a **barrel** or **hub,** and the **thimble.**

The **ratchet** and the **lock nut** are convenient accessories which are available on some micrometers. With the use of the ratchet, a consistent pressure can be applied on the spindle when measurement is made, regardless of who uses the tool. A consistent pressure is important in making accurate measurements. Without the ratchet, one must develop the right feel for accurate measurement. The lock nut locks the spindle in position after measurements are made.

Kinds of Micrometers

Many kinds and sizes of micrometers have been developed for various measuring applications. The following are the principal kinds:

Outside micrometers, also called **micrometer calipers,** Fig. 7-1, are used for measuring outside diameters or thickness. Outside micrometers are available in various sizes which are limited to 25 mm (1″) measuring ranges, such as 0-25 mm (0″ to 1″), 25-50 mm (1″ to 2″), and so on. A 1″ to 2″ micrometer is shown in Fig. 7-7 with a 1″ reference standard.

Large micrometers are available for measuring within various 25 mm (1″) ranges. The micrometer in Fig. 7-8 may be used for measurements in the range from 225-300 mm (9″ to 12″) simply by changing and installing the appropriate anvil.

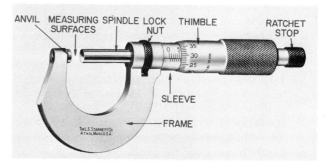

(L. S. Starrett Company)

Fig. 7-6.   Principle micrometer parts.

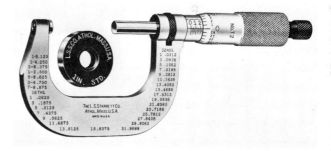

(L. S. Starrett Company)

Fig. 7-7.   1″ to 2″ micrometer caliper with 1″ reference standard.

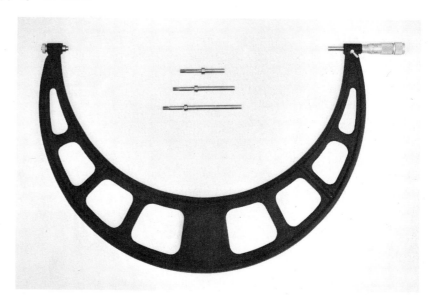

Fig. 7-8.   Micrometer with interchangeable anvils for 225-300 mm range.

Inside micrometers are used for measuring inside diameters, parallel surfaces, or other inside dimensions. There are several types and sizes of inside micrometers available. The inside micrometer caliper in Fig. 7-9 may be used to measure within the range from 5 mm to 25 mm. Figure 7-10 shows a **tubular inside micrometer.** It may be used for measuring inside diameters from 40 mm to 300 mm (1-1/2" to 12"), in range increments of 13 mm (1/2"). Measuring rods are added to either or both ends of the micrometer head to increase its range. A tubular inside micrometer can be used for measuring large diameters, Fig. 7-11.

Mike hole gages, Fig. 7-12, are used for accurately measuring the diameter of relatively small holes. They are available in various size ranges from 6 mm to 200 mm (in inch sizes from .275" to 8"). Setting rings, Fig. 7-13, are available in various sizes for testing and setting the accuracy of inside micrometers.

Depth micrometers are used for measuring the depth of holes, grooves, shoulders, and projections, as shown in Fig. 7-14. The measuring range for depth micrometers can be increased in multiples of 25 mm (1"), as desired, by installing interchangeable measuring rods.

Thread micrometers are used to measure the pitch diameter of screw threads.

(Yamazen USA, Inc.)

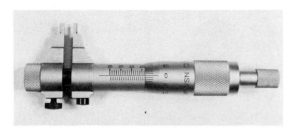

Fig. 7-9.   Inside micrometer caliper.

Care of Micrometers

Micrometers are delicate instruments and are easily damaged. If properly cared for they will give a lifetime of service. They should be kept clean and lightly oiled with instrument oil. When not being used, they should be kept in a clean instrument case or tool box drawer. If dropped, they should be immediately checked for damage and, if necessary, be readjusted to read properly, or sent back to the factory for repair.

(MTI Corporation)

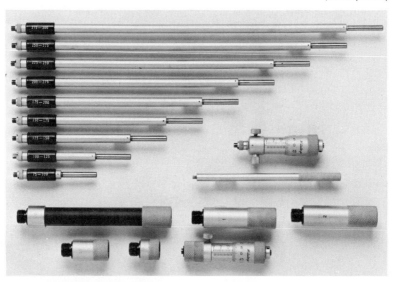

Fig. 7-10. Tubular inside micrometer.

To check a 25 mm (1″) micrometer for accuracy, apply light pressure to a clean piece of paper while pulling it between the anvil and the spindle. This cleans the measuring surfaces so that an accurate reading can be obtained. Then bring the anvil and spindle together with normal pressure. Observe whether the zero line on the thimble is aligned with the longitudinal index line on the sleeve. If the two lines are not aligned, they should be adjusted according to the manufacturer's directions, as supplied with the micrometer.

Larger micrometers are checked in a similar manner with the aid of a reference standard. See Fig. 7-7.

(MTI Corporation)

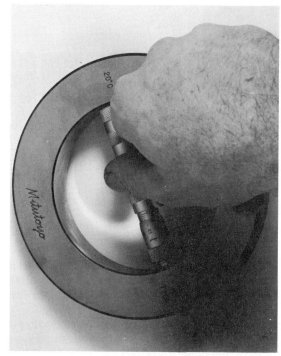

Fig. 7-13. Setting ring for testing and setting inside micrometers.

Fig. 7-11. Measuring with a tubular inside micrometer.

(L. S. Starrett Company)

Fig. 7-14. Measuring depth of a shoulder with a depth micrometer.

Fig. 7-12. Mike hole gage.

Unit

8

How to Read Micrometers

Metric Micrometers

The metric micrometer spindle has 50 threads per 25 mm. Therefore, two revolutions of the spindle are required to move the spindle one millimeter. Since the thimble is graduated with 50 divisions, and two revolutions of the thimble are required to move one millimeter, the degree of precision obtained is 0.01 mm (one-hundredth of a millimeter). Note that the sleeve of the metric micrometer has both an upper and a lower set of graduations, Fig. 8-1. The upper set consists of 25 graduations one millimeter apart, with each fifth millimeter numbered. The lower set of graduations divides each upper set in two, providing one-half millimeter graduations.

The metric micrometer reading of 5.78 mm illustrated in Fig. 8-2, is obtained as follows:

1. Upper sleeve reading
   (whole millimeters)   5.00 mm
2. Lower sleeve reading
   (half millimeters)   0.50 mm
3. Thimble reading
   (hundredths of a millimeter)   <u>0.28 mm</u>
          Total Reading   5.78 mm

Additional metric micrometer readings are given in Fig. 8-3.

Vernier metric micrometers are designed for measuring to 0.001 mm or 0.002 mm. The vernier scale is placed on the sleeve above the usual graduations, Fig. 8-4, but note that the millimeter and half millimeter graduations are both marked

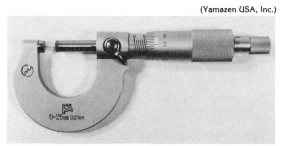
(Yamazen USA, Inc.)

Fig. 8-1.   A 0-25 mm micrometer opened to show graduations.

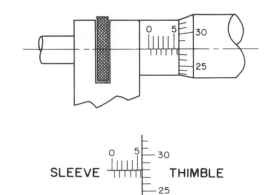

Fig. 8-2.   Metric micrometer reading of 5.78 mm.

on the lower part of the sleeve. This allows the vernier to be read without having to twist the micrometer.

The example in Fig. 8-4 shows a vernier metric micrometer graduated for 0.001 mm readings. (A 0.002 mm vernier micrometer scale would have only the numbered graduations and thus could not be read directly to 0.001 mm). It is read as follows:

1. Reading from sleeve
   (to nearest 0.5 mm)   2.500 mm
2. Reading from thimble
   (to nearest 0.01 mm)   0.360 mm
3. Reading from vernier
   (to nearest 0.001 mm)   0.008 mm
                   Total Reading   2.868 mm

Inch Micrometers

Inch micrometers are the same as metric micrometers except that they are designed to measure accurately to 0.001″ (one-thousandth of an inch). Their spindles are made with 40 threads to the inch, requiring the thimble to be rotated 40 complete revolutions in order to move the spindle one inch. Each revolution of the spindle, then, moves the spindle 1/40th of an inch or .025″. By dividing and marking the thimble into 25 divisions, each mark on the thimble becomes 1/25th of a revolution or 0.001″ (one-thousandth of an inch).

Markings on the sleeve correspond to one full revolution of the thimble, placing them .025″ apart. Every fourth line is a little longer than the others and is marked 1, 2, 3, etc., which stands for 0.100″, 0.200″, 0.300″, etc. See Fig. 8-5.

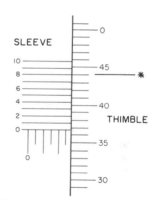

Fig. 8-4.   Metric vernier micrometer graduations.

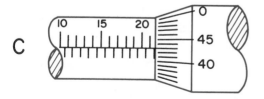

Fig. 8-3.   Additional metric micrometer readings.
   A. 13.25 mm
   B. 8.89 mm
   C. 21.43 mm

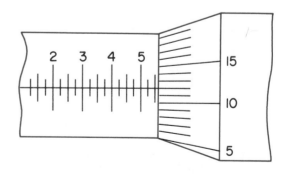

Fig. 8-5.   Inch micrometer graduations. The reading is 0.562″.

Following is an explanation of how the reading in Fig. 8-5 is obtained:

Five numbered graduations represent	0.500"
Two additional lines are exposed, each representing 0.025"	0.050"
The twelfth line on the thimble is aligned with the index line on the sleeve, each line representing 0.001"	0.012"
Total Reading	**0.562"**

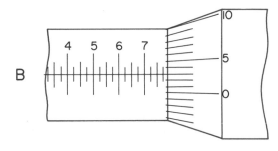

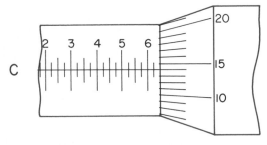

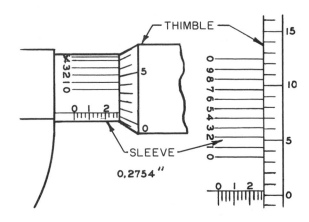

Fig. 8-6.   Additional inch micrometer readings.
   A. 0.385"
   B. 0.778"
   C. 0.639"

Several additional micrometer readings are provided in Fig. 8-6.

The addition of a vernier to the inch micrometer, Fig. 8-7, provides a direct reading accuracy of 0.0001" (one ten-thousandth of an inch). The vernier consists of ten graduations on the sleeve which run parallel to the index line. To read a ten-thousandth micrometer, first read the measurement to the nearest one-thousandth in the usual way. Then check to see which of the vernier lines is aligned with a line on the thimble. If the line **1** is aligned, add one ten-thousandth to the reading. If the second line is aligned, add two ten-thousandths to the reading, and so on.

Fig. 8-7 provides an example of how to read an inch vernier micrometer:

1. The "2" line on the sleeve is visible, representing	0.200"
2. Three additional lines are fully exposed, representing	0.075"
3. The fifth zero on the thimble has passed the index line, representing	0.000"
4. Line **4** on the vernier lines up with a line on the thimble, representing	0.0004"
Total Reading	0.2754"

Fig. 8-7.   Inch vernier micrometer graduations.

Unit

9

Vernier Measuring Tools

(L. S. Starrett Company)

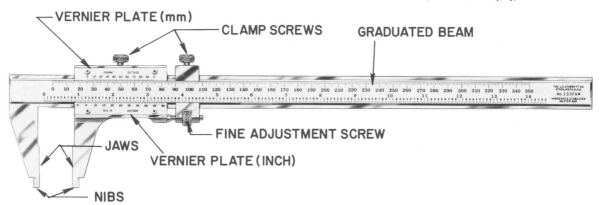

Fig. 9-1. Vernier caliper, showing principal parts.

A vernier is a short rule or scale that is mounted on a measuring instrument so that its graduations subdivide the divisions on the main scale. See Fig. 9-1. Verniers increase the degree of precision which can be obtained from both linear and angular measuring tools. The French mathematician, Pierre Vernier, invented the vernier scale about 1630 A.D.

Vernier Calipers

The vernier caliper is made up of a graduated beam with a fixed measuring jaw, a movable jaw which carries a vernier scale, and a mechanism for making fine adjustments, Fig. 9-1. Vernier calipers are capable of making both outside and inside measurements. See Fig. 9-2. In addition, some are also provided with a depth measuring rod. Some vernier calipers require that when outside measurements such as length, thickness, or

(L. S. Starrett Company)

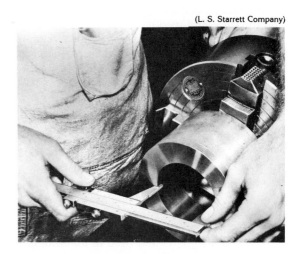

Fig. 9-2. Making an inside measurement with a vernier caliper.

(Brown and Sharpe Manufacturing Company)

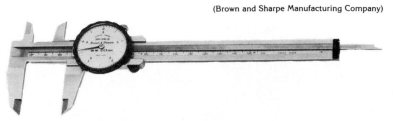

Fig. 9-3.   Metric vernier dial caliper.

outside diameters are made, the size must be read on the scale marked **outside.** When inside measurements are made, such as the width of slots or inside diameters, the size is read on the scale marked **inside.** Other vernier calipers use a single scale for both outside and inside measurements.

Making accurate measurements with vernier tools requires the same care and delicate touch as when using micrometers. First, the slide assembly should be moved along the beam until the jaws almost contact the workpiece. Then the part of the slide assembly that carries the fine adjusting screw should be locked to the beam. The jaws are now brought into contact with the workpiece by moving the fine adjusting nut. The jaws should make definite contact with the workpiece but should not be tight. The main slide assembly is then locked to the beam. The caliper is carefully removed from the workpiece to prevent springing the jaws. The reading is then made.

Vernier dial calipers, Fig. 9-3, have a direct reading dial. It is both faster and easier to read than the traditional vernier scale.

Vernier calipers are made in sizes from 150-1200 mm (6″ to 48″) in length.

(Brown and Sharpe Manufacturing Company)

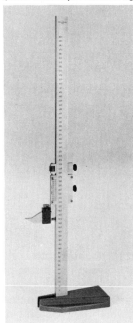

Fig. 9-4.   Vernier height gage.

Vernier Height Gage

The vernier height gage, Fig. 9-4, is a precision tool which has a graduated steel beam mounted upright on a steel base. It is equipped with a vernier plate mounted on a sliding assembly, similar to the vernier caliper. Measurements are read in exactly the same manner as the vernier caliper.

A vernier height gage is used to measure or to mark off accurate vertical distances. A hardened, ground, and lapped scriber is mounted on the movable jaw, as in Fig. 9-4, for marking off heights. The vernier height gage is being used for layout work in Fig. 5-23.

The vernier height gage also may be used on the flat surface of a machine table to measure the height of machined surfaces. A depth gage attachment may be mounted on the movable jaw, Fig. 9-5, for measuring the depth of holes or recesses. An indicator also may be clamped on the movable jaw for inspection work.

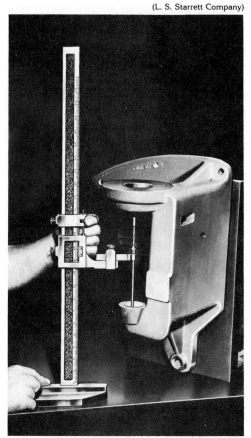

The beam may be graduated on one or on both sides (for outside and inside measurements). Vernier height gages generally are available in sizes ranging from 300 to 1200 mm (12″ to 48″).

Vernier Depth Gage

The vernier depth gage, Fig. 9-6, is used for measuring depths of holes, slots, and recesses. It commonly is available in sizes from 150 to 300 mm (6″ to 12″) in length. Readings are taken in the same manner as for the vernier calipers and height gages.

Vernier Gear Tooth Caliper

This special tool with two vernier scales, Fig. 9-7, is used for measuring the size of gear teeth. To measure a gear tooth, the vertical scale is first set to the **corrected addendum** of the gear tooth

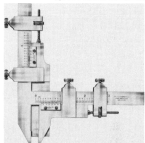

Fig. 9-5.   Measuring depth of recesses with vernier height gage.

Fig. 9-7.   Vernier gear tooth caliper, also available with metric measurements.

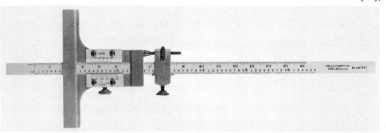

Fig. 9-6.   Vernier depth gage.

to be measured. This adjusts the bar which projects into the space between the jaws, effectively placing the end of the jaws at the pitch circle of the gear tooth. The horizontal vernier is then adjusted to measure the width or **chordal thickness** of the gear tooth. Gear tooth dimensions are listed in charts in handbooks for machinists.

Reading Metric Vernier Calipers

Metric vernier calipers are made to measure to a fineness of either 0.05 mm (.002″) or 0.02 mm (0.0008″), depending on how they are calibrated. The graduations on the main scale are always one

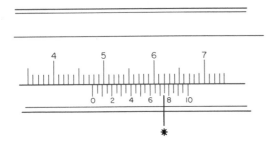

Fig. 9-8.   A 20-division metric vernier caliper reading of 47.75 mm.

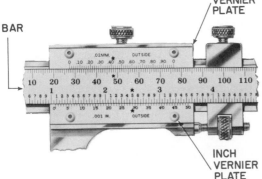

Fig. 9-9.   This 50-division dual vernier scale shows an inch reading of 1.079″ and a metric reading of 27.42 mm.

millimeter apart. If the main scale is numbered 10, 20, 30, etc., it is read directly in millimeters. If the main scale is numbered 1, 2, 3, etc., then it is numbered in centimeters. Thus "1" means 10 millimeters, "2" means 20 millimeters, etc. The calipers which measure to 0.05 mm precision have a vernier scale with 20 divisions, Fig. 9-8, the numbered lines identifying tenths of a millimeter. Fig. 9-8 is read as follows:

1. Main scale reading
    (to nearest millimeter)   47.00 mm
2. Vernier scale reading
    (to nearest 0.05 mm)   0.75 mm

        Total Reading   47.75 mm

The calipers which measure to 0.02 mm precision have a vernier scale with 50 divisions, Fig. 9-9. The numbered lines identify the number of 0.02 mm graduations at that point on the scale. Thus, "5" indicates 5 × 0.02 mm = 0.10 mm (one-tenth of a millimeter), "10" indicates 10 × 0.02 mm = 0.20 mm (two-tenths of a millimeter), etc. Figure 9-9 is read as follows:

1. Main scale reading
    (to nearest millimeter)   27.00 mm
2. Vernier scale reading
    (to nearest 0.02 mm)   0.42 mm

        Total Reading   27.42 mm

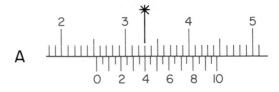

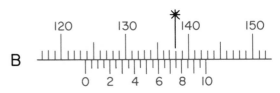

Fig. 9-10. Readings on a 20-division metric vernier scale.
A. 25.4 mm
B. 123.75 mm

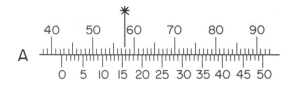

A

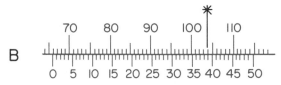

B

Fig. 9-11. Readings on a 50-division metric vernier
    scale.
    A. 42.32 mm
    B. 65.78 mm

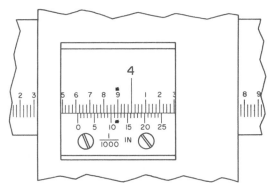

Fig. 9-12. A 25-division inch vernier reading of 3.612".

Additional metric vernier caliper readings are
given in Fig. 9-10 and Fig. 9-11.

Reading Inch Vernier Calipers

Inch vernier calipers measure to a fineness of
0.001" (one-thousandth of an inch). They are
made with either a 25-division vernier or a 50-
division vernier.

25-Division Vernier

This type has every inch of the main beam
graduated into 40 parts, each of which is 0.025"
(twenty-five thousandths of an inch). Every fourth
line is a little longer than the others and is
marked 1. 2, 3. etc., which stands for 0.100",
0.200", 0.300", etc., Fig. 9-12. The 25 divisions
on the vernier plate each represent 0.001" (one-
thousandth of an inch). Every fifth line on the vernier scale is numbered for convenience in making readings.

To read the vernier, first note how many
inches, tenths (0.100", 0.200", etc.), and 40ths
(0.025") the zero on the vernier plate is from the
zero on the beam. Then add to this amount the
number of thousandths indicated by the matching lines on the vernier plate and beam. (Note
starred lines in Fig. 9-12.)

The reading in Fig. 9-12 is explained as
follows:

1. The zero on the vernier
    plate is between the
    3" and 4" marks, thus
    the number of whole
    inches is   3.000"
2. The zero on the vernier
    plate has passed six
    1/10th graduations, or   0.600"
3. The zero on the vernier
    plate has not passed
    any 1/40th gradua-
    tions, thus   0.000"
4. The twelfth line on the
    vernier plate matches
    a line on the beam,
    adding   0.012"

Total Reading   3.612"

50-Division Vernier

The 50-division vernier, Fig. 9-13, is read in the
same manner as the 25-division vernier. However, it is easier to read with the unaided eye,
because the division lines are spaced farther
apart on the vernier plate. Also, the 1/10th divisions on the beam have only two divisions representing 1/20th of an inch (0.050") instead of the
four divisions representing 1/40th of an inch
(0.025").

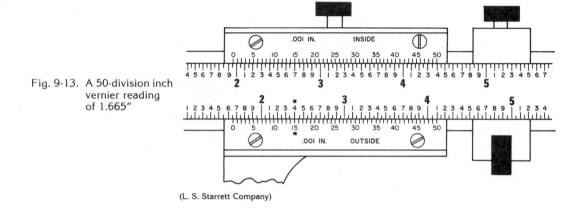

Fig. 9-13. A 50-division inch vernier reading of 1.665"

(L. S. Starrett Company)

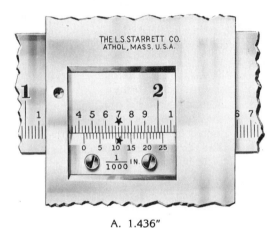

A. 1.436"

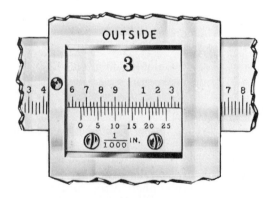

B. 2.659"

Fig. 9-14. Readings on a 25-division inch vernier scale.

The reading in Fig. 9-13 is explained as follows:

1. The zero on the vernier plate is between the 1" and 2" marks on the beam, making .... 1.000"
2. The zero on the vernier plate has passed the 6 on the beam, indicating 6/10ths or .... 0.600"
3. The zero on the vernier plate has passed the midpoint division between the 6th and 7th marks, adding 1/20th" or .... 0.050"
4. The 15th line on the vernier matches a line on the beam, adding .... 0.015"

Total Reading ..... 1.665"

Additional inch vernier caliper readings are given in Fig. 9-14.

Universal Bevel Protractor

The universal bevel protractor equipped with a vernier, Fig. 9-15, measures angles accurately to 5 minutes or one-twelfth of a degree. This tool also is called a **vernier protractor.** It may be used to lay out, measure, or check angles, Fig. 9-16.

The principal parts of the protractor are shown in Fig. 9-15. A 150 or 300 mm (6" or 12") blade may be inserted in the graduated dial and locked in position with the blade clamp nut. The blade

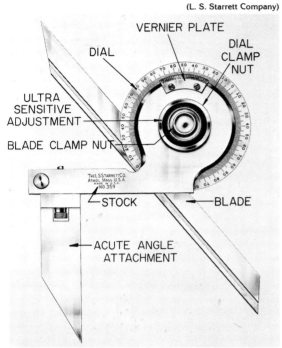

VERNIER PLATE

DIAL

DIAL CLAMP NUT

ULTRA SENSITIVE ADJUSTMENT

BLADE CLAMP NUT

STOCK

BLADE

ACUTE ANGLE ATTACHMENT

Fig. 9-15. Universal bevel protractor with vernier.

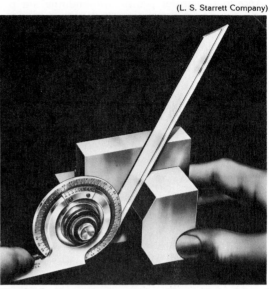

Fig. 9-16. Measuring an angle with a universal bevel protractor.

and dial are swiveled to the angle desired, and the dial is locked with the dial clamp nut. An acute-angle attachment is provided for use in measuring angles of less than 90°.

The protractor dial is graduated 360°, reading in whole degrees from 0 to 90, 90 to 0, 0 to 90, and 90 to 0. Each 10 degrees is numbered, and a long graduation divides each 5 degrees, as shown in Fig. 9-15. The vernier plate is graduated with 12 spaces. Thus, each line here represents 5 minutes or one-twelfth of a degree. Every third line on the vernier plate is numbered to represent 15, 30, 45, and 60 minutes. Both the protractor dial and the vernier plate have numbers in both directions from zero.

Reading the Vernier

When angles are read in whole degrees, the zero line on the vernier plate coincides with a graduation line on the protractor dial. Also, the graduation for 60 minutes will coincide exactly with a graduation on the dial.

When angles are read which are not in whole degrees, the following procedure is used: Note how many degrees can be read from the zero line on the dial up to the zero line on the plate. Then, reading in the **same direction** (and this is important) note the number of minutes indicated by the line on the vernier which coincides exactly with a line on the dial. Add this amount to the number of whole degrees. Remember, each graduation line on the vernier represents 5 minutes.

An explanation of the protractor reading in Fig. 9-17 follows:

1. The zero line on the vernier plate has passed line 50 on the dial, indicating 50 degrees   50°00′

2. Reading the vernier plate in the same direction as the numbers on the dial increase (this is very important), line 20 coincides with a line on the dial (shown with stars), adding a reading of 20 minutes   20′
   _____

   Total Reading   50°20′

Additional vernier protractor readings are provided in Fig. 9-18.

A magnifying glass is a useful aid in reading the fine graduations of all vernier scales more accurately.

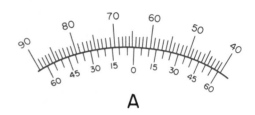

A

(L. S. Starrett Company)

Fig. 9-17 Universal bevel protractor vernier with reading of 50°20′.

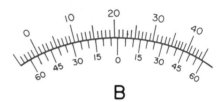

B

Fig. 9-18. Vernier protractor readings.
A. 64°10′
B. 20°45′

Unit

10

Dial Indicating Instruments

Machinists and toolmakers frequently use dial indicators. Dial indicators make it possible to quickly check correct alignment of one machine component with another and to correctly position workpieces in machine tools prior to machining. Dial indicating gages provide parts inspectors with a rapid and accurate way to check large quantities of nearly identical parts.

Dial Indicators

A dial indicator, Fig. 10-1, shows visually the amount of error, in size or alignment, for a part

(MTI Corporation)

Fig. 10-1.   Dial indicator with extended range counter.

being measured or positioned as in Fig. 10-2. Metric dial indicators are calibrated in hundredths of a millimeter (0.01 mm), or in two-thousandths of a millimeter (0.002 mm) units. Most inch dial indicators are calibrated in thousandths of an inch (0.001″), and ten-thousandths of an inch (0.000 1″), but instruments with calibrations as fine as fifty-millionths of an inch (0.000 050″) are available.

There are two common types of dial indicators. The **balanced type** has figures in both directions from the zero, as in Fig. 10-3. The **continuous reading type** is numbered continuously, as in Fig. 10-9. The balanced type is the more widely used. Generally, smaller dial indicators are made with a smaller range of travel than larger dial indicators. Travel is frequently limited to one rotation of the dial pointer. Extended travel dial indicators are equipped with a second pointer which counts the number of revolutions of the main dial pointer, Fig. 10-1. The dial reading may be set at zero at any time simply by rotating the entire dial face.

Dial Test Indicator Sets

Dial test indicator sets are made for use in a wide variety of setup and inspection work. The set shown in Fig. 10-3 has a steel base with T-slots, and a column clamped to the base for holding the dial indicator. The dial indicator is first set to the proper gauging height. The gauging height is the basic dimension of the part to be gauged or tested. The gauging height may be established with gage blocks (Fig. 11-3), with a planer gage which is set to the proper height, or with any appropriate gauging device. The dial should be set at zero. It should be under sufficient spring tension to allow the dial to rotate in either direction through the desired measuring range. In Fig. 10-2, the gauging height has been established between the contact point and the surface of the machine table. After the correct gauging height has been set and the dial indicator has been brought into contact with the workpiece, the difference between the gauging height and the height of the workpiece can be read directly from the dial indicator.

Universal dial indicator sets, as shown in Fig. 10-4, are used widely. With the accessories provided in this set, the dial indicator may be mounted on a surface gage for use on a surface

(L. S. Starrett Company)

(L. S. Starrett Company)

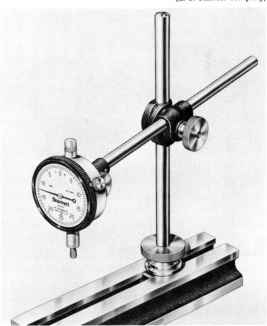

Fig. 10-2. Checking height of machined part with dial test indicator.

Fig. 10-3. Dial test indicator set.

(L. S. Starrett Company)

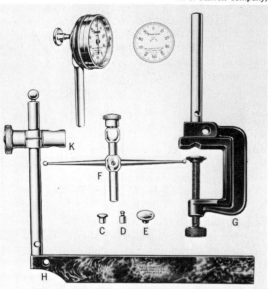

Fig. 10-4. Universal dial indicator set. C, D, and E are contact points; F is hole attachment; G, clamp; H, tool post holder; K, sleeve.

(Brown and Sharpe Manufacturing Company)

Fig. 10-5. Hole attachment permits accurate internal tests with dial indicator.

plate or on a machine table. It may be mounted in the tool post of a lathe. It is particularly useful for aligning work in a four-jaw chuck, as in Fig. 10-5, where it is being used with a hole attachment. It is also provided with a clamp so that it may be attached wherever needed for special applications.

Universal dial test indicators of the type shown in Fig. 10-6 are small and versatile, and are popular with machinists and inspectors. Their range of travel is restricted to one revolution of the pointer. A full range of accessories permits them to be held in surface gages, vernier height gages, and in lathe tool posts. With the aid of a magnetic base, the universal dial test indicator can be quickly attached to any machine, as in Fig. 10-7. It can also be held in a drill chuck or collet in a jig borer or vertical milling machine for use in checking alignment of vises, fixtures, or workpieces. In Fig. 10-8, the dial test indicator is mounted on a milling machine arbor and is used to align a vise on the machine table.

Dial Indicating Gages

Numerous special gages are equipped with dial indicators. These gages are used widely for determining whether parts are within required size limits.

(L.S. Starrett Company)

Fig. 10-6. Small dial test indicator of a type popular with machinists.

Dial indicating depth gages, Fig. 10-9, are used for gauging or testing the depth of holes, slots, shoulders, recesses, and keyways. Extension points increase the measuring depths at which the gage may be used.

Dial indicating hole gages, Fig. 10-10, are used to gauge or test holes for size, taper, out-of-roundness, or other irregular conditions. They are available in a wide range of sizes.

(L. S. Starrett Company)

Fig. 10-7.   Aligning a workpiece with the aid of a dial indicator.

(L.S. Starrett Company)

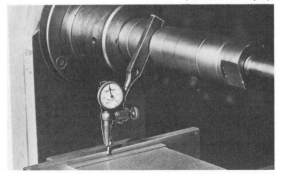

Fig. 10-8.   Dial test indicator mounted on milling machine arbor. It is used to align vise on the machine table.

(MTI Corporation)

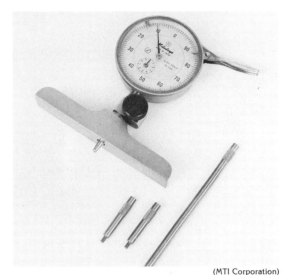

Fig. 10-9.   Dial indicating depth gage with extension points.

(MTI Corporation)

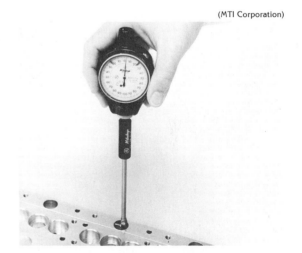

Fig. 10-10. Dial indicating hole gage.

Dial indicating snap gages, Fig. 10-11, are used for gauging diameters of parts to determine whether they are within the size limits specified. In use, the gage is snapped over the diameter of the part being gauged. Snap gages are efficient for checking parts which are produced in large numbers, and they are available in a wide range of sizes. Each gage may be used for measuring

(Federal Products Corporation)

Fig. 10-11. Dial indicating snap gage used to gauge work in cylindrical grinder.

(Federal Products Corporation)

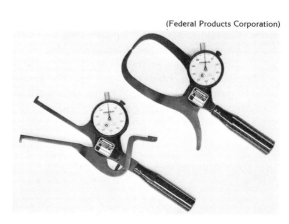

Fig. 10-13. Dial indicating caliper gages. Inside caliper at left; outside caliper at right.

(Federal Products Corporation)

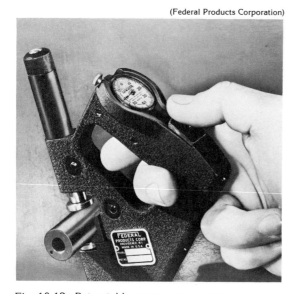

(Federal Products Corporation)

Fig. 10-14. Setting disc and setting ring.

Fig. 10-12. Retractable contact snap gage.

A retractable contact snap gage is shown in Fig. 10-12. The contactor point is opened with a button located conveniently for thumb operation. This type of gage is available in sizes which gauge, within certain ranges, from 0-112 mm (0″-4½″).

Dial indicating caliper gages, Fig. 10-13, have revolution counters which make it possible to measure directly through their complete range of 75 mm (3″). Calipers of this type are available with 0.02 mm (0.0008″) graduations.

Setting discs and setting rings, Fig. 10-14, are used for checking and setting indicating caliper gages, snap gages, hole gages, comparators, and other types of gages. Setting discs and rings are available in many sizes.

within a 25 mm (1″) range, such as 0-25 mm (0″-1″), 25-50 mm (1″-2″), etc. Size is set by adjusting the frame itself with the knurled wheel near the indicator. The gage may be used at an inspection bench where it can be mounted in a bench stand or right at the machine.

Dial Comparators

Dial comparators, Fig. 10-15, are used primarily for measurement of thickness and outside diameters. They are used as a gage for checking large quantities of parts to determine whether they are within required size limits. The lifting lever at the top retracts the contact point. A handwheel raises or lowers the table to the desired height.

Dial indicating comparators are available in a wide range of models and sizes. Metric dials with either 0.01 mm or 0.002 mm graduations, and inch dials with .001″, .0005″, or .0001″ graduations are commonly used.

(Federal Products Corporation)

Fig. 10-15. Dial comparator.

Unit

11

Gage Blocks and Gages

The purpose of gauging is not to determine the specific size of parts, but to determine whether the parts are within specified size limits. The devices or instruments designed for this purpose are called **gages**.

Toolmakers, machinists, and inspectors in production plants use gage blocks, gages, and gauging instruments of many types. They must know how to use and care for them properly. Many types of gages are very expensive and are not readily available to the machine shop student. Nevertheless, the student should be familiar with the basic types and their uses.

Gages, such as the snap gage in Fig. 10-11 and the comparator in Fig. 10-15, are designed for measuring large numbers of parts accurately and quickly. Measurements can be checked more rapidly with gages than with the use of conventional measuring tools, such as micrometers and verniers. Gages generally may be used with less training and skill. There also is less possibility for human error in using them.

Gage Blocks

Precision gage blocks, Fig. 11-1, are available either rectangular or square in cross section. They are made of hardened steel, chrome or tungsten carbide, or a special ceramic and are finely finished. Grade 1 gage blocks (formerly Grade AA) of 10 mm length or less are made to plus or minus 0.05 microns (1 micron = 0.001 mm). Inch blocks of the same grade, of one inch or less, are made to plus or minus two millionths of an inch (0.000 002″). These tolerances are valid at 20°C (68°F) — the standardized temperature for precision measurement in the United States.

(MTI Corporation)

Fig. 11-1.   An 88-piece gage block set.

(MTI Corporation)

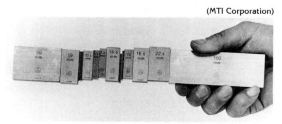

Fig. 11-2.   Gage blocks wrung together.

Gage blocks are recognized throughout the world as a practical **reference standard** for measuring length. They are used in precision measurement laboratories, inspection departments, toolrooms, and machine shops for calibrating and setting many types of inspection gages, measuring tools, and measuring instruments. Gage blocks, therefore, are the connecting link between the national standard of measurement and measurement in the shop.

Manufacturers produce gage blocks in relation to master gage blocks which must be accurate to plus or minus 0.05 microns (0.000 002″). In many cases, their master gage blocks are accurate to within plus or minus 0.025 microns (0.000 001″). These master gage blocks are

checked periodically for accuracy with an **interferometer.** This instrument measures gage blocks in units of light wave length to within a fraction of the required tolerance. (Light wave length is not significantly affected by changes in temperature and atmospheric conditions.)

Gage blocks may be used together to make up various gage block combinations of greater length. When gage blocks are properly combined, they are said to be **wrung** together. Their surfaces are so flat and smooth that when properly wrung, they stick together as though magnetized, Fig. 11-2. Some have been known to support a dead weight of over 90 kilograms (200 pounds) in tension when wrung together.

The reason for the adhesion of gage blocks is not clearly understood, but it is believed to be due to a combination of factors, including atmospheric pressure, molecular attraction, and a thin film of moisture or oil. It has been found that thoroughly cleaning the surfaces with certain solutions causes the blocks to fall apart easily. Gage blocks are wrung together by sliding them together under slight pressure. They also should be separated by sliding them apart.

Uses

Gage blocks are used for many purposes involving precision measurement. They are used to establish the height of dial indicating gages, Fig. 11-3. They are used to establish the original height setting for precise electronic measuring instruments. They are used in conjunction with a sine bar for establishing precise angular measurements, Fig. 12-1. Gage blocks also may be equipped with gage block jaws for determining precision length measurements, as in Fig. 11-4.

Measuring tools lose their original accuracy through wear, warpage, shrinkage, or damage in handling. Therefore, they must be readjusted or recalibrated to gage block standards. Gage blocks make possible accurate recalibration of micrometers, comparators, height and depth measuring tools, parallels, and squares.

Classification of Gage Blocks

Gage blocks normally are classified according to three accuracy classifications: Class 1 (formerly AA), Class 2 (formerly A +), and Class 3 (a blend of former A and B). These are U.S. Federal quality classifications based on permissible tolerances for length, flatness, parallelism, and surface finish. The length tolerances for metric blocks 10 mm or less, and inch blocks 1″ or less, at a temperature of 20°C (68°F), are as follows:

	Metric	Inch
Class 1 ±	0.05 microns	± 0.000 002″
Class 2 +	0.10 microns	+ 0.000 004″
−	0.05	− 0.000 002″
Class 3 +	0.20 microns	+ 0.000 008″
−	0.10	− 0.000 004″ *

Some manufacturers produce gage blocks with the special classification, **laboratory master gage blocks,** to tolerances of half that for Class 1 blocks.

These gage blocks, as well as regular **master gage blocks** (Class 1), are intended for special purposes in temperature-controlled gauging and measurement laboratories. They are used for experimental work, research work, and as grand master gages for measurement and inspection of other gages.

Class 2 gage blocks often are called **inspection gage blocks.** They are used primarily for inspection of finished parts and for inspecting and setting working gages. They also may be used as masters in inspection departments and toolrooms.

Class 3 gage blocks often are called **working gage blocks.** They are used for many applications requiring accurate measurement throughout the shop. They are used on surface plates for accurate layout, on machines for setting cutting tools accurately, and for ordinary inspection work.

Gage Blocks Sets

Gage blocks are available in sets or as individual blocks. Sets are available with from 5 to more than 100 gage blocks.

A commonly used standard set of 88 metric gage blocks consists of:

3 blocks:	0.5, 1.00, 1.005 mm
9 blocks:	1.001 through 1.009 by 0.001 mm
49 blocks:	1.01 through 1.49 by 0.01 mm
17 blocks:	1.5 through 9.5 by 0.5 mm
10 blocks:	10 through 100 by 10 mm

A similar set of 81 inch gage blocks consists of:

4 blocks:	0.062 5, 0.078 125, 0.093 75, 0.109 375″

*U.S. Federal specification GGG-G-15B.

(Brown and Sharpe Manufacturing Company)

Fig. 11-3.   Gage blocks used to establish the height of dial indicating gages.

(MTI Corporation)

Fig. 11-4.   Using gage blocks with jaws to determine precise length measurements.

3 blocks:	0.100 025, 0.100 050, 0.100 075″
9 blocks:	0.100 1 through 0.100 9 by 0.000 1″
49 blocks:	0.101 through 0.149 by 0.001″
19 blocks:	0.050 through 0.950 by 0.050″
4 blocks:	1.000 through 4.000 by 1″

Over 120,000 different gage sizes may be made with either of the above sets.

Gage blocks of greater length also are available individually or in sets. Various kinds of fixtures and accessories can be used to clamp or hold long combinations together. With appropriate accessories, these combinations are used as precision height gages, snap gages, and as gages for other special purposes.

Procedure for Wringing Gage Blocks

To assemble gage blocks as in Fig. 11-2, use the following procedure:

1. Clean the gage blocks as necessary by rubbing with a clean, soft, cotton cloth or chamois.
2. Hold the desired gage blocks as shown at **A**, Fig. 11-5.
3. Wipe the contacting surfaces on the wrist or palm of the hand, and place together as at **B**, Fig. 11-5.
4. With a slight contact pressure, slide the blocks together, as shown at **C**, Fig. 11-5. If the contact surfaces are clean, the blocks will adhere to each other as though magnetized.

Continue the above steps until the proper combination of blocks has been completed. The blocks are self-checking. Equal combinations may be checked against each other or against a single block. At **D** in Fig. 11-5, seven blocks totaling 25 millimeters are checked against a 25-millimeter block. Gage blocks should be disassembled by removing the end blocks with a sliding pressure.

Precautions in Care and Use of Gage Blocks

1. Do not put gage blocks away while still wrung together. Moisture between the surfaces may cause them to rust.
2. Keep gage blocks free from dust, dirt, or moisture.
3. After use, wipe the gage blocks free of finger marks with a clean cloth or chamois. Then, using a clean cloth or chamois, apply a thin film of white petrolatum to prevent rusting. Gage block manufacturers may recommend other substances as a protective film.
4. Nicks may be removed with a very fine deburring stone, available from the gage block manufacturer. Examine the block with a magnifying glass, and be careful to remove only the high spots. Do not wear the surface.
5. Gage blocks which are used regularly should be inspected at least once per year. Gage block manufacturers are equipped to inspect gage blocks accurately.

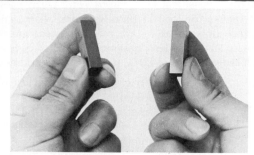

A. Hold gage block in each hand.

B. Wipe contacting surfaces on wrist or palm of the hand and place together.

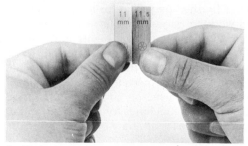

C. With slight pressure, slide the blocks together.

D. Blocks totaling 25 mm are checked against a 25 mm block.

(MTI Corporation)

Fig. 11-5.   Building gage block combinations.

Effect of Temperature Change on Gage Blocks

The accuracy of gage blocks is affected by temperature changes. They are accurate within specified classification tolerances at a temperature of 20°C (68°F).

Expansion or contraction, per degree of temperature variation from 20°C (68°F), essentially is the same for both the gage blocks and the steel parts being gauged. Consequently, if the gage blocks are at the same temperature as the steel parts being tested or gauged, size compensations need not be made. The measurements will be essentially equal.

Gage blocks which have raised in temperature due to contact with the hands should be placed flat on a cast-iron surface plate for 15 minutes or more. This will help reduce the difference in temperature between the gage blocks and the parts being inspected or gauged. A soft cloth should be used to insulate gage blocks against heat from the hand while holding them for setting snap gages and other gauging instruments.

The coefficients of expansion for nonferrous metals are different than for steel. Therefore, mathematical compensations may need to be made when nonferrous metal parts are gauged at temperatures which vary significantly from 20°C (68°F). However, such compensation is required only when the parts must be produced and inspected **within gage block tolerances.**

Procedure for Building Gage Block Combinations

In building up a specific gage block combination, use as few blocks as possible. You should know the sizes of the gage blocks in the set available. Start by selecting gage blocks which will remove the right-hand figure in the decimal size which you wish to build. Then select blocks which will remove the next right-hand decimal, and so on. Several examples will illustrate the procedure.

Example 1: (Metric)

Build up	39.985 mm	
Subtract	1.005	— First block
	38.980	
Subtract	1.48	— Second block
	37.50	
Subtract	7.5	— Third block
	30.00	— Fourth block

Example 2: (Inch)

Build up	2.8417″	
Subtract	0.1007	— First block
	2.7410	
Subtract	0.141	— Second block
	2.600	
Subtract	0.600	— Third block
	2.000	— Fourth block

Angle Gage Blocks

Angle gage blocks are precision tools used for accurate measurement of angles. A set of 16, Fig. 11-6, may be used for measuring 356,400 angles in steps of 1 second up to 99 degrees. These angle gage blocks may be wrung together in various combinations, just as rectangular gage blocks are. Angle blocks can also be wrung together for inspection of a simple angle on a part. In Fig. 11-7, they are used to establish the proper grinding angle on a revolving magnetic chuck.

(L. S. Starrett Company)

Fig. 11-6.   Set of one-second angle gage blocks.

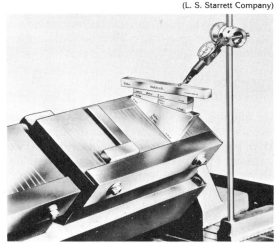

Fig. 11-7.   Setting a revolving magnetic chuck with angle gage blocks.

Angle blocks are manufactured in two accuracy classifications: **laboratory master angle gage blocks,** which are the most expensive, have an accuracy classification of plus or minus 1/4 second; **toolroom angle gage blocks** have an accuracy of plus or minus 1 second.

Angle gage blocks are so designed that they may be combined in plus or minus positions. One end of each angle block is marked **plus,** while the opposite end is marked **minus.** Several examples will illustrate how the blocks may be combined in either position, thus forming different angles. The plus end of a 15° angle block may be wrung together with the plus end of a 5° block to form a 20° angle. Wringing the plus end of the 15° block together with the minus end of the 5° block forms an angle of 10°. The angle blocks may be wrung together to form angles in steps of degrees, minutes, seconds, or in any combination of these units.

The procedure for setting up or inspecting precision angles with angle gage blocks is less complex than the procedure used with a sine bar. The latter method, which is explained in Unit 12, involves more complicated mathematical procedures. Also, for many angles, angle blocks are more accurate than the sine bar.

Other Types of Gages

Many kinds of gages are designed for many purposes. Some are designed for general use. Others are designed for special or limited measuring applications within a specific size range.

Adjustable-type gages usually are designed for general use and may be used within various size ranges. Several common gages of this type (described previously in Unit 10) include the following dial indicating instruments: depth gage shown in Fig. 10-9, hole gage in Fig. 10-10, snap gage in Figs. 10-11 and 10-12, caliper gages in Fig. 10-13 and the comparator in Fig. 10-15.

Fixed-type gages usually are designed for special or limited measuring applications within specific size limits. A few of the more common ones include plug gages, ring gages, snap gages, and thread gages.

Reference gages, also called **master gages,** are used for checking other gages. Inspection gages are tested for accuracy with a reference gage.

Go and Not-Go Gages

Go and not-go gages are also called **limits gages.** Because they have two gauging surfaces or points, they sometimes also are called **double gages.** One gauging surface is used for testing the upper size limit; the other, the lower size limit. The common types of limits gages include snap gages, ring gages, and plug gages.

The principles involved in testing parts with go and not-go gages are illustrated in Fig. 11-8. The figure shows how cylindrical parts are checked with a limits **snap gage.** The upper gauging point is the **go** point, while the lower is the **not-go** point. At **A,** the part is too small, since it passed between both the upper and the lower gauging points. At **B,** the part is satisfactory, since it passed the upper point, but hangs on the lower gauging point. At **C,** the part is too large, since it hangs on the upper gauging point. Most such gages are labeled "go" and "not go", meaning that the part will **go** at the first point, but it will **not go** at the second.

Limits gages never should be forced under high pressures when checking parts. The pressures applied to the gauging surfaces should be slight. Limits gages have contact surfaces which are hardened, precision-ground, and lapped.

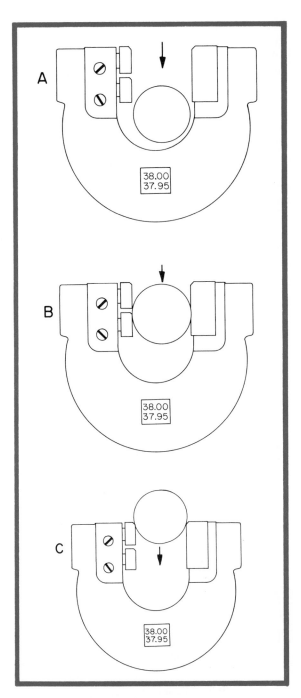

Although they are designed for accuracy and wear resistance, they should be handled and used carefully.

Snap Gages

One of the most widely used limits gages is the snap gage, Fig. 11-9. This caliper-type gage is used to gauge thicknesses, lengths, and outside diameters. Snap gages are available in several styles and sizes. That shown in Fig. 11-9 is an adjustable limits gage. It has one stationary anvil and two button anvils which are adjustable. The outer button is set to the go size, and the inner button to the not-go size. The procedure for using this type of snap gage was described earlier and is shown in Fig. 11-8.

The size limits on a snap gage may be checked with gage blocks, Fig. 11-10. Limits snap gages may be supplied by the manufacturer **set** and **sealed** at specified size limits, or **unset** and **unsealed**. When preset, the adjustment screws usually are sealed with sealing wax and the size limits are stamped on the gage.

Snap gages of special types are fitted with special anvils, buttons, or rolls for gauging special forms or external threads. A roll thread snap gage, Fig. 26-4, is used to check the size limits for the pitch diameter of screw threads. Dial indicating snap gages are described in Unit 10.

Fig. 11-8.   Parts tested in a limits gage.
   A. Part is undersize.
   B. Part is satisfactory.
   C. Part is too large.

Fig. 11-9.   Adjustable limits snap gage.

Ring Gages

Three types of ring gages commonly are used for checking the external diameters of parts: plain ring gages, taper ring gages, and thread ring gages.

Plain ring gages are designed in the form of a cylindrical ring, as in Fig. 11-11. These gages are used for checking the external diameters of straight round parts. The **not-go** ring (identified by the groove around the outside diameter) is used to check the minimum size limit. The **go** ring is used to check the maximum size limit.

The go ring will pass over a part which is within specified size limits with little or no interference. The not-go ring will not pass over the work. If both rings pass over, the part is undersize. If neither does, the part is oversize.

Taper ring gages are used for checking the size and the amount of external cylindrical taper on parts. The upper three items in Fig. 11-12 are cylindrical taper ring gages. Gages of this type are used for checking the taper shanks on drills, reamers, lathe centers, and other machine accessories.

In using a taper ring gage, first draw three equally spaced chalk lines lengthwise on the external tapered surface which is to be checked. Then slip the ring over the external taper by applying a light pressure for good surface contact.

Rotate the ring forward and backward a small amount while continuing to apply light pressure. Remove the gage and observe the external tapered surface. If all three chalk lines have been uniformly rubbed and distributed, the taper is correct. If the chalk lines have been rubbed harder at one end than at the other, then a correction should be made on the taper. The correct amount of taper should be established before the part is machined to finished size. Size may be determined by measuring the small diameter with a micrometer or by noting the distance to which the taper enters the ring gage.

Thread ring gages of the go and not-go type, Fig. 26-3, are used for checking the pitch diameter of external threads. The use of these gages is explained in Unit 26.

Plug Gages

Three basic types of plug gages are used for checking the accuracy of holes: plain cylindrical

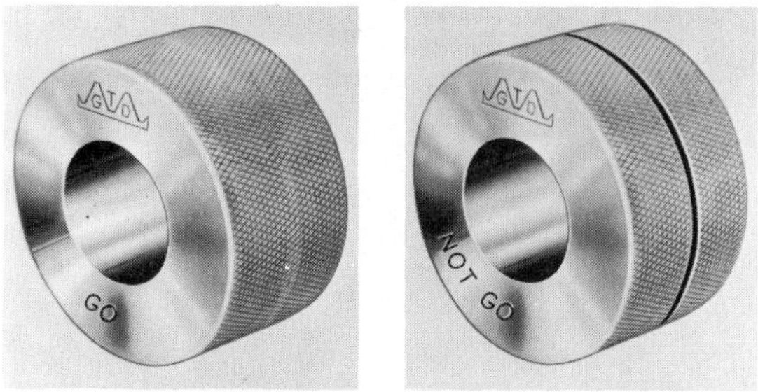

Fig. 11-11. Plain ring gages.

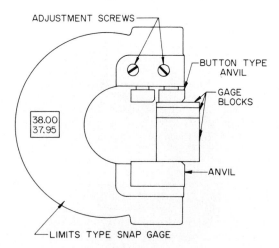

Fig. 11-10. Testing size of snap gage with gage blocks.

Fig. 11-12. Cylindrical taper rings and plus gages.

plug gages, cylindrical taper plug gages, and thread plug gages. Plug gages of special types also are available for use in checking holes of special shape, such as square holes and rectangular holes.

Plain cylindrical plug gages, as shown in Fig. 11-13, are accurate cylinders which are used for checking the size limits of straight cylindrical holes. The gage is provided with a handle for convenient use. The gage may be either the single-end type or the double-end type. The go gage is longer than the not-go gage; it should enter the hole with little or no interference. If great pressure is required, the hole is undersize and is not acceptable. The not-go gage should not enter the hole. If it does, the hole is too large.

A **progressive-type plug gage** has both the go and the not-go gages on the same end of the handle. It is efficient for checking through holes, but it cannot be used for shallow, blind holes.

Tapered cylindrical plug gages, as shown in the lower portion of Fig. 11-12, are used for checking the size and amount of taper in tapered cylindrical holes in drill sleeves, in machine tool spindles, and in adapters for use with taper shank tools.

In using the tapered cylindrical plug gage to check a tapered hole, use the same procedure described above for using a tapered cylindrical ring gage. The hole is finished to size when the gage enters to the end of the plug or to a depth indicated by a mark on the gage.

Thread plug gages, Fig. 26-1, are used for checking the size limits for the pitch diameter of internal screw threads. The use of thread plug gages is explained in Unit 26.

Optical Height Gage

The optical height gage shown in Fig. 11-14 is used for measuring the height of parts accurately to 0.127 microns (0.000 005″). This instrument is designed for use on a surface plate only. The optical height gage is basically a stack of gage blocks wrung together to an accuracy of 0.127 microns (0.000 005″). The gage is available in several model sizes for use in measuring heights which range from 0–2125 mm (0–85″).

The optical height gage measures through a range of 25 mm (1″). Measurements may start with any desired block setting within the range of the particular gage. Settings on the height gage are made by means of a lever-operated cam and vernier handwheels and are read in the optical eyepiece. Measurements may be taken from the numbered blocks and may be transferred easily to the part being measured on the surface plate.

(Greenfield Tap and Die Division of TRW Inc.)

Fig. 11-13. Plain cylindrical plug gages. Top—single-end gages. Bottom—double-end gage.

(L. S. Starrett Company)

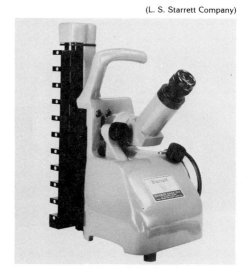

Fig. 11-14. Optical height gage.

Unit

12

The Sine Bar and Its Use

The Sine Bar

The **sine bar,** Fig. 12-1, is a precision tool used to establish or check angles to within one minute of arc. Sine bars must be used in conjunction with some true surface, such as a surface plate, from which accurate measurements may be taken. They are used to establish and check angles for layout work and inspection work. They may be used for making machining setups such as those often required for surface grinding. They also may be used to accurately determine unknown angles.

A sine bar is a hardened and precision-ground steel bar which has two hardened and precision-ground steel rolls of the same diameter attached. The edge of the bar is parallel with the center line of the rolls. For convenient mathematical calculation, it is available in lengths which provide a distance of 127 mm, 254 mm, or 508 mm (5″, 10″, or 20″) between the centerlines of the rolls.

The sine bar is named after the sine trigonometry function, which states that the sine of an angle is equal to the length of the side opposite the angle divided by the length of the hypotenuse. Applying this rule to the right triangle in Fig. 12-2, either unknown angle can be found when the side opposite the angle, and the hypotenuse are known: sine of angle $C = \dfrac{\text{side } c}{\text{hypotenuse } a}$; sine of angle $B = \dfrac{\text{side } b}{\text{hypotenuse } a}$. Also, when the

(Brown and Sharpe Manufacturing Company)

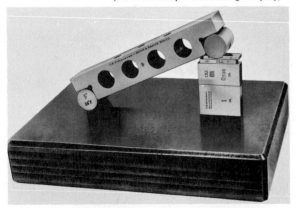

Fig. 12-1. Sine bar set at desired angle with gage blocks. Sine bars can readily convert to metric, but metric sine bars are also available.

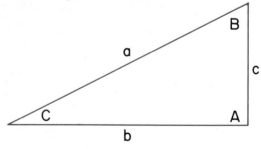

Fig. 12-2. A right triangle. Angle A is always 90°.

angle and the hypotenuse are known, the length of either side can be found: side c = sine of angle C × hypotenuse a; side b = sine of angle B × hypotenuse a. Thus, with the sine bar serving as a fixed length for the hypotenuse, Fig. 12-3, any unknown side or angle can be quickly calculated.

The sines for various angles are included in a table of **natural trigonometric functions,** Table A-15 in the Appendix. While this brief four-place table is desirable for instruction, the sines should be listed to at least five decimal places for accuracy to 1 minute. Such tables in machinists or mathematics handbooks, for example, indicate that the sine for an angle of 21° 36′ (21 degrees 36 minutes) is 0.368 125. (Table A-15 gives the sine of 21° 30′ as 0.3665 and of 21° 40′ as 0.3692, so by interpolation, 0.6 of the difference places the sine of 21° 36′ at 0.3681.)

Using the above angle as an example, the distance X in Fig. 12-3 for a sine bar of 127 mm (5″) length is calculated as follows:

Dimension X = Length of sine bar ×
  sine of angle
  X = 127 mm (5″) × sine 21° 36′
  X = 127 mm (5″) × 0.368 125
  X = 46.752 mm (1.840 625″)

To use the sine bar to determine an unknown angle, align the bar carefully with the unknown angle by establishing the distance X with gage blocks. The unknown angle is determined by calculating its sine and then consulting a sine table for the angle. The sine of the unknown angle is equal to the distance X divided by the length of the sine bar.

An example will illustrate how an unknown angle may be determined with a 127 mm (5″) sine bar when the distance X is known to be 59.847 mm (2.356 2″).

$$\text{Sine of Angle} = \frac{\text{Distance X}}{\text{Length of Sine Bar}}$$

$$\text{Sine of Angle} = \frac{59.847 \text{ mm } (2.356\ 2″)}{127 \text{ mm } (5″)}$$

Sine of Angle = 0.471 24
  Angle = 28° 7′ (to the nearest whole minute)

A sine bar or sine plate may be used with a master planer gage for setting grinding angles, as shown in Fig. 12-4. The height of the planer gage may be measured with a micrometer or vernier caliper.

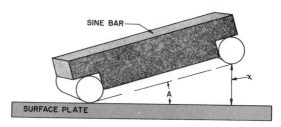

SINE BAR

SURFACE PLATE

Fig. 12-3.   Dimension "X" establishes the sine bar angle.

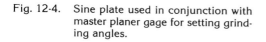

Fig. 12-4.   Sine plate used in conjunction with master planer gage for setting grinding angles.

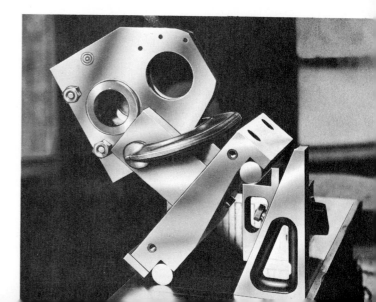

Sine Plate

The **sine plate** operates on the same principle as a sine bar and is very similar to a sine bar. The type shown in Fig. 12-5 is a **simple inspection sine plate,** because it permits only simple angles to be obtained. It is designed primarily for simple angular measurement, layout, and inspection work.

The sine plate has hardened precision rolls, hinges, and gage block surfaces. The roll on the right-hand end (Fig. 12-5) is located underneath the sine plate and cannot be seen in the picture. The roll at the left end is hinged to the base with a firm-fitting hinge.

When the sine plate is used, angles are established by wringing gage blocks together and placing them on the smooth gage block surface underneath the right-end roll. The sine plate may be used on a surface plate or it may be mounted on a machine table by clamping the flanged base. Tapped holes are provided on the sides and top of the plate for various clamping arrangements. The end and side holes may be used to fasten end or side plates.

Compound sine plates are used to establish compound angles. See Fig. 12-6.

Perma Sine

A **simple perma sine,** Fig. 12-7, is a sine plate equipped with a magnetic chuck for holding workpieces securely at any desired angle. A perma sine is well suited for use on surface grinders. The base is equipped with a flange for clamping to the machine table. The holding power of the chuck is turned on or off with a simple lever. Compound perma sines are also made.

(Brown and Sharpe Manufacturing Company)

Fig. 12-6.   Compound sine plate.

(Brown and Sharpe Manufacturing Company)

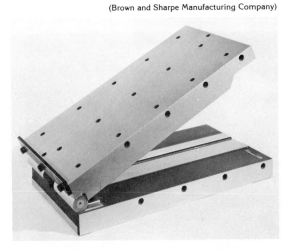

Fig. 12-5.   Simple sine plate.

(Brown and Sharpe Manufacturing Company)

Fig. 12-7.   Simple perma sine.

Unit

13

Inspection of Surface Finish, Flatness and Shape

Control of Surface Finish

The smoothness of a machined surface is determined by a combination of factors involved in the machining process. Some of the most significant of these include: type and condition of cutting tool used, rigidness of the machine and setup, type of material being cut, depth of cut, rate of feed, cutting speed, and kind of cutting fluid used. Figure 13-1 shows surface characteristics involved in measurement of surface finish quality.

The type of surface required on a given product is determined by the designer. Such items as bearings, gear teeth, and pistons must have controlled surface quality. For example, a surface on a bearing can be excessively rough or smooth. If it is too rough, it will wear rapidly, resulting in limited life. If it is too smooth, it will not have adequate provision for oil pockets and will be difficult to keep lubricated, thus again resulting in limited life.

To require a high surface quality where it is not necessary is expensive and unprofitable. Where detailed specifications concerning surface quality are not indicated, it means that the surface normally produced by that particular kind of machine operation is adequate.

The machinist must produce machined surfaces which meet specified standards of quality, and must be able to interpret the surface quality specifications indicated on drawings and blueprints. See Fig. 13-2. The machinist also must know how to determine whether machined surfaces meet surface quality specifications.

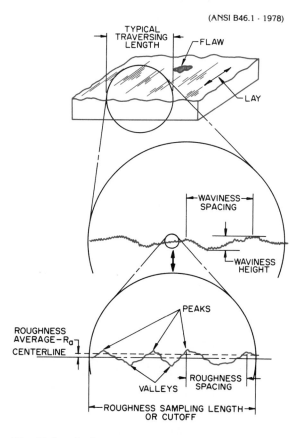

(ANSI B46.1 - 1978)

Fig. 13-1.   Surface characteristics.

All machined surfaces, including those which appear to be very flat and smooth, have surface irregularities. Under high magnification, scratches or grooves in the form of peaks and valleys are revealed. These irregularities may or may not be superimposed on larger waves. Such complex factors as height, width, and direction of surface irregularities determine surface texture. They are specified with standard symbols on drawings, Fig. 13-2.

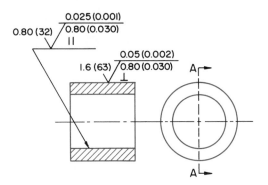

Fig. 13-2.   Application of surface roughness symbols.
Section AA Interpretation:
Roughness average (OD) . . . . 1.6 μm (63 μin.)
Roughness average (ID) . . . . 0.80 μm (32 μin.)
Roughness sampling length . . . . . . 0.80 mm (0.030″)
Waviness height (OD) . . . . 0.05 mm (0.002″)
Waviness height (ID) . . . . 0.025 mm (0.001″)
Lay (OD) . . . . Circumferential
Lay (ID) . . . . Axial

(ANSI B46.1 - 1962)

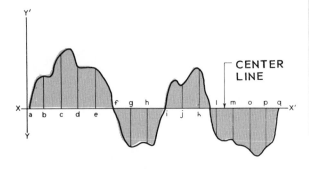

Fig. 13-3.   Short section of hypothetical profile divided into increments.

Terms Related to Surface Texture

The following terms related to surface texture have been defined by the American National Standards Institute (ANSI) and have been extracted from ANSI B46.1-1978 with the permission of the publisher:*

Surface texture. Repetitive or random deviations from the nominal surface which form the three-dimensional topography of the surface. Surface texture includes roughness, waviness, lay, and flaws.
Profile. The profile is the contour of a surface in a plane perpendicular to the surface, unless some other angle is specified.
Centerline. The centerline is the line about which roughness is measured and is a line parallel to the general direction of the profile within the limits of the sampling length, such that the sum of the areas contained between it and those parts of the profile which lie on either side are equal. See Fig. 13-3.
Micrometer. A micrometer is one millionth of a meter (0.000 001 m), and may be abbreviated μm.
Microinch. A microinch is one millionth of an inch (0.000 001 inch), and may be abbreviated μin. One microinch equals 0.0254 micrometer, (1 μin. = 0.0254 μm).
Roughness. Roughness consists of the finer irregularities in the surface texture, usually including those irregularities which result from the inherent action of the production process. These are considered to include the traverse feed marks and other irregularities within the limits of the roughness sampling length. See Fig. 13-1.

Roughness Average. Roughness average is the arithmetic average of the absolute values of the measured profile height deviations taken within the sampling length and measured from the graphical centerline, Fig. 13-1. Roughness average is expressed in micrometers (microinches). Table 13-1 gives preferred roughness average values.

Roughness Spacing. The average spacing between adjacent peaks of the measured profile within the roughness sampling length. See Fig. 13-1.
Roughness Sampling Length. The roughness sampling length is the sampling length within which the roughness average is determined. This length is chosen, or specified, to separate the profile irregularities which are designated as roughness from those irregularities designated as waviness. Standard values are given in Table 13-2. Roughness sampling length is measured in millimeters (inches). See also Fig. 13-1. (Continue to refer to Fig. 13-1 throughout the ANSI definitions.)

*American National Standards Institute. **Surface Texture**, ANSI B46.1-1978. The American Society of Mechanical Engineers. United Engineering Center, 345 East 47th Street, New York, NY 10017.

Cutoff. The cutoff is the electrical response characteristic of the roughness average measuring instrument which is selected to limit the spacing of the surface irregularities to be included in the assessment of roughness average. The cutoff is rated in millimeters (inches).

Waviness. Waviness is the more widely spaced component of surface texture. Unless otherwise noted, waviness is to include all irregularities whose spacing is greater than the roughness sampling length and less than the waviness sampling length. Waviness may result from such factors as machine or work deflections, vibration, chatter, heat treatment, or warping strains. Roughness may be considered superimposed on a 'wavy' surface.

Table 13-1
Preferred Series Roughness
Average Values (R$_a$)

| Micrometers– µm | | Microinches– µin. | |
µm	µin.	µm	µin.
0.012	0.5	1.25	50
0.025*	1*	1.60*	63*
0.050*	2*	2.0	80
0.075	3	2.5	100
0.10*	4*	3.2*	125*
0.125	5	4.0	160
0.15	6	5.0	200
0.20*	8*	6.3*	250*
0.25	10	8.0	320
0.32	13	10.0	400
0.40*	16*	12.5*	500*
0.50	20	15	600
0.63	25	20	800
0.80*	32*	25*	1000*
1.00	40		

*Recommended

Table 13-2
Standard Roughness Sampling Length
(Cutoff) Values*

| Millimeters–mm | | Inches—in. | |
mm	in.	mm	in.
0.08	0.003	2.5	0.1
0.25	0.010	8.0	0.3
0.80	0.030	25.0	1.0

*When no value is specified, the value 0.8 mm (0.030 in.) applies.

Waviness Height. The waviness height is the peak-to-valley height of the modified profile from which the roughness and flaws have been removed by filtering, smoothing, or other means. The measurement is to be taken normal to the normal profile within the limits of the waviness sampling length and expressed in millimeters (inches). Table 13-3 gives preferred waviness height values.

Waviness Spacing. The waviness spacing is the average spacing between adjacent peaks of the measured profile within the waviness sampling length.

Lay. Lay is the direction of the predominant surface pattern, ordinarily determined by the production method used.

Flaws. Flaws are unintentional irregularities which occur at one place or at relatively infrequent or widely varying intervals on the surface. Flaws include such defects as cracks, blow holes, checks, ridges, scratches, etc. Unless otherwise specified, the effect of flaws shall not be included in the roughness average measurements. Where flaws are to be restricted or controlled, a special note as to the method of inspection should be included on the drawing or in the specifications.

Application of Surface Finish Symbols

Surface quality is designated with a surface finish symbol and ratings. The symbol is similar to a check mark, but with a horizontal extension line added, Fig. 13-2. The long leg of the check-like symbol is to the right as the drawing is read. If only the roughness height is designated, the horizontal extension line may be omitted.

The point of the surface symbol is located on the line indicating the surface specified. It also may be located on an extension line or leader pointing to the surface specified, as in Fig. 13-2. Symbols used with the surface symbol to indicate lay are shown in Fig. 13-4.

Table 13-3
Preferred Series Maximum Waviness
Height Values*

mm	in.	mm	in.	mm	in.
0.0005	0.00002	0.008	0.0003	0.12	0.005
0.0008	0.00003	0.012	0.0005	0.20	0.008
0.0012	0.00005	0.020	0.0008	0.25	0.010
0.0020	0.00008	0.025	0.001	0.38	0.015
0.0025	0.0001	0.05	0.002	0.50	0.020
0.005	0.0002	0.08	0.003	0.80	0.030

*Waviness is not currently shown in ISO standards. It is included here to follow present industry practice in the United States.

Fig. 13-4.   Lay symbols.   (ANSI Y14.36 · 1978)

Lay Symbol	Meaning	Example Showing Direction of Tool Marks
—	Lay approximately parallel to the line representing the surface to which the symbol is applied.	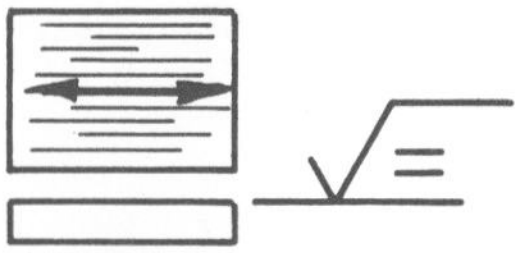
⊥	Lay approximately perpendicular to the line representing the surface to which the symbol is applied.	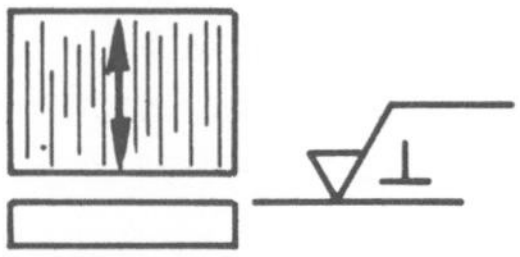
X	Lay angular in both directions to line representing the surface to which the symbol is applied.	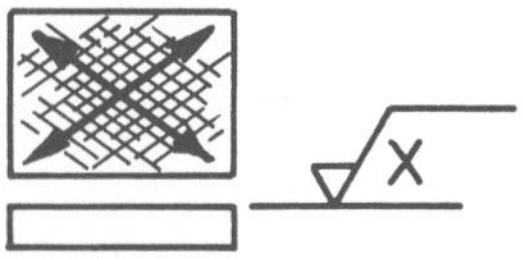
M	Lay multidirectional.	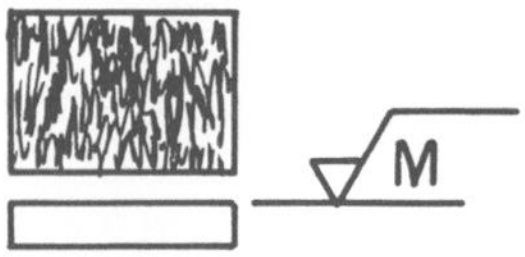
C	Lay approximately circular relative to the center of the surface to which the symbol is applied.	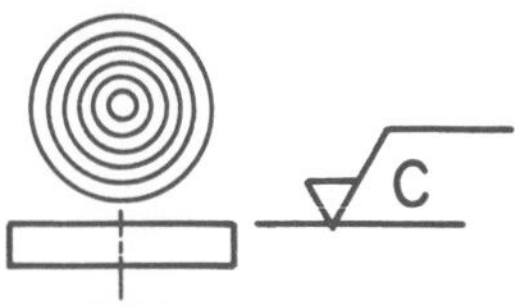
R	Lay approximately radial relative to the center of the surface to which the symbol is applied.	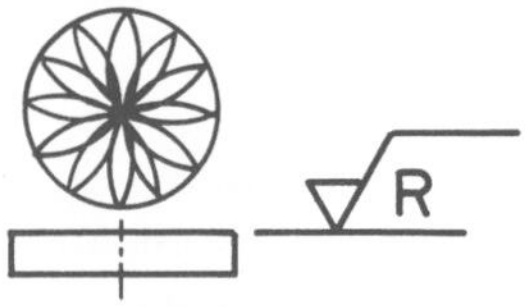
P*	Lay particulate, nondirectional, or protuberant.	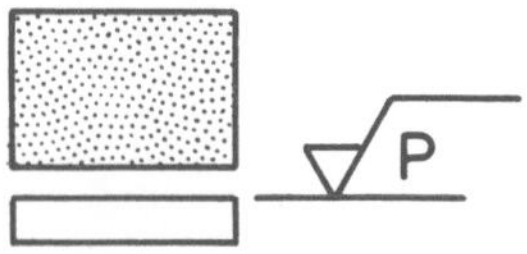

*The "P" symbol is not currently shown in ISO Standards. American National Standards Committee B46 (Surface Texture) has proposed its inclusion in ISO 1302—"Methods of Indicating Surface Texture on Drawings."

Surface quality ratings for various characteristics such as roughness, waviness, and lay are positioned specifically in relation to the surface symbol. The relative location of these specifications and ratings is indicated in Fig. 13-5. An example of surface quality symbols with specifications is shown and interpreted in Fig. 13-2.

The quality of a surface is determined to a large extent by the machining method used to produce the surface. The range of surface roughness made by various machining methods is illustrated in Table 13-4.

Table 13-4
Surface Roughness Produced by Common Production Methods

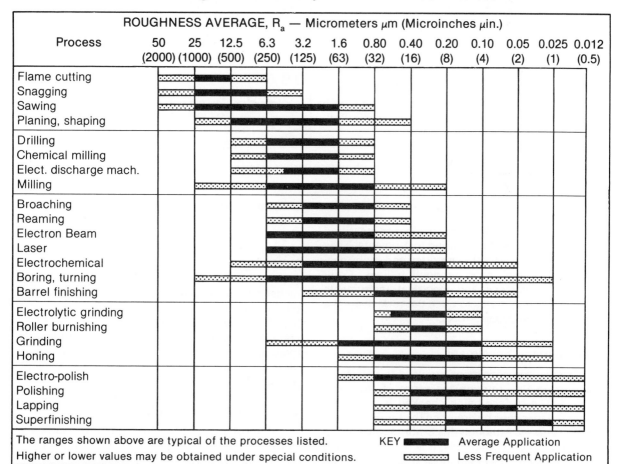

| ROUGHNESS AVERAGE, R_a — Micrometers μm (Microinches μin.) |
| Process: 50 (2000), 25 (1000), 12.5 (500), 6.3 (250), 3.2 (125), 1.6 (63), 0.80 (32), 0.40 (16), 0.20 (8), 0.10 (4), 0.05 (2), 0.025 (1), 0.012 (0.5) |
| Flame cutting, Snagging, Sawing, Planing, shaping |
| Drilling, Chemical milling, Elect. discharge mach., Milling |
| Broaching, Reaming, Electron Beam, Laser, Electrochemical, Boring, turning, Barrel finishing |
| Electrolytic grinding, Roller burnishing, Grinding, Honing |
| Electro-polish, Polishing, Lapping, Superfinishing |

The ranges shown above are typical of the processes listed.
Higher or lower values may be obtained under special conditions.

KEY ▬ Average Application   ▭ Less Frequent Application

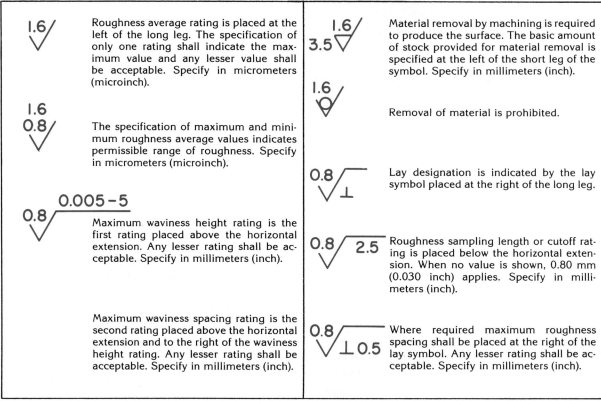

(ANSI Y14.36 · 1978)

Fig. 13-5.   Application of surface texture values to symbol.

Surface Finish Measurement

Surface irregularities are checked or measured by various types of measuring instruments. They are used in research laboratories and by inspectors or machinists on production lines. Comparison specimens and electrical instruments with a sensitive stylus to trace the surface are the most commonly used tools.

Comparison Specimens

Comparison specimens are used to determine surface quality through the sense of sight and touch. The surface being checked is compared with the specimen by dragging the fingernail over both surfaces. Although this is not as reliable and accurate as other methods, it is economical and satisfactory for checking those surfaces which do not require extremely close tolerances.

Commercial comparison specimens, also called **master standards,** are available in sets, Fig. 13-6. The sets include blocks with various types of lay and with surface qualities ranging from rough to very smooth. They are available in various materials, including stainless steel, nickel alloys, aluminum alloys, and hard black plastic. The black plastic aids in elimination of possible error in judgment which may result in comparing bright specimens with less brilliant materials. An example of a specimen set complete in one solid piece is the comparator shown in Fig. 13-7.

Surface Roughness Indicator

The most widely used instrument for surface finish measurement is an electrical instrument of the **stylus type.** As the perpendicular stylus is passed over the surface, its motion is amplified,

(GAR Electroforming)

Fig. 13-7.   Microfinish comparator.

and the average roughness of the surface is indicated in micrometers (microinches). An instrument of this type is shown in Fig. 13-8.

Surface measuring instruments of the stylus type are available for either hand- or machine-powered movement of the stylus. More accurate readings are obtained with the power-driven stylus. If the surface being measured has a high degree of waviness, and if the specifications require an accurate roughness-width-cutoff setting, hand-guided operation of the tracer head and stylus is not recommended.

The meter scale on the instrument can be switched to read any one of several roughness average ranges. The instrument also may be provided with several roughness sampling length values, including 0.08 mm (0.003"), 0.25 mm (0.010"), and 0.80 mm (0.030"). Unless otherwise specified on the drawing, a roughness-width cutoff of 0.80 mm (0.030") should be selected. With this wider cutoff range, a more accurate average reading of surface irregularities may be determined.

Designations for Roughness Height

The roughness average, according to the 1978 ANSI standard, is expressed in micrometers (microinches) as the simple **arithmetical average (AA)** deviation, measured normal to the centerline. The centerline through the profile of a rough surface is shown in Fig. 13-3. In previous standards,

(GAR Electroforming)

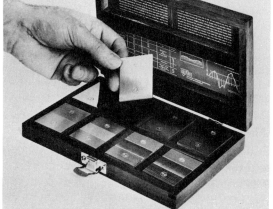

Fig. 13-6.   Surface finish master standards.

(AirTronics)

Fig. 13-8.   Surface roughness indicator, motor drive, and tracer.

the roughness height was expressed in micrometers (microinches) as the **root mean square average** (RMS) deviation, measured normal to the centerline. Certain instruments are equipped with a selector switch for selecting either the RMS or the AA reading.

Roughness measuring instruments calibrated for AA values will indicate approximately 11 percent lower for a given surface than those calibrated for RMS average values. However, because the absolute limit of roughness for satisfactory functioning of a surface is indefinite, many manufacturers adopt AA ratings without changing the RMS values indicated on older drawings. For most surface measurement applications, the difference between the two values is of no consequence.

In order to eliminate error or confusion in the use of various stylus instruments, standards are included in ANSI B46.1-1978. For instruments indicating a numerical value only, a spherical-tip stylus with a 10 micrometer (400 microinch) radius tip is standard. The accuracy of instruments for surface roughness measurement should be checked periodically by measuring a precision reference specimen.

Flatness Measurement

Flatness is critically important to the accuracy of gage blocks and certain other gages, micrometers, parallels, and other precision tools. It is equally important to proper functioning of flat metal-to-metal assemblies which must be leak-free without use of sealing materials. This degree of flatness goes well beyond the capacity of dial indicators or other conventional measuring tools to detect. **Interferometry,** using the interference of two beams of light for measurement, is capable of measuring flatness to 25 millionths of a millimeter (one millionth of an inch). Optical flats utilize interferometry and are relatively inexpensive tools used for measuring flatness.

Optical Flats

Optical flats, Fig. 13-9, are quartz or glass lenses that have been polished accurately flat on one or both surfaces. When both surfaces are polished, they are known as **optical parallels.** They range in size from 25 mm (1″) diameter, 13 mm (1/2″) thick, to 300 mm (12″) diameter, 70 mm (2-3/4″) thick, and larger. A choice of three grades provides accuracies of 25, 50, or 100 millionths of a millimeter (1, 2, or 4 millionths of an inch).

The precision surface of the optical flat facing the work is both transparent and light reflecting. Because of this, light waves striking this surface are effectively split into two light waves, one passing through and one reflecting back. When two reflected light waves cross or interfere with each

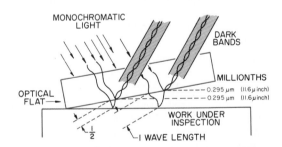

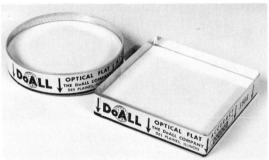

(DoAll Company)

Fig. 13-9. Optical flats.

Fig. 13-10. The surface that faces the work is the functioning part of an optical flat. It is transparent and can also reflect light. Therefore, all light waves that strike this surface are, in effect, split in two longitudinally. One part is reflected back by the surface of the flat. The other part passes through and is reflected back by the surface under inspection, as illustrated. Whenever the reflected split portions of two light waves cross each other (interfere), they become visible and produce dark bands. This happens whenever the distance between the reflecting surfaces is one-half of a wave length or multiples thereof.

other, they become visible as dark bands, Fig. 13-10 and Fig. 13-11. This occurs when the surface of the optical flat and the surface being measured are out of parallel by one half of a wavelength, or multiples thereof, of the light being used. Helium light, with a predominant wavelength of 589 millionths of a millimeter (23.2 millionths of an inch) is ordinarily used with optical flats. See Fig. 13-12. Thus, each dark band provides a measuring unit of 294.5 millionths of a millimeter (11.6 millionths of an inch).

Flat surfaces out of parallel in one axis produce a pattern of dark bands at right angles to that axis. Diagonal bands result when the surfaces are out of parallel in both axes. Convex or concave surfaces produce a pattern of curved bands, Fig. 13-13.

(DoAll Company)

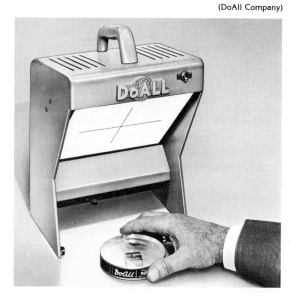

Fig. 13-12. Helium light source.

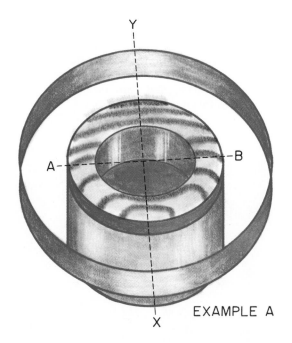

EXAMPLE A

Fig. 13-11. Pattern of dark bands produced by a rotating seal that is not flat. The greatest variation from flatness is across the approximate center, where the center is higher than the outer edges by 1½ scale lines or 442 millionths of a millimeter (17.4 millionths of an inch).

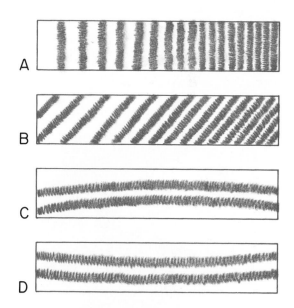

Fig. 13-13. Band patterns reveal surface characteristics.
A. Flat surfaces out of parallel in one axis.
B. Flat surfaces out of parallel in both axes.
C. Convex surface.
D. Concave surface.

Shape Measurement

Where only close approximation of shape is required, template type gages such as screw pitch and radius gages (see Unit 5), and custom-made templates are often adequate. Templates, however, cannot provide the degree of accuracy required for precision measurement of thread and gear tooth profiles, cams, miniature instrument parts, and other small parts calling for precise shape. Highly precise inspection of shape can be obtained with **toolmaker's microscopes** and **optical comparators.**

Toolmaker's Microscope

The toolmaker's microscope, Fig. 13-14, is designed especially for use in toolmaking. Images can be enlarged from 10 to 200 times, depending on equipment and job requirements. Images are not reversed as in ordinary microscopes. The platform, called a **stage,** on which workpieces are mounted for measurement, can be moved both crossways and sideways with vernier micrometer accuracy. A vernier protractor built into the microscope head, reads to five minutes of arc through the magnifying lens. The microscope optics provide 90° cross hairs and concentric circles with diameters from 0.25–5 mm by 0.25 mm (0.10″–0.200″ by 0.010″). An attachment allows cylindrical workpieces to be supported between centers. Direct angular measurement, thread dimensions, radii, hole roundness, and even squareness can be easily checked with a high degree of accuracy.

Optical Comparators

An optical comparator, Fig. 13-15, projects an accurately enlarged shadow-like profile of the part being measured onto a screen. Here, both its size and shape are compared to a master drawing or template. Magnification of 10, 20, and 50 diameters are commonly used. A stage provides sideways and crossways movement of the part to vernier micrometer accuracy. Angular measurement can be directly from templates or from a protractor attachment.

Optical comparators are especially useful for checking small, irregularly shaped objects which cannot be easily measured with conventional tools. Flexible parts, such as springs and rubber or plastic objects, which would distort under the pressure of conventional measuring tools, can easily be inspected. The accuracy of thread forms, gear tooth shape, and cam profiles are easily checked in this manner.

(Brown and Sharpe Manufacturing Company)

(Jones and Lamson TEXTRON)

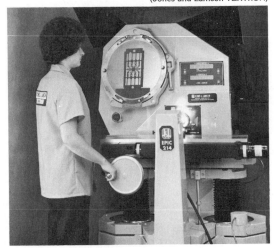

Fig. 13-14. Toolmaker's microscope.

Fig. 13-15. Optical comparator with an image of a part shown projected on its screen.

Unit

14

Setup Tools

Setup tools include machine vises, parallels, hold-downs, angle plates, and other assorted work-holding accessories. These tools and accessories are used for mounting workpieces in machine vises or for fastening the workpieces directly to the machine table.

Machine Vises

Plain, swivel, and tilting machine vises are used for mounting work on drill presses, shapers, milling machines, and other machine tools, Fig. 14-1. The jaws of the machine vise may be made of soft or hardened steel. Care must be taken to avoid scarring or otherwise damaging them, so as to preserve their finish and accuracy.

Parallels

Parallels often are used to make setups for machining and for layout and inspection work. See Fig. 14-2. In machining setups, parallels are used to raise the work above the jaws of the vise and to provide a solid seat for it, Figs. 14-3 and 14-5. Parallels are strips of hardened steel which have been machined accurately to the desired size. They are made in many sizes and are ordinarily sold in pairs. Parallels which are numbered according to pairs should be used in pairs to insure accuracy of the setup.

For holding workpieces at an angle, **angular parallels** are used, as in **B**, Fig. 14-3. These are similar to **rectangular parallels,** except that one

(Kurt Manufacturing Company)

(Universal Vise & Tool Company)

A. A plain vise with a flanged base

B. A swivel vise

Fig. 14-1.   Machine vises.

C. A tilting vise

Fig. 14-2.   Hardened and ground steel parallels.

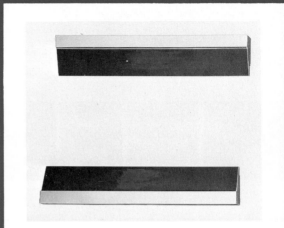

Fig. 14-4.   Hold-downs.

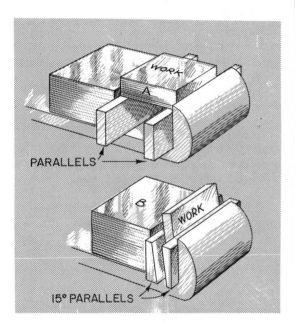

Fig. 14-3.   Using parallels.

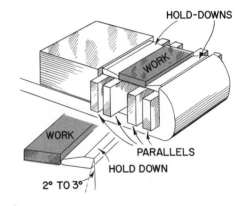

Fig. 14-5.   Using hold-downs.

side has been machined at an angle with the base, for example, 15 degrees.

Hold-Downs

Sometimes parallels, or combinations of parallels, of the size required to raise the work above the jaws of the vise are not available. This is particularly true when machining thin stock. Hold-downs of the type shown in Fig. 14-4 are

used to grip the work and hold it in position, as shown in Fig. 14-5. Hold-downs are wedge-shaped in cross section with the thick edge being beveled 2° or 3°. This causes the hold-down to press downward at the thin edge when it is tightened against the work. Thus, the work is held sufficiently rigid to permit machining with light cuts.

Angle Plates

One common type of angle plate is L-shaped, with two surfaces accurately finished at 90° to each other, Fig. 14-6. Work that cannot be held in a vise for machining may be clamped to an angle plate which is bolted to the machine table. In Fig. 5-25, an angle plate is shown resting on a

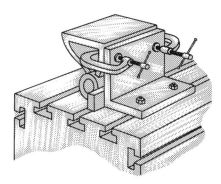

Fig. 14-6.   Using angle plates for machining setups.

Fig. 14-8.   Holding irregularly shaped casting with U-clamps and adjustable bolts.

(Armstrong Bros. Tool Co.)

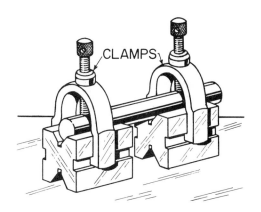

Fig. 14-7.   Matched pair of V-blocks.

Fig. 14-9.   Typical setup tools and accessories.

magnetic chuck and supporting a magnetic V-block. An angle plate is also shown in Fig. 5-23 supporting a workpiece for precision layout.

Angle plates are precision tools which should be used carefully. They should not be nicked or abused, since nicks, scratches, or rust will affect their accuracy.

V-Blocks

V-blocks are valuable work-holding tools used mainly for holding round stock for drilling and other light machining operations. They are made in many styles and are often sold in matched pairs. Some styles permit the V-block clamps to be installed or removed while the blocks can remain clamped in a machine vise. Other styles are made to be clamped directly to the machine table, Fig. 5-25, or to be held in the hand during use, as in Fig. 14-7.

Work-Holding Accessories

Workpieces which cannot be mounted in a machine vise often are bolted directly to the worktable with the use of **strap clamps** and parallels, as shown in Fig. 14-8. The parallels are used to support the part, hold it level, or provide a bearing surface for the clamps. Various clamps, jacks, and other accessories are used in supporting and clamping the large casting shown in Fig. 14-9.

T-slot bolts, Fig. 14-10, are used to fasten strap clamps or machine vises to the machine table. The bolts are designed so that they will not turn in the T-slot of the worktable when tightened. The bolts, nuts, and washers are heat-treated for strength and durability, and they are available in a variety of sizes.

Fig. 14-10. T-slot, bolt, nut, and washer.

Fig. 14-12. T-slot clamp.

Fig. 14-11. T-nut and stud set.

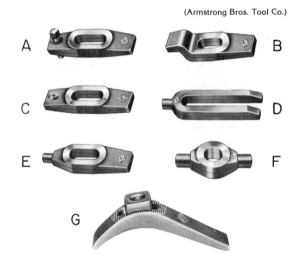

Fig. 14-13. Machine strap clamps.
- A. Screw head
- B. Goose neck
- C. Plain
- D. U
- E. Finger
- F. Double finger
- G. Universal adjustable

T-nuts and studs serve the same purpose as T-bolts, but are more versatile. If necessary, studs of several lengths can be combined to get the necessary length. Figure 14-11 shows a typical set of T-nuts and studs.

T-slot clamps of the type shown in Fig. 14-12 are equipped with a T-slot bolt and are used for holding work on all types of machine tables equipped with T-slots. The clamp is tightened to the table with the T-slot bolt. The work is held down by tightening the hand screw.

Machine strap clamps of several kinds are shown in Fig. 14-13. To get maximum holding power with strap clamps, they should always be supported parallel to the table surface, and the clamping bolt should be as near the workpiece as possible. Figures 14-8 and 14-9 show correct applications.

Planer jacks, Fig. 14-14A, are useful in leveling and supporting castings for setups, as shown

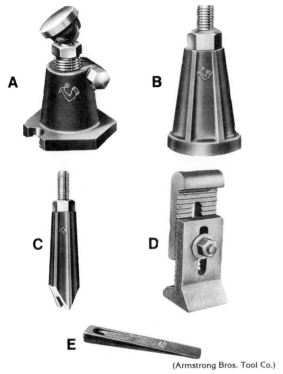

A

B

C

D

E

(Armstrong Bros. Tool Co.)

Fig. 14-14. Setup tools.
    A. Planer jack
    B. Vertical jack
    C. Bracing jack
    D. Adjustable step block
    E. Setup wedge

in Fig. 14-9. **Vertical jacks,** Fig. 14-14B, also are used to level and support castings. **Bracing jacks,** Fig. 14-14C, are used to prevent the workpiece from sliding, as shown in Fig. 14-9.

Step blocks, Fig. 14-14D, provide support at various heights required for setups. They are useful particularly as support blocks for strap clamps, as shown in Fig. 14-9. Figure 14-15 shows a step block and clamp set designed for rapid setup. The system is shown in use in Fig. 14-16.

The **wedge,** Fig. 14-14E, is useful in leveling or supporting large castings which are bolted to machine tables.

(Northwestern Tools, Inc.)

Fig. 14-15. Step block and clamp set.

Fig. 14-16. Step block and clamp system provides secure holding power.

(Northwestern Tools, Inc.)

Test Your Knowledge of Section 3

Unit 4: Principles of Linear Measurement

1. What is the name of the modern form of the metric system that is being adopted world-wide?
2. What is the standard metric unit of linear measurement, and how is it related to the standard inch unit of linear measurement?
3. List the units into which the meter is commonly divided.
4. List the units into which the inch is commonly divided.
5. List the decimal equivalents of 3/4", 5/8", 9/16", 15/32", 25/64".
6. What is the name of the United States agency that is responsible for maintaining standards of measurement?
7. What is used as the practical standard of measurement in machine shops throughout the world?
8. Why is it impossible to produce parts to absolute or exact sizes?
9. Define the following terms: (a) actual size; (b) basic size; (c) limits of size.
10. Define the meaning of the term **tolerance.**
11. Define the following terms: (a) fit; (b) clearance fit; (c) interference fit.
12. Define the meaning of the term **allowance.**
13. List several ways in which tolerances may be specified on a drawing.

Unit 5: Measuring and Layout Tools

1. In what lengths are metric steel rules made? Show approximate inch equivalents in parentheses.
2. What graduations are provided on metric steel rules? Inch?
3. List the four principal parts of a combination set.
4. For what purpose is a center head used?
5. Why should precision steel squares be handled so carefully?
6. What is a cylindrical square? Describe how it is used.
7. What kind of square can be adjusted to measure angles up to 10°?
8. List two uses for hermaphrodite calipers.
9. For what purposes is a center gage used?
10. Explain how a screw pitch gage is used.

11. Describe a radius gage and its use.
12. What is meant by the term **layout work?**
13. Explain the difference between dividers and calipers.
14. Explain the difference between a center punch and a prick punch.
15. Describe a surface plate and list several uses for it.
16. For what purpose is a straightedge used?
17. Explain what V-blocks are and describe how they are used.
18. List several uses for surface gages.
19. Describe a telescoping gage and explain how it is used.
20. List several uses for adjustable parallels.
21. List several uses for a master planer and shaper gage.

Unit 7: Micrometers

1. List six principal parts of a micrometer.
2. To what degree of accuracy does a plain metric micrometer measure? A plain inch micrometer?
3. Vernier-equipped metric micrometers measure to what degree of accuracy? Vernier inch micrometers?
4. What is the typical range of measurement for metric micrometers? Inch micrometers?
5. How are tubular inside micrometers and depth micrometers able to measure through a wider range than outside micrometers?
6. How do mike hole gages differ from other inside micrometers?
7. Describe how to check a 25 mm or 1 inch micrometer for accuracy.
8. How are micrometers larger than 25 mm or 1 inch checked for accuracy?

Unit 10: Dial Indicating Instruments

1. To what degree of accuracy are metric dial indicators calibrated? Inch dial indicators?
2. Explain the difference between the balanced dial indicator and the continuous reading type.
3. List several uses for dial indicators.
4. List several kinds of dial indicating gages and briefly describe their use.
5. What is a dial comparator? For what and how is it used?

Unit 11: Gage Blocks and Gages

1. Name two main advantages of using gages over using micrometers or vernier measuring tools.
2. What measuring instrument measures accurately to within 25 micromillimeters (one millionth of an inch), using light waves?
3. List several uses for gage blocks.
4. List the grades in which gage blocks are made and give their corresponding accuracies in both millimeters and inches.
5. What is the standardized temperature for precision measurement in the United States? Celsius? Fahrenheit?
6. Describe the procedure for wringing gage blocks together.
7. List several precautions in caring for and using gage blocks.
8. Explain the effects of temperature change on the accuracy of gage blocks.
9. Explain the procedure for building up a gage block combination measuring: (a) 16.125 mm; (b) 1.7326".
10. List the two classes of angle gage blocks and give the degree of accuracy to which they are made.
11. Explain the use of go and not go, or limits, gages.
12. How are the size limits for a snap gage checked or accurately set to size?
13. Explain how tapered ring or plug gages are used.
14. Describe an optical height gage. For what is it used?

Unit 12: The Sine Bar and Its Use

1. Describe how a sine bar is used.
2. In what lengths are sine bars usually made?
3. Why is a sine bar so named?
4. Give the formula for calculating the altitude (the distance **X**) for a given angle.
5. Give the formula for determining an unknown angle with a sine bar, when the altitude is known.
6. List several purposes for which a sine plate may be used.
7. How do the simple and compound sine plates differ?
8. What is a perma sine?

Unit 13: Inspection of Surface Finish, Flatness, and Shape

1. List several factors which determine the quality of a machined surface.
2. How is it possible for a finish to be too smooth?
3. Why should a designer avoid requiring a high-quality surface when a surface of lesser quality would be satisfactory?
4. What unit of measure is used for measuring surface finish quality?
5. What is the meaning of waviness? How is it caused?
6. Explain the meaning of **lay.**
7. Write the surface symbol for the following conditions: (a) 0.40–80 micrometer (16–32 microinch) finish; (b) 0.025 mm (0.001") waviness height; (c) perpendicular lay; (d) 0.76 mm (0.030") roughness width cutoff.
8. What are the average roughness height ranges produced by the following machining operations: (a) drilling; (b) milling; (c) turning; (d) grinding; (e) lapping.
9. Explain what comparison specimens are and how they are used to check surface texture.
10. What is the advantage in using a surface roughness indicator which has a power operated stylus?
11. What **roughness width cutoff** rating should be selected for measuring with a stylus instrument when no specific figure is indicated with the surface texture symbol?
12. Explain how an optical flat works.
13. Name two instruments used for precision measurement of parts with irregular shapes or contours.
14. Explain the technique of measurement used in optical comparators.

Unit 14: Setup Tools

1. Name three types of machine vises.
2. What are parallels and for what are they used?
3. Describe how hold-downs are used.
4. How are T-nuts and studs more versatile than T-bolts?
5. What is the function of step blocks?
6. List three kinds of jacks used in supporting or bracing workpieces.
7. Make a sketch to show correct use of strap clamps, step blocks, and T-bolts to hold a rectangular workpiece to a machine table.

Section 4

Threads and Threading

Unit

15

Screw Threads and Their Use

A screw thread is a helical or spiral ridge of uniform section on the surface of a cylinder or a cone, either external or internal. Threads on bolts and screws are external threads, Fig. 15-1. Threads on nuts are internal threads, Fig. 15-2. Threads on a cylindrical surface (such as bolts, machine screws, and nuts) are **straight** or **parallel** threads. Threads on a conical surface are **tapered** threads.

Threads may be **right-hand** (RH) or **left-hand** (LH), Fig. 15-3. A right-hand thread advances away from the observer when turned clockwise. A left-hand thread advances away from the observer when turned counterclockwise. A grinder with two grinding wheels, one mounted on each end of the arbor, has a right-hand thread and nut on one end and a left-hand thread and nut on the other end. Taps and dies are available for use in cutting right-hand and left-hand threads. Unless a thread is otherwise designated, it is assumed to be right-hand.

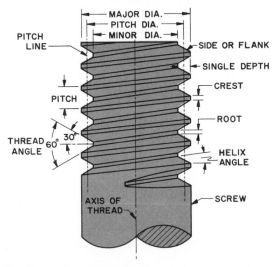

Fig. 15-1. Principal parts of an external screw thread.

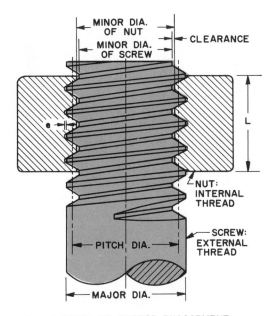

L = LENGTH OF THREAD ENGAGEMENT
e = EXTENDED MAJOR DIAMETER OF TAP
    FOR CLEARANCE

Fig. 15-2.   Comparison between the minor diameters of a screw and a nut, showing clearance. External threads and internal threads have the same basic pitch diameters.

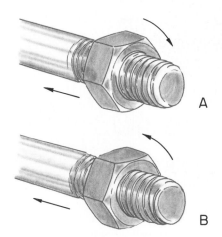

Fig. 15-3.   Right-hand thread advances with clockwise rotation as in A; left-hand thread advances with counterclockwise rotation as in B.

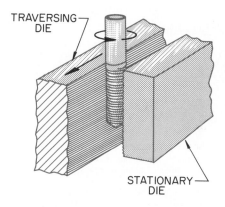

Fig. 15-4.   Technique of thread rolling used in cold heading machines.

Uses of Screw Threads

Screw threads are widely used on fasteners such as nuts, bolts, and screws. They permit easy assembly and dismantling for replacement of parts.

Screw threads are also widely used to transmit motion, transmit power, increase mechanical advantage, control movement accurately and uniformly, and permit adjustments on machines. The lead screw on a lathe transmits power. The screw on a vise provides for increasing mechanical advantage. The screw on a micrometer provides accurate and uniform control of movement, thus making accurate measurement possible. Screw threads allow for adjustments on tools, machines, instruments, and control devices.

How Screw Threads Are Produced

Screw threads are produced manually with hand tools or power-machine tools. Hand taps and dies are used to cut threads manually.

Machine tools that produce screw threads are the lathe, turret lathe, automatic screw machine, tapping machine, milling machine, specialized thread rolling machine, and thread grinding machine. Tapping attachments allow threads to be tapped efficiently with a drill press or with a portable power drill.

Most standard threaded fasteners are made on high-speed machines called **cold headers**, which automatically make the fasteners from wire. It forms the threads by **thread rolling**, Fig. 15-4.

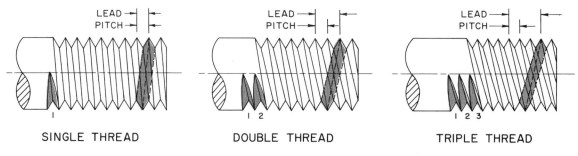

Fig. 15-5. Relationship of pitch and lead on multiple threads.

Turret lathes or automatic screw machines produce quantities of special fasteners and can either cut or roll threads with equal efficiency.

Screw Thread Terms

The following **screw thread terms** and definitions are adapted from ANSI definitions. Use these terms when discussing screw threads or when calculating or cutting screw threads:*

External thread: The thread on the external surface of a cylinder or cone, Figs. 15-1 and 15-2.

Internal thread: The thread on the internal surface of a cylinder or cone, Fig. 15-2.

Major diameter: The largest diameter of a straight external or internal thread, Figs. 15-1 and 15-2.

Minor diameter: The smallest diameter on a straight external or internal screw thread, Figs. 15-1 and 15-2.

Pitch diameter: On a straight thread, the diameter of an imaginary cylinder which passes through the thread profile at points where the width of the groove and the width of the thread are equal, Figs. 15-1 and 15-2. The pitch diameter may be measured with a thread micrometer, Fig. 26-2. The amount of clearance permitted between two mating threads is controlled by maintaining close tolerances on their pitch diameters.

The pitch diameter on a taper thread, at a given position on the thread axis, is the diameter of the pitch cone at that position.

Pitch: The distance from a point on one screw thread to a corresponding point on the adjacent thread, measured parallel to the thread axis, Fig. 15-1. The pitch of a thread is a measure of the size of the thread form. For metric threads, pitch is expressed in millimeters. For inch-based threads, pitch is equal to 1 divided by the number of threads per inch.

*__Unified Screw Threads__ (ANSI B1.1 — 1960), Published by the American Society of Mechanical Engineers, New York. N.Y.

Lead: The distance a thread moves along its axis, with respect to a mating part, in one complete revolution. On a single thread, the lead and the pitch are the same. On a double thread, the lead is equal to twice the pitch. On a triple thread, the lead is equal to three times the pitch. See Fig. 15-5. (A single thread has one groove; a double thread, two grooves; a triple thread, three grooves; and so on, Fig. 15-5).

Multiple thread: A thread having the same form produced with two or more helical grooves, such as a double, triple, or quadruple thread, Fig. 15-5.

Angle of thread: The included angle between the sides or flanks of the thread, measured in an axial plane, Fig. 15-1.

Lead angle: (Sometimes called helix angle.) Angle made by the helix of a thread at the pitch diameter, measured in a plane perpendicular to the axis of the thread, Fig. 15-1.

Axis of a screw thread: The axis of the pitch cylinder or cone on which the screw thread appears, Fig. 15-1.

Crest: The top surface which joins the two sides of a thread. The crest of an external thread is at its major diameter. The crest of an internal thread is at its minor diameter, Figs. 15-1 and 15-2.

Root: The bottom surface which joins the sides of two adjacent threads. The root of an external thread is at its minor diameter. The root of an internal thread is at its major diameter, Figs. 15-1 and 15-2.

Flank: The surface which connects the crest with the root on either side of the thread, Fig. 15-1.

Clearance: The distance between the crest of a thread and the root of the mating thread, measured perpendicular to the thread axis, Fig. 15-6.

Depth of engagement: The depth to which one thread is engaged with a mating thread, measured perpendicular to the thread axis.

Length of engagement: The contact distance between an external and internal thread, measured parallel to the axis along the pitch cylinder or cone, Fig. 15-2.

Height of thread: (Sometimes called depth of thread.) The distance between the major and minor cylinders or cones of the thread, measured perpendicular to the axis of the thread.

Form: The profile for a length of one pitch in an axial plane, Fig. 15-6.

Maximum material limits: The maximum limit of size for an external dimension, or the minimum limit of size for an internal dimension.

Minimum material limits: The minimum limit of size for an external dimension, or the maximum limit of size for an internal dimension.

Design form of thread: The form of the thread (either internal or external) under the maximum material condition, i.e., under maximum material limits (Fig. 17-2). Specified tolerances generally permit the thread roots to be cut deeper than those illustrated in Fig. 17-2.

Fig. 15-6.   Sharp V-thread and original American (National) thread form.

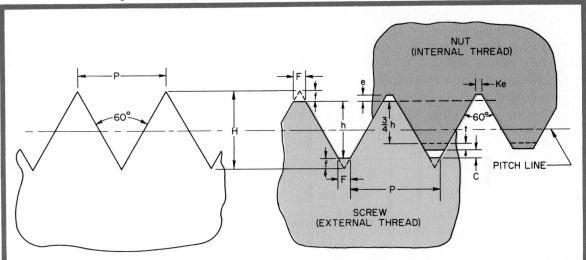

$H = $ Depth of Sharp Vee thread $= .866 \times$ Pitch

$P = $ Pitch $= \dfrac{1}{\text{No. Threads Per Inch}} = \dfrac{1}{n}$

$n = $ Number of Threads Per Inch

$h = $ Basic Depth (or Height) $= 0.6495\,P = \dfrac{0.6495}{n}$
$\qquad\qquad = \dfrac{6}{8}\,H$

$F = $ Flat $= 0.1250\,P = \dfrac{1}{8}\,P$

$f = $ Truncation $= \dfrac{1}{8}\,H = \dfrac{1}{6}\,h = 0.10825\,P$

$e^{\cdot} = $ Extended Major Dia. on Tap for Clearance

$Ke = $ Not Less Than $\dfrac{1}{3}\,F$ or $\dfrac{1}{24}\,P$

$t = $ Tolerance $= \dfrac{1}{12}$ of Basic Depth $= \dfrac{1}{12}\,h$

$C = $ Clearance $= \dfrac{1}{6}$ of Basic Depth $= \dfrac{1}{6}\,h$

$t + C = \dfrac{1}{4}$ of Basic Depth $= \dfrac{1}{4}\,h$

$\dfrac{3}{4}\,h = \dfrac{3}{4}$ of Basic Depth (or $\dfrac{3}{4}$ Full Thread in Nut)

$\dfrac{3}{4}\,h + t = \dfrac{5}{6}\,h = \dfrac{5}{8}\,H = \dfrac{0.54127}{n} = 0.54127\,P = $
Maximum Depth of Thread Engagement, Design Form (Maximum Material Condition)

Minor Dia. of Internal Thread (Design Form) = Major Diameter of Internal Thread — 1.08253 P

Terms Related to Thread Classes or Fits

Nominal size: A designation used for general identification purposes.

Actual size: The size of that dimension as measured on the individual part.

Basic size: The theoretical or designated size from which the limits of size are derived through the application of allowances and tolerances.

Allowance: On mating threads or parts, an allowance is the intentional difference in size between the maximum material limits of the parts; the minimum clearance between the parts is called **positive** allowance, and the maximum interference is called **negative** allowance. The allowance on mated threads represents the difference between the largest external thread and the smallest internal thread, thus the tightest possible fit.

The following examples illustrate allowances for external threads mated with internal threads:

Example 1: M15 × 1.5 thread, class 2A external thread and class 2B internal thread:

Minimum pitch diameter of nut	14.026 mm
Minimum pitch diameter of screw	13.990 mm
Allowance (positive)	0.036 mm

Example 2: 5/8"-11 Unified National coarse thread, class 3A external thread and class 3B internal thread:

Minimum pitch diameter of nut	0.5660"
Maximum pitch diameter of screw	0.5660"
Allowance	0.0000"

Tolerance: The tolerance is the total permissible variation in the size of a thread pitch or other dimension. It may be expressed as plus or minus, or both. The total tolerance is the sum of the plus and minus tolerances. Tolerances on threads may be specified for pitch diameter, major diameter, minor diameter, thread angle, half-thread angle, and lead.

The following is an example of the tolerance or amount of variation permitted on the pitch diameter of a M15 × 1.5 thread, class 2A:

Maximum pitch diameter	13.990 mm
Minimum pitch diameter	13.871 mm
Tolerance	0.119 mm

For Unified and American National screw threads, the tolerance is applied **minus** to external threads and plus to internal threads.

Limits: The maximum and minimum sizes permissible for a given dimension. The limits for the pitch diameter of a M15 × 1.5 thread, class 2A are 13.990 mm and 13.871 mm, as shown in the example above.

Fit: The range of tightness existing between two mating parts as a result of clearance or interference when they are assembled. The fit of threads may range from a loose fit with some play to an interference fit requiring a screwdriver or wrench for assembly. Where a specific class of thread or fit is designated, it means that certain allowances and tolerances are specified.

Classes of threads: One class of thread is distinguished from another by the amount of tolerance (or allowance and tolerance) specified for the thread.

History and Development of Screw Threads

It is not known who first discovered the use of the screw thread, but it occurred many centuries ago. A very early application was made by Archimedes (278-212 B.C.) who developed a screw which was enclosed in a cylinder for the purpose of drawing water. During the Middle Ages, nuts and bolts were used to fasten metal suits of armor together. Such early screws were made by hand and were very crude. Until about 1800, screws were forged to shape and the threads were layed out and filed by hand. Since there were no standards concerning size or pitch, these early bolts and nuts could not be used interchangeably.

Screw Threads Made by Machine

Henry Maudslay developed the first practical screw-cutting lathe in England in 1797. (See Fig. 1-5.) His lathe was equipped with a master lead screw and change gears which made it possible to cut uniform threads of different sizes and pitches.

Maudslay was interested in threads and methods of cutting threads. He developed his own system of thread sizes and pitches, and also made his own taps and dies for hand threading.

Screw Threads in Great Britain

In 1841, Sir Joseph Whitworth, who had worked several years for Henry Maudslay, developed a standard for screw threads. During the next 20 years, his standard was generally adopted in Great Britain. Whitworth's original standard, with various modifications, formed the basis of the British Standard Whitworth (BSW) thread, which still is used to some extent in Great Britain. The BSW thread has a 55° V-form with rounded crests and rounded roots, Fig. 15-7.

Screw Threads in the United States

The first progress toward the standardization of screw threads in the United States occurred in 1864. During that year, a committee appointed by the Franklin Institute in Philadelphia, PA, presented a report recommending proposed screw thread standards, many of which were developed by William Sellers. The report was adopted, and the thread system became known as the **Sellers Standard**.

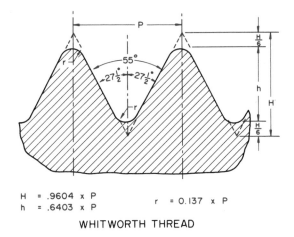

H = .9604 x P
h = .6403 x P   r = 0.137 x P

WHITWORTH THREAD

Fig. 15-7.   Cross section of Whitworth thread form.

Sellers accepted many of the diameter-pitch combinations of the Whitworth system. His thread had a 60° included angle, instead of the 55° angle used on the Whitworth thread. Hence, the Sellers thread was not interchangeable with the Whitworth thread. The 60° V-form was easier to produce and measure. Sellers modified the sharper V-form of thread which was customary at the time by utilizing flats at the crest and roots.

USS Thread

In 1868 the Sellers Standard was adopted by the U.S. Navy and became known as the **United States Standard** (USS) thread. It became widely adopted for government services, railroads, and industry. The USS thread was a system of coarse threads which served the needs of industry in the U.S. at that time.

SAE Thread

With expanding industry and the development of the automobile, the airplane, and other modern equipment, other diameter-pitch combinations with finer threads were needed. This brought forth the adoption of the SAE (Society of Automotive Engineers) thread standard about 1911. The SAE standard included a system of fine threads, which became rather widely adopted in the U.S.

Following the adoption of the USS and SAE thread standards, efforts were made to further develop and improve these standards. There was a need for the establishment of more specific dimensional limits, tolerances, and classes of thread fits. The established standards were concerned only with thread form, pitches, and external diameters.

Original American (National) Screw Thread Standard (1924)

The National Screw Thread Commission was established by Congress in 1918 for the purpose of adopting new screw thread standards. Its aim was to eliminate unnecessary sizes and to use predominant existing screw sizes as much as possible. At the same time, industries and engineering societies were interested in the development of new thread standards. Groups cooperating with the Screw Thread Commission included the American Standards Association, the Society of Automobile Engineers, and the American Society of Mechanical Engineers.

As a result of these cooperative efforts, the existing USS and SAE series of threads were used as the basis and were further developed as a new thread standard. This resulted in the **American (National) Screw Thread Standard** which was approved in 1924. The thread profile was designated as the **American (National) Form** of thread (Fig. 15-6).

Two thread series were adopted in the American (National) Screw Thread Standard — the coarse thread and the fine thread. The coarse thread corresponded closely with the USS thread, and the fine thread corresponded with the SAE thread.

Four **classes of fits** were adopted under the 1924 American (National) Screw Thread Standard — the **loose fit**, the **free fit**, the **medium fit**, and the **close fit**. Tables of tolerances and allowances were provided for the four classes of fits which could be used on either the coarse- or the fine-thread series.

American (National) Thread System (1935)

In 1933 the National Screw Thread Commission made several modifications in the American (National) Screw Thread Standard of 1924. The coarse-thread series, previously designated USS, was changed to NC (National Coarse). The fine-thread series, previously designated SAE was changed to NF (National Fine). These modifications were included in the revised American Standard for screw threads approved by the American Standards Association in 1935. Other series of threads which are less commonly used included the following:

Extra-fine series was designated NEF.
8-thread series was designated 8N. This series has 8 threads per inch for all diameters from 1″ to 6″ in diameter.
12-thread series was designated 12N. This series has 12 threads per inch for all sizes from 1/2″ to 6″ in diameter.
16-thread series was designated 16N. This series has 16 threads for all sizes from 3/4″ to 6″ in diameter.
Special threads were designated NS.

The original American (National) thread profile was retained on all of the thread series classified under the American (National) thread system. The form of the thread profile and the formulas used in calculating the required dimensions are shown in Fig. 15-6.

The 1935 standard for American (National) screw threads, including the four classes of fits, still is being used to a very limited extent by American industry. It has been largely replaced by the **Unified and American (National) Screw Thread Standards** published in 1949 and 1960.* These new Unified Standards include six thread classes designed to replace the four classes of fits in the 1935 standard.

*__Unified and American Screw Threads__ (ANSI B1.1-1949) and later edition, __Unified Screw Threads__ (ANSI B1.1-1960), Published by the American Society of Mechanical Engineers, New York, New York.

Unit

16

ISO Metric Threads

Metric standards of measurement have long been used by a majority of countries, but lack of international cooperation led the major industrialized nations to develop their own national standards. This resulted in the creation of several different metric thread systems which were not interchangeable between countries. In 1949, the International Standards Organization (ISO) recommended worldwide adoption of three series of metric threads based on the same 60° thread form adopted in 1948 for the ISO Inch (Unified) threads. This thread form is shown in detail in Fig. 17-2. The three series are referred to as (1) ISO Metric coarse pitch, (2) ISO Metric fine pitch, and (3) ISO Metric constant pitch. While the ISO

Metric and the ISO Inch threads share the same thread form, they are not interchangeable due to differences in diameters and pitches. Figure 16-1 compares the ISO Metric coarse series thread sizes with the Unified National (ISO Inch) coarse series thread sizes. A complete listing of ISO Metric coarse and fine pitch threads is given in Table A-7, Appendix.

Only the ISO Metric coarse pitch threads are in common use on fasteners. ISO Metric fine pitch threads are used primarily on precision tools and instruments. The constant pitch series are used mainly on machine parts, but are also found on all spark plugs.

Classes of ISO Metric Threads

Basically, there are three classes of fit: **fine, medium,** and **coarse.** The classes of fit are more accurately identified by specification of the **tolerance grade** and **tolerance position** of the mating external and internal threads. Tolerance grades are specified by a number, and may be applied to both the major diameter and pitch diameter. Figure 16-2 lists the range of tolerance grades for external and internal threads. Grade 6 is recommended for medium fits on general purpose threads, and is closest to Unified class 2A and 2B fits. (See Unit 17.)

Tolerance position is specified with a lower-case letter for external threads, and a capital letter for internal threads as follows:

External threads:
    e = large allowance
    g = small allowance
    h = no allowance

Internal threads:
    G = small allowance
    H = no allowance

The combination of tolerance grade and tolerance position constitutes the **tolerance class** of the thread. Figure 16-3 shows the tolerance classes for external and internal threads which are identified with the three classes of fit.

Thread Designations

Basic designations for all ISO Metric threads begin with the capital letter "M". Next, the nominal size (basic major diameter) in millimeters is given. This is followed by the pitch in millimeters, separated by an "X". ISO practice calls for the pitch to be omitted when designating coarse series threads. Therefore, an ISO Metric

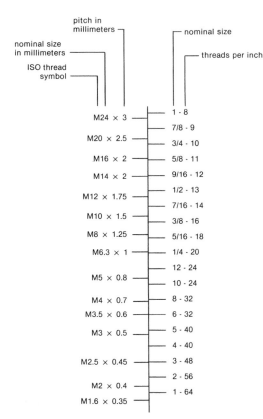

Fig. 16-1. Comparison of common sizes of ISO Metric coarse thread series with Unified National (ISO inch) coarse thread series.

Fig. 16-2.
Tolerance grades for ISO Metric threads.

External Thread		Internal Thread	
Major Diameter	Pitch Diameter	Major Diameter	Pitch Diameter
	3		
4	4	4	4
	5	5	5
6	6	6	6
	7	7	7
8	8	8	8
	9		

Fig. 16-3.
ISO Metric thread tolerance classes identified with class of fit.

Class of Fit	Tolerance Class	
	External Threads	Internal Threads
Fine	4h	5H
Medium	6g	6H
Coarse	8g	7H

10 mm coarse series thread with a pitch of 1.5 mm would simply be designated M10, whereas the same diameter in the fine series would be designated M10 X 1.25.

Complete designations for ISO Metric threads includes identification of the tolerance class. The tolerance class follows the basic designation, separated by a dash. The tolerance grade and position for the pitch diameter are given first, followed by the tolerance grade and position for the major diameter. If the pitch and major diameter tolerance are the same, then the symbols need only be given once. Examples:

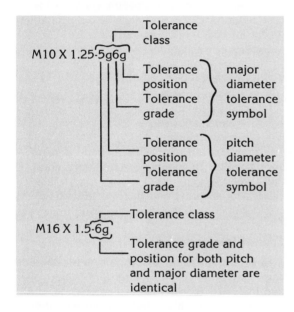

Optimum Metric Fastener System (OMFS)

Beginning in 1971, a special committee appointed by the American National Standards Institute began a study to develop a metric threaded fastener system for United States industries which would use the fewest possible number of sizes, grades, series, types, and styles of fasteners. The major goal was to reduce the 116 diameter and pitch combinations originally provided in the coarse and fine ISO Metric and ISO Inch series threads. The study was carried out by the Industrial Fasteners Institute with the support of United States fastener manufacturers.

Initial recommendations called for a single system of 25 diameter-pitch combinations, with sizes ranging from M1.6 through M100. M6 × 1 and M7 × 1 sizes were to be replaced with an M6.3 × 1 size, which is a close approximation of the popular 1/4-20 UNC thread. Slight differences in thread profile were proposed in order to strengthen the thread. Other differences proposed included a change to two classes of fit, slightly different material grades standards, and the method of describing the thread.

A series of international meetings held in 1975 resulted in agreements which made it unnecessary for the United States to have a separate standard for the OMFS. ISO Metric standards were revised to accept the single series of 25 diameter-pitch combinations, including the M6.3 × 1 size. See Fig. 16-4. Only one tolerance class for internal threads (6H) and two tolerance classes for external threads (6g for general purpose and 5g6g for close tolerance) were called for. Also, a second thread system using a stronger thread profile was approved for all aerospace fasteners.

Additional information on ISO Metric threads can be found in handbooks for engineers and machinists.

M1.6 × 0.35	M20 × 2.5
M2 × 0.4	M24 × 3
M2.5 × 0.45	M30 × 3.5
M3 × 0.5	M36 × 4
M3.5 × 0.6	M42 × 4.5
M4 × 0.7	M48 × 5
M5 × 0.8	M56 × 5.5
M6.3 × 1	M64 × 6
M8 × 1.25	M72 × 6
M10 × 1.5	M80 × 6
M12 × 1.75	M90 × 6
M14 × 2	M100 × 6
M16 × 2	

Fig. 16-4.   The single series of 25 ISO Metric thread sizes which make up the Optimum Metric Fastener System.

Unit

17

Unified and American (National) Screw Threads (1948)

Before 1948, screw threads used by the United States, Canada, and Great Britain were not interchangeable. The British Whitworth screw thread used a 55° angle, and the American (National) screw thread used a 60° angle.

During World Wars I and II, many difficult problems developed because repairs involving screw threads on American-made military equipment could not be made with British tools. Dual inventories of tools and parts had to be carried — one British and one American. For this reason, efforts were made after World War II, to develop a unified screw thread system which could be used interchangeably by the United States, Canada, and Great Britain.

Unification of Screw Threads

On December 18, 1948 an agreement was reached on standardization of screw threads by representatives of the United States, the United Kingdom, and Canada. The following quotation presents the chief features of the agreement:

The present unification agreement provides a 60° angle and a rounded root for screw threads. The crest of the external thread may be flat, as preferred in American practice, or rounded, as preferred by the British. The number of threads per inch for the various series of thread diameters has been unified, and the limiting dimensions, for three grades of fit have been agreed upon; thus, interchangeability of screw thread parts, based on the accord, now becomes feasible.

There is, however, a further degree of interchangeability attained by agreements on the numerical values for allowances and tolerances, thereby setting limits to the least and greatest amounts of looseness between mating parts. Such agreement provides for identity of sizes (or interchangeability of use) of screw thread gages used in the different countries for controlling the limits of size of the threads. It also standardizes the grade or grades of fits between mating parts.*

Maximum and minimum major diameters, pitch diameters, and minor diameters for both external and internal threads were agreed upon. A in Fig. 17-1 shows the old American (National)

*Technical Report 1315. National Bureau of Standards, U.S. Department of Commerce, Washington, D.C.

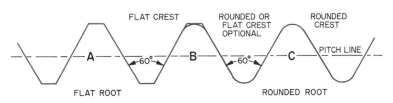

Fig. 17-1.   Comparison between (A) the American (National) and (B and C) the Unified Standard form of thread.

form of thread, and **B** and **C** show the Unified standard form.

Careful study of Figs. 15-7 and 17-1 will reveal that the new Unified standard screw thread profile is a compromise between the British Whitworth thread profile and the American (National) thread profile. With the Unified thread, British practice changed to the 60° thread angle. The root forms are round by design, as preferred by British practice. However, crest clearances permit interchangeability with the optional flat root forms, as preferred by previous American practice. Except for the rounded roots and crests, the Unified thread form is essentially the same as the American (National) thread form.

The basic height or depth of the external Unified thread in the maximum metal condition is slightly less than the height of the American (National) standard thread. The basic height of the external Unified thread is equal to 0.61343 × pitch, Figs. 17-2A and B. The basic height of the American (National) standard thread is 0.64952 × pitch, Fig. 15-6. Because of the application of tolerances and allowances, the two types of threads with the same diameter and pitch are mechanically interchangeable.

The American standard for screw threads, **American Standard Unified Screw Threads** (ANSI B1.1-1960)*, includes both systems of screw threads used in the United States — (1) the Unified system of threads and (2) the particular pitch-diameter combinations in the 1935 American (National) standard thread system, which have not been unified.

The transition from the American (National) to the Unified thread standards is essentially completed. The use of standard Unified threads is required on most equipment purchased by the federal government, including military and other equipment. It is expected that use of Unified threads in the United States will gradually diminish as manufacturers voluntarily change to use of ISO Metric threads.

*Unified and American Screw Threads** (ANSI B1.1-1949) and later edition, **Unified Screw Threads** (ANSI B1.1-1960). Published by the American Society of Mechanical Engineers, New York, New York.

Classes of Unified Threads

In the Unified thread system, the term **class of fits,** which referred to the assembly characteristics of mating threaded parts, has been discontinued. The new term used is **class of thread.** It refers to the tolerance and allowance, for one thread component. It does not imply that both the nut and the screw must have the same class of tolerance. In fact, any class of external thread, old or new, may be mated with any class of internal thread, so long as the resulting product meets assembly requirements. Fit is determined by the particular combination of classes of thread selected for the mating parts.

Six thread tolerance classes are included in the Unified thread system — three for screws and three for nuts. The external classes of threads are designated 1A, 2A, and 3A. The internal classes of threads are designated 1B, 2B, and 3B. Associated with these thread classes in the **Unified Screw Thread Standard** (1960) are the old American (National) thread classes 2 and 3 (formerly called fits). The old class 4 was dropped under the new standard. The old class 2 and class 3 threads are interchangeable with any of the new classes. Thread classes 1A and 1B are intended to replace the old class 1 fit.

The following are the new thread classes for Unified threads:

1A and 1B: Provide a fit with some play or looseness, even for rusty or slightly damaged threads. Some allowance is provided on class 1A.

2A and 2B: Provide a free fit suitable for a large majority of commercially threaded fasteners. Some allowance is provided on class 2A.

3A and 3B: Provide a close fit for applications which require more than the usual accuracy of thread angle and lead. Accurate equipment and frequent inspection are required to maintain consistent production of these thread classes. No allowance is provided on class 3A.

Basic thread dimensions, limits, allowances, and tolerances for Unified and American (National) screw threads are given in the standards, ANSI B1.1-1949 and ANSI B1.1-1960, footnoted earlier. This information is also available in several of the standard machine shop handbooks.

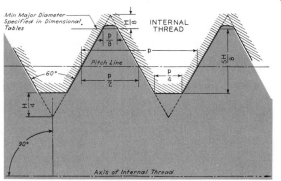

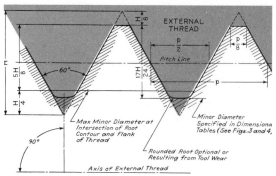

Angle of thread . . . . $2\alpha = 60°$

Half angle of thread . . . . $\alpha = 30°$

Number of threads per inch . . . . $n = \dfrac{1}{p}$

Pitch of thread . . . . $p = \dfrac{1}{n}$

(B)

Height of sharp V thread . . . . $H = 0.86603p$
$$= \dfrac{0.86603}{n}$$

[1]Height of external thread, design form . $h_s = 0.61343p$
$$= \dfrac{0.61343}{n}$$
$$= \dfrac{17}{24}H$$

Height of internal thread, design form . $h_n = 0.54127p$
$$= \dfrac{0.54127}{n}$$
$$= \dfrac{5}{8}H$$

Depth of thread engagement . . . . $h_e = 0.54127p$
$$= \dfrac{0.54127}{n}$$

Flat at crest of external thread . . . . $F_{cs} = 0.125p$
$$= \dfrac{0.125}{n}$$
$$= \dfrac{p}{8}$$

Truncation of external-thread crest . . . . $f_{cs} = 0.10825p$
$$= \dfrac{0.10825}{n}$$
$$= \dfrac{H}{8}$$

[1]Truncation of external-thread rounded
root . . . . $s_{rs} = 0.14434p$
$$= \dfrac{0.14434}{n}$$
$$= \dfrac{H}{6}$$

Flat at crest of internal thread,
design form . . . . $F_{cn} = 0.25p$
$$= \dfrac{0.25}{n}$$
$$= \dfrac{p}{4}$$

Truncation of internal-thread
crest . . . . $f_{cn} = 0.21651p$
$$= \dfrac{0.21651}{n}$$
$$= \dfrac{H}{4}$$

Flat at root of internal thread . . $F_{rn} = 0.125p$
$$= \dfrac{0.125}{n}$$
$$= \dfrac{p}{8}$$

Truncation of internal-thread
root . . . . $f_{rn} = 0.10825p$
$$= \dfrac{0.10825}{n}$$
$$= \dfrac{H}{8}$$

Addendum of external thread . . . $h_{as} = 0.32476p$
$$= \dfrac{0.32476}{n}$$
$$= \dfrac{3}{8}H$$

Major diameter of external
thread . . . . D_s

[2] Pitch diameter of external
thread . . . . $E_s = D_s - 2h_{as}$
$$= D_s - 0.64952p$$
$$= D_s - \dfrac{0.64952}{n}$$

Minor diameter of external
thread, design form . . . . $K_s = D_s - 2h_s$
$$= D_s - 1.22687p$$
$$= D_s - \dfrac{1.22687}{n}$$

Major diameter of internal thread . $= D_n$

Pitch diameter of internal thread . . $= E_n$

Minor diameter of internal
thread, design form . . . . $K_n = D_n - 2h_n$
$$= D_n - 1.08253p$$
$$= D_n - \dfrac{1.08253}{n}$$

---

[1] For calculating minor diameter values in tables.
[2] $2h_{as} = hb =$ the basic height, h, of the original American National form.

Fig. 17-2. Unified internal and external screw thread (maximum material condition) design forms (A) and thread formulas, Unified and American National (B). (Extracted from American Standard Unified Screw Threads (ANSI B1.1-1960) with permission of the publisher, The American Society of Mechanical Engineers.)

Designation of Screw Threads

Thread Sizes

The sizes (major diameters) of Unified National screw threads are indicated by fraction or by screw gage number. Thread diameters 1/4" and larger are designated by fractional size, such as 1/4-20 UNC; thread diameters under 1/4" generally are designated by screw gage number such as 10-24 UNC. A No. 10 machine screw has a diameter of 0.190". (See Table A-8, Appendix.) Certain screw threads in the special series, UNS or NS, under 1/4" diameter are designated by fractional size, such as 3/16-24 NS or 3/16-24 UNS.

Thread Series Included

The following series of screw threads are included in the American standard, **Unified Screw Threads** (ANSI B1.1-1960):

UNC — Unified National Coarse
UNF — Unified National Fine
UNEF — Unified National Extra-Fine

The following constant-pitch series of threads also are included in the Unified screw thread standard: UN4, UN6, UN8, UN12, UN16, UN20, UN28, and UN32. A constant-pitch thread has the same number of threads per inch for all diameters included. Constant-series threads are used where the coarse, fine, or extra-fine series do not meet specified requirements. When a constant-series thread is selected, preference should be given to the 8-, 12-, or 16-thread series.

Thread Symbols

The following symbols are used in designating Unified and American (National) screw threads on drawings and blueprints:

A indicates external thread.
B indicates internal thread.
U indicates Unified thread, and the particular thread size is recognized as a standard thread in the United Kingdom, Canada and the United States.

N indicates American (National) screw thread.
LH indicates left-hand thread.

When the letter U does not appear in the thread designation, but the letter **A** or **B** does appear, it means that all of the thread limits and tolerances conform to the principles involved in the establishment of Unified threads, though the particular thread size is not unified.

When none of the letters **A, B,** or **U** appears in the thread designation, the thread conforms to the earlier American (National) thread standard.

Thread Specifications

Unified and American (National) threads are completely specified on drawings with a note. The note always follows the same order: the nominal size is indicated first, followed by the number of threads per inch, thread series designation, thread class, and LH if the thread is left-hand. For example, a thread specified by the note 5/8-18 UNF-3A-LH designates a 5/8" diameter thread, with 18 threads per inch, the Unified National Fine series, with a class 3A (external) thread, and with left-hand threads.

Thread Formulas and Calculations

When threads are cut on a lathe with a single-point tool, it often is necessary to use thread formulas to make the required calculations for such thread elements as external thread height, internal thread height, and internal minor diameter. The formulas required for the calculation of Unified screw threads, according to present standards, are included in **A** and **B** of Fig. 17-2. The formulas used for the calculation of the original American (National) screw thread are included in Fig. 15-6. The basic formulas required for other thread forms are included with the figures showing the desired thread.

Unit

18

Other Forms of Screw Threads

Square Thread

The depth, width, and space between square threads are equal. See Fig. 18-1A. The square thread sometimes is used on vise screws, heavy jack screws, and similar items. The square thread cannot be made efficiently with dies, taps, or milling cutters unless the thread form is modified. It must be cut with a single-point tool on a lathe. Acme threads have largely replaced the use of square threads in machine design and construction.

Acme Thread

Acme threads have a 29° thread angle, as shown in Fig. 18-1B. Originally they were designed as a modification of the square thread. They are used to produce traversing movements on machine tools, steam valves, vises, etc. Although acme threads are not quite as strong as the square thread, they are much easier to machine. They may be cut with dies, taps, milling cutters, grinders, and a single-point tool in a lathe.

There are two basic thread standards for 29° acme threads — the old **standards** and the **new standards** (introduced in 1952). Acme threads produced according to these two standards are not interchangeable. The basic dimensions for old standard general-purpose American (National) acme threads may be determined from the formulas in Fig. 18-1B. The basic dimensions, limiting dimensions, and tolerances for these threads may be secured in standard handbooks for machinists. The machinist, particularly the maintenance machinist, should be familiar with both the old and new acme thread standards.

New standards for general-purpose acme threads were established in 1952. The American standard, **Acme Screw Threads***, gives complete data including basic dimensions, limiting dimensions, and tolerances for three classes of general-purpose acme threads. The three classes (designated 2G, 3G, and 4G) are provided with clearances on all diameters for free movement. Class 2G is most widely used for general-purpose work. The other classes are used where backlash or end play must be reduced in the mating threads. Data concerning the new standards for acme threads also are available in standard handbooks for machinists.

An **acme thread gage,** Fig. 18-2, is used in grinding an acme threading tool to the correct thread angle and to the correct tool width. The width of the tool point must conform to the numbered notch in the gage. For example, the point of a tool ground to cut five threads per inch must fit in the notch numbered five. The gage also is used to set the tool square with the work.

The number of acme threads per inch for various diameters (on both the old and the 1952 standards) for standard acme general-purpose threads, is included in Table 18-1.

---

*Acme Screw Threads (ANSI B1.5-1952). Published by the American Society of Mechanical Engineers, New York.

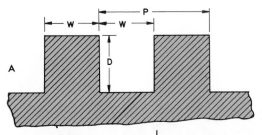

P = PITCH = $\dfrac{1}{\text{NO. THREADS PER INCH}}$

D = DEPTH = .500 X PITCH

W = .500 X PITCH

WIDTH W OF THREAD GROOVE IN NUT = (.500 X PITCH) + .001 TO .002 INCH CLEARANCE MAKE .001 TO .003 OVERSIZE TO FIT

SQUARE THREAD

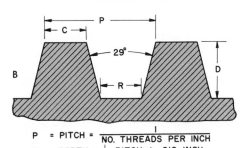

P = PITCH = $\dfrac{1}{\text{NO. THREADS PER INCH}}$

D = DEPTH = $\frac{1}{2}$ PITCH + .010 INCH

C = FLAT ON TOP OF THREAD = P X .3707

R = FLAT ON BOTTOM = (P X .3707) − .0052

ACME THREAD
(Old Standard Acme General-Purpose) Thread

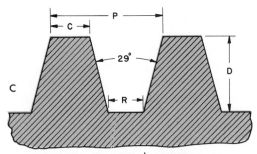

P = PITCH = $\dfrac{1}{\text{NO. THREADS PER INCH}}$

D = DEPTH = .6866 X PITCH

C = CREST FLAT = .335 X PITCH

R = ROOT FLAT = .310 X PITCH

29° BROWN AND SHARPE WORM THREAD

Fig. 18-1.   Cross sections of several forms of screw threads.

Brown and Sharpe Worm Thread

Worm threads are used in worm gear drive systems which have three principal uses: (1) to effect a large reduction in shaft R.P.M.; (2) to provide a sharp increase in mechanical advantage; (3) for steady and efficient power transmission.

The general form and the basic dimensions for the Brown and Sharpe worm thread are shown in Fig. 18-1C. This thread is similar to the acme thread, Fig. 18-1B. Both threads have a 29° included angle, but the Brown and Sharpe thread is deeper. The widths of the crests and roots of the thread are also different.

(Brown and Sharpe Manufacturing Company)

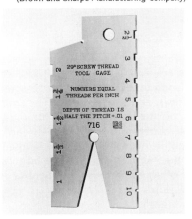

Fig. 18-2.   Acme Standard 29° thread tool gage.

Table 18-1
Standard General-Purpose Acme Threads

Size	No. of threads per inch	Size	No. of threads per inch
1/4	16	1-3/8	4
5/16	14	1-1/2	4
3/8	12	1-3/4	4
7/16	12	2	4
1/2	10	2-1/4	3
5/8	9	2-1/2	3
3/4	6	2-3/4	3
7/8	6	3	2
1	5	3-1/2	2
1-1/8	5	4	2
1-1/4	5	4-1/2	2
		5	2

(Brown and Sharpe Manufacturing Company)

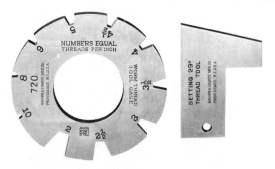

Fig. 18-3. Brown and Sharpe 29° worm-thread gage (left) and setting tool.

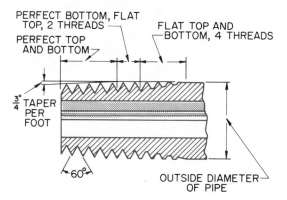

Fig. 18-4. Taper pipe threads (NPT).

The Brown and Sharpe worm-thread gage in Fig. 1ᴾ-3 is used to measure the correct tool angle and point width for a single-point tool used in cutting the worm thread. The setting tool, also shown in Fig. 18-3, is used to set the tool square with the work.

Pipe Threads

The most commonly used system of pipe threads is the American Standard or American National pipe thread. Two types of pipe threads are included in this system — the American Standard taper pipe thread (NPT) and the American Standard straight pipe thread (NPS). A third type which is a variation of the NPT is the American standard Dryseal pipe thread (NPTF).

Table 18-2
American Standard Pipe Dimensions and Tap Drill Sizes

Pipe Diameters			Threads Per Inch	Tap Drill Size
Nominal Size	Actual Inside	Actual Outside		
1/8	0.270	0.405	27	11/32
1/4	0.364	0.540	18	7/16
3/8	0.494	0.675	18	19/32
1/2	0.623	0.840	14	23/32
3/4	0.824	1.050	14	15/16
1	1.048	1.315	11-1/2	1-5/32
1-1/4	1.380	1.660	11-1/2	1-1/2
1-1/2	1.610	1.900	11-1/2	1-23/32
2	2.067	2.375	11-1/2	2-3/16
2-1/2	2.468	2.875	8	2-5/8

Pipe Thread Sizes

Pipe diameter designations are different from the actual diameter of the pipe. For example, a 1/8″ pipe has an outside diameter of 0.405″ with an actual inside diameter of 0.270″. Table 18-2 gives tap drill sizes and dimensions of pipes to 2 1/2″. More complete tables in standard handbooks for the machinist include the basic pipe thread dimensions, nominal pipe sizes with the number of threads per inch, outside diameters, pitch diameters, and other dimensions.

Tapered Pipe Threads (NPT)

The tapered pipe thread is similar to the American National thread. It has a 60° angle between the sides of the threads, and it has flattened crests and roots. It differs in that the threads are tapered 3/4″ per foot of length from the small end of the thread toward the large end. See Fig. 18-4.

Both the external and the internal threads are tapered. This permits drawing the joint up tightly for a rigid joint. However, a pipe compound must be used to seal the clearance space which exists between the crests and roots of the mating threads in order to prevent leakage of liquid or gas under pressure.

Straight Pipe Threads (NPS)

Straight and tapered pipe threads are similar in form and dimensions. The straight pipe thread more closely resembles the regular National Form thread in general form, but pitches and nominal diameter size designations are significantly different.

Straight pipe threads (NPS) sometimes are used in couplings which join pipes having tapered pipe threads (NPT). This type of pipe assembly often is satisfactory for low-pressure lines when a pipe compound is used and when the line is free of vibration.

Dryseal Pipe Threads

American Standard Dryseal pipe threads (NPTF) are used for those pressure joints where the use of a pipe sealing material is objectionable. This type of thread is used on both internal and external threads. On Dryseal threaded pipe joints, the external thread is tapered, while the internal thread may be tapered or straight. Joints with both internal and external threads tapered are superior.

The basic dimensions of most American Standard Dryseal pipe threads are the same as those for the American Standard tapered pipe threads, except that the crests and roots of the Dryseal threads are modified to cause a pressure-tight seal. The crest flats are equal to or less than the root flats of the mating thread, thus causing physical contact between the crests and the roots when turned up hand-tight. When turned wrench-tight, the threads mash together causing a pressure-tight seal, Fig. 18-5.

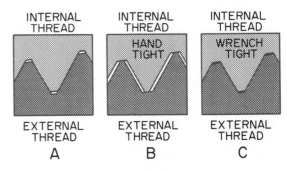

Fig. 18-5.   Pipe thread joints.
A. Regular taper pipe joint.
B. Dryseal pipe thread, hand-tight.
C. Dryseal pipe thread, wrench-tight.

Unit

19

Taps and Accessories for Hand Tapping

Taps are used to cut internal threads in holes. The process of cutting internal threads is called **tapping.** External threads are cut with a die, and the process is called **threading.**

Taps are made of hardened tool steel, either carbon steel or high-speed steel. Carbon steel taps usually are used for hand-tapping operations. High-speed steel taps are used for both hand- and machine-tapping operations.

The flutes on the tap provide for the cutting edges and also provide space for the chips. Standard hand taps usually have four flutes, but some taps have two or three flutes, Fig. 19-1. The square end on the tap is used for holding the tap with a wrench or other holding device. Taps are hard and brittle, and they break quite easily when excessive force is applied to them.

Hand Taps

The three basic types of hand taps are the **taper tap,** the **plug tap**, and the **bottoming tap,** Fig. 19-2. The only basic difference in the three taps is the number of threads which are ground to a taper on the end of the tap. Taps are ground tapered in order to start and cut the threads more easily. Most of the actual cutting of the thread is performed by the chamfered threads on the end of the tap.

As the name implies, **hand taps** are designed for hand-tapping operations. They can be purchased singly or in sets of three, including the **taper, plug,** and **bottoming** taps for each thread size and pitch.

Taper Tap

The taper tap has 8 to 10 threads tapered. It is used to tap **open or through** type holes, as in Fig. 19-3A. It also is used as the first step in tapping closed or blind holes in hand-tapping operations, Fig. 19-3B and C. Tool life and breakage largely depend on the length of the tapered threads on the tap. Less power is required and less strain is exerted on a tapered tap than on a plug or bottoming tap. Therefore, the tapered tap should be used whenever possible.

Plug Tap

The plug tap has 3 to 5 threads tapered. Whenever possible, all three hand taps should be used in tapping closed or blind holes, Fig. 19-3B and

C. The tapered tap is used first in tapping to the bottom of the hole, and it is followed with the plug tap and the bottoming tap. For most materials, plug taps can be successfully used in place of taper taps. Plug taps are preferred for machine tapping because through holes and blind holes can be tapped in one operation, as in Fig. 19-3C, thus effecting considerable cost savings.

(Greenfield Tap and Die Division of TRW Inc.)

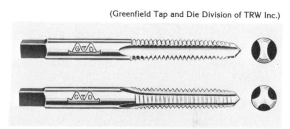

Fig. 19-1.   Two-fluted and three-fluted taps.

(Greenfield Tap and Die Division of TRW Inc.)

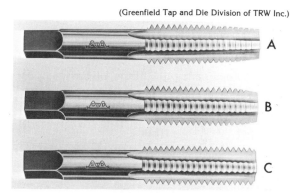

Fig. 19-2.   Set of hand taps.
A. Taper tap
B. Plug tap
C. Bottoming tap

Fig. 19-3. Types of holes commonly tapped
A. Open or through
B. Blind bottoming
C. Blind but not bottoming

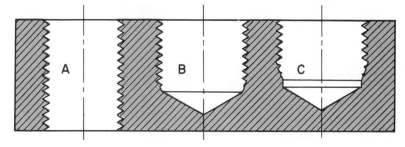

Bottoming Tap

The bottoming tap has 1 to 1½ threads tapered. It is used after the plug tap for tapping to the bottom of a hole as in Fig. 19-3B.

Serial Taps

Serial hand taps usually are produced in sets of three (numbered 1, 2, and 3) and are identified with one, two, or three rings on the shank of the tap near the square, Fig. 19-4. Serial taps are used for hand tapping deep holes in tough metals. They are similar to the taper, plug, and bottoming tap in general appearance, but they differ in size of the pitch diameter and major diameter.

Each tap is designed to remove a certain portion of the metal in cutting a thread. This procedure eliminates excessive strain on the tap and aids in preventing tap breakage. The No. 1 tap is used first for cutting a shallow thread. This is followed by the No. 2 tap, which cuts deeper, and finally, the No. 3 tap is used to cut the thread to full depth.

Tap Size Identification

The diameter and pitch of the tap are stamped on its shank. A tap labeled M12 × 1.75 has an ISO Metric thread of 12 mm diameter and 1.75 pitch. A tap labeled ½-13 UNC is ½" diameter,

has 13 threads per inch, and is a Unified National Coarse thread. Left-hand taps are identified with the additional letters LH, such as ½-13 UNC-LH.

UNC machine screw diameters under ¼" are designated by gage number, such as 0, 6, 8, 10, and 12. A typical designation would be 10-24 UNC. Major diameters of machine screw sizes are listed in tables of tap drill sizes, such as Table A-8, Appendix.

Tap Wrenches

Tap wrenches are used to turn the tap for hand-tapping operations. The wrenches have adjustable jaws which enable the tap to be tightened in the wrench and held firmly. There are two common types of tap wrenches — the **T-handle** tap wrench (Fig. 19-5) and the **straight handle** tap wrench (Fig. 19-6). Both types are available in several sizes. The T-handle tap wrench usually is used for holding smaller diameter taps than the

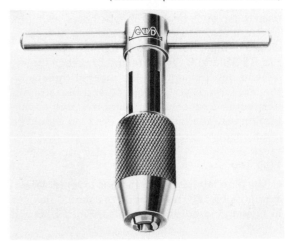

Fig. 19-5.   T-handle tap wrench.

Fig. 19-6.   Straight-handle tap wrench.

SERIAL SIZE

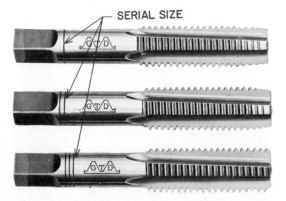

Fig. 19-4.   Serial hand taps, including No. 1, No. 2, and No. 3 taps, identified by rings in shanks.

straight handle tap wrench. Taps of a small size may be broken easily when used with tap wrenches which are too large. Tap wrenches are also used to drive reamers, screw extractors, and other tools which are turned by hand.

Screw Plates

Taps, tap wrenches, dies, and die stocks may be purchased in sets called **screw plates,** as shown in Fig. 19-7. Screw plates are available in several sizes and combinations. They may include either a small or large range of thread sizes and either the fine, the coarse, or both series of threads.

Screw Pitch Gage

The pitch of a screw thread may be checked with a screw pitch gage, Fig. 19-8. The gage is made of a series of sheet metal plates into which is cut the form of screw threads of various pitches. Pitch is checked by determining which plate fits accurately into the given thread. The pitch also may be measured with a rule, Fig. 19-9.

Tap Extractor

If a tap is broken off in a hole so that it cannot be reached with a pliers, it can sometimes be removed with a tap extractor, Fig. 19-10. In use, the steel prongs of the extractor are pushed down into the flutes of the broken tap. The steel bushing is then pushed down against the broken tap to hold the prongs firmly. A tap wrench is

(Greenfield Tap and Die Division of TRW Inc.)

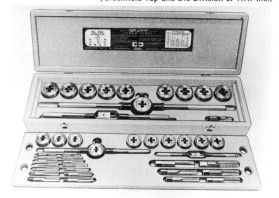

Fig. 19-7.   Taps, dies, and tap wrenches included in screw plate.

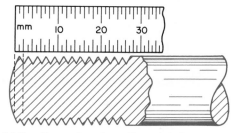

Fig. 19-9.   Measuring shows 2 mm pitch.

Fig. 19-10. Tap extractor used to remove broken taps.

(L. S. Starrett Company)

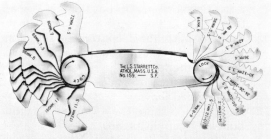

Fig. 19-8.   Screw pitch gage.

(Walton Company)

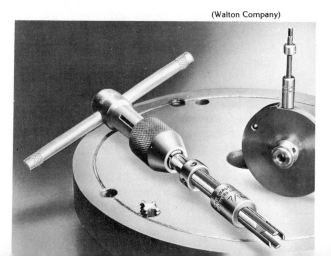

used to turn the extractor in removing the broken tap. Taps which cannot be removed with tap extractors can be burned out by electrical discharge machining.

Screw Extractor

A bolt or machine screw which is broken off below the surface, or which is broken off too short to grasp with a pliers, may be removed with a screw extractor, Fig. 19-11. A hole slightly larger than the small end of the extractor must be drilled into the broken screw. The screw extractor is inserted into the hole and turned counterclockwise with the use of a tap wrench. The largest size

screw extractor possible for the size of the screw should be used. Screw extractors are available in a full range of sizes. A screw extractor can also be used to remove short pieces of pipe which are broken off below the surface of a pipe coupling.

(Cleveland Twist Drill — an Acme-Cleveland Company)

Fig. 19-11. Screw extractor.

Unit
20

Procedure for Hand Tapping

1. Determine the size of the thread from information provided on the part drawing.
2. Look up the proper tap drill size on a drill chart or in Table A-8 or A-9, Appendix.
3. Center punch, center drill, and drill the hole (s) to be tapped.
4. Mount the work in a bench vise so that the hole is in a vertical position.
5. Select the proper tap and tap wrench, and tighten the tap in the wrench. Check the tap to be sure it is sharp.
6. Use the dominant hand to grasp the tap and tap wrench. Cup your hand over the center of the wrench, as in Fig. 20-1 and place the tap in the hole. Keeping the tap aligned as parallel as possible to the hole, start the tap by turning two or three turns to the right, at the same time keeping a steady pressure

downward on the tap. Harder metals require more pressure.
7. After several threads have been cut, remove the tap wrench without disturbing the tap. Place the blade of a small square against the solid shank of the tap to check for squareness. Check from two positions 90° apart. See Fig. 20-2. If the tap is not square with the work, it will cut too deeply on one side of the hole, will ruin the thread, and possibly will break in the hole. If the tap is not started square with the work, back it out and restart (as described above).
8. Except on cast iron, always use a cutting fluid when cutting threads.
9. Continue tapping the hole. If an ordinary hand tap is used, turn the tap two or three turns. Then reverse the tap one turn to break

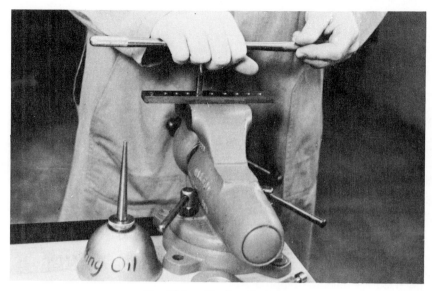

Fig. 20-1. Starting the tap.

the chips. This aids in chip removal, prevents damage to the threads, and helps prevent tap breakage. If a machine tap is being used, it is not necessary to reverse the tap to break the chips.

10. When hand tapping to the bottom of a deep blind hole, it is good practice to back the tap out and clean the chips from the tap and the hole periodically. The taper tap, plug tap, and bottoming tap are used, in that order, to cut the thread to the bottom of deep holes. In shallow blind holes, only the plug tap and bottoming tap are used. Be careful to avoid breaking the tap when the bottom of the hole is reached.

11. Clean the chips from the hole, and check the fit of the thread with a thread plug gage. If a gage is not available, check the fit with a bolt or screw of the type to be used in the hole.

Fig. 20-2. Checking tap for squareness.

Unit
21

Other Styles of Taps for Machine and Hand Tapping

Taps are made in various styles and designs for hand-tapping and machine-tapping operations. While many taps used for machine tapping are called hand taps, some taps are designed with special features which cause them to operate more efficiently for machine tapping on specific materials.

Two- and Three-Fluted Taps

For average hand-tapping operations, taps with the standard number of flutes produce satisfactory results. Larger size hand taps usually have four flutes, while small size machine screw taps usually have three flutes, Fig. 19-1. Certain sizes of hand taps in machine screw sizes and in sizes up to M12 (½") also are regularly produced with two or three flutes.

For tapping deep holes in stringy metal or in blind holes where the chips fall to the bottom of the hole, more flute space is needed for the chips. A tap with two or three flutes usually has deeper flutes and more chip space than one with four flutes. However, with deeper flutes, the strength of the tap is reduced. A three-fluted tap generally should be tried first, since it is stronger than a two-fluted tap.

Gun Taps

The gun tap is very similar to the standard hand tap, except that it is designed for machine tapping. It is called a gun tap because the angular design of the cutting point causes the chips to **shoot** ahead of the tap as the thread is cut. See Fig. 21-2. This is a desirable feature when tapping stringy metals, because the chips do not readily clog the flutes, thus minimizing tap breakage. The gun tap has fewer flutes than a standard hand tap, but the flutes are not as deep; therefore, the tap is stronger and will withstand the greater strains required in production tapping. Gun taps also work nicely as hand taps.

There are three common types of gun taps — the **plug** gun tap (Fig. 21-1A), the **bottoming** gun tap (Fig. 21-1B), and the **gun flute only** plug tap (Fig 21-1C). The plug gun tap is designed primarily for tapping open holes. The bottoming gun tap is designed for blind holes. The point of the bottoming gun tap has a somewhat different

(Greenfield Tap and Die Division of TRW Inc.)

A. Plug gun tap   B. Bottoming gun tap   C. Gun flute only tap

Fig. 21-1.   Three common types of gun taps.

design than the plug gun tap. It breaks the chip up finely so that the chips may escape more easily. The flutes are also deeper and larger for chip removal. The gun flute only plug tap is designed for tapping shallow through holes which are not more than one diameter in depth. Because there are no flutes in the body, the tap is stronger than the plug gun tap. It is recommended for tapping soft, stringy materials.

Spiral Fluted Taps

Helical fluted taps, commonly called spiral fluted taps, have spiral flutes similar to a drill. The flutes may be the low-angle or high-angle type, Fig. 21-3. The spiral flutes aid in drawing the chips out of the hole being tapped, as shown in Fig. 21-4. This type of chip removal also aids in preventing tap breakage when the tap is backed out of the hole at high speeds.

Low-angle spiral fluted taps are especially useful in machine tapping of stringy materials such as aluminum, copper, magnesium, brass, and die cast metals. High-angle spiral fluted taps work well with tough alloy steel in deep blind holes. They also may be used in materials which produce powder or granular chips, such as die castings. Spiral fluted taps also work well in tapping holes where a gap or slot must be bridged in a hole.

Spiral fluted taps are available with two, three, or four flutes (depending on the diameter) and in plug and bottoming styles.

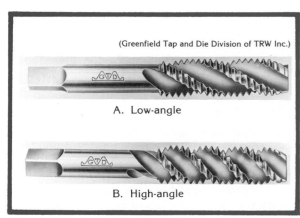

(Greenfield Tap and Die Division of TRW Inc.)

Fig. 21-2.   Plug gun tap shoots chips ahead of it as the thread is cut.

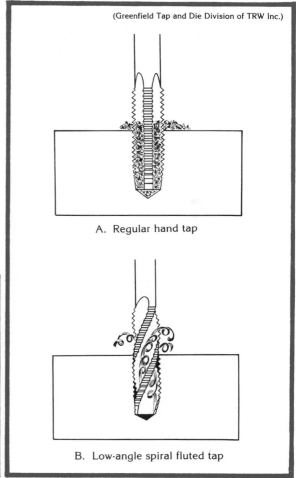

(Greenfield Tap and Die Division of TRW Inc.)

A.   Regular hand tap

B.   Low-angle spiral fluted tap

Fig. 21-4.   Cutting action of taps.

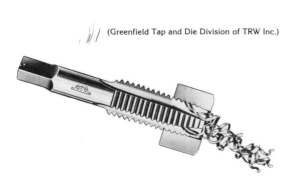

(Greenfield Tap and Die Division of TRW Inc.)

A.   Low-angle

B.   High-angle

Fig. 21-3.   Spiral fluted taps.

Fig. 21-5.   Thread forming tap, plug style.

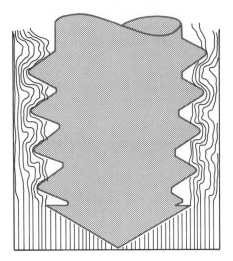

Fig. 21-6.   Displacement of material which results when thread-forming taps are used.

Thread Forming Taps

Thread forming taps, Fig. 21-5, have no cutting edges or flutes. They are simply forced into the hole, cold forming the threads by displacement of material, Fig. 21-6. Formed threads are stronger and have better surface finish than cut threads. Problems associated with chip clogging are eliminated. The taps work well in ductile metals such as lead, zinc, copper, aluminum, brass, and even leaded steels.

CAUTION

Be sure to follow the manufacturer's recommended tap drill sizes for thread forming taps. Tap drill sizes for thread cutting taps are too small and will cause thread forming taps to break.

Pipe Taps

Metric pipe taps currently being offered in the U.S. are 55° modified Whitworth pipe threads.

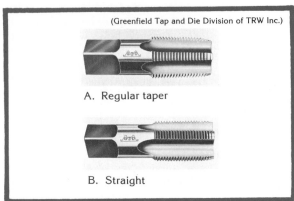

(Greenfield Tap and Die Division of TRW Inc.)

A. Regular taper

B. Straight

Fig. 21-7.   Pipe taps.

These are available in left or right hand, and straight or taper types. Because of differences in thread angle, pitch, and pitch diameter, metric and American pipe threads are not interchangeable. Presently, however, most pipe threads in use in the U.S. are American standard or American National pipe threads. The following three types of American standard pipe threads commonly are used:

American standard taper pipe thread (NPT)

American standard straight pipe thread (NPS)

American standard Dryseal pipe thread (NPTF)

Further information concerning American standard pipe threads may be found on pages 101 and 102.

Taper pipe taps are used to tap NPT threads, Fig. 21-7A. Straight pipe taps are used to tap NPS threads, Fig. 21-7B. American standard Dryseal pipe threads are tapped with ground thread taper taps which have the American standard Dryseal pipe form.

Acme Taps

Internal acme threads may be cut with taps, or they may be cut with a single-point threading tool on a lathe. When acme threads are tapped, special consideration must be given to the selection of the proper taps. A specific tap is designed for each type and each class of acme thread.

Acme taps often are manufactured as serial taps in sets of two or three, Fig. 21-8. The roughing tap is used first to remove some of the material, thus cutting the thread to partial depth. The finishing tap then is used to cut the thread to finished depth. In some cases involving open or through holes, one tap of special design may be used to tap the thread to full depth.

Other factors to be considered in the selection of acme taps include the depth of the thread, the hardness and toughness of the material to be tapped, and the depth of the hole. Fine-pitch threads tap more easily than coarse-pitch threads.

To order acme taps, furnish the manufacturer with complete data concerning the type and class of acme thread. Include the following: thread dimensions, hole type, material to be tapped, and type of machine used, if for machine tapping.

(Greenfield Tap and Die Division of TRW Inc.)

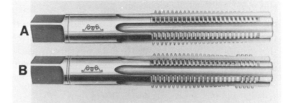

Fig. 21-8.   Acme taps.
A. Roughing tap
B. Finishing tap

Unit

22

Tap Size Limits

Taps are manufactured according to specific size limits and tolerances and are available with either **cut** threads or **ground** threads. Those with ground threads are more expensive, but they generally produce threads with closer size limits and a smoother finish. Both carbon steel taps and high-speed taps are made with cut threads. High-speed steel taps also are made with ground threads.

Pitch Diameter Size Limits

The size of the pitch diameter on a tap largely determines the size and fit of the tapped thread produced. If the pitch diameter is oversize, a screw will fit the tapped thread loosely. If the pitch diameter is undersize a slight amount, the thread will fit more tightly. Taps are made with standard pitch diameter size limits, oversize limits, and undersize limits.

Taps with cut threads usually have standard pitch diameter size limits. Taps with ground threads may have standard size, oversize, or undersize pitch diameter size limits, indicated by code letters and numbers on their shank. When ground thread taps are purchased, the limits code number and the class of thread should be specified. Otherwise, the manufacturer generally recommends a tap with pitch diameter size limits appropriate for threads with Grade 6 tolerance for ISO Metric threads, a class 2 fit for National form threads, or a class 2B thread for Unified threads.

Tap standards now include only one classification of ground thread taps — the **ground** thread.

Ground threads with pitch diameter tolerances above basic size are designated as **High** and are identified by the letter H. Those ground below basic are designated **Low** and are identified by the letter L. Numerals are used to identify the pitch diameter limits of the tap as compared with the basic pitch diameter. The following are pitch diameter limit numbers for taps through 1″ diameter:

L1 — Basic to basic minus .0005″
H1 — Basic to basic plus .0005″
H2 — Basic plus .0005″ to basic plus .0010″

H3 — Basic plus .0010″ to basic plus .0015″
H4 — Basic plus .0015″ to basic plus .0020″
H5 — Basic plus .0020″ to basic plus .0025″
H6 — Basic plus .0025″ to basic plus .0030″

For example, a ½-13 UNC ground thread hand tap selected for a class 3B thread would be identified with the additional label H3. A ½-13 NC ground thread tap selected for a class 2 fit would be identified with the additional label H5. Some manufacturers use the letter G with these labels (such as GH3 or GH5), to indicate that the tap is ground.

Unit
23

Selection and Calculation of Tap Drill Size

The size of the drill required for a given threaded hole is called the **tap drill size.** Each thread size and pitch has a specific tap drill size to produce the necessary percentage depth of the thread to be cut. The tap drill sizes for average applications are based on threads which are approximately 75 percent of a full-depth thread. Thus, the hole is larger than the minor diameter for a full-depth thread. Tap drill sizes for ISO Metric, and Unified and American (National) screw threads requiring approximately 75 percent thread depth are listed in Tables A-7, A-8 and A-9, Appendix.

If a particular tap drill size is not on hand, the next larger size may be selected. For example, the recommended tap drill size for an M12 × 1.75 thread is 10.25 mm. A letter "Y" drill, which measures 10.262 mm, is actually closer to the theoretically correct drill size of 10.295 mm. Use of tap drills smaller than the recommended size should be avoided due to the increased danger of tap breakage.

An internal thread with a 75-percent depth has approximately 95 percent of the strength of a full-depth thread. The power required to turn a tap for a thread with 100-percent depth is three times greater than that required for one with 75-percent depth, and excessive tap breakage will result.

For a majority of tapping requirements, a tap drill producing a minor diameter which provides 55- to 75-percent thread depth is adequate. Thread strength tests indicate that a nut made with 50-percent thread depth, according to standard specifications, will break the bolt or screw before the threads strip. These tests also indicate that any increase in the internal thread depth over 60 percent does not increase the strength significantly.

For very tough metals, often it is desirable to use a tap drill which provides less than 75-percent thread depth, but usually not less than 55-percent thread depth. The larger hole size will reduce tap breakage without significant loss of thread strength. When holes deeper than one-and-one-half times the diameter of the tap are tapped (particularly for small machine screw sizes), a tap drill size larger than the 75-percent type may be used. In such cases, drills which provide for a 55-percent thread depth give satisfactory results. Tap drill sizes for threads of various percentage depths, such as 70, 65, 60, 55, or 50 percent, may be found in handbooks for machinists.

Calculation of Tap Drill Size

The tap drill size for ISO Metric threads requiring a 75-percent thread depth may be calculated with the formula: Nominal Outside Diameter minus (0.97 × Pitch).

Example: Determine the tap drill size for an M25 × 2 thread.

Formula:

$$
\begin{aligned}
\text{Tap Drill Size} &= \text{O.D.} - (0.97 \times \text{Pitch}) \\
&= 25 - (0.97 \times 2) \\
&= 25 - 1.94 \\
&= 23.06 \text{ mm} \\
\text{Tap Drill Size} &= 23 \text{ mm}
\end{aligned}
$$

The tap drill size for Unified and American (National) form threads requiring a 75-percent thread depth may be calculated by subtracting the pitch from the major diameter. The pitch is equal to 1 divided by the number of threads per inch.

Example: Determine the tap drill size for a 3/8-16 UNC thread of approximately 75-percent thread depth.

Formula:

$$
\begin{aligned}
\text{Diameter of tap drill} &= \text{major diameter} - \text{pitch} \\
&= \quad 3/8 \quad - 1/16 \\
\text{Diameter of tap drill} &= 5/16 \text{ inch}
\end{aligned}
$$

Unit
24

Hand Threading Dies

Threading dies cut external threads on round rods or bolts. Dies of various designs are available, usually made of carbon tool steel or high-speed steel. They are designed with internal threads, much like a nut, but with flutes or grooves intersecting the thread to provide space for chips to escape. See Fig. 24-1.

Threading dies generally are provided with some means for adjusting the depth of cut for a snug or free fit. Adjustment of thread depth usually is made with one or more small screws in the die or die holder. The principal types of threading dies are the adjustable **round split die,**

(Greenfield Tap and Die Division of TRW Inc.)

Fig. 24-1.   Adjustable round split die.

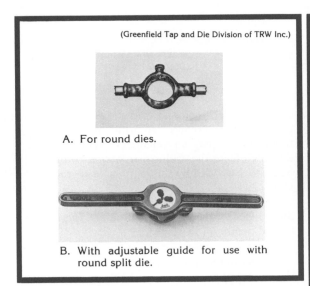

A. For round dies.

B. With adjustable guide for use with round split die.

Fig. 24-2.   Diestocks.

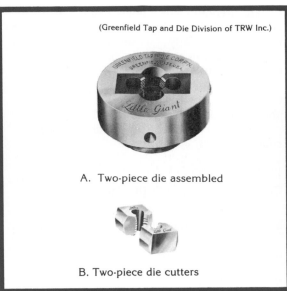

A. Two-piece die assembled

B. Two-piece die cutters

Fig. 24-3.   Two-piece die.

the adjustable **two-piece die,** and the **solid bolt thread die.**

Round Split Die

The round split die is a single-piece, adjustable die, Fig. 24-1. For hand-threading operations, it is mounted in a diestock, Fig. 24-2. When used on a screw machine, it is mounted in a special die holder.

The round split die can be adjusted to cut a thread with a snug or loose fit through use of the small screw in the die. When the screw is turned in, the die opens and a snug-fitting thread results. When a loose-fitting thread is desired, the screw is backed out slightly. For hand-threading purposes, the screw in the diestock also must be turned in snugly to hold the die closed against the die adjustment screw, so that the thread will cut to the adjusted depth.

Two-Piece Die

Adjustable two-piece dies are contained in collets consisting of a cap and a guide. See **A** in Fig. 24-3. The die halves, **B** in Fig. 24-3, are inserted in the beveled cap and are held in place by the guide which is screwed into the cap. The dies may be adjusted for depth of cut with the small screws on either end of the slot. The dies must be inserted in the cap with the tapered threads down-

Fig. 24-4.   Solid square die.

ward toward the guide and with the label side of the die toward the top. The guide used on the collet serves as an aid in starting and holding the thread square with the work being threaded. After several threads have been cut, they should be checked for proper fit. If necessary, further adjustments should be made until the proper fit is achieved. Gages and methods used in measuring threads and determining thread fits are included in Unit 26.

Solid Bolt Thread Die

The solid square die is designed primarily for dressing or cleaning up bruised or damaged threads, Fig. 24-4. No special holder is used to turn the die. Any wrench of sufficient size may be used.

Unit 25

Procedure for Using Hand Threading Dies

1. Chamfer the end of the rod to be threaded. This may be done with a file or by grinding. The depth of the chamfer should be about equal to the depth of the thread to be cut, Fig. 25-1.
2. Select the proper threading die, and mount it in the diestock, usually with the labeled side up. The opposite side is tapered for starting and cutting the thread.
3. Mount the rod or workpiece in a bench vise. Short pieces should be mounted in a vertical position; long pieces, usually in a horizontal position.
4. If the guide is adjustable, adjust it to slide freely over the rod.
5. Place the die over the rod, cup the hand over the die, apply pressure evenly, and turn the die to get the thread started, Fig. 25-2. After several threads have been cut, the hands may be shifted to the ends of the diestock as in Fig. 25-3.

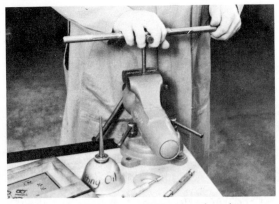

Fig. 25-2.   Cup hand over die to start thread.

(Greenfield Tap and Die Division of TRW Inc.)

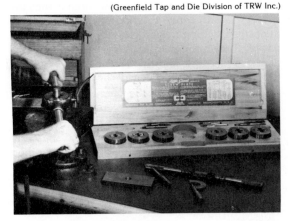

Fig. 25-3.   Threads being cut with a die mounted in a diestock.

Fig. 25-1.   Chamfer end of rod before threading.

6. Apply a cutting fluid to the die and work-piece. This will cause the die to cut more easily, stay sharp longer, and produce a better quality finish on the thread.

7. Check to see that the die is started square with the work. If the die is **out-of-square,** a crooked thread will result; this is called a **drunken thread.**

8. Continue cutting the threads by turning two or three turns forward. Then reverse the die about one-half turn to break the chips. When several turns of the thread have been completed, back the die off and check the fit of the thread.

9. Check the thread fit with a thread gage, a thread micrometer, a nut, or the mating part.

If the thread is not correct, adjust the screw(s) in the die for the desired fit. (As the die is closed, the thread is cut deeper. As the die is opened, the thread depth is reduced.) Run the die over the thread, and check the fit again. Continue this adjustment until the proper fit is achieved. Sometimes it is desirable to cut a thread on a practice piece first, until the correct fit is achieved, before cutting on a final workpiece.

10. Continue cutting the thread to the correct length. If the thread must be cut close to a shoulder or close to the head of a bolt, the die may be backed off, inverted, and the last several threads completed while cutting in the inverted position.

Unit 26

Screw Thread Measurement

The size and accuracy of screw threads may be measured with screw thread plug gages, thread micrometers, ring thread gages, roll thread snap gages, thread comparators of various types, optical comparators, and by the three-wire method.

The emphasis in thread measurement is always on the measurement of the pitch diameter. The limits and tolerances on the pitch diameter largely determine the fit of screw threads. Since the pitch diameter of thread ring gages and snap roll gages is difficult to determine accurately by other methods, these gages are set or fitted with the use of accurate plug gages.

Thread Plug Gage

Tapped holes are checked for the correct fit with thread limits plug gages. The gage in Fig.

26-1 is a double-end gage, providing a go gage on one end and a not-go gage on the opposite end. The go gage is always the longer of the two, and it has a chip groove for cleaning the threads in the tapped hole being measured.

The thread limits plug gage is used to determine whether the pitch diameter of a tapped thread is within limits for a specified class of thread. The diameter of the go gage is minimum diameter, usually basic, and the not-go gage is maximum diameter.

(Greenfield Tap and Die Division of TRW Inc.)

Fig. 26-1.   Thread plug gage.

The thread is gaged by having the go gage enter the tapped hole the full length of the gage. The not-go gage may or may not enter. If it does, it should have a snug fit on or before the third thread, thus indicating that the hole is the maximum size permitted for the specified fit. If this gage enters farther, the thread is oversize and will not fit properly.

Thread Micrometer

The pitch diameter of 60° V-threads may be measured directly with a thread micrometer. See Fig. 26-2. The spindle of the micrometer has a 60° conical point, and the anvil has a 60° groove. The anvil point swivels to enable measurement of different pitches.

A given thread micrometer is designed to measure a specific range of screw threads. Care should be taken to select a micrometer with the correct thread range when measuring a specific thread. The micrometer always should be checked for a zero reading before measuring threads. See inset of Fig. 26-2.

Thread Ring Gage

The accuracy of an external thread may be checked with a pair of thread ring gages. The pair includes a go gage and a not-go gage, Fig. 26-3. The **go gage** is designed to check the maximum pitch diameter, flank angle, lead, and the clearance at the minor diameter simultaneously. The **not-go gage** is designed to check only the pitch diameter to determine whether it is below minimum limits.

To check a thread, both gages are used. If the **go gage** does not turn on freely, one of the thread elements is not accurate, and the thread will not assemble with the mating part. If the **not-go gage** turns on the thread, the pitch diameter of the thread is under the specified minimum limits, and the thread will not fit properly with its mating part.

Roll Thread Snap Gage

External threads may be checked rapidly for accuracy with a roll thread snap gage, Fig. 26-4. The gage illustrated is the open-face type which may be used close to shoulders. Right- and left-hand threads can be checked with the same gage.

(L. S. Starrett Company)

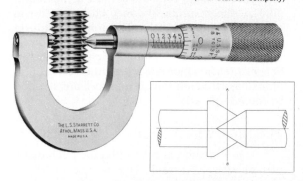

Fig. 26-2.   Thread micrometer measuring the pitch diameter of a screw thread directly. Inset shows anvil and spindle position at Line AB which corresponds to zero reading.

(Greenfield Tap and Die Division of TRW Inc.)

Fig. 26-3.   Thread ring gages — go and not-go gages with holder.

(Greenfield Tap and Die Division of TRW Inc.)

Fig. 26-4.   Roll thread snap gage.

The outer or **go rolls** are set to the maximum pitch diameter limits, and they check all thread elements simultaneously. The inner or **not-go rolls** are set to minimum pitch diameter limits and check only the minimum pitch diameter. Screw threads which are within the correct size limits pass through the go rolls and are stopped by the not-go rolls.

The Three-Wire Method

This is a method of measuring the pitch diameter of external threads. The method is recommended by the National Bureau of Standards in Washington, D.C. The three-wire method requires the use of an ordinary outside micrometer and three accurately sized wires. A different **best wire size** is recommended for each pitch and diameter combination.

The best wire size is determined by calculation or by selection from a chart of recommended wire sizes. The wires are placed in the thread grooves as in Fig. 26-5, and a micrometer measurement across the wires is made. The correct dimensions for measurement over wires can be found in handbooks for machinists, or the pitch diameter can be calculated.

Because the three-wire method is more cumbersome and time consuming to use, many machinists prefer to check the pitch diameter of external threads with a ring gage, thread micrometer, or other instrument. However, the three-wire method is considered to be more accurate than the use of many gages designed for this purpose.

Special Thread Measuring Devices

Numerous special thread gaging and measuring devices are available for use in measuring thread elements. Some of these devices measure only one element, such as the pitch diameter. Others measure several thread elements simultaneously.

External Thread Comparator

An external thread comparator is used to inspect external threads by means of a single visual reading between indicator tolerance hands. See Fig. 26-6. The comparator checks for errors in lead, thread angle, and pitch diameter. The reading on the indicator dial shows whether the cumulative error of all these elements, combined, falls between the high and low limits for a class-of-thread tolerance. The comparator is set

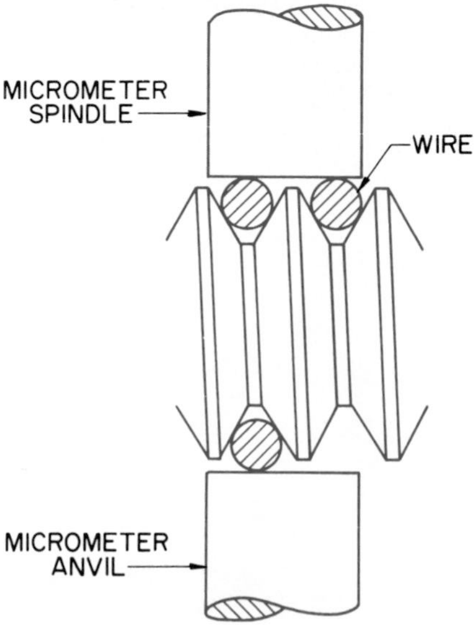

(Hanson-Whitney Co.)

MICROMETER
SPINDLE

WIRE

MICROMETER
ANVIL

Fig. 26-5.   Three-wire thread measurement.

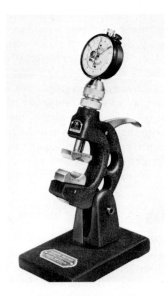

Fig. 26-6.   External thread comparator.

to a given size with the use of a master plug thread gage, and the tolerance hands are set for the desired class of thread.

Internal Thread Comparator

The internal thread comparator, Fig. 26-7, works on the same principle as the external comparator. It is used to inspect all classes of internal threads ranging in diameters from 9.5-50 mm (3/8"-2"). The comparator is set to a given size with the use of a master ring gage, and the tolerance hands are set for the desired class of thread.

Optical Comparator

Optical comparators, Fig. 13-15, may be used to check the accuracy of thread forms and dimensions. An accurately enlarged profile of the thread is projected onto a screen where it is compared to a master drawing. Magnification of up to 50 diameters may be used, depending on the degree of accuracy required.

(Hanson-Whitney Co.)

Fig. 26-7.   Internal thread comparator.

Unit
27

Machine Threading with Taps and Dies

Special accessories are used for machine threading with taps and dies on drill presses, tapping machines, turret lathes, and automatic lathes. One type of accessory is designed for use with conventional taps and dies. It requires the workpiece or the toolholder to reverse rotation in order to back the tap or die off the workpiece. A second type of accessory includes self-opening die heads and collapsing taps. They are designed to automatically trip clear of the workpiece so they can be rapidly withdrawn without reversing toolholder or workpiece rotation.

Die Holders

Die holders are of two types: nonreleasing and releasing. Both types must be used on machines with reversing spindles. The nonreleasing type simply allows the die holder to move lengthwise, independent of the shank in which it is mounted. Once started on the workpiece, the die can lead itself, eliminating the need to feed the die at exactly the same rate as the lead of its thread. The die continues to advance on the workpiece up to the moment spindle rotation is stopped or reversed.

With releasing holders, the die holder automatically disengages after it has pulled ahead of its shank a certain amount, stopping its advancement on the workpiece. When the workpiece rotation is reversed, the die holder again engages with the shank, which causes it to back off the thread. Care must be taken to apply very light tension when retracting the die from the thread. This avoids damaging the first thread on the end of the workpiece. A releasing die holder for spring-type dies is shown in Fig. 27-1. This type of die is used mainly for diameters up to about 3 mm (1/8″).

Self-Opening Die Heads

A large percentage of external threads are made with die heads on various production machines. Die heads are of two basic types: the sta-

(Greenfield Tap and Die Division of TRW Inc.)

Fig. 27-1.   Spring die mounted in releasing die holder.

tionary or nonrevolving type, and the revolving type. Either type is available with retractable cutters called **chasers** for cutting threads, Fig. 27-2, or with rollers for making rolled threads, Fig. 27-3. Various diameter and pitch combinations may be made with a given die head. The chasers and rollers are of the insert type and are available in sets of a given diameter and pitch. Die heads are provided with a means for adjustment of thread depth to conform to required limits and tolerances.

Stationary Die Head

A stationary die head used widely on turret lathes is shown in Fig. 27-2. It is equipped with a lever for resetting the chasers manually. When stationary die heads are used on automatic screw machines, they are reset by an automatic mechanism, making them totally automatic in operation.

Revolving Die Head

Revolving die heads, Fig. 27-4, are used on drill presses and other machines which require the die head to revolve about its axis. This type of die head is equipped with a yoke or other mechanism which is actuated by the machine itself and causes the head to open and close automatically for each threading cycle.

(Eastern Machine Screw Corp.)

Fig. 27-2.   Stationary or nonrotating die head with thread-cutting chasers.

(Geometric Tool)

Fig. 27-3.   Stationary or nonrotating die head with rollers for thread rolling.

(Eastern Machine Screw Corp.)

Fig. 27-4.   Revolving die head.

Machine Tapping

A wide variety of tapping accessories is made to suit the operating features of different machines. Tap holders, tapping heads, and collapsible taps are among the more commonly used accessories. Skilled machine operators can perform tapping operations with nothing more than a chuck or solid tap holder. However, there is less risk of tap breakage when tapping accessories are used.

Tap Holders

Tap holders are virtually identical to die holders except for the method of holding the tools. The explanation of nonreleasing and releasing die holders given earlier applies as well to tap holders. Figure 27-5 shows a tap holder of the nonreleasing type. Split bushings are used for holding different sizes of taps in the tap holders.

Tapping Heads

Tapping heads are reversing or nonreversing. Machines with reversing spindles may use reversing heads which extract the tap by reversal of the

(Boyar-Schultz)

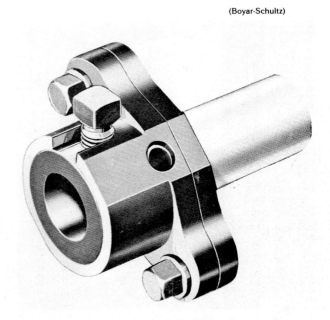

Fig. 27-5.   A nonreleasing tap holder.

spindle. Machines not equipped with a reversing spindle must use a nonreversing tapping head. The tapping head shown in Fig. 27-6 is a nonreversing tapper. When the thrust is down toward the work, the forward clutch engages to drive the tap in. When the forward pressure is released, torque on the tap is released, and tap rotation stops. When the drill press lever is raised, the reversing gear in the tapping head is engaged, causing the tap to back out of the hole.

Some types of tap drivers are provided with friction clutch mechanisms which aid in preventing tap breakage. A torque setting device provides for setting the torque to coincide with the strength of the tap being used. If the preset torque is exceeded, the clutch mechanism slips, thereby protecting the tap from breakage.

Collapsible Taps

Collapsible taps have cutters that can be set to retract when the thread is the desired length, enabling the tap to be rapidly withdrawn without reversing the tap holder or the workpiece rotation. They are of two types: revolving, for use on machines with revolving spindles, and stationary, for use on machines that revolve the workpiece. Both types are actuated when the tap has penetrated the workpiece far enough to bring it into contact with a ring or yoke surrounding the tap. A hand lever is provided for resetting stationary collapsible taps, Fig. 27-7. Collapsible taps are made in sizes from about 32 mm to 125 mm (1¼" to 5").

Test Your Knowledge of Section 4

Unit 15: Screw Threads and Their Use

1. What tools are used to cut threads by hand?
2. List five machine tools which may be used to cut threads.
3. Explain the difference between the major diameter and the minor diameter of a screw thread.
4. How is the amount of clearance between two mating threads controlled?
5. Explain the difference between the pitch and the lead of a screw thread.
6. What is meant by multiple threads?
7. What is meant by the nominal size of a screw thread? Basic size?
8. List three thread factors which may have tolerances specified.
9. What is meant by class of thread?

(Supreme Products Corporation)

Fig. 27-6.   Nonreversing taper mounted on drill press.

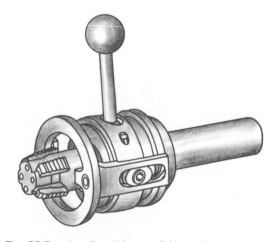

Fig. 27-7.   A collapsible tap of the stationary type.

10. What machine was first used to produce precision screw threads?
11. Who developed the first screw thread standards used in the United States?
12. Explain the basic differences between the first standardized thread system used in Great Britain and that used in the United States.
13. When was the original American (National) Screw Thread Standard approved, and what threads were included?
14. Explain the thread designations used for coarse- and fine-thread series under the American (National) thread standard adopted in 1935.

    List several less commonly used thread series included under the 1935 standard.
15. Of what significance are the Unified and American (National) screw thread standards adopted in 1949 and 1960?

Unit 16: ISO Metric Threads

1. Why were the metric threads originally made by different countries not interchangeable with each other?
2. What thread form or profile was adopted for the ISO Metric thread?
3. Why is it that ISO Metric and ISO Inch threads are not interchangeable?
4. What three series of ISO Metric threads were recommended for worldwide adoption? Which series is in common use on fasteners?
5. Name the three classes of fit provided for ISO Metric threads.
6. Explain the basic designation of ISO Metric coarse and fine series threads.
7. What information beyond the basic designation is required for a complete ISO Metric thread designation?
8. How is the Optimum Metric Fastener System different from the three ISO Metric thread series?

Unit 17: Unified and American (National) Screw Threads (1948)

1. What countries cooperated in the development of the Unified screw thread system of 1948?
2. Describe the general form of the Unified screw thread.

3. Are American (National) standard threads and Unified standard threads interchangeable? Explain.
4. Explain the thread tolerance classes used in the Unified thread system.
5. Explain how Unified screw threads are designated on drawings or blueprints.
6. What series of threads are included under the American standard, **Unified Screw Threads** (NASI B1.1-1960)?
7. When is it necessary for the machinist to use thread formulas in making thread calculations?

Unit 18: Other Forms of Screw Threads

1. Why have square threads been largely replaced with Acme threads?
2. Describe the Acme thread and list several places where it is used.
3. Explain how an acme thread gage is used.
4. Describe the Brown and Sharpe worm thread and list three main uses for it.
5. Describe the tapered pipe thread, NPT, and explain its use.
6. Describe the characteristics of the straight pipe thread, NPS.
7. Describe the features of the Dryseal pipe thread.

Unit 19: Taps and Accessories for Hand Tapping

1. Describe the three basic types of hand taps and tell when each should be used.
2. Describe how tap sizes with Unified National threads are identified and labeled.
3. Describe two types of tap wrenches.
4. Describe a screw plate.
5. Explain how a screw pitch gage is used.
6. Explain how a tap extractor is used.
7. Explain how a screw extractor is used.

Unit 20: Procedure for Hand Tapping

1. How can the proper tap drill size for a tap be found?
2. How can the tap be checked to see if it is aligned parallel with the hole?
3. A cutting fluid should always be used when tapping, except for tapping what kind of metal?
4. What kind of tap requires backing up every two or three turns in order to break the chip?

Unit 21: Other Styles of Taps for Machine and Hand Tapping

1. For what tapping applications are gun taps used?
2. Describe the characteristics and uses of spiral fluted taps.
3. Describe the characteristics of threads made by thread forming taps.
4. What types of taps are available for tapping pipe threads?
5. What special factors must be considered in selecting taps for cutting acme threads?

Unit 22: Tap Size Limits

1. What advantages are gained in using ground thread taps rather than cut thread taps?
2. If the limits code number or class of thread fit is not specified when ordering a ground tap, what pitch diameter size limits will the manufacturer usually provide?
3. How are pitch diameter size limits designated on ground thread taps?

Unit 23: Selection and Calculation of Tap Drill Size

1. Why are internal threads usually tapped only 75 percent of full depth for most thread applications?
2. For what tapping applications should tap drill sizes be increased to produce less than 75-percent thread depth?
3. Calculate the tap drill size for an M24 × 3 thread with 75% depth.
4. Calculate the tap drill size for a ½-13 UNC thread with approximately 75-percent thread depth.

Unit 24: Hand-Threading Dies

1. Why must threading dies be provided with some means of adjusting the depth of cut?
2. Describe the characteristics of the round split die.

3. Describe the characteristics of the adjustable, two-piece threading die.
4. For what purpose is the solid bolt thread die used?

Unit 25: Procedure for Using Hand-Threading Dies

1. Why should the end of a rod be chamfered before it is threaded with a die?
2. In what ways can the starting side of a die be identified?
3. What benefits are gained by using a cutting fluid when threading with a die?
4. If a die gets started out-of-square on the rod, what kind of thread results?

Unit 26: Screw Thread Measurement

1. Why is measurement of the pitch diameter of screw threads always important?
2. Explain how thread limits plug and ring gages are used.
3. Describe the characteristics of a thread micrometer, and explain how it is used.
4. Explain how a roll thread snap gage is used.
5. How can an ordinary outside micrometer be used to measure the pitch diameter of a thread?
6. Explain how thread comparators are used to inspect threads.

Unit 27: Machine Threading with Taps and Dies

1. Name several machines on which machine threading with taps and dies is routinely done.
2. Explain the difference in operation between nonreleasing and releasing tap and die holders.
3. Explain how a stationary die head operates on a turret lathe or automatic screw machine.
4. Explain the difference between a reversing and a nonreversing tapping head.
5. What is the advantage in having a preselective torque setting device on a tapping head?

Section 5

Tool and Utility Grinder Operation

Unit 28

Tool and Utility Grinder Operation

A **utility grinder** is essentially an electric motor driving one or two grinding wheels which are provided with wheel guards, safety shields, and simple work supports. Those intended for dry grinding, Fig. 28-1, are often provided with a water pot into which the workpiece is dipped for cooling. Wet grinders are equipped with a coolant system which pumps a continuous supply of coolant to the wheel and workpiece.

In some shops, utility grinders are used mainly for coarse grinding operations, such as deburring, preparing edges for welding, and removing flashing and other imperfections from castings. When fitted with the proper grinding wheels, however, the utility grinder can be used for sharpening many kinds of sharp edge tools. In the machine shop, bench- or floor-mounted utility grinders are commonly used for sharpening lathe and shaper tool bits, drills, cold chisels, and

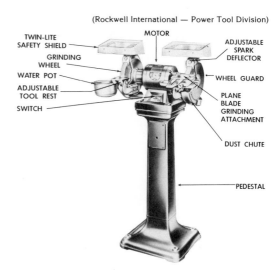

(Rockwell International — Power Tool Division)

TWIN-LITE SAFETY SHIELD
MOTOR
ADJUSTABLE SPARK DEFLECTOR
GRINDING WHEEL
WATER POT
WHEEL GUARD
ADJUSTABLE TOOL REST
PLANE BLADE GRINDING ATTACHMENT
SWITCH
DUST CHUTE
PEDESTAL

Fig. 28-1.   A typical utility grinder.

scribers. Utility grinders may be fitted with attachments designed to aid sharpening of tools which are difficult to sharpen accurately by hand. A drill sharpening attachment is shown in Fig. 41-3. Grinding operations performed with the tool or workpiece hand-held are called **offhand grinding** operations.

Tool grinders, Fig. 28-2, are a type of utility grinder. They are equipped with a tilting table to enable accurate angles to be ground on tool bits for lathes, shapers, and planers. The table also has a tilting angle jig, or miter gage used for grinding edge or end angles accurately. This grinder usually is equipped with a silicon-carbide wheel or a diamond wheel for grinding tungsten-carbide cutting tools. It can also be equipped with aluminum-oxide wheels for grinding high-speed steel cutting tools.

Operating the Utility Grinder

The utility grinder is a simple machine to use, but care must be taken in order to operate it effectively, efficiently, and safely. When the grinder is in operation, the tool rests should be as close to the wheel as possible. This prevents the work from becoming wedged between the rest and the wheel. After adjusting the rest, the wheel should be rotated by hand before turning on the power, to determine if the wheel will clear the rest. If the tool rest is too close to an out-of-round wheel, the high spot may strike the rest.

If too much pressure is used when grinding, the tool or metal being ground will become overheated. The pressure may cause the face of the wheel to wear away more rapidly at some points than at others and thus become out of round. In order to distribute the wear evenly on the face of the wheel, the work should be moved back and forth alternately.

The harder the stock being ground, the more quickly the grains of the wheel become dull. Thus, the harder the material, the softer should be the wheel. This is not true, however, when grinding very soft metals such as brass, copper, or aluminum. In this case, a medium-soft wheel having coarse grains should be used. If the wheel becomes glazed because the cutting particles have become dull, or the pores between the cutting particles have become clogged or loaded with the material being ground, or the wheel becomes out of round, it should be dressed and trued with a wheel dresser.

Grinding Wheel Dressing Tools

A loaded grinding wheel is shown in Fig. 28-3. After the wheel is dressed and trued with a wheel dressing tool, the wheel appears as shown in Fig. 28-4.

Grinding wheel dressers are used for dressing and truing grinding wheels. **Dressing** removes the ground particles in the pores of the wheel. It also causes the dull abrasive grains on the surface of the wheel to fracture away, thus exposing sharp new abrasive grains. The clean, sharp abrasive surface improves the cutting action of the wheel.

Truing a grinding wheel restores the original shape to the wheel. In some cases, truing involves forming a new shape on the surface of the wheel for the purpose of grinding a surface of special shape.

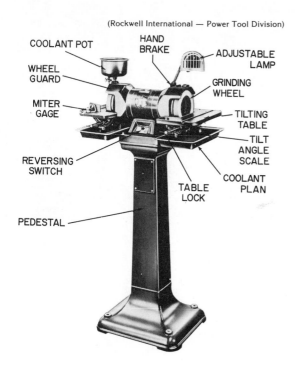

(Rockwell International — Power Tool Division)

COOLANT POT
WHEEL GUARD
MITER GAGE
HAND BRAKE
ADJUSTABLE LAMP
GRINDING WHEEL
TILTING TABLE
TILT ANGLE SCALE
COOLANT PLAN
REVERSING SWITCH
TABLE LOCK
PEDESTAL

Fig. 28-2.   Tool grinder with tilting table.

Grinding wheel dressers often are classified as **mechanical** dressers, **abrasive** dressers, or **diamond** dressers. A widely used mechanical dresser is the **Huntington** type shown in Fig. 28-5. When dressing a wheel with this tool, place the tool on the grinder tool rest and press it firmly against the revolving wheel. Move the tool across the wheel evenly in both directions until the wheel is dressed and true. Care should be taken to avoid running off the edge of the wheel, since this will cause the corner of the wheel to wear off unevenly.

An abrasive wheel dresser is shown in Fig. 28-6. This type usually has a silicon-carbide abrasive wheel which is mounted at a 15° to 30° angle with the axis of the wheel which is to be dressed. The abrasive tool dresses and trues grinding wheels with a shearing action when pressed against the wheel.

(Desmond-Stephan Mfg. Co.)

Fig. 28-3. Loaded grinding wheel before dressing.

(Desmond-Stephan Mfg. Co.)

Fig. 28-4. Grinding wheel after dressing.

(Desmond-Stephan Mfg. Co.)

Fig. 28-5. Huntington wheel dressers.

(Desmond-Stephan Mfg. Co.)

Fig. 28-6. Abrasive wheel dresser.

(Desmond-Stephan Mfg. Co.)

Fig. 28-7.   Abrasive stick wheel dressing tool.

The abrasive stick wheel dressing tool in Fig. 28-7 consists of a steel tube filled with hard abrasive such as silicon carbide or boron nitride. This tool works well on the smaller sizes of medium- and soft-grade grinding wheels. In using the dressing tool, place it on the tool rest, as in Fig. 28-8. Apply pressure to the tool while **rolling** it across the face of the grinding wheel, thus dressing and truing the wheel. As the abrasive is worn off on the end, the steel tube is ground back, thus exposing new abrasive.

Diamond wheel dressing tools, Fig. 28-9, are used for dressing and truing grinding wheels where an extremely accurate finish is required. They are used for dressing and truing wheels on surface grinders, cylindrical grinders, and tool grinders. On these machines, a short diamond-point **nib** is mounted in a guide or holding device at a drag angle of about 15°.

Fig. 28-8.   Correct use of abrasive stick wheel dressing tool.

(Desmond Stephan Mfg. Co.)

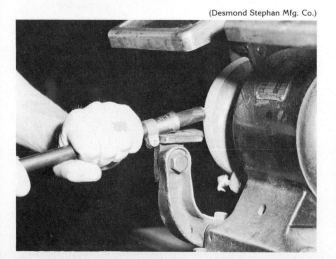

A diamond dressing tool with a handle, Fig. 28-9, is made for manual dressing or truing of small grinding wheels. The depth of cut should be limited to about 0.025 mm (0.001″), and the tool should be fed slowly and uniformly across the surface of the wheel. The tool should be held at a **drag** angle of about 15° in relation to the centerline of the wheel, Fig. 28-10. If there is doubt as to the location of this centerline, the point should be lowered about 3 mm (1/8″) below the line for safety. The tool also should be held at a **lead** angle of about 30° in relation to the face of the grinding wheel. This procedure will aid in preventing chatter and will help keep the diamond point sharp.

Grinding Wheel Selection

Complete information concerning both abrasives and grinding wheels is included in Section 12. Information concerning the classification and selection of grinding wheels is included in Unit 99. Table 99-2 in Section 12 may be used as an aid in selecting the proper wheel. A list of ten precautions for using grinding wheels also is included in Unit 99, and these precautions should be studied at this time, in addition to those given below.

Safety Precautions for Grinding

The following safety precautions should be followed when using the tool grinder:

1. Always wear approved safety goggles when grinding, even though the grinder is equipped with safety glass shields.
2. Avoid grinding on the side of the wheel. Always grind on the face of the wheel.
3. Keep the tool rest adjusted to within 1.5 mm (1/16″) from the face of the wheel.

(Desmond-Stephan Mfg. Co.)

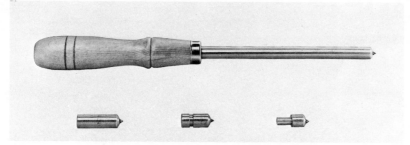

Fig. 28-9. Diamond wheel dressing tool and diamond nibs.

4. Hold the workpiece securely while grinding.
5. The wheel must be kept true and balanced at all times.
6. Be sure that the grinding wheel is not operated at a speed in excess of the rpm marked on the wheel label.
7. See that all safety guards or hoods are secure and in the proper place.
8. Avoid personal contact with the grinding wheel.
9. Wear safe clothing while grinding. Avoid loose neckties or shirt sleeves hanging loosely.
10. If the grinding wheel or grinder appear to be unsafe, report it to the person in charge.

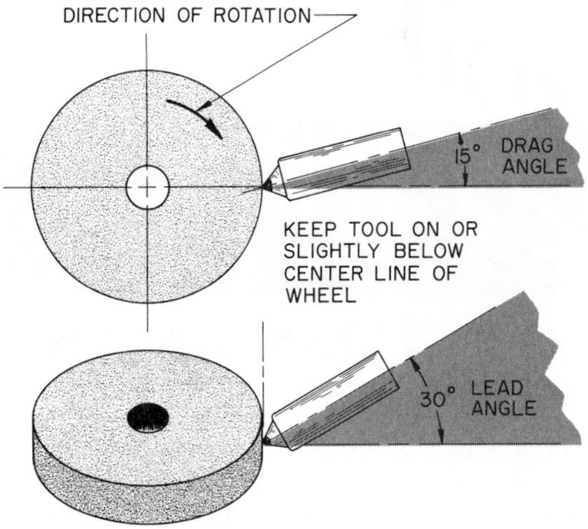

DIRECTION OF ROTATION

15° DRAG ANGLE

KEEP TOOL ON OR SLIGHTLY BELOW CENTER LINE OF WHEEL

30° LEAD ANGLE

Fig. 28-10. Angle of diamond-point tool when dressing or truing a grinding wheel.

Test Your Knowledge of Section 5

Unit 28: Tool and Utility Grinder Operation

1. Describe a typical utility grinder and list the range of grinding operations for which they are used.
2. How does a tool grinder differ from a utility grinder?
3. What kind of grinding wheel is used for grinding tungsten carbide cutting tools?
4. What kind of grinding wheel is used for grinding high-speed steel-cutting tools?
5. How close should the tool rest be to the grinding wheel?
6. Explain the principal difference between **dressing** and **truing** a grinding wheel.
7. List three common types of grinding wheel dressing tools.
8. Explain how a hand-held diamond wheel dresser should be positioned for truing or dressing a grinding wheel.
9. How should a worker be dressed for safely using a tool or utility grinder?
10. Describe how to use and properly care for grinding wheels.

Section 6

Stock Cutoff Machines

Unit 29

Types of Stock Cutoff Machines

Four kinds of stock cutoff machines are in common use: (1) power hacksaws, (2) horizontal bandsaws, (3) cold saws, and (4) abrasive cutoff saws. These saws are used primarily for cutting standard stock to working lengths. The machines must be capable of rapidly making straight, square cuts on a wide variety of workpiece shapes. Some machines are equipped with vises that swivel through 45°. This added versatility is useful in job or fabricating shops where angles other than 90° are often encountered.

Power Hacksaws

Power hacksaws operate on the principle of the reciprocating stroke. On the cutting stroke, the blade teeth are forced or fed into the metal either by gravity or by controlled hydraulic pressure. The pressure is automatically removed on the return stroke in order to prolong the service life of the blade. Most machines are equipped with a chip tray and a cabinet base which houses the coolant reservoir and pump for circulating the coolant, Fig. 29-1.

For production work, heavy duty power hacksaws are provided with automatic bar feeds, Fig. 29-2. With this type of machine, the stock to be cut is loaded on a carriage which, at the completion of a cut, automatically moves forward the distance required. Hydraulic pressure automatically operates the vise jaws, gauges the material, and raises and lowers the saw blade. Once the machine has been set up for cutting stock to a given length, it will operate automatically without the attention of an operator until all stock loaded on the carriage has been cut. Machines of this type are available with a capacity for round stock as large as 800 mm (30½").

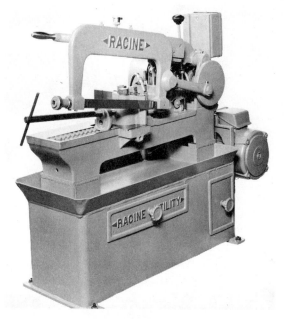

Fig. 29-1.   Wet-cutting power hacksaw.

Fig. 29-2.   Heavy-duty automatic hydraulic saw.

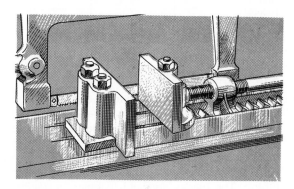

Fig. 29-3.   Toothed rack provides quick positioning of power hacksaw vise jaw.

Vises

On manually operated machines, the vises may be plain or the swivel type. The movable jaw can be positioned quickly by means of a toothed rack, Fig 29-3, which is part of the table plate. Notice in Fig. 29-1 that the movable jaw of the vise has been advanced along the toothed rack until it is close to the fixed jaw, thus requiring little advance of the clamping screw to close the jaws completely. Most machines have adjustable stops which, when set to a particular dimension, enable the operator to cut a number of pieces identical in length without having to measure each piece.

Feed and Cutting Speed

Both of the machines shown are equipped with hydraulic feed. This provides accurate feed and pressure control throughout the cut regardless of the type of material. When the cut has been completed, a knockout disengages the clutch; this

deactivates the saw, and the saw frame automatically rises to its highest inactive position. The machine in Fig. 29-1 is available with two or four cutting speeds. The two-speed type has 100 to 140 strokes per minute. The four-speed type operates at speeds of 35, 70, 100, and 140 strokes per minute. On machines equipped with four speeds, a four-speed drive motor is used. Changes in cutting speed are made by moving the speed change lever to the position indicating the desired speed. Hard and tough metals, such as high carbon steel and alloy tool steel, should be cut at lower cutting speeds than low carbon steel. Table 29-1 gives recommended cutting speeds for cutting different metals with power hacksaws.

Table 29-1
Cutting Speeds Recommended
for Power Hacksaws

Material	Strokes Per Minute Dry	Strokes Per Minute Wet
Low-carbon steel	70-100	100-140
Medium-carbon steel	70	100-140
High-carbon steel	70	100
High-speed steel	70	100
Drill rod	70	100
Alloy steel	70	100
Cast iron	70-100	(cut dry)
Aluminum	100	140
Brass	70	100-140
Bronze	70	100

(Kasto-Racine)

Fig. 29-4.   Automatic band saw cutoff machine.

(Kasto-Racine)

Fig. 29-5.   Working area of automatic band saw cutoff machine.

Saw Blades and Their Selection

The blades recommended for the saw shown in Fig. 29-1 are 350 mm (14") in length, 32 mm (1¼") wide, and normally are available with tooth pitches of 2.5, 4, and 6 mm (10, 6, and 4 teeth per inch). Blades with more teeth can be used when necessary, for example, when cutting thin-walled tubing. The blades are held in position against the aligning surfaces of the blade holder with socket head screws, and they are tensioned by means of hardened bolts. The blades may be selected as recommended in Table 29-2.

The following general principles apply to the selection of power hacksaw blades:

1. A minimum of two teeth should be kept in contact with the workpiece at all times.
2. Soft and easily machined materials and large cross sections require coarse teeth for adequate chip clearance.
3. Small sections and thin-walled sections require fine teeth.
4. Hard materials which are difficult to machine require finer teeth, thus providing more cutting edges.

Horizontal Band Saws

Most stock cutoff band saws are horizontal machines designed especially for this purpose. They range from small manually operated machines that are intended for intermittent toolroom operation to large, fully automatic production machines. Figure 29-4 shows an automatic machine of medium size. A closeup of its working area showing the blade guides, coolant hoses, and automatic feed rolls built into the vise is shown in Fig. 29-5.

Table 29-2
Tooth Selection for Power Hacksaw Blades

Material Being Cut	Machinability Group*	Pitch of Teeth Minimum Material Thickness				
		6 mm (1/4")	12 mm (1/2")	18 mm (3/4")	25 mm (1")	50 mm (2") or over
Easily machined	1 (above 70%)	2.5 mm (10)**	4 mm (6)	4 mm (6)	6 mm (4)	6 mm (4)
Moderately difficult to machine	2 (50 - 70%)	2.5 mm (10)	2.5 mm (10)	4 mm (6)	4 mm (6)	6 mm (4)
Difficult to machine	3 (40 - 50%)	2.5 mm (10)	2.5 mm (10)	2.5 mm (10)	4 mm (6)	4 mm (6)

*For general machinability ratings of metals, see Selection Chart for Cutting Fluids, Table A-11, Appendix.
**Teeth per inch.

Band saws offer a number of advantages over other kinds of stock cutoff machines. Their saw blades cut a kerf of only 1.5 mm (1/16") compared to 3 mm (1/8") for power hacksaws and abrasive disc saws, and 6 mm (1/4") for cold saws. This represents a considerable saving, especially when cutting large stock or expensive material. Cutting rates are higher because their endless blades cut continuously, unlike the intermittent cutting of power hacksaws. And because the blades are thinner, they require less power to cut through the workpiece, making them more economical to operate.

Blade Selection

Band saw blades of solid carbon tool steel, alloy tool steel, and high-speed steel are available. Also, composite blades with high-speed steel or tungsten-carbide teeth welded to a carbon steel back are available. High-speed steel blades are preferred for cutoff work because they provide long service life at high cutting speeds. Blades 25 mm (1") or wider are normally used because their high strength can tolerate the heavier loading encountered in high feed rates.

Blades with **regular** tooth form are used for workpieces up to 25 mm (1") thick, with **hook** tooth blades recommended for thicker workpieces. Tooth form profiles are shown in Fig. 31-6. Table 29-3 gives blade width and pitch recommendations for cutting various thickness ranges. Permissable cutting speeds and feeds vary with kind of blade material, size and condition of blade, kind and thickness of material to be cut, and whether or not a coolant will be used.

Table 29-3
Band Saw Blade Widths and Pitches
for Stock Cutoff

Workpiece Thickness	Blade Width	Pitch
up to 16 mm (5/8")	25 mm (1")	2.5 mm (10*)
16 mm (5/8") to 25 mm (1")	25 mm (1")	3 mm (8)
25 mm (1") to 100 mm (4")	25 mm (1")	4 mm (6)
100 mm (4") to 250 mm (10")	25 mm (1") to 32 mm (1-1/4")	6 mm (4)
250 mm (10") to 500 mm (20")	32 mm (1-1/4") to 50 mm (2")	8 mm (3)

*Teeth per inch.

Cold Saws

Cold saws are low rpm circular saws designed especially for cutting metals. They range in size from hand-operated bench-top models with 200 mm (8″) blades to fully automatic machines with blades of 900 mm (3′) diameter and larger. Light duty manual or automatic machines are usually provided with a swivel head that enables cuts to be made at any angle. See Fig. 29-6. These machines are most efficient for cutting structural shapes such as I-beams, angles, and channels because the circular blades can complete their cuts with less travel than straight blades. Heavy-duty machines with automatic feeds are efficient for cutting solid bars up to about 250 mm (10″). Beyond this size, blade diameters, which must be more than double the cutting capacity, become excessively costly as does the machine necessary to drive them. Several speed ranges are provided for cutting metals of different toughness, and built-in coolant systems help produce good finishes and prolong blade life.

Cold Saw Blades

Blade widths increase with diameters, varying from 2 mm (5/64″) to as much as 8 mm (5/16″). Blades 350 mm (14″) and over are made with inserted teeth of high-speed steel. Only highest quality, precision-ground blades will produce accurate cuts. Solid blades which are properly used may be resharpened up to 20 times. Hollow ground, staggered tooth, and swaged tooth blades are available. Tooth pitches range from very fine at 3 mm (1/8″) to very coarse at 16 mm (5/8″). In general, fine pitch blades are used for cutting solid stock up to 12.7 mm (1/2″) and tubing, medium pitch blades for cutting solids from 12.7 mm (1/2″) to 37.5 mm (1-1/2″), and coarse pitch blades for cutting solids above 37.5 mm (1-1/2″).

Abrasive Cutoff Saws

Abrasive cutoff saws employ an abrasive disc to separate material by grinding. Many types of abrasive cutoff saws are made. They are built either for manual operation or with power feeds, and with either fixed or oscillating wheel heads. Oscillating wheel head machines are recommended for cutting thick sections of tough

(Kasto-Racine)

(Jet Equipment and Tools)

Fig. 29-6.   Manually operated cold saw with swivel head.

Fig. 29-7.   A dry cut 406 mm (16″) abrasive cutoff machine.

materials such as titanium and high alloy steels. Sizes range from small bench-top machines with 200 mm (8″) wheels to machines with 500 mm (20″) and larger wheels. Figure 29-7 shows a manually operated, dry cutting machine designed for either portable or stationary use.

Abrasive cutoff saws are useful for rapid cutting of small sizes of bar stock, tubing, and structural shapes. They are most valuable for cutting extremely tough or hardened materials that cannot be cut efficiently with other types of saws. Figure 29-8 shows a wet cutting machine used for cutting metallurgical samples.

Abrasive Cutoff Wheels

Most abrasive cutoff wheels for cutting metals are made with aluminum oxide or silicon carbide abrasives. For wet cutting, rubber-bonded wheels of medium grit should be used. Dry cutting is done with resinoid-bonded wheels of fairly coarse grit. The 1 mm (1/32″) and 1.5 mm (1/16″) small diameter wheels used for cutting metallurgical samples are not reinforced and are thus very fragile. Larger wheels used for general purpose cutoff are 3 mm (1/8″) thick and are reinforced with layers of fiberglass for added strength and safety.

Safety Precautions for Using Cutoff Saws

1. Mount or remove work only when the saw is stopped and is raised well above the vise.
2. Clamp the workpiece securely on both sides of the blade if possible. Injury can result when brittle blades shatter due to rolling or shifting of the workpiece under the pressure of cutting.
3. Use only blades that are in good condition and which are the correct kind for the material being cut.
4. Operate the machine at a cutting speed and feed rate that does not strain the machine or the blade.
5. Keep hands away from the saw blade when it is running.
6. Stand out of line with the blade when starting an abrasive cutoff saw.
7. Do not bend over in front of a power hacksaw frame when the machine is running.
8. Support long pieces of stock so that they will not fall and cause injury, and hang a cloth on the protruding ends to warn others of the potential hazard to them.
9. Broken blades can result on power hacksaws with hydraulic feeds if blades are brought closer to the workpiece than 6 mm (1/4″) at the start of a cut.
10. Take due care when handling freshly cut pieces. They have sharp edges and can be hot enough to cause burns.
11. Always wear approved eye protection when operating cutoff saws.

(Buehler, Ltd.)

Fig. 29-8.   Wet cut abrasive cutoff machine.

Unit

30

Stock Cutoff Machine Operation

The following procedure applies to the operation of all types of cutoff machines:

1. Wear safety glasses or goggles.
2. On machines with adjustable vises, check that the vise is set at the desired angle and is securely fastened.
3. Obtain the stock and position it properly in the vise. See Fig. 30-1. Several pieces may be cut at the same time, provided they can be held securely, Fig. 30-2.
4. If the piece of stock being cut is short, another piece of the same size must be placed on the opposite side of the vise so that the vise jaws remain parallel. See Fig. 30-3. If the stock is long, support the overhanging end(s) with a work support.
5. Change the saw blade if it is in poor condition or if it is the wrong kind of blade for the metal to be cut.
6. Lower the saw blade so that the length of the piece to be cut can be measured. If additional pieces of the same size will be cut, set the stock stop to eliminate the need for further measuring, Fig. 30-4.
7. Set the machine for the correct cutting speed and feed, if necessary.

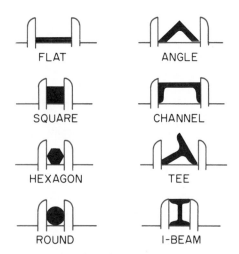

Fig. 30-1.   Holding different metal shapes in stock cutoff machine vise.

(DoAll Company)

Fig. 30-2.   Angles nested in vise for cutting multiple pieces of identical length.

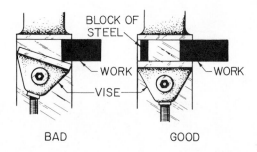

Fig. 30-3.   Holding short piece of metal in power saw vise.

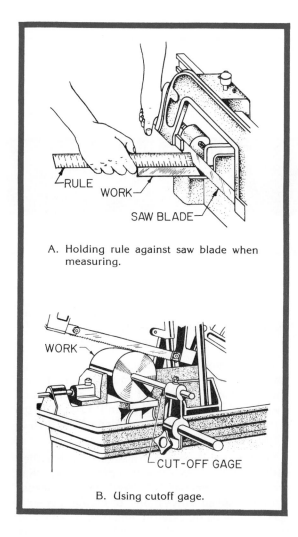

A. Holding rule against saw blade when measuring.

B. Using cutoff gage.

Fig. 30-4.   Measuring metal to be cut.

8A. On machines with power feed systems, start the machine, turn on the coolant, and engage the feed system. **Be prepared to slow down the feed rate if the machine begins to labor, or to shut off the machine in case the blade breaks or the workpiece shifts.**

8B. On machines with manual feed, start the machine, turn on the coolant, if any, and grasping the feed handle firmly, feed the blade into the workpiece at a rate that allows it to cut freely.

9. When the cut has been completed, raise the saw frame and shut the machine off. Some hydraulically powered machines will automatically stop the blade and raise the saw frame in preparation for the next cut.

10. Measure the cut piece to determine whether it is the correct length. If not, adjust the stock stop accordingly.

11. Loosen the vise, position the stock for the next cut, if any, then proceed as in steps 6 through 9.

Test Your Knowledge of Section 6

Unit 29: Kinds of Stock Cutoff Machines

1. Name the four kinds of stock cutoff machines.
2. How do stock cutoff machines designed for production work differ from those that are intended for intermittent job shop use?
3. List the general principles that apply to the tooth selection of power hack saw blades.
4. List several advantages of band saw cutoff machines when compared to other kinds of cutoff machines.
5. Give the tooth form, blade width, and pitch of the band saw blade recommended for cutting a workpiece 18 mm (3/4") thick.
6. Blades made of what kind of metal give best results for cutoff band sawing?
7. For what kind of work are light duty cold saws best suited? Heavy duty cold saws?
8. Abrasive cutoff saws are especially useful for cutting what kinds of materials?
9. What kinds of abrasives are used in metal cutting abrasive cutoff wheels?

Section 7

Band Machining

Unit

31

Metal Cutting Band Saws

Two general types of band saws are available for cutting metals: the **horizontal** type on which the blade passes through the work in a horizontal plane, and the **vertical** type which presents the blade to the workpiece vertically. Horizontal machines, Fig. 29-4, are designed for stock cutoff work and are discussed in Section 6. Some light duty machines are so designed that they may be changed easily from either a horizontal or vertical position. On both types of band saws, the saw blade travels around saw carrier wheels which usually have a heavy layer of hard rubber cemented to their rims to protect the saw teeth.

The vertical band saw is available in a number of sizes ranging in work capacity from 200 mm (8″) to 600 mm (24″) in thickness, and in throat capacity from 400 mm (16″) to 1500 mm (60″).

A heavy duty type is shown in Fig. 31-1. This machine has a throat capacity of 400 mm (16″) and a work thickness capacity of 300 mm (12″).

Vertical band saws frequently are called **contour machines.** They are used for sawing, filing, and polishing contours of regular or irregular shape. See Figs. 32-4, 35-2, and 35-3. A large production sawing machine, called a **band mill,** is shown in Fig. 31-2.

Vertical band saws are available with either variable speed drives or with a drive system that provides a selection of fixed blade speeds. Variable speed drives are desirable because speeds may be quickly set, and may be adjusted to match cutting conditions while the machine is running.

(DoAll Company)

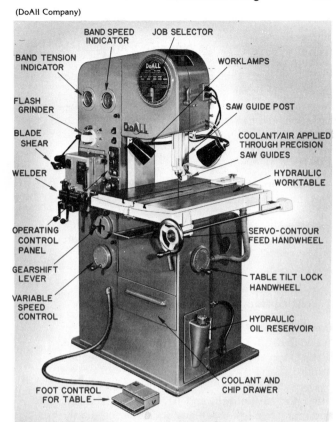

BAND SPEED INDICATOR

JOB SELECTOR

BAND TENSION INDICATOR

WORKLAMPS

FLASH GRINDER

SAW GUIDE POST

BLADE SHEAR

COOLANT/AIR APPLIED THROUGH PRECISION SAW GUIDES

WELDER

HYDRAULIC WORKTABLE

OPERATING CONTROL PANEL

SERVO-CONTOUR FEED HANDWHEEL

GEARSHIFT LEVER

TABLE TILT LOCK HANDWHEEL

VARIABLE SPEED CONTROL

HYDRAULIC OIL RESERVOIR

Fig. 31-1.   Conventional (vertical) heavy-duty band saw.

COOLANT AND CHIP DRAWER

FOOT CONTROL FOR TABLE →

(DoAll Company)

Fig. 31-2.   Band milling using a 50 mm (2″) width blade for cutting large diameter work.

Variable-Speed Band Saws

The variable-speed machine shown in Fig. 31-1 is equipped with a job selector, Fig. 31-3. This is a circular chart that gives recommended blade speeds and other data pertinent for cutting a large number of commonly used materials. The desired speed can be obtained by merely turning a hand crank located at the left side of the machine.

CAUTION

The machine must be running when speed adjustments are made.

A gear shift, also located at the side, provides easy shifting of the gears to produce either high or low speeds.

CAUTION

The machine must be stopped when shifting from one speed range to another.

Some saws are equipped with a tachometer or speed indicator which indicates the speed the blade is traveling in meters per minute or feet per minute.

The table on most of these machines can be tilted 45° to the right and 10° to the left, front, and rear. On a few machines, it can be tilted only 45° to the right and 5° to the left.

Some machines are equipped with power feed. The power-feed mechanism shown on the machine in Fig. 31-4 is actuated by an adjustable weight. A power-feed control handwheel, located at the front end of the machine, regulates the pull on the work. Pressure on the work can be released instantly by pressing on the power-feed pedal.

The machine illustrated in Fig. 31-1 is equipped with an infinitely variable hydraulic-feed mechanism which has a feed range from 0 to 300 mm (12") per minute. The feeding force also may be adjusted within an infinitely variable range from 0 to 160 kg (350 lbs.). Machines are also made which use compressed air cylinders to provide power to the table.

Vertical metal-cutting band saws are used both for cutting off stock and for contour machining — internal and external. When used for internal sawing, as in Fig. 32-3, the saw blade must be cut, threaded through a pilot hole drilled in the material, and then butt welded. For this purpose,

(DoAll Company)

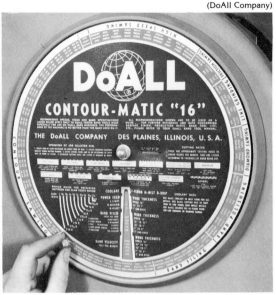

Fig. 31-3.   Job selector chart for a variable-speed band saw.

(DoAll Company)

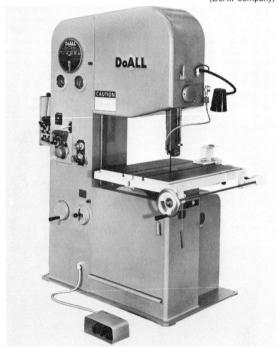

Fig. 31-4.   Band saw equipped with power feed.

(DoAll Company)

Fig. 31-5. Butt-welding and annealing unit with grinder for joining and dressing saw blade.

machines are equipped with blade shears, and butt welding and annealing attachments. Such a unit is shown in Fig. 31-5 combined with a flash grinder and saw-thickness gage. Thus the blade may be welded, annealed, reduced to uniform thickness, and returned to service with a minimum of effort and interruption of production.

Many metal-cutting band saws, particularly those which run at high speed, are equipped with safety brakes. Such brakes are automatically set and the driving power cut off should the blade happen to break.

Saw Blades

Band saw blades are made of solid carbon tool steel or solid high-speed steel. Composite blades of high-speed steel or tungsten-carbide teeth welded to a carbon-steel band are also made. Carbon-steel blades are the most widely used for general applications requiring moderate cutting speeds and a moderate production rate. High-speed steel blades (which stay sharp longer) may be used at higher cutting speeds and are recommended where higher rates of production are required. The cutting rate with high-speed steel blades may double the rate possible with carbon-steel blades; however, they are more expensive.

Tungsten-carbide blades generally are recommended for work thicknesses greater than 75 mm (3"). They usually have a coarse pitch of 10-16 mm (2½-1½ teeth per inch). They are recommended only for heavy-duty saws where the cutting speed and rate of production must be greater than for either carbon-steel or high-speed steel blades. The carbide blades operate at higher cutting speeds than high-speed steel blades, but they are also more expensive.

Carbon-steel blades for metal cutting have hardened teeth and a flexible back. The latter feature resists fatigue and facilitates continuous bending. Blades are available in various widths, lengths, and pitches.

Common pitches are 1.5, 2, 2.5, 3 and 4 mm (18, 14, 10, 8, and 6 teeth per inch). Common blade widths are 6, 9, 12.7, 16, 18, and 25 mm (1/4", 3/8", 1/2", 5/8", 3/4", and 1"). The length of saw blade required for a particular machine is determined by the construction of the machine and its blade capacity. The life of the saw blade is affected in no small degree by the manner in which the operator uses it. Improper speed, feed, or tension will shorten the life of a blade. Consequently, it is the responsibility of an operator to see that these factors are correct for each job. The vertical band saws illustrated in this unit are equipped with a blade tension indicator, which shows proper tension for each width of blade. On saws not so equipped, the manufacturer's recommendation should be followed concerning proper blade tension. Manufacturers usually supply with each machine delivered an operator's manual which includes this type of information.

Tooth Form

Band saw blade tooth forms are shown in Fig. 31-6. Although blade manufacturers sometimes use different names, the forms are identified here as **regular, hook, skip,** and **carbide.** The regular tooth form is used for all blades of 4 mm (6 teeth per inch) pitch. The hook form is used for blades of 4 mm (6 teeth per inch) and coarser. Most contour cutting is done with regular tooth form blades. They provide good chip capacity for making fast cuts with good finishes. Skip tooth blades use the same rake and clearance angles as regular tooth blades, but the teeth are shorter and spaced farther apart. The wider tooth spacing provides greater chip clearance needed for making cuts through thick sections of soft metals.

Blade Set

The teeth of band saw blades are bent or offset slightly to the right and to the left to provide clearance for the blade while cutting. This offsetting is called **set**, Fig. 31-7. The distance from the outside of one tooth to the outside of the tooth on the opposite side of the blade is the **set dimension.** The set on a band saw enables the blade to cut contours. With a small-width blade and a wide set, small-radius contours may be cut.

The common types of set on metal-cutting band saw blades are the **raker** set and the **wave** set, Fig. 31-7. The raker set has one straight tooth at the center, one offset to the right, and one offset to the left, continuing in that order. The wave set gradually weaves from left to right and back again. A third type of set, the straight set which has teeth alternating left and right, no longer is used for metalworking.

The raker set is recommended for cutting most contours. The wave set is recommended wherever the work has a varying cross-sectional thickness and wherever one band saw blade must be used for a wide range of material sizes and shapes. It is recommended for cutting pipe, angle iron, and other irregular structural shapes.

Blades for Special Applications

Metal cutting band saws can be used to cut or shape a great variety of materials and products, including practically all metals, and nonmetals such as wood, plastics, glass, and stone.

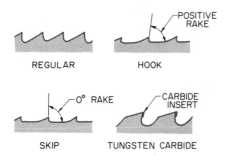

Fig. 31-6.   Tooth forms for metal-cutting band saw blades.

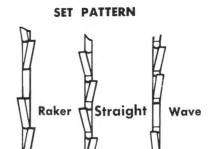

(DoAll Company)

SET PATTERN

Raker   Straight   Wave

Set exaggerated for clarity

Fig. 31-7.   The three set patterns. Straight set is no longer used for metalworking.

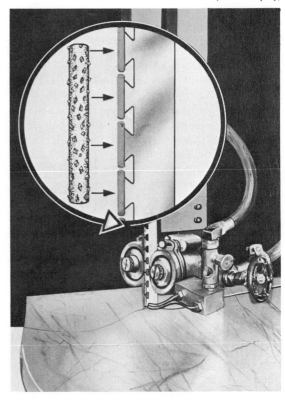

(DoAll Company)

Fig. 31-8.   Diamond-tooth band saw blade.

A **diamond-tooth blade,** Fig. 31-8, makes possible the cutting of refractory materials such as glass, tile, marble, granite, china, porcelain, and silicon carbide. The blade is constructed of highly fatigue-resistant steel, and hundreds of small rods consisting of natural or synthetic diamond particles bonded with a patented tungsten alloy and electronically brazed to the steel band.

Diamond or **aluminum oxide edge blades** can successfully cut hardened steels and the cobalt and nickel-based super alloys. The **spiral tooth blade,** Fig. 31-9, has a round cross section and a cutting edge that spirals continuously around the blade, enabling it to cut in any direction. It is made in several sizes from 0.5 mm (0.020″) to 2 mm (0.074″). Its ability to cut in any direction makes it especially useful for cutting complex contours in thin metal.

Friction Sawing

In **friction sawing,** cutting is accomplished by momentary contact between the material and a rapidly moving blade, which produces enough friction to heat the metal immediately ahead of the saw to its softening point. It is much faster than conventional sawing, and will cut metal that can otherwise only be cut with great difficulty or not at all. Relatively little heat is conducted into the side walls of the cut, leaving the metallurgical properties of the material undisturbed. Metals which become soft and sticky at low temperatures, such as aluminum, copper, and their alloys, cannot be cut satisfactorily by friction sawing.

This method is capable of both straight and contour sawing, with radii as short as 16 mm (5/8″) possible using a 6 mm (1/4″) blade. Only saw blades recommended for friction sawing should be used, Fig. 31-10. The blade need not be sharp as a dull blade increases friction, thus producing higher heat and higher cutting speed.

Armor plate as thick as 12.7 mm (1/2″) can be friction cut successfully. Thicker material can be cut by using a rocking technique. In this procedure, the rear end of the material is raised, thus

(DoAll Company)

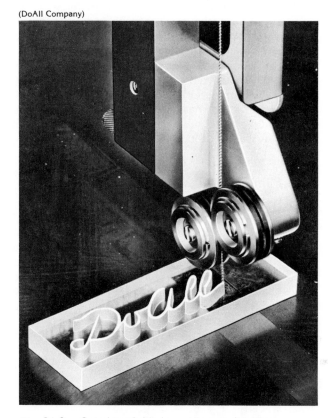

Fig. 31-9.   Spiral-tooth blades cut in any direction.

(DoAll Company)

Fig. 31-10. Dull edge for friction sawing.

presenting the top edge of the material to the saw. When the saw starts to cut, the rear is lowered to the table, and then raised and lowered alternately throughout the cut.

Friction cutting employs speeds ranging from 1,000-5,000 meters per minute (3,000 to 15,000 feet per minute), depending upon the composition and thickness of the material. Blade speed and pitch recommendations for cutting different materials are found in Table A-1, Appendix.

Filing and Polishing

Both filing and polishing can be performed effectively on metal-cutting band saws by means of endless filing and polishing bands. Excellent work can be done with filing bands and in much shorter time than by hand filing (Fig. 35-2). This is also true of polishing bands (Fig. 35-3). File bands are available in various styles and cuts: flats, half-round, and oval. Common sizes are 6 to 12.7 mm (1/4″ to 1/2″) widths.

Safety Precautions for the Band Saw

1. Always be sure that the saw guide is set to within 6 to 9 mm (1/4″ to 3/8″) above the thickest portion of the workpiece to be cut.
2. When starting the saw, stand to one side of the saw frame; then adjust the speed as desired.
3. Support protruding ends of long pieces so they will not fall and cause injury.
4. Hang a cloth over protruding ends of long pieces, and see that others do not run into them.
5. Be sure that the saw blade is in good condition, and that the machine is set to the correct cutting speed.
6. Avoid getting the hands too close to the blade.
7. Use a stick to remove short pieces of work from the area close to the blade.
8. Be careful in handling parts with sharp burrs on the ends. The burrs may be removed with a file.
9. Wear approved safety goggles at all times in the shop.

Unit 32

Vertical Band Saw Operation

Vertical band saws are used for straight, angular, and contour cutting, both external and internal.

Procedure for Cutting with Table Horizontal

1. Locate the position on the work at which the cut is to be made. Then, if necessary, draw guidelines or use appropriate machine guides, as in Fig. 32-1. The attachment illustrated can be used for cutting off, ripping, or mitering operations.
2. Determine kind and size of saw blade required by consulting the job selector chart. If necessary change the saw blade.
3. Set the machine to the cutting speed required; consult the job selector chart.
4. Protect the eyes by wearing a pair of properly fitted goggles.
5. Lower the saw guide until it is within 6 mm (1/4″) of the thickest or highest part of the work surface.

(DoAll Company)

Fig. 32-1. Cutoff and mitering attachment used to cut stock at any angle. Manual or power feed may be used.

(DoAll Company)

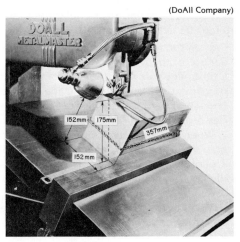

Fig. 32-2. Making cut with table tilted.

6. With the machine running, bring the stock slowly into contact with the traveling blade. Advance the work by exerting a light, but steady, pressure. Do not crowd the saw; allow it time to remove the metal. When the machine is so equipped, power feed should be used for cutting heavy pieces.

7. Round or irregularly shaped pieces should be held in a vise, jig, or fixture. This then should be held against a suitable guide.

The use of a cutting fluid is recommended when cutting steel. When used, it should be directed at the teeth of the saw where they contact the workpiece (or slightly above).

CAUTION
In placing the hands on the material, be sure they will not come in contact with the traveling saw blade.

8. At the end of the cut, stop the machine and remove the workpiece. Remove the waste stock and clean the machine, unless other pieces are to be machined.

CAUTION
Use a piece of wood or other suitable material to remove short workpieces or waste stock from the vicinity of the saw blade. Use a brush to remove metal chips from the saw table.

Procedure for Cutting with Table at an Angle

1. Proceed as when making a straight cut, steps 1 through 4 inclusive.
2. Tilt the table to the angle desired, and clamp it in position. When cutting a heavy piece of material, use a guide or fixture such as shown in Fig. 32-2.
3. Place the work on the saw table, and move it close to the saw blade. Then adjust the saw guide, bringing it as close to the work surface as practicable.
4. Start the machine, move the work against the saw, and make the cut in the usual manner.

Procedure for Cutting External Contours

1. Lay out the required shape on the surface of the workpiece.
2. Determine the size and kind of saw blade needed. When the shape involves sharp curves, a suitably narrow blade should be used.
3. Proceed as when making straight cuts, steps 3 through 6 inclusive.
4. When it is necessary to cut a very sharp curve, it may be advisable to bypass it temporarily. Then, when the waste material has been removed from other parts of the work, the cut can be made in the usual manner.

At the end of the cut, stop the machine, remove the workpiece, and clean the machine, unless other pieces are to be machined.

Procedure for Cutting Internal Contours

Rectangular, circular, or irregular internal contours may be cut with a band saw.

1. Lay out the required contour or outline on the surface of the workpiece, as in Fig. 32-3.
2. Drill a pilot hole near one edge of the contour or outline. Be sure to make the hole large enough to permit free entry of the saw blade. See Fig. 32-3.

   When the enclosure is rectangular, a pilot hole should be drilled at each corner. Drilling a pilot hole wherever a contour changes its direction sharply will facilitate turning or reversing the direction of the cut.
3. Determine the size and kind of saw blade required to perform the operation; consult the job selector chart.
4. Cut the saw blade. Then draw one end through the pilot hole in the workpiece, place the ends of the blade in the automatic butt welder, Fig. 31-5, and weld the ends together.
5. Place the welded joint in the flash grinder, and grind the joint smooth and even.
6. Place the welded blade on the wheels of the machine, and tension it correctly.
7. Proceed as when making straight cuts, steps 3 through 6 inclusive.
8. When it is necessary to cut a very sharp curve, it may be advisable to bypass it temporarily; then after waste material has been removed, the cut can be made in the usual manner.
9. When the cut has been completed, Fig. 32-4, stop the machine, remove the waste material and the blade, cut the saw blade, and remove it.
10. Re-weld and re-install the saw blade or set it aside for later use on a similar operation.
11. Brush waste stock and metal chips from the saw table.

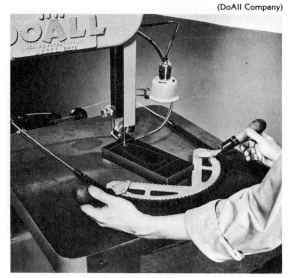

(DoAll Company)

Fig. 32-3.   Band saw inserted through drilled hole for internal contour sawing.

(DoAll Company)

Fig. 32-4.   Internal contour sawed out.

Unit

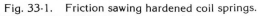

33

Friction Band Sawing

Straight or contour sawing can be done by friction sawing. Care must be taken that the recommended saws, Fig. 31-10, and cutting speeds are used.

Procedure

1. Locate the position on the workpiece at which the cut is to be made; then draw appropriate guidelines or use appropriate machine guides.
2. Install the size and kind of saw required.
3. Protect the eyes, hands, and clothing by wearing properly fitted glasses, leather gloves, and appropriate coveralls.

4. Adjust the machine to the correct cutting speed.

5. Start the machine. When it has attained full speed, bring the work gently into contact with the blade, as in Fig. 33-1. Advance the work steadily. Use the rocking technique, if necessary; that is, alternately slightly raise and lower the rear end of the work.

At the completion of the cut, stop the machine, remove the work, and clean the worktable.

Fig. 33-1.   Friction sawing hardened coil springs.

(DoAll Company)

Unit

34

Band Sawing Refractory Materials

A variety of refractory materials (such as glass, porcelain, china, marble, granite, and silicon carbides) can be cut with diamond-tooth blades.

Procedure

1. Lay out the design on the workpiece or set a saw guide as in Fig. 34-1.
2. Secure a diamond-tooth saw blade, Fig. 34-1. Mount it on the saw, and tension it properly.
3. Set the machine for a cutting speed of 600 to 1200 meters per minute (2000 to 4000 feet per minute).
4. Protect the eyes by wearing goggles.

5. Start the machine, turn on the cutting fluid, and direct it at the teeth of the blade where they contact the workpiece. A spray lubricator is recommended. Figure 34-2 shows the spray tubes.

   Note: When sawing cylindrical pieces, hold them in a vise, a jig, or other suitable fixture. The same applies when sawing irregular shapes.

6. Bring the workpiece gently against the traveling blade, and feed it steadily forward. Do not crowd the saw.
7. At the completion of the cut, stop the machine, remove the work, and clean the machine.

(DoAll Company)

Fig. 34-1.   Cutting glass with a diamond-tooth blade.

(DoAll Company)

Fig. 34-2.   A spray lubricant is recommended for refractory material.

Unit

35

Band Filing and Polishing

Endless filing and polishing bands are available for use with metal cutting band saws. When properly handled, they will do accurate work, much more rapidly than can be done by hand. An endless file is fitted with a joint which permits its use for internal filing, Fig. 35-1 and 35-2.

Procedure for Band Filing

1. Select a file band of appropriate grade and shape. Also select a suitable band guide.
2. Mount the guide in the guidepost.
3. If internal filing is to be done, uncouple the file band, Fig. 35-1, insert one end through the opening in the workpiece, and recouple the band.
4. Mount the band on the wheels of the saw, and tension it properly, usually about one quarter of a turn after the blade comes under tension.
5. Protect the eyes by wearing properly fitted goggles.
6. Set the machine for the correct cutting speed.
7. Start the machine, and bring the workpiece gently against the traveling file band, Fig. 35-2. Move the work in a manner that will cause the file to remove the excess stock from the surface to be filed.

(DoAll Company)

(DoAll Company)

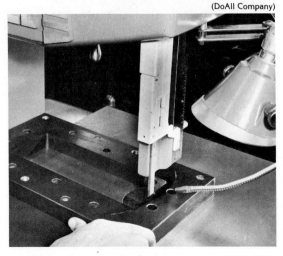

Fig. 35-1.   Uncoupling the file band.

Fig. 35-2.   Band filing application.

8. When the filing has been completed, remove the filing band and the guide; then clean the machine.

Procedure for Polishing

1. Select a suitable polishing band and guide.
2. Mount the guide on the guidepost and the band on the wheels of the saw.
3. Protect the eyes by wearing properly fitted goggles.
4. Set the machine for the correct band speed.
5. Start the machine; then bring the workpiece gently against the traveling band, Fig. 35-3. Move the work in a manner that will bring the polishing band into contact with all parts of the surface to be polished.
6. When the operation has been completed, remove the polishing band and clean the machine.

(DoAll Company)

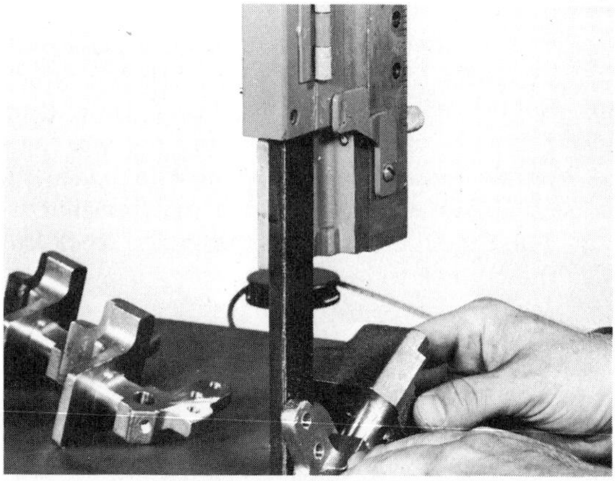

Fig. 35-3.   Polishing with an abrasive band on the band saw.

Test Your Knowledge of Section 7

Unit 31: Metal Cutting Band Saws

Note: Additional information concerning some of the following questions may be found in Units 32 through 35.

1. List the two general types of band sawing machines and indicate which type is also known as a contour machine.
2. What two dimensions determine the size of a band sawing machine?
3. Why are variable speed drive systems an advantage on band saws?
4. What information is provided by a job selector chart?
5. Describe three power feed systems used to feed the workpiece or worktable on vertical band machines.
6. List three types of material from which band saw blades are made, and tell what kind of cutting is done with each.
7. What operational factors will shorten the life of a band saw blade?
8. List four kinds of band saw blade tooth forms and tell briefly the kind of work for which each is best suited.
9. Name two kinds of blade sets used on metal cutting band saw blades and tell the kind of cutting for which each is recommended.
10. List three kinds of special purpose band saw blades and tell for what kind of cutting they are used.
11. List several safety precautions concerning the personal safety of the operator when operating a band saw.
12. Describe the procedure for band machining internal cuts.

Section 8

Drills and Drilling Machine Operations

Unit 36

The Drill Press

The drill press, Fig. 36-1, is one of the most important metalworking machine tools. In addition to its principal task of drilling holes, the drill press is widely used for hole-machining operations such as reaming, boring, counterboring, countersinking, and tapping. See Fig. 36-2. The drill press also may be used for honing or lapping, as well as for spot-finishing as shown in Fig. 45-1.

The six operations illustrated in Fig. 36-2 are also commonly performed on metalworking lathes, turret lathes, hand and automatic screw machines, and vertical milling machines. They are also performed on modern numerically controlled machines, such as the turret drilling machine shown in Fig. 37-6. The principles involved in performing these six operations are

essentially the same, whether done on a drill press or on other basic or specialized production machine tools.

Sensitive Drilling Machines

Sensitive drilling machines are equipped with a hand feed which enables the operator to **feel** the progress of the cutting tool. The floor-model drill press in Fig. 36-1 and the bench model in Fig. 36-3 are sensitive drilling machines. This type machine also may be equipped with power feed, as shown in Fig. 36-1. These machines generally are used for light-duty drilling operations in tool and die shops and in maintenance shops. They are available with drilling capacities ranging up to about 38 mm (1-1/2″) in diameter. Often they are arranged in a line to make up a **gang drilling** machine as shown in Fig. 37-2.

(Clausing Div. — Atlas Press Co.)

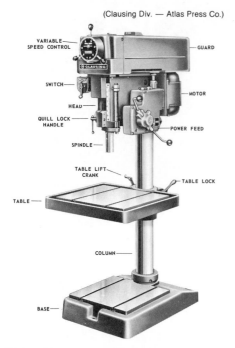

VARIABLE SPEED CONTROL
GUARD
SWITCH
MOTOR
HEAD
QUILL LOCK HANDLE
POWER FEED
SPINDLE
TABLE LIFT CRANK
TABLE LOCK
TABLE
COLUMN
BASE

Fig. 36-1.   Floor model of a sensitive drill press.

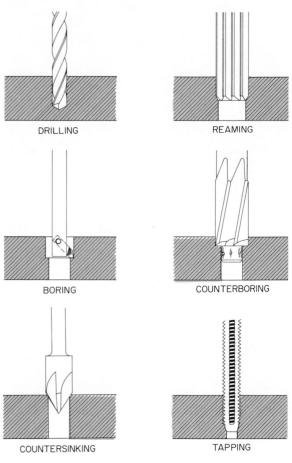

DRILLING

REAMING

BORING

COUNTERBORING

COUNTERSINKING

TAPPING

Fig. 36-2.   Six common operations that can be performed on a drill press.

Principal Parts and Characteristics

The principal parts of a drilling machine, as shown in Fig. 36-1, include the following: base and column, spindle, motor and head, table, feed mechanism, and quill. Other important parts include a drill chuck, on-off switch, table-raising crank, table lock or clamp, and a depth stop as identified in Fig. 36-3.

Speed Variation

Variation of spindle speed is accomplished through several types of drive systems. One common method involves the **step-pulley drive** with a V-belt, as shown in Fig. 36-4. With this system, spindle speed is increased by stepping the V-belt up to a larger diameter on the motor pulley. Similarly, spindle speed is decreased by stepping the belt down to a smaller diameter on the motor pulley. When changing speeds either way, begin at the stepped pulley which will put its belt onto a smaller pulley. Then move the opposite end to the correspondingly larger step. (The belt must not run diagonally nor be stretched.)

A second drive system is the **variable-speed drive.** With this system, the speed is infinitely variable throughout each speed range. The drill press in Fig. 36-1 has a variable-speed countershaft drive of the type shown in Fig. 36-5. **Usually the speed must be changed only while the machine is running.**

A third method for variation of spindle speed involves a **gear drive system.** Here speed-selector handles are used to shift the gears for the desired spindle rpm. **Usually the machine must be stopped while speed changes are made.** This type of drive system normally is used on heavy-duty drilling machines.

A two-speed motor may be used in conjunction with any of the above drive systems. This conve-

(Rockwell International — Power Tool Division)

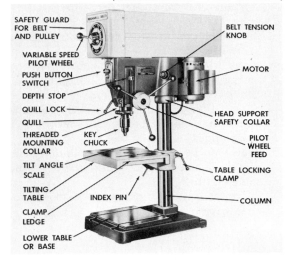

SAFETY GUARD FOR BELT AND PULLEY

VARIABLE SPEED PILOT WHEEL

PUSH BUTTON SWITCH

DEPTH STOP

QUILL LOCK

QUILL

THREADED MOUNTING COLLAR — KEY CHUCK

TILT ANGLE SCALE

TILTING TABLE — INDEX PIN

CLAMP LEDGE

LOWER TABLE OR BASE

BELT TENSION KNOB

MOTOR

HEAD SUPPORT SAFETY COLLAR

PILOT WHEEL FEED

TABLE LOCKING CLAMP

COLUMN

Fig. 36-3.   Bench model of a sensitive drill press.

(Clausing Div. — Atlas Press Co.)

Fig. 36-4.   Step-pulley countershaft drive.

(Clausing Div. — Atlas Press Co.)

Fig. 36-5.   Variable-speed countershaft drive.

(Atlas Press Co.)

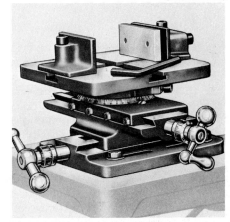

Fig. 36-6.   Universal compound vise.

niently provides both a **low**- and a **high**-speed range for the spindle.

The speed ranges on drill presses may vary from about 30 to 5000 rpm. Correct spindle speed is determined mainly by the kind and size of tool being used and the kind of material being cut.

Feed

All drill presses are equipped with a **hand feed.** In addition, many are equipped with an **automatic feed.** The drill press illustrated in Fig. 36-1 has a mechanical automatic-feed mechanism. **Mechanical** automatic feeds usually are set at a certain feed per revolution of the drill press spindle, such as 0.10 mm (0.004″) per revolution. The range of feeds varies from 0.05 to 0.64 mm (0.002″ to 0.025″) per revolution.

An **air-hydraulic** feed mechanism may be set at infinitely variable feeds from 12.7 to 1524 mm (1/2″ to 60″) per minute. It operates from air pressure and may be either automatic or semi-automatic in operation. **Liquid hydraulic** feed mechanisms also are widely used on large production drill presses.

Attachments

Several attachments are available for use on drill presses. The **universal compound vise,** Fig. 36-6, is useful in clamping work securely for many drilling operations. The attachment should be bolted to the drill press table. With slides providing movement in both directions, workpieces can be easily and accurately positioned. A plain drilling vise is shown in Fig. 42-4. A **tapping attachment** is shown in Fig. 27-6.

Figure 36-7 shows a small **multiple spindle drilling head** arranged for drilling two vertical holes at the same time. A drilling jig is normally used with this type drilling head to assure accurate hole location. Figure 36-7 also shows a **portable drilling head** positioned for drilling a horizontal hole at the same time the vertical holes are being drilled. Several styles of multiple spindle drilling heads are made, the largest being capable of drilling 30 or more holes simultaneously.

Size and Capacity

The size of a drill press is given as twice the distance from the column to the center of the spindle. For example, a drill press that measures 200 mm (8") from the column to the center of the spindle is called a 400 mm (16") drill press because it can drill a hole in the center of a 400 mm (16") circle. The vertical capacity of the machine is determined by the distance from the table (in its lowest position) to the bottom of the jaws of the chuck when fully elevated, less the amount the drill projects. Other factors determining capacity include the distance of quill travel and the size of drill which the spindle or chuck will accommodate. Drill presses are made in many sizes and for many special purposes.

Safety with Drilling Operations

1. Always wear appropriate safety goggles.
2. Never leave the chuck key in the drill chuck or the drill drift in the spindle.
3. Mount the work securely before drilling. Do not hold thin or small pieces in the hands.
4. Remove chips with a brush or piece of wood. Never use the hands.
5. Ease up on the feed pressure as the drill begins to break through a hole. This will prevent the drill from catching or breaking, or pulling the work loose.
6. Do not attempt to stop the spindle with the hands after turning the machine off.
7. Never drill copper alloys, including brass and bronze, with a drill which is ground for steel. Request the instructor to show you how to grind the drill for this purpose. A drill ground for steel may dig in, break, ruin your work, or cause injury.
8. Keep long sleeves, other loose clothing, and especially long hair away from the revolving spindle or belts.
9. Do not operate the drill with the covers or guards removed.

(Zager, Inc.)

Fig. 36-7. Multiple-spindle drill head mounted on single-spindle drill press.

Unit
37

Industrial Drilling Machines

Several types of drilling machines are in common use. Some are designed for general use, while others are designed for mass production. **Sensitive** drilling machines were explained in the previous unit.

Upright Drilling Machine

The upright drilling machine, Fig. 37-1, is designed for general-purpose and heavy-duty work on small parts which may be mounted on the table. It can be used with drills of larger diameter than those normally used on standard bench- and floor-model drill presses. Upright drilling machines are available in various sizes and with various drilling capacities (ranging up to 75 mm (3″) diameter drills on the largest machines). Machines of this type often have a gear-driven mechanism for changing spindle speeds and feeds. Wide ranges of speeds and feeds are provided.

Gang Drilling Machine

The gang drilling machine, Fig. 37-2, basically is a series of single-spindle drilling heads on a long table. It is designed for mass-production purposes where a number of drilling operations must be performed in a certain sequence. Each drilling head may be equipped with a different

(Clausing Div. — Atlas Press Co.)

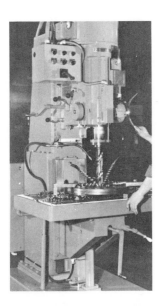

Fig. 37-1.   Upright drilling machine with power feed.

Fig. 37-2.   Gang drill.

tool (such as a center drill, drill, reamer, countersink, boring tool, counterbore, tap, etc.), and the part being machined is moved from one drilling head to the next. Gang drills may be equipped with from 2 to 10 or more spindles.

Radial Drilling Machine

The radial drilling machine, Fig. 37-3, is the most versatile of all drilling machines. It is designed primarily for use on parts which are too large for the operator to move and handle for each drilling operation. The drilling head slides along a large radial arm, thus providing for adjustment of the drilling distance from the column. The arm also swings about the column for locating the drill horizontally. The arm is raised or lowered with power by means of a lead screw. These machines are provided with a wide range of spindle speeds and automatic feeds.

Radial drilling machines are available in various sizes and capacities. Arm lengths range up to about 4 meters (14'), column diameters up to about 750 mm (30") or more, and drilling capacity for drills up to 125 mm (5") in diameter. The machine illustrated in Fig. 37-3 is the two-station type. It is equipped with revolving indexing fixtures and jigs for drilling a typical job.

Multiple-Drill-Head Machine

Multiple-drill-head machines, Fig. 37-4, are designed for drilling a number of holes at the same time for mass-production purposes. They are used for drilling parts such as engine blocks or parts with bolt circles. Besides saving time, such machines insure accurate spacing between holes. For light-duty work, a multiple-drill-head attachment may be mounted on the spindle of a standard drill press, as shown in Fig. 36-7.

Drills in the drill head may be located randomly, as desired, within the range of the drilling head and the machine table. Drilling jigs or fixtures generally are used in conjunction with this type drilling head to maintain accurate hole location. Machines of this type have been produced with from 2 to more than 200 spindles. They range in horsepower from 368W (1/2 hp) to more than 73 550W (100 hp). Most of these machines are designed for drilling in the vertical position, but some drill horizontally.

(Carlton Machine Tool Co.)

Fig. 37-3.   Two-station radial drilling machine is equipped with revolving indexing fixtures and jigs for drilling a typical job.

(Precision Tool)

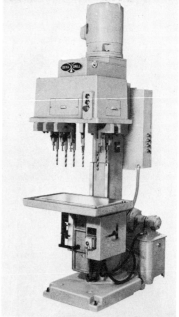

Fig. 37-4.   Multiple-drill-head machine drills several holes at the same time.

Turret Drilling Machine

The turret drilling machine has a multiple spindle turret with 6, 8, or 10 spindles. See Figs. 37-5 and 37-6. The turret may be indexed to the desired tool in the sequence necessary. That shown in Fig. 37-5 is a small hand-operated machine with a six-spindle turret. It has capacity for drilling 12.7 mm (1/2") diameter in steel. The turret head may be tooled with six tools in any desired order. Turret drilling machines provide for increased production by eliminating time needed for tool changes, or by reducing the number of standard drilling machines required.

Turret drilling machines are available in a variety of sizes. The large turret drilling machine in Fig. 37-6 is capable of performing all common drilling operations, as well as light face-milling and end-milling operations (such as face milling flat surfaces or end milling shoulders and slots). This type of machine may be operated either manually through the use of the electrical controls or automatically through the use of its **numerical control** system. Section 16 deals with the operation and programming of numerically controlled machine tools.

(Burgmaster Division — Houdaille Industries, Inc.)

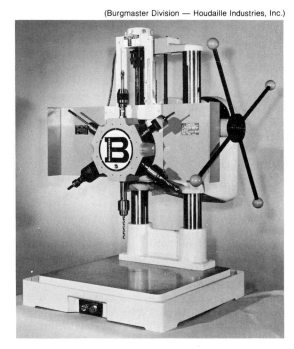

Fig. 37-5.   Small turret drilling machine.

(Cincinnati Lathe and Tool)

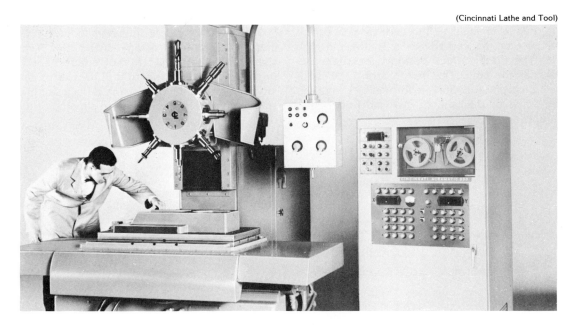

Fig. 37-6. Numerically controlled turret drilling machine.

Unit

38

Drilling, Boring, Counterboring, and Countersinking

Drills are used for cutting holes into or through material. There are many kinds of drills. However, twist drills are by far the most commonly used.

Parts of a Drill

Generally speaking, a drill has three principal parts: the point or dead center, the body, and the shank, Fig. 38-1. The spiral grooves that wind around the body of the drill are called **flutes.** They provide a means whereby (1) a suitable lip or cutting edge may be formed on the point of the bit; (2) the chip removed by the cutting lip may be carried by a channel to the surface; and (3) a lubricant can be carried easily to the cutting edge. The body surface between the flutes is known as the **land.** The narrow strip of metal, labeled **margin,** Fig. 38-1, is formed by grinding away some of the land to give the drill body clearance.

The thin wall between the flutes is called the **web.** It is this part that gives rigidity and strength to the drill. As the web approaches the shank, it thickens. Cutting the flutes somewhat shallower but slightly wider permits free passage of the chips.

The **shank** (that part of the drill which fits the spindle or chuck of the drill press) varies in shape according to its size or the purpose for which it was designed. **A** in Fig. 38-2 is an ordinary straight shank found on drills up to 12.7 mm (1/2″) diameter and is intended for use with a chuck; **B** represents a taper-shank drill. Taper-shank drills have standard Morse tapers and will fit the spindles of standard drill presses or auxiliary sleeves. The tang on shank **B** fits a slot in the spindle to prevent the drill from slipping or turning in the spindle.

(Cleveland Twist Drill — an Acme-Cleveland Company)

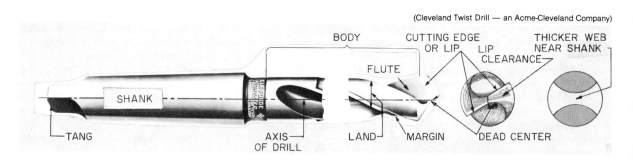

Fig. 38-1. Parts of a drill.

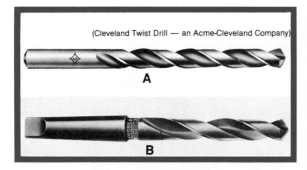

Fig. 38-2. Types of drill shanks.

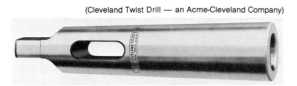

Fig. 38-3. Sleeve or shell socket.

Fitting Taper-Shank Drills

Morse-taper shanks are standard on taper-shank drills. They also are used on a variety of other tools such as reamers, milling cutters, counterbores, and spot-facing tools. Morse tapers are made in various sizes ranging from Nos. 0 through 7. The No. 2, 3, and 4 tapers are used most commonly on drills from 9.5 mm (3/8") to 37.5 mm (1-1/2") diameter.

Since a 12.7 mm (1/2") drill with a No. 2 Morse-taper shank will not fit a drill press spindle with a larger No. 3 or 4 taper hole, a reducing fitting called a **sleeve** or **shell socket** must be used, Fig. 38-3. These sleeves are available in several standard sizes. The sleeve is placed over the drill shank and tapped lightly with a hammer. It fits the taper shank securely because of a close friction fit. The drill together with its sleeve then is mounted in the drill press spindle by tapping the end of the drill lightly with a lead hammer for a snug fit.

To remove the drill from the spindle or from the sleeve, a drill drift is used, Fig. 38-4. The drift is inserted in the hole in the sleeve and tapped lightly with a hammer, thus forcing the taper shank to separate from the taper hole.

Fig. 38-4. Plain drill drift (above) and safety drill drift (below).

Fig. 38-5. Fitted socket.

The safety drill drift in Fig. 38-4 has a sliding handle which is attached permanently to the drift, eliminating the need to use a hammer.

Fitted sockets, Fig. 38-5, are used to adapt tool shanks to fit into machine sockets which they otherwise would not fit. Fitted sockets are available with a variety of standard external and internal tapers so that a large shank may be adapted to a small hole or vice versa. (They also can be used simply to extend the reach of a drill.) For example, an adapter with a No. 2 hole and a No. 3 shank may be used to adapt the shank of a drill or other tool having a No. 2 taper to a spindle having No. 3 internal taper. Similarly, an adapter with No. 3 hole and No. 2 shank may be used to

(Cleveland Twist Drill — an Acme-Cleveland Company)

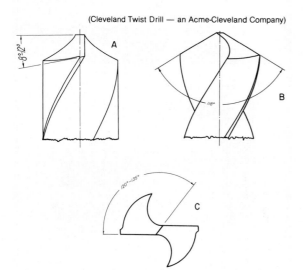

(Cleveland Twist Drill — an Acme-Cleveland Company)

CUTTING LIP

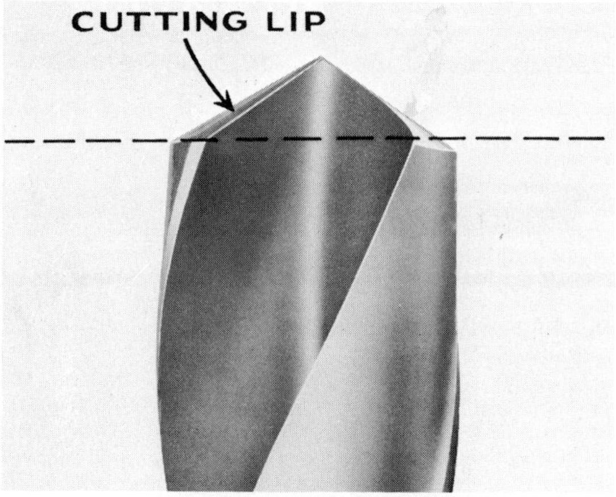

Fig. 38-7.   Drill without lip clearance will not cut.

Fig. 38-6.   Correct angles for standard general-purpose drills.
A. Lip clearance
B. Point angle
C. Chisel edge angle

adapt a tool with a No. 3 taper to a machine spindle having a No. 2 internal taper. A socket with a No. 3 hole and a No. 3 shank is used as an extension.

Drill Sizes

Drill sizes are indicated in four ways: (1) in millimeters, (2) by number, (3) by letter, and (4) by fractional parts of an inch. Metric drill sizes range from 0.1499 mm (0.0059″) to 100 mm (4″). Drill sizes by number are given in terms of wire gage, and they range from No. 80, which has a diameter of 0.0135″, up to No. 1 with a diameter of 0.2280″. The series continues with lettered sizes from A up to Z. An **A** drill has a diameter of 0.2340″ and Z, a diameter of 0.4130″. Drill sizes given in fractional parts of an inch are available as small as 1/64″ in diameter to 3-1/2″ or larger on request. Drill sizes designated in fractions of an inch increase in size by 1/64″ up to 3″, after which stock drills increase by 1/32″ or 1/16″.

Decimal equivalents for metric, letter, number, and fractional sizes of twist drills are listed in Tables A-16 and A-17, Appendix.

Grinding Drills

Most of the difficulties encountered in drilling may be attributed to improper grinding. See Table 38-1. When grinding, three factors are important: (1) correct **lip clearance** as at **A,** Fig. 38-6; (2) correct **point angle** as at **B;** (3) correct **chisel edge angle** as at **C.** Correct lip clearance for standard general-purpose drills used for drilling most steels is 8° to 12°. Within this range, provide smaller drills with greater clearance than larger drills. Clearance may be increased 50% when drilling soft or free-machining material. Lip clearance permits the cutting edge of the lip to penetrate the workpiece.

A drill with no lip clearance will not cut. Note that in Fig. 38-7 the cutting lip and the heel are at the same level horizontally. This drill cannot possibly cut because the cutting lips cannot penetrate the workpiece.

The correct lip angle for general-purpose drills used for drilling most machine steels and many other metals is 59° measured to the center line of the drill, **B,** Fig. 38-6. Thus, the total included angle of the drill point is 118°. To check the lip

Table 38-1
Causes of Drill Breakage, Damage, or Inaccuracy

Symptoms	Probable Cause	Remedy
Breaking of drill.	Spring or back lash in press or work. Too little lip clearance. Too low speed in proportion to the feed. Dull drill.	Test press and work for rigidity and alignment. Regrind properly. Increase speed or decrease feed. Sharpen drill.
Breaking down of outer corners of cutting edges.	Material being drilled has hard spots, scale, or sand inclusions. Too much speed. Improper cutting compound. No lubricant at point of drill.	Reduce speed. Use proper cutting compound and correct application.
Breaking of drill when drilling brass or wood.	Chips clog up flutes.	Increase speed. Use drills designed for these materials.
Broken tang.	Imperfect fit of taper shank in the socket — due to nicks, dirt, burrs, or worn out socket.	Get a new socket or ream old one to prevent recurrence.
Chipping of margin.	Oversize jig bushing.	Use proper size bushing.
Chipping of lip or cutting edges.	Too much feed. Too much lip clearance.	Reduce feed. Regrind properly.
Chipping or checking of a high-speed drill.	Heated and cooled too quickly while grinding or while drilling. Too much feed.	Warm slowly before using. Do not throw cold water on hot drill while grinding or drilling. Reduce feed.
Change in character of chips while drilling.	Change in condition of the drill such as chipping of cutting edge, dulling, etc.	Regrind drill properly.
Hole too large.	Unequal angle or length of the cutting edges, or both. Loose spindle.	Regrind properly. Test spindle for rigidity.
Only one lip cutting.	Unequal length or angle of cutting lips or both.	Regrind drill properly.
Splitting up center.	Too little lip clearance. Too much feed.	Regrind with proper lip clearance. Reduce feed.
Rough hole.	Dull or improperly ground drill. Lack of lubricant or wrong lubricant. Improper set-up. Too much feed.	Regrind properly. Lubricate or change lubricant. Reduce feed.

angle, use a **drill point gage,** Fig. 38-8. An angle of 59° is recommended because drills ground at that angle cut more rapidly and with less exertion of power than when ground at any other angle. The chisel edge of a drill should have an angle of 120° to 135° and be centered exactly in line with the center of the drill, **C,** Fig. 38-6.

When grinding drills for drilling hard materials such as manganese steel, the lip angle should be 75°; total point angle of 150°. This material is very hard and tough. Consequently, the shorter lip secured when ground at the flatter angle takes less power to operate and does not cause as great a strain on the drill.

When drills are ground for use in drilling softer materials such as bakelite, hard rubber, molded plastics, fiber, and wood, the lip angle should be ground to an angle of 45°, a total angle of 90°. More information concerning drill grinding and the procedures used in grinding drills is included in Unit 41.

A drill with a standard 118° point is used for drilling brass and bronze. However, the rake angle should be modified as shown in Fig. 38-9. Drills with standard rakes tend to screw themselves into free machining brass and some plastics, sometimes breaking the drill, the material, or both. This can be avoided if the drills are reground to provide a neutral or slightly negative rake. This also is recommended for drilling very hard steel, because it reduces the angle of the cutting edge of the drill. It increases the strength of the cutting edge, thus preventing chipping.

Spiral-Point Drill

A spiral point may be ground on a standard twist drill with a drill-grinding machine of special design. The spiral-point drill offers many operating advantages over the conventional chisel-point drill.

With the conventional drill, a center-drilled hole is required to start the drill. Without the center-drilled hole, the drill has a tendency to **walk** to one side or another, thus drilling a slanted hole or a hole in the wrong location. The spiral-point drill has a sharp point on the end, rather than a wide chisel point, Fig. 38-10. This eliminates the need for center punching or center-drilling to start the drill accurately.

Fig. 38-8. Drill point gage. The metric version of this tool would be calibrated with mm and ½ mm graduations.

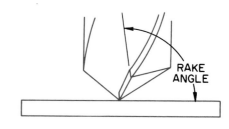

Fig. 38-9. Drill point modified to provide negative rake.

Fig. 38-10. Comparison of spiral-point drill (left) and chisel-point drill (right).

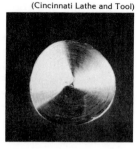

(Cincinnati Lathe and Tool)

Fig. 38-11. Comparison of hole roundness with spiral-point drill (left) and chisel-point drill (right).

With the conventional chisel-point drill, jigs and drill bushings normally are required to start and guide the drill most accurately for production setups. In most instances, the spiral-point drill will start accurately and drill a true hole without jigs and fixtures.

Most **two-lipped** drills produce holes several hundredths of a millimeter (thousandths of an inch) oversize, depending on the material being drilled, the accuracy of the ground point, the rigidness of the setup, and whether or not fixtures with bushings are used to guide the drill. With the use of spiral-point drills, the average range of **hole oversize** is reduced by more than 50 percent.

Spiral-point drills also produce holes which are more accurate in roundness. See Fig. 38-11. Where accurate size and shape are required for holes drilled with chisel-point drills, it is common practice to drill the hole slightly undersize, followed by reaming or boring. In many instances, spiral-point drills hold size and shape sufficiently to eliminate the need for reaming and boring.

The use of the spiral-point drill is particularly desirable on numerically controlled drilling machines. The spiral points generally eliminate the need for center-drilling operations, thereby saving valuable machine time.

Using Drills

For best results, use a drill at the correct speed and feed. The speed of a drill refers to the rate at which it travels at the circumference. This is called **peripheral** or **outside speed** and is given in terms of meters (or surface feet) traveled per minute. The feed is the rate at which the drill advances into the work per revolution, measured in hundredths of a millimeter or thousandths of an inch. For correct speeds and feeds, see Unit 40.

Types of Drills

Numerous styles and types of drills are made. Each type of drill is designed for certain applications. The types of drills used most are the **straight-shank** and the **taper-shank twist drill,** Fig. 38-2.

Less common types of drills, Fig. 38-12, include:

A. **Drills over 12.7 mm (1/2″) but with a 12.7 mm (1/2″) shank** are designed for use either in portable electric drills or in drill chucks on smaller drill presses. Care should be exercised in drilling over 19 mm (3/4″) diameter, since the shanks may not be strong enough for heavy feeds on tough materials. A pilot hole should be drilled first.

B. **Straight-shank drills with carbide tips** are recommended for production drilling of cast iron, cast steel, and nonferrous materials. They are not recommended for drilling steel.

C. **Carbide-tipped die drills** of the straight-shank type are recommended for use in drilling hardened steel in the range from 48 to 65 Rockwell-C hardness. Holes may be drilled without annealing the metal. A steady hand feed with a good flow of cutting fluid should be used.

D. **Three-fluted core drills** are used for enlarging cored holes or previously punched or drilled holes. Because of their wide use in drilling cored holes in castings, they are known as **core drills.** This type of core drill can enlarge holes as small as 60 percent of the drill diameter. The advantages of a multiflute drill include increased rate of metal removal, increased accuracy in hole size and location, and improved finish. Four-fluted core drills are also made.

E. **Subland drills** are special multicut drills which can drill several diameters in one operation. Drilling operations of this type are called **step-drilling** operations. Many variations of step-drilling operations are possible with multicut drills, Fig. 38-13.

F. **Oil hole drills** are used on high-production screw machines. Oil is forced through the machine spindle and through the oil holes to the

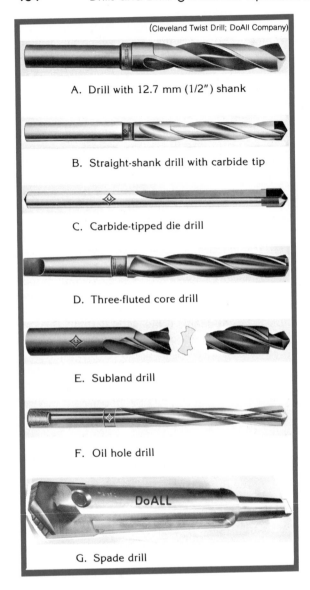

(Cleveland Twist Drill; DoAll Company)

A. Drill with 12.7 mm (1/2″) shank

B. Straight-shank drill with carbide tip

C. Carbide-tipped die drill

D. Three-fluted core drill

E. Subland drill

F. Oil hole drill

G. Spade drill

Fig. 38-12. Several types of drills.

cutting edges. The oil not only serves as a lubricant and coolant, but it also helps force the chips out of the drilled hole.

G. **Spade drills** are made up of a tool shank and a replaceable spade bit. They are most widely used for drilling holes 25 mm (1″) to 125 mm (5″) in diameter. Straight and taper shank holders are available in several lengths. For deep hole drilling, an oil hole is provided through the shank to allow cutting fluid to be delivered to the drill point under pressure. Spade drills are much less expensive than twist drills of like size.

Boring Tools

When a drill of a particular dimension is not available, or when a very straight, accurate hole is desired, a boring tool may be used. In such cases, a hole large enough to permit entry of the boring tool is drilled with a standard drill. A simple boring tool is shown in Fig. 38-14.

The boring head in Fig. 38-15 is adjustable to within 0.0127 mm (0.0005″) or less. This type is available in a wide range of sizes. Boring capacity

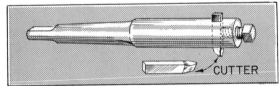

Fig. 38-14. Boring tool.

(Illinois State University Photographic Service)

Fig. 38-13. Types of single operations possible with multiple-diameter drills.

Fig. 38-15. Adjustable boring head and boring tools.

may range from 19 mm (3/4") to 50 mm (2") offset from center, or a hole capacity of over 100 mm (4").

Adjustable boring heads eliminate the need for a complete inventory of expensive, large-size drills. When used with care on sturdy machines with power feed and rigid setups, extremely accurate holes may be machined with this type of tool.

Counterboring Tools

The counterboring tool, Fig. 38-16, is used to spot-face or counterbore for bolts and screws, to enlarge holes to receive the head of fillister head screws, Fig. 38-17, and for similar purposes. It has a pilot or guide on the point which frequently is interchangeable. Ordinarily, the pilot has a diameter about 0.05 mm (0.002") smaller than that of the drilled hole.

Counterboring tools are made in many sizes. They are available in high-speed steel and with carbide insert teeth. The former type is designed

Fig. 38-16. Counterboring tool.

Fig. 38-17. Counterboring for fillister head screws.

for counterboring steel, while the latter is recommended for use with cast iron.

Counterboring tools should be run at lower cutting speeds than a drill of corresponding diameter. A cutting fluid should be used freely.

Countersinks

Countersinks are used to machine a cone-shaped enlargement at the end of a hole. They are made in many styles and sizes, and with point angles of 60° for lathe centers, 82° for flat-headed machine screws, 90° for deburring, and 100, 110, and 120 degrees for some types of rivets. Other angles are made for special applications. Figure 38-18 shows six types of countersinks: one-, three-, and four-flute; add-on; chatterless; and piloted. Other types are made for special applications. The single-flute is essentially chatter-free but is limited to relatively shallow countersinking. The chatterless style provides additional cutting edges offset from the main cutting edges, which effectively defeats the tendency

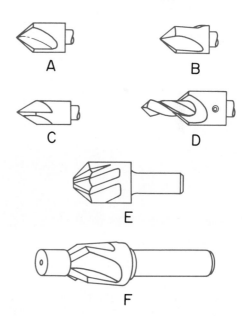

Fig. 38-18. Six types of countersinks.
    A. Single-flute
    B. Three-flute
    C. Four-flute
    D. Add-on
    E. Chatterless
    F. Piloted

to chatter. Add-on countersinks are often used for countersinking on the back side of a surface which would be inaccessible otherwise. Piloted countersinks help attain close concentricity between the hole and the countersink. For best results, countersinks should be run at 1/2 to 2/3 the rpm of drills of like size.

Combination drill and countersinks, also called **center drills,** are used for two main purposes: (1) for providing a starting hole for drills, and (2) for drilling center holes in stock to be held in a lathe or cylindrical grinder. They are made in many sizes, Fig. 38-19, and of either high carbon or high-speed steel. The **plain** style has a single 60° countersink angle. The **bell** style has a secondary angle of 120°. See Fig. 38-20. The hole made by the bell style is desirable because it protects the edges of the 60° center hole from being nicked during handling, thus preserving the accuracy of the workpiece.

(Cleveland Twist Drill — an Acme-Cleveland Company)

Fig. 38-19. Combination drills and countersinks are made in many sizes.

(Cleveland Twist Drill — an Acme-Cleveland Company)

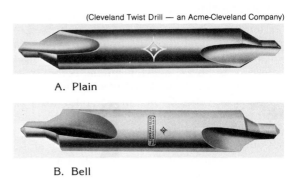

A. Plain

B. Bell

Fig. 38-20. Combined drill and countersinks.

Unit

39

Reamers

Reamers are multiple cutting edge tools used to enlarge and finish a drilled hole to exact dimension and smoothness, Figs. 36-2 and 39-1. Drilled holes usually have a rough finish and are inaccurate in size and roundness. Drilled holes are usually satisfactory for bolts and screws, however, where round, straight, accurately sized holes with a smooth finish are required for use with bushings or bearings, the holes often are reamed, either by hand or by machine.

General Classification of Reamers

Reamers may be made of carbon tool steel, high-speed steel, or may have carbide-tipped cutting edges. They are available in a wide variety of types and sizes.

AMERICAN STANDARD

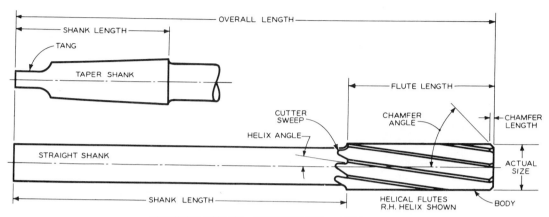

CHUCKING REAMER, STRAIGHT AND TAPER SHANK

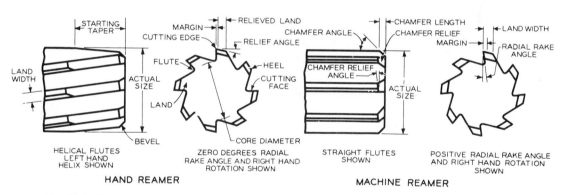

HAND REAMER   MACHINE REAMER

Fig. 39-1. Terms applying to reamers. (Extracted from American Standard Reamers, ANSI B94.2-1964, with the permission of the publisher, The American Society of Mechanical Engineers.)

Reamers may be classified into two general groups — **machine reamers,** Fig. 39-2, and **hand reamers,** Fig. 39-3. **Hand reamers** are designed for hand operation and light cuts. They are equipped with a straight shank having a square tang which is turned with a tap wrench.

Machine reamers are designed for use on drill presses, engine lathes, turret lathes, vertical milling machines, and other special production machines.

Machine reamers are made with either Morse taper shanks or straight shanks, Fig. 39-1.

Cutting Action

Machine reamers have a bevel of from 40° to 50° (usually 45°) on the cutting end, Fig. 39-1. They cut differently than hand reamers. Machine reamers are **end-cutting** reamers which cut on the beveled end in the same manner as a drill cuts as it enters a hole. When the reamer becomes dull, it is resharpened on the 45° beveled end only. It must be sharpened on a tool-and-cutter grinder so that all cutting edges are exactly even.

Hand reamers, on the other hand, cut on the **periphery** of the reamer. The flutes are ground

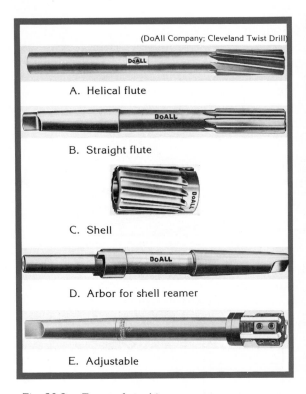

(DoAll Company; Cleveland Twist Drill)

A. Helical flute

B. Straight flute

C. Shell

D. Arbor for shell reamer

E. Adjustable

Fig. 39-2. Types of machine reamers.

straight for the whole length, except near the cutting end which is ground with a **starting taper,** Fig. 39-1. This taper permits the reamer to enter the hole easily. Thus, hand reamers are essentially scraping tools, and nearly all cutting takes place along the starting taper at the cutting end of the reamer. Hence, the hand reamer is designed for removing only very small amounts of material. When the hand reamer becomes dull, it must be resharpened on the starting taper portion only. Like a machine reamer, it must be sharpened on a tool-and-cutter grinder.

Types of Machine Reamers

Rose Reamers

Rose reamers are intentionally designed to rough-ream cored or drilled holes 0.07 to 0.25 mm (0.003" to 0.010") undersize. They are designed for two-step reaming operations. After rough reaming, the hole is finish-reamed to final size with a fluted reamer.

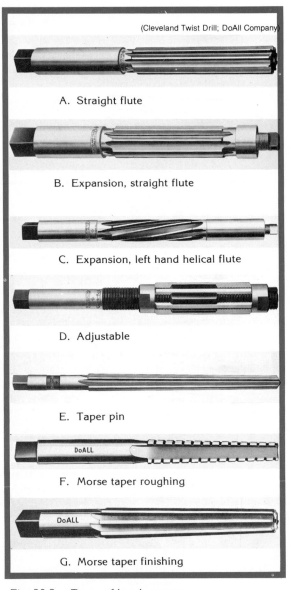

(Cleveland Twist Drill; DoAll Company)

A. Straight flute

B. Expansion, straight flute

C. Expansion, left hand helical flute

D. Adjustable

E. Taper pin

F. Morse taper roughing

G. Morse taper finishing

Fig. 39-3. Types of hand reamers.

Rose reamers have a 45° end-cutting angle which does all of the cutting and which is capable of cutting rapidly. The reamer also has a **back taper;** that is, it is tapered a maximum of 0.025 mm per 25 mm (0.001" per inch) of length for the length of the flutes. The back taper prevents binding in deep holes with heavy cuts. The outside diameter of the reamer is ground cylindrically, thus leaving a wide circular margin. There are no cutting edges along the flutes, which merely provide space for cutting fluid and chip ejection.

Fluted Reamers

Fluted reamers generally have more flutes than rose reamers and are designed for reaming holes to finished size. They may be used for one-step reaming operations or after using a rose reamer in a two-step operation.

The cutting action takes place at the 45° beveled cutting end. A narrow circular margin, from 0.127 to 0.508 mm (0.005″ to 0.020″) in width, runs along the entire length of the flute. The lands are slightly beveled along the length of the flute to provide a body clearance angle. The reamer is not provided with significant back taper.

Several common types of machine reamers are shown in Fig. 39-2. The **helical fluted** reamer has a straight shank and is designed for reaming materials which ordinarily are considered difficult to ream. The helical flutes provide a free-cutting action which aids in producing smooth, accurate holes.

Straight fluted machine reamers are designed for use with materials which possess average reaming properties.

Shell reamers are made with helical flutes, Fig. 39-2, or with straight flutes. They are also made in the rose type or the fluted type. Shell reamers are designed with the hole for economy reasons. The hole is tapered to fit snugly and accurately on a special arbor, which is available with either a straight or taper shank. Several sizes of shell reamers may fit the same arbor. When the reamer is worn out, it may be discarded, and a new reamer may be used on the old arbor.

Adjustable machine reamers, as shown at E in Fig. 39-2, are made with either high-speed steel blades or carbide-tipped high-speed steel blades, and with either straight or taper shanks. These reamers are easily adjusted for size within a range of about 1 mm (1/32″) diameter. As the blades become dull, they can be expanded and reground several times. When worn out, the blades may be replaced.

Carbide tipped reamers are more abrasion-resistant and can withstand higher temperatures

and higher cutting speeds than high-speed steel reamers. They are particularly useful for reaming castings, both ferrous and nonferrous, which have sand or scale inclusions. For more information concerning high-speed steel and cemented carbides, see Unit 117 concerning cutting tool materials.

Special machine reamers of many types and sizes are available for special purposes. One important type is the **combination drill and reamer,** which has a drill at the end and a reamer farther back. This design makes it possible to drill and ream a hole in one operation.

Types of Hand Reamers

Several basic types of hand reamers, including those shown in Fig. 39-3, are described below.

Straight-fluted hand reamers of the solid type are made of either carbon tool steel or high-speed steel, see Fig. 39-3A. The cutting end is ground with a starting taper for easy entry into a hole. This type reamer is recommended for general-purpose reaming of holes to finished size.

Helical-fluted hand reamers are similar to the straight-fluted hand reamer. This type reamer is recommended for reaming holes with interruptions or keyways. The helical flute produces a smooth cutting action with minimum chatter.

Expansion hand reamers may be of the straight-flute type, Fig. 39-3B, or the helical-fluted type, Fig. 39-3C. The amount of expansion possible depends on the diameter of the reamer. It may vary from about 0.152 mm (0.006″) for a 6.35 mm (1/4″) reamer to about 0.305 mm (0.012″) for a 38 mm (1-1/2″) reamer. These reamers are provided with an adjusting screw for expansion. An undersize pilot is provided on the end to aid in alignment.

Adjustable hand reamers, Fig. 39-3D, may be adjusted for any size, above or below basic size, within the range of the reamer. These reamers are available in standard sizes from 6.35 mm (1/4″) to about 85 mm (3-11/32″) diameter. Each reamer may be expanded to the smallest size of the next-size reamer.

Blades are available in carbon steel or high-speed steel. They slide in accurately tapered slots and may be adjusted by loosening one nut and tightening the other, thus moving the blades in the slots.

(Cleveland Twist Drill — an Acme-Cleveland Company)

Fig. 39-4.   Burring reamer with spiral flutes.

Taper pin reamers, Fig. 39-3E, have a taper of 6.35 mm per 300 mm (1/4" per foot). They are used for reaming holes for standard taper pins. Best results are achieved when the drilled hole is slightly larger than the small end of the taper pin. These reamers are made with straight- or left-hand helical flutes, and in carbon steel or high-speed steel.

Taper socket reamers are made for reaming standard Morse or Brown and Sharpe taper holes. Straight-shank hand reamers are intended for maintenance of taper holes in machine tools and accessories. See Fig. 39-3F and G. Taper-shank roughing reamers and finishing reamers are made for production reaming of standard taper holes.

Burring reamers, Fig. 39-4, are used for removing burrs from cut pipe and conduit. They also may be used for enlarging holes in thin materials.

Reaming Practices

Reaming speeds may vary considerably, depending on the type of material to be reamed, the type of machine used, the type of finish required, and the accuracy required. As a general rule, machine reaming is done between half and two-thirds the speed used for drilling the same material. (Drilling speeds are shown for various materials in Table 40-1.)

Reaming feeds are two to three times higher than for drilling. If the feed is too slow, excessive reamer wear will result. If the feed is too fast, the hole will be inaccurate. The feed should be sufficient to cause each flute to cut a chip rather than burnish or rub the material. A good starting point is to use a feed from 0.04 to 0.10 mm (0.0015" to 0.004") per flute per revolution. Then the feed can be adjusted as required for desired results.

Reamer alignment is one of the most important factors in reaming accuracy. When reaming, the spindle, reamer, reamer bushing, and the hole to be reamed should be in perfect alignment. Any variation in these factors detracts from reaming accuracy and results in excessive reamer wear. The effects of reamer misalignment may be reduced through the use of a **floating** reamer holder which allows the reamer to be self-centering. When possible, holes should be drilled and reamed in the same setup without moving the workpiece.

Chatter can affect the accuracy and finish of reamed holes. The following are possible causes for chatter:

1. Excessive speed.
2. Too light a feed.
3. Setup not rigid.
4. Spindle too loose.
5. Excessive clearance on reamer.
6. Excessive looseness in floating holder.

Cutting fluids aid in producing good finishes when reaming. Mineral-lard oils and sulfurized oils are desirable for most reaming applications. See Table A-11, Appendix, for specific cutting fluid recommendations. Gray cast iron generally should be reamed dry, or the reamer may be cooled with a jet of compressed air.

Care of reamers affects the accuracy of reamed holes. Reamers should be handled without bumping, nicking, or otherwise abusing them. They should be stored in containers or holders which will prevent nicks along the cutting edges and flutes.

Allowance for Reaming

The usual procedure for producing a reamed hole is to produce an **undersize** hole first. The hole may be produced in one of three ways — drilling, boring, or both drilling and boring. The method employed depends on the material allowance which is to be left for reaming or the accuracy required for the hole location. Very little material allowance is made for hand reaming compared to machine reaming.

Holes for machine reaming should be produced undersize an amount which will provide the following material allowances:

6.35 mm (1/4″) holes . . . 0.254 mm (0.010″)
12.7 mm (1/2″) holes . . . 0.406 mm (0.016″)
25.4 mm (1″) holes . . . . 0.508 mm (0.020″)
50.8 mm (2″) holes . . . . 0.813 mm (0.032″)
76.2 mm (3″) holes . . . . 1.194 mm (0.047″)

For one-step machine reaming, it is common practice to drill holes under 12.7 mm (1/2″) 0.40 mm (1/64″) undersize, followed by reaming with a fluted reamer. For holes 12.7 mm (1/2″) and over, 0.813 mm (1/32″) undersize is not excessive.

Where extreme accuracy is required, hand reaming often is required. A cut of 0.05 mm (0.002″) usually is recommended. **Never leave over 0.127 mm (0.005″) of material for hand reaming.** The usual procedure for hand reaming is to drill and bore (or drill and rough ream) the

hole to 0.05 mm (0.002″) undersize and then ream the hole by hand with a hand reamer.

Hints for Reaming

1. See that the reamer is free from nicks.
2. Utilize a setup and procedures which will prevent the causes of chatter.
3. Never reverse a reamer or the workpiece when reaming. To do so will cause chips to damage the margins of the reamer, thus causing an inaccurate hole with a poor-quality finish.
4. Avoid starting a reamer on a slanted or uneven surface. To do so will cause the reamer to drift out of alignment.
5. Follow the procedure outlined in Unit 44 when reaming.

Unit 40
Cutting Speeds and Feeds for Drilling

The cutting speeds and the feeds at which metal should be drilled vary for different metals. The feeds also vary according to the size of the drill. The terms **cutting speed** and **rpm** (revolutions per minute) are interrelated.

Cutting Speed

Cutting speed for drills and other rotating tools is expressed in terms of **meters per minute (mpm)** or **feet per minute** (fpm). It is the distance that a point on the circumference of the drill will travel in one minute. If the cutting speed for drilling is too high, the drill will become overheated and will dull easily. If the cutting speed is too low, the production rate will be low and the drill may break easily.

Cutting speed has a different meaning than **revolutions** per minute (rpm). A 6.35 mm (1/4″) diameter drill turning at 1222 rpm has a cutting speed of 24.38 mpm (80 fpm). A 12.7 mm (1/2″) diameter drill at 24.38 mpm (80 fpm) would turn at only 611 rpm. Consequently, the drill press spindle must be set at the proper rpm in order to obtain the correct cutting speed for each different size tool.

There is no one correct speed for drilling all materials. It is common practice to select an **average** cutting speed for each type of material. The cutting speed selected may then be increased or decreased according to conditions which affect the particular job setup and the material drilled. The following factors affect the cutting speed selected for drilling operations:

1. The kind of material being drilled. Softer materials generally are drilled at higher cutting speeds than harder materials. See Table 40-1.
2. The kind of cutting tool material. Carbon-steel drills are used with cutting speeds which are about one-half those used with high-speed steel drills. Tungsten-carbide drills generally are run at higher speeds and lighter feeds than high-speed steel drills.
3. Whether or not a cutting fluid is used.
4. The size and the type of drilling machine used, and the rigidness of the work setup.
5. The quality of finish desired in the hole.

Table 40-1
Cutting Speeds for High-Speed Steel Drills

Material	Meters Per Minute	Feet Per Minute
Low-carbon steel (0.05-0.30% carbon)	24.4-33.5	80-110
Medium-carbon steel (0.30-0.60% carbon)	21.4-24.4	70-80
High-carbon steel (0.60-1.70% carbon)	15.2-18.3	50-60
Steel forgings	15.2-18.3	50-60
Alloy steel	15.2-21.4	50-70
Stainless steel	9.1-12.2	30-40
Cast iron, soft gray	30.5-45.7	100-150
Cast iron, hard-chilled	21.4-30.5	70-100
Cast iron, malleable	24.4-27.4	80-90
Ordinary brass and bronze	61.0-91.4	200-300
High-tensile bronze	21.4-45.7	70-150
Monel metal	12.2-15.2	40-50
Aluminum and its alloys	61.0-91.4	200-300
Magnesium and its alloys	76.2-122.0	250-400
Slate, marble, and stone	4.6-7.6	15-25
Bakelite and similar plastics	30.5-45.7	100-150
Wood	91.4-122.0	300-400

(The Cleveland Twist Drill Co.)

Carbon-steel drills should be run at speeds of from 40% to 50% of those given above. Carbide-tipped or solid-carbide drills may be run two to three times faster than high-speed steel drills.

Calculating rpm

The rpm for a given cutting speed may be calculated with the following formulas:

A. Metric:

$$rpm = \frac{CS\ (mpm) \times 1000}{D\ (mm) \times \pi}$$

Where: CS (mpm) = Cutting speed in meters per minute

D (mm) = Diameter in millimeters

π = Pi or 3.1416

B. Inch:

$$rpm = \frac{CS' \times 12}{D'' \times \pi}$$

Where: CS = Cutting speed in feet per minute

D = Diameter in inches

π = Pi or 3.1416

A short cut formula which yields the approximate rpm satisfactory for most applications follows:

$$rpm = \frac{4 \times CS'}{D''}$$

Example: Calculate the rpm for a 6.35 mm (1/4″) diameter drill to cut at a cutting speed of 30.48 mpm (100 fpm).

Metric Solution:

$$rpm = \frac{30.48 \times 1000}{6.35 \times 3.1416}$$

$$= \frac{30480}{19.94916}$$

$$= 1527.88\ or\ 1528$$

Inch Solutions:

$$rpm = \frac{100 \times 12}{0.250 \times 3.1416}$$

$$= \frac{1200}{.7854}$$

$$= 1527.88\ or\ 1528$$

Using the shortcut formula:

$$rpm = \frac{4 \times 10}{0.250}$$

$$= \frac{400}{0.250}$$

$$= 1600$$

1600 rpm is about 5% faster than the precise answer of 1528 rpm; not enough to harm the drill in most cases.

Calculating Cutting Speed

When the diameter of the drill and the rpm are known, the cutting speed for drilling may be calculated by using the following formulas:

A. Metric:

$$CS \text{ (mpm)} = \frac{D \text{ (mm)} \times \pi \times rpm}{1000}$$

B. Inch:

$$CS \text{ (fpm)} = \frac{D'' \times \pi \times rpm}{12}$$

and the shortcut formula:

$$CS \text{ (fpm)} = \frac{D'' \times rpm}{4}$$

Example: Calculate the cutting speed for drilling with a 12.7 mm (1/2") diameter drill at 688 rpm:

Metric Solution:

$$CS \text{ (mpm)} = \frac{12.7 \times 3.1416 \times 688}{1000}$$

$$= \frac{27450.044}{1000}$$

$$= 27.45$$

Inch Solutions:

$$CS \text{ (fpm)} = \frac{0.5 \times 3.1416 \times 688}{12}$$

$$= \frac{1080.71}{12}$$

$$= 90$$

Using the shortcut formula:

$$CS \text{ (fpm)} = \frac{0.5 \times 688}{4}$$

$$= \frac{344}{4}$$

$$= 86$$

Again, the shortcut formula gives an approximation that is accurate enough for most applications.

When the diameter of the drill and the cutting speed are known, the rpm may be determined simply by referring to a **table of cutting speeds** (Table A-10, Appendix). Tables of this type very often are posted in machine shops. However, the competent machinist or machine operator also should be able to calculate rpm and cutting speeds when necessary.

The above formulas also can be used to calculate the cutting speed and rpm for all types of cylindrical cutting tools, including taps, reamers, counterboring tools, and milling cutters. In addition, they may be used to calculate cutting speeds for turning or boring on a lathe. However, in this case, the **diameter** (D) refers to the diameter of the workpiece being turned (or the diameter of the hole being bored).

Feeds

The **feed** for a drill refers to the rate at which the drill advances into the work in one revolution. Hence, with a feed setting of 0.10 mm (0.004"), the drill advances into the work 0.10 mm (0.004") deeper each revolution.

Feeds are governed by the size of the drill and the material to be drilled. The general rule when drilling mild steel is to use a feed of 0.025 to 0.058 mm (0.001" to 0.002") per revolution for drills smaller than 3 mm (1/8"); 0.058 to 0.10 mm (0.002" to 0.004") for drills 3 mm (1/8") to 6.35 mm (1/4"); 0.10 to 0.178 mm (0.004" to 0.007") for drills 6.35 to 12.7 mm (1/4" to 1/2"); 0.178 to 0.38 mm (0.007" to 0.015") for drills 12.7 to 25.4 mm (1/2" to 1"); 0.38 to 0.635 mm (0.015" to 0.025") for drills larger than 25.4 mm (1"). Alloy and hard steels should be drilled with a lighter feed than given, while cast iron, brass, and aluminum usually may be drilled with a heavier feed.

Extreme speed or feed will cause drills to chip or break at the cutting edges or to split the web. Similar damage also may result from improper grinding. Rapid wearing at the outer corners of the cutting edges usually is an indication of too much speed.

Cutting Fluids

To maintain the feeds and speeds recommended, it is necessary to use good cutting fluids. The following are recommended for drilling operations:

Hard refractory steel — turpentine, kerosene, emulsifiable oils (soluble oils), mineral-lard oils.

Soft steel and wrought iron — emulsifiable oils (soluble oils), sulfurized oils, mineral-lard oils.

Malleable iron — emulsifiable oils, mineral-lard oils.

Brass and bronze — emulsifiable oils, or dry.

Aluminum and aluminum alloys — mineral oils, emulsifiable oils, mineral-lard oils.

Gray cast iron — Dry or with a jet of compressed air.

The selection of cutting fluid also varies with the **machinability** of the material and the **severity** of the operation being performed. For more specific recommendations concerning these factors in relation to cutting fluid selection, see Table A-11, Appendix.

Unit

41

Sharpening Twist Drills

Drills produce round holes in metal. They are effective tools for this purpose when properly sharpened. Three things are important when grinding a drill: (1) lip clearance, (2) length and angle of the cutting lip, and (3) location of the point.

Lip clearance is produced by grinding away the surface back of the cutting edge or lip, Fig. 41-1. If lip clearance were not given the drill, it would be impossible for it to enter the metal. Correct lip clearance for ordinary work is 8° to 12° (A, Fig. 38-6). Clearance greater than this weakens the cutting edge.

By experimentation and experience, it has been found that the most effective **lip angle** for metalworking drills is 59° with the axis (B, Fig. 38-6). The **length of the lips** must be equal, as must the lip angle. Otherwise, one lip will do most of the cutting, and an oversize hole will be produced, Fig. 41-2. Furthermore, the point will be off-center, which may cause the drill to chip or break.

To produce accurate work and to avoid undue strain on the spindle and bearings of the drill press, **locate the point** in the center of the drill. Figure 41-2 shows the effect of the point being

located off-center, even though the lip angles are equal. To insure correct length of lip and correct lip angle, test with a drill point gage, Fig. 38-8.

Grinding machines made especially for drill sharpening are produced by many different manufacturers. One need only follow the simple directions provided with these machines to quickly obtain properly sharpened drill points. Drill grinding attachments for utility grinders,

(Cleveland Twist Drill — an Acme-Cleveland Company)

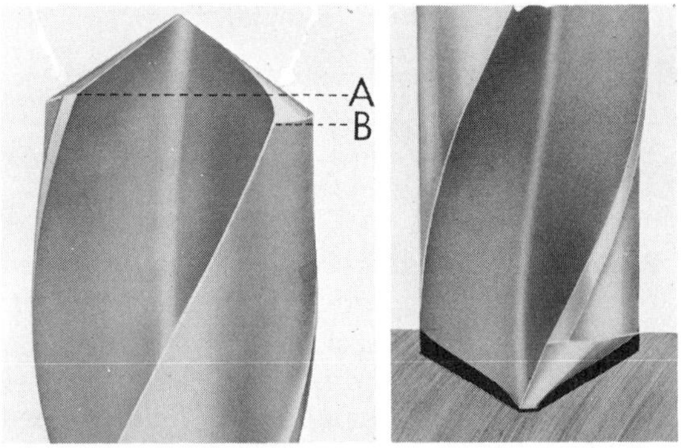

Fig. 41-1.   Proper lip clearance. Note how much lower heel line B is than cutting line A.

(Cleveland Twist Drill — an Acme-Cleveland Company)

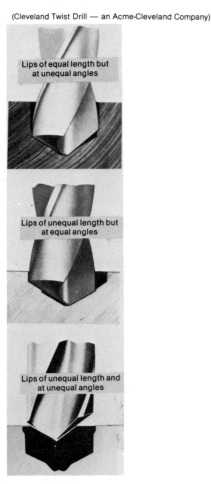

Lips of equal length but at unequal angles

Lips of unequal length but at equal angles

Lips of unequal length and at unequal angles

Fig. 41-2.   Incorrect points.

(Atlas Press Co.)

Fig. 41-3.   Drill-grinding attachment.

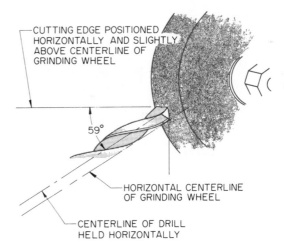

CUTTING EDGE POSITIONED HORIZONTALLY AND SLIGHTLY ABOVE CENTERLINE OF GRINDING WHEEL

59°

HORIZONTAL CENTERLINE OF GRINDING WHEEL

CENTERLINE OF DRILL HELD HORIZONTALLY

Fig. 41-4.   Correct position of drill for hand sharpening at the beginning of a grinding stroke.

Fig. 41-3, are a low cost alternative, and generally give good drill points. The V-block holder on the attachment in Fig. 41-3 centers the drill and holds it in place. A special lip stop and micrometer feed insures accurate grinding of both lips.

Procedure for Hand Sharpening

1. **Put on goggles.** Do this even when the wheel is protected with a shield.
2. Examine the grinding wheel to make sure the face is straight and true. If it is not, dress it with a wheel dresser.
3. If a coolant is used, see that there is sufficient liquid in the reservoir; then start the machine.

4. Examine the angle at which the drill was last ground. Check with a drill point gage.
5. Assuming the angle is correct, hold the drill so that its centerline is horizontal, rotate the drill until the cutting edges are horizontal. Then position the cutting edge that is pointing up parallel with the face of the grinding wheel, see Fig. 41-4.

6. Bring the drill lightly into contact with the wheel at a point slightly above the centerline of the wheel. Then slowly drop the heel of the drill while at the same time pushing the drill forward just enough to keep it in contact with the wheel. If more clearance is needed, pushing forward harder while the heel of the drill is being lowered will grind more metal away below the cutting edge, thus providing more clearance. **It is not necessary to rotate the drill.**

Repeat the operation until the cutting edge is sharp and the correct lip clearance is achieved.

Be sure to hold the drill at the correct angle throughout the operation and do not force the rate of grinding. Otherwise, there is danger of drawing the temper by overheat-ing, thus damaging the cutting qualities of the drill.

7. Grind the other cutting edge in a similar manner.

8. Test the drill with a gage. Be sure that both cutting edges are the same length and are ground at the correct angle. Correct if necessary.

9. If the drill is being sharpened for drilling free machining brass, use the side of the grinding wheel to grind the rake off the cutting edges as shown in Fig. 38-9.

10. Test the drill by drilling a hole in a piece of scrap. A properly sharpened drill running at the correct rpm will cut without excessive pressure and will produce identical chips from the cutting edges. See Fig. 2-1.

Unit

42

Hole Drilling Procedures

Holes may be drilled in most metals with ordinary twist drills, providing they are correctly ground and operated at appropriate cutting speeds and feeds.

Procedure for Drilling Work Clamped Flat on the Table or Held in a Vise

1. Locate the position of the hole with light layout lines, Fig. 42-1A.
2. Center-punch a distinct impression at the intersection of the lines, Fig. 42-1B.

If the hole must be accurately located, scribe a circle the size of the hole with a pair of dividers, Fig. 42-2. With a prick punch, make four or more marks along the cir-

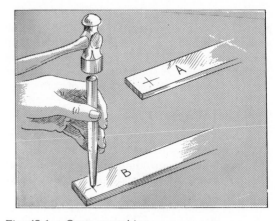

Fig. 42-1. Center punching.

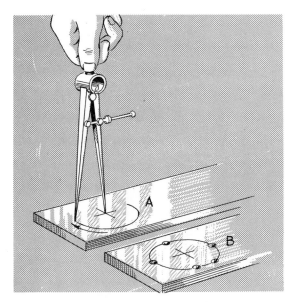

Fig. 42-2. Scribing the circumference of a hole.

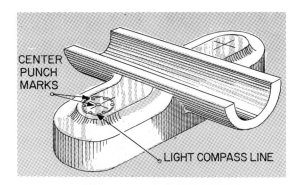

Fig. 42-3. Second concentric circle scribed and punched.

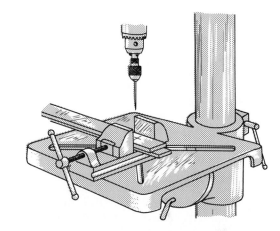

Fig. 42-4. Vise used to hold work.

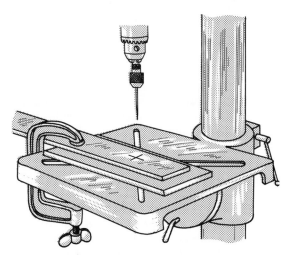

Fig. 42-5. C-clamp used to hold work.

cumference line, as shown in view B. These marks enable the operator to see if the hole is being drilled at the position desired, even though the circumference line becomes erased. For even greater accuracy, a concentric circle 1 to 2 mm (1/32" to 1/16") larger in diameter sometimes is drawn and punch marked, Fig. 42-3. The second circle serves as a check in case the drill should happen to lead off-center and eliminate some of the punch marks on the first concentric circle.

3. Examine the center punch mark to see that the center is located at the intersection of the lines. Correct if necessary.
4. Adjust the spindle speed to run about 600 rpm.
5. Install a **wiggler** in the drill chuck, Fig. 42-4. Adjust the wiggler so that its point runs perfectly true with the machine running.
6. Clamp the workpiece in a vise. Support the workpiece either on parallels or on a block of wood so as to avoid drilling into the vise, Fig. 42-4. If a vise is not available, or if the vise is too small for the workpiece, then support it on a piece of wood on the drill press table as in Fig. 42-5.
7. Position the workpiece so that the punch mark is under the wiggler point. Turn the

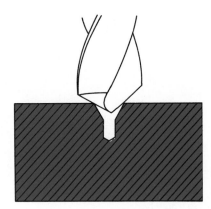

Fig. 42-6.   Center-drilled hole helps provide accurate location for twist drills.

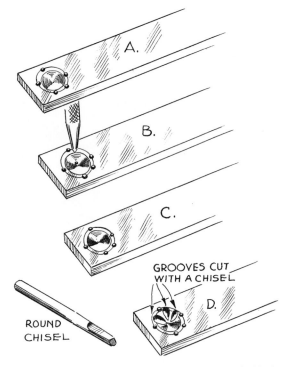

GROOVES CUT WITH A CHISEL

ROUND CHISEL

Fig. 42-7.   Moving the center of a partially drilled hole.

machine on, and while gently moving the point of the wiggler in and out of the hole, observe it closely from several directions for sidewise movement. Sidewise movement of the wiggler means the punch mark is off center. Adjust the position of the workpiece until the wiggler can be moved in and out of the punch mark without visible sidewise movement. Then clamp the vise or the workpiece to the work table if possible.

8.   Remove the wiggler from the drill chuck and install a small center drill (combination drill and countersink).

9.   Determine the speed at which the center drill should run (Table 40-1 or A-10, Appendix). Then adjust the drive system so as to produce the rpm desired.

CAUTION

With a variable-speed drive, the speed usually must be changed only while the machine is running.

10.   Drill the center hole deep enough to produce a cone-shaped hole, Fig. 42-6.

11.   Replace the center drill with a twist drill. Determine the correct speed for the twist drill and set the machine accordingly.

12.   Start the machine and advance the drill until it has drilled almost its full diameter.

13.   Raise the drill and examine the impression to be sure that the hole is concentric with the scribed circle.

   If the impression is not in the center, as in **A** of Fig. 42-7, the center may be drawn over by cutting one or more grooves on the side toward which the center is to be drawn, as in **D.** If a center drill has not been used, center punching in the direction the center must be moved, as in **B,** will often work.

14.   When the drilled depression is concentric with the scribed circle, continue to feed the drill into the work. If the machine is equipped with a power feed, engage the automatic feed. See Unit 40 for selection of the proper rate of feed.

15.   Apply a cutting fluid appropriate for the kind of metal being drilled.

16.   When hand feeding, and the point of the drill is about to break through the metal, decrease the pressure on the feed mechanism, but continue to feed slowly until the hole is completed.

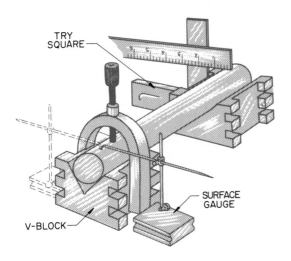

Fig. 42-8. Checking the center punch mark.

CAUTION

Be sure the stock is held securely throughout the operation, particularly when the point breaks through the underside. Otherwise the piece is likely to spin around with the drill and damage the drill or the work or injure the operator.

When it is necessary to drill a hole a certain depth, set the quill **depth stop,** Fig. 36-3, the desired distance, and feed the drill to the limit of travel. When drilling deep holes, be sure to withdraw the drill frequently to clear the chips from the flutes of the drill. Small diameter drills are particularly prone to clogging and breakage. Some machines with automatic feed systems have a **peck cycle** which automatically withdraws the drill to clear it of chips after a certain amount of travel.

17. When the operation is completed, stop the machine. Unclamp the workpiece, clean up the machine, and return the tools to their proper storage place.

Procedure for Drilling Work Clamped in a V-Block

1. With a center punch, mark the position on the stock at which the hole is to be drilled.

Fig. 42-9. Work clamped in position on the table.

2. Select a suitable center drill and insert it in the drill chuck.
3. Place the stock in a V-block, and adjust it until the center of the punch mark is exactly under the point of the drill.

   If necessary, test the accuracy of the setting by testing the center from both sides with a surface gage, or by measuring with a rule and try square as indicated in Fig. 42-8.

   Another alignment method is simply to center a drill at the bottom of the V in the empty V-block. When the workpiece is carefully placed in the V-block, and the center punch mark is aligned with the point of the drill, or with a wiggler, the hole will automatically go through the center of the workpiece.

4. When the center has been located correctly under the point of the drill, clamp the piece in position with the V-block clamp as in Fig. 42-9. For best results, mount the V-block in a vise and fasten it to the machine table.
5. Adjust the machine for the correct speed and feed, and center-drill the workpiece.

6. Remove the center drill and install the twist drill. Reset the machine for the correct rpm for the twist drill.
7. Start the machine. Then draw the drill down against the stock and continue feeding it forward until the hole is drilled to the depth desired. Use a cutting fluid, if appropriate.
8. When the operation is completed, stop the machine. Unclamp the workpiece, clean up the machine, and return the tools to their proper storage place.

Procedure for Drilling Work Held Against an Angle Plate

1. Accurately locate the point at which the hole is to be drilled, and mark it distinctly with a center punch.
2. Bolt an angle plate on the table of the drill press in such position that when the work to be drilled is clamped against it, a hole may be drilled in the work at the position desired, Fig. 42-10.
3. Clamp the work securely to the angle plate.
4. Locate the workpiece with the aid of a wiggler.

5. Center-drill the hole location.
6. Install the twist drill, start the machine, and draw the drill down against the work. Use the power feed if the machine is so equipped. Continue advancing the drill until the hole is drilled the depth desired, using a cutting fluid if appropriate.
9. When the operation is completed, stop the machine. Unclamp the workpiece, clean up the machine, and return the tools to their proper storage place.

Procedure for Drilling a Large Hole

1. Prepare the stock in the usual manner, and fasten it securely on the table of the drill press.
2. Select a drill of the size required; also select a small drill having a diameter larger than the thickness of the web of the large drill.
3. Locate the workpiece and drill a center hole.
4. Insert the small drill in the spindle or chuck and drill the pilot hole, Fig. 42-11.
5. Remove the small drill, insert the large drill in the spindle, and drill the hole as in Fig. 42-11.

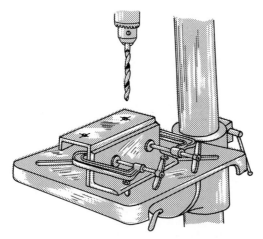

Fig. 42-10. Work held against an angle plate.

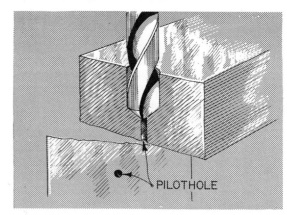

Fig. 42-11. Drilling a large hole with the aid of a pilot hole.

Boring a Hole with a Drill Press

When a very round, straight hole with an excellent finish must be produced, it is usually drilled undersize and finished with a boring tool. See Fig. 43-1. Boring also is resorted to when a drill of the size desired is not available. For ordinary boring, the hole is first drilled from 1.5 to 3 mm (1/16" to 1/8") undersize.

For good results, boring requires a rigid setup on a sturdy machine with power feed to the quill. Attempting to bore precision holes on a small, sensitive drill press with hand feed will produce mediocre results at best **and may even be dangerous.**

Procedure

1. Clamp the stock on the drill press, and drill an undersize hole of suitable size.

CAUTION
Do not loosen the clamps or change the position of the stock on the table of the drill press, unless the workpiece has moved out of position during drilling.

2. Select a boring head or boring bar of suitable size, and insert it in the spindle of the drill press. Tap the protruding end of the boring bar with a lead hammer, thus seating it firmly in the socket or spindle.
3. Examine the cutter to make sure it is sharp. Grind it if necessary.
4. Adjust the cutter in the boring head or boring bar so that it will make a cut of the depth necessary to produce a hole of the size desired. For greater accuracy, a roughing cut should be made first, followed by a light finishing cut of 0.127 to 0.254 mm (0.005" to 0.010") depth. Be sure the cutter is fastened securely.
5. Set the machine at the speed recommended for drilling a hole of similar size in the same material, Table 40-1 or A-10, Appendix.
6. Start the machine, and feed the tool slowly into the hole.
7. When the tool has entered a distance of about 3 mm (1/8"), stop the machine, and caliper or measure the diameter of the hole accurately.
8. If the hole is not being bored the diameter desired, make the necessary adjustment. Then continue boring until the job is completed. If the machine is equipped with a power feed, its use is recommended.

CAUTION
When using a light boring tool, be sure to take a light cut and feed.

9. When the operation is completed, remove the clamps, workpiece, and boring tool and return them to their proper storage place.

SPINDLE OF DRILL PRESS

Fig. 43-1.   Using a boring tool in a drill press.

Unit

44

Hand and Machine Reaming Procedure

A reamer in good condition is capable of quickly producing a straight, smooth hole, far superior to that obtained by drilling. A hole may be machine reamed, hand reamed, or both.

Procedure for Machine Reaming

1. Drill the hole 0.40 to 0.80 mm (1/64″ to 1/32″) undersize, depending on the size of the reamer to be used.
2. If precision location is required, drill the hole 1.5 to 3 mm (1/16″ to 1/8″) undersize. Then bore the hole to straighten it out, leaving the usual reaming allowance.
3. If power feed is to be used, set the feed rate at 0.04 to 0.10 mm (0.0015″ to 0.004″) per flute per revolution.
4. Ream the hole at between 1/2 and 2/3 the rpm used for drilling a hole of like size in the same kind of material. See Table 40-1 for drilling speeds.
5. Use of an appropriate cutting fluid usually improves the quality of the finish, see Table A-11, Appendix.

Procedure for Hand Reaming

1. Drill, bore, or machine ream the hole 0.05 to 0.13 mm (0.002″ to 0.005″) undersize.
2. Clamp the workpiece to be reamed in a vise.
3. Obtain the reamer and a suitable tap wrench to drive it.

4. If an adjustable or expansion reamer is being used, adjust it so that it will just enter the hole snugly.
5. Insert the reamer in the hole, Fig. 44-1, and align it carefully with the hole axis. Apply a cutting fluid. Then turn the reamer slowly clockwise while applying enough pressure to the end of the reamer to advance it into the hole.

If hand reaming is to be done immediately after the hole is drilled or bored in a lathe or

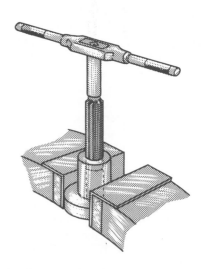

Fig. 44-1.   Starting a hand reamer.

drill press, it is best to leave the workpiece clamped in the machine. Remove the drilling or boring tool and replace it with a lathe center. Fit the tap wrench on the end of the reamer. Then insert the point of the reamer in the hole. Advance the lathe center in the machine spindle until its point enters the center hole in the shank of the reamer, Fig. 44-2. This setup assures perfect alignment of the reamer with the hole in the workpiece. Ream the hole as described above, keeping the lathe center in the reamer center hole until the reamer is far enough into the hole to assure correct alignment.

6. When the hole has been reamed its full length, continue turning the reamer clockwise while withdrawing it from the hole.

CAUTION
Never turn a reamer backwards. Doing so will damage the cutting edges and damage the finish in the hole.

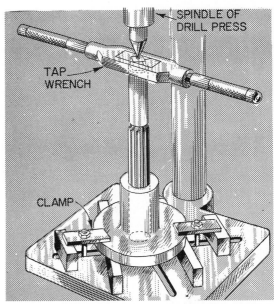

Fig. 44-2. Using the drill press spindle to keep reamer aligned.

Unit
45

Spot Finishing Procedure

A **spot finish,** also known as an **engine turned finish,** consists of a series of small circular spots ground into the surface of the metal in a regular pattern. See **A** in Fig. 45-1. The pattern usually consists of overlapping spots, each row overlapping the previous row. Either a drill press or a vertical milling machine can be used for applying the finish. Use of a vertical milling machine is preferred because of its ability to precisely space the spots. Spot finishing applied in the manner described below works best on steel and steel alloys, and can only be applied to flat surfaces.

Procedure

1. Machine the workpiece to its final size and shape.
2. Surface-grind or hand-polish the surface (s) to be spot finished. If hand polishing is necessary, use a fine aluminum oxide abrasive

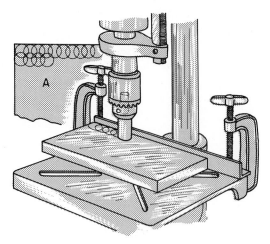

Fig. 45-1.   Spot-finishing with a drill press.

10. Again bring the dowel into contact with the surface, and make a second spot. Continue thus to the end of the work.
11. Move the workpiece back two-thirds of the dowel diameter; proceed as in steps 9 and 10.
12. Repeat the above steps until the whole surface is spot-finished.
13. Remove the workpiece and tools. Then thoroughly clean the machine with an old rag or a shop towel.
14. With a soft cloth, remove the remaining abrasive and oil from the finished surface of the object. Then apply a coat of commercial anti-rust compound. Unless protected, steel rapidly oxidizes or rusts.

or emery cloth and a light oil. Keep the surface as flat as possible.

3. Cut a piece of hardwood dowel 50 to 75 mm (2″ to 3″) long. Popular diameters for spot finishing are 9.5 and 12.7 mm (3/8″ and 1/2″), but any diameter may be chosen which is proportional to the size of the surface on which the finish will be applied.
4. Mount the dowel securely in a chuck or collet.
5. Start the drill press. Then square the end of the dowel by drawing it down against a sheet of abrasive paper placed on the table of the machine.
6. Prepare a pasty mixture of fine abrasive and oil. Coat the surface to be spot-finished with abrasive and oil. Valve grinding compound may be used instead.
7. If a drill press is to be used, clamp a straightedge on the table of the drill press in such a position that when the dowel is drawn down on the surface of the work, a spot will be made near one edge. If a vertical milling machine is to be used, mount the workpiece in a vise or fasten it to the work table.
8. Locate the work so that the first spot will be made at one corner, Fig. 45-1.
9. Bring the revolving dowel into contact with the surface of the work, and hold it there for about 10 seconds. Then raise the dowel and move the piece over about two-thirds of the dowel diameter.

Test Your Knowledge of Section 8

Unit 36: The Drill Press

1. List six hole-machining operations which can be performed on a drill press.
2. Describe sensitive drilling machines and list their basic uses.
3. Name three types of drive systems which make it possible to vary the speed of a drill press spindle. Which type should only be adjusted with the machine running?
4. How is the size of a drill press designated?
5. List five safety precautions which you think are most important when operating a drill press.

Unit 37: Industrial Drilling Machines

1. Describe an upright drilling machine, and explain the purpose for which it is designed.
2. Describe a gang drilling machine, and explain its principal uses.
3. Describe a radial drilling machine, and the type of work for which it is designed.
4. For what purpose are multiple-spindle drilling machines designed?
5. Describe a turret drilling machine and list major advantages.
6. What types of operations other than drilling may be performed on a large turret drilling machine?

Unit 38: Drilling, Boring, Counterboring and Countersinking

1. Name the three principal parts of a drill.
2. Describe how sleeves or shell sockets are used.
3. Describe how a taper shank drill is removed from a drill press spindle.
4. Describe how you could mount a drill with a No. 3 Morse taper shank in a drill press with a No. 2 Morse taper hole.
5. List four ways in which drill sizes may be indicated.
6. List three factors which are important in grinding a drill properly.
7. What should the lip-clearance angle be for a general-purpose drill for drilling most steels?
8. What should the total included angle be for the point of a general-purpose drill?
9. What should the total included angle be for a drill used to drill molded plastics and fibrous materials?
10. What rake angle should be ground on the front of the cutting lip of a drill which is used to drill brass? Why?
11. What are the advantages of a spiral point on a drill?
12. List several purposes for which carbide-tipped drills are used.
13. For what purposes are three- or four-fluted core drills used?
14. For what principal purpose are subland drills used?
15. Describe spade drills and list their advantages.
16. For what purposes are boring tools used?
17. For what purpose are counterboring tools used?
18. For what purpose are holes countersunk at an angle of 82°? 90°? 60°?

Unit 39: Reamers

1. Describe a reamer and list several advantages of reaming.
2. Describe the difference in cutting action between hand and machine reamers.
3. How do rose reamers differ from fluted machine reamers?
4. Of what advantage are shell reamers?
5. Describe an adjustable machine reamer and discuss its advantages.
6. Of what materials are reamers generally made?
7. For what kinds of reaming operations are straight-fluted solid hand reamers recommended? Helical-fluted?
8. How do expansion reamers differ from adjustable reamers in design and range of size adjustment?
9. How are reaming speeds determined for machine-reaming operations?
10. How do reaming feeds generally compare with drilling feeds?
11. List several factors which cause chatter while reaming.
12. What cutting fluid should be used for reaming steel?
13. List several precautions to be observed in caring for reamers.
14. What is common practice for one-step machine reaming allowances on holes less than 12.7 mm (1/2") diameter? Over 12.7 mm (1/2") diameter?
15. Why should reamers never be reversed?

Unit 40: Cutting Speeds and Feeds for Drilling

1. Describe the difference between cutting speed and rpm (revolutions per minute).
2. List several factors which affect the cutting speed selected for drilling operations.
3. Name suggested cutting speeds for some of the more common metals drilled in machine shops.
4. Give the formula which may be used for calculating the rpm for a drill when the cutting speed is known.
5. Define the meaning of feed as related to drilling operations.
6. List suggested feeds for drilling mild steel with drills 4.8 mm (3/16") and 15.9 mm (5/8") in diameter.
7. What damage can result from excess speed and feed in drilling operations?
8. List several cutting fluids which may be used for drilling soft steel.

Section

9

The Lathe and Lathe Operations

Unit

46

The Metalworking Lathe

The metalworking lathe is the most basic of all the metalworking machine tools. Lathes are built primarily for making cylindrical and conical parts such as pins, bolts, shafts, discs, pulleys and gear blanks, and for boring holes more precisely or larger than can be made by drilling and reaming. Some lathes can be equipped with attachments which enable them to perform milling, grinding, and broaching operations. When equipped with a full range of accessories, the lathe is said to be the only machine tool capable of reproducing itself.

Figure 46-1 shows a manually operated lathe typical of those found in maintenance, tool and die, and job shops. This type of lathe is used primarily for making either single parts, or small quantities of identical parts which cannot be made economically on lathes designed for higher volume production. Unit 47 is an introduction to

(DoAll Company)

Fig. 46-1.   Operator turning work mounted in a chuck on a lathe.

the production machines which have been developed from the basic metalworking lathe.

Most lathes are capable of performing both external and internal machining operations. External cylindrical machining operations include straight turning, taper turning, turning grooves, facing flat surfaces on the end of stock, threading, knurling, and cutting off stock. Some of these operations and the types of cutting tools used to perform them are shown in Fig. 49-1.*

Internal cylindrical machining operations include all common hole-machining operations such as drilling, boring, reaming, counterboring, countersinking, tapping, and threading with a single-point tool (Fig. 36-2). Hence, all of the common operations performed on a drill press also can be performed on a lathe.

The same skills required for performing basic lathe operations apply to the operation of other machine tools, most of which are modifications or adaptations of the lathe. Therefore, a basic understanding of the metalworking lathe and its operation is fundamental to the development of a broad knowledge of machine shop theory and practice.

Fig. 46-2. Primitive lathe.

Historical Background

The lathe is thought to be one of the first machines invented by man. By whom or when the first lathe was built no one knows. One of the earliest-known illustrations of a lathe is shown in Fig. 46-2. It is what is known as a tree lathe. This machine was intended for turning wood, and, as may be seen, consisted of two centers, a tool rest (the board in the rear), and a means of rotating the work.

From the time of this first invention until 1797 when an Englishman, Henry Maudslay, designed and built the screw-cutting lathe shown in Fig. 1-5, there doubtless were many improvements in lathe construction. None, however, were strikingly outstanding or efficient. The outstanding feature of Maudslay's lathe was a lead screw geared to the spindle of the lathe. This invention made possible the advancement of the tool at a constant rate of speed, which made possible the production of interchangeable screw threads and uniformly smooth finishes so important to good bearing surfaces.

Modern Lathes

Modern lathes are efficient, accurate, and complex devices, capable of doing a great quantity and variety of work. A well-constructed lathe, when properly operated, can produce work accurate to 0.01 mm (0.0004") or less. Figure 46-3 gives an idea of the complexity of a modern lathe and shows many of its principal parts.

Power for driving a lathe comes from an individual motor. On some machines, the motor is direct mounted, while on others, it is connected by means of a short belt, usually of the V-type. Figure 46-4 illustrates a type of lathe in which the motor is mounted under the head in the hollow machine base.

The size of a lathe is given in terms of swing and length of bed. The **swing** refers to the diameter of work that can be rotated in the lathe. Thus a 400 mm (16") lathe will swing work as large as 400 mm in diameter. The length of a lathe bed should not be confused with the maximum distance between centers when the tailstock is moved to the far end of the lathe bed. For example, a 1 524 mm (5') bed may only provide 914 mm (3') between centers.

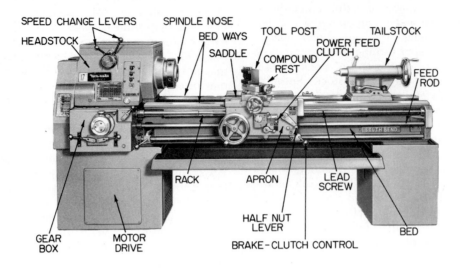

Fig. 46-3.   Parts of a modern lathe.

Lathes are made in a wide variety of types and sizes, from the small precision lathe found in watch repair shops to the immense machines used in manufacturing big guns or mill rolls. A heavy-duty lathe with a geared-head drive is shown in Fig. 46-8.

Modern lathes perform basically the same operations which were performed on lathes at the turn of this century. However, because of improved design features and improved cutting tools, they accomplish these operations more accurately and efficiently today. Without the metalworking lathe, modern industrial machines and equipment could not be produced.

How a Modern Lathe Operates

Before attempting to operate a lathe, you should become familiar with its principal parts, controls, and accessories. The principal component parts of a metalworking lathe are shown in Fig. 46-5 on page 190. By studying this figure carefully you will learn the function of each basic part of a lathe and its relationship to other parts.

The principal controls on a lathe are shown in Fig. 46-3. Although the controls are not identical on all lathes, most lathes are equipped with similar controls.

Changing Spindle Speeds

With a **flat-belt drive,** spindle speeds are changed by first loosening the belt with the belt tension lever and then moving the flat belt to the desired step on the **cone pulley.** See Fig. 46-4. **The lathe must be stopped** before moving the belt to another pulley.

The cone pulley has four steps which provide four speeds while in direct drive. The lathe is also equipped with a **back-gear** mechanism which provides four lower speeds. Hence, this lathe has eight speeds available, ranging from about 40 to 940 rpm, with a single-speed motor. When equipped with a two-speed motor, 16 speeds are available. The speeds may then range up to 1600 rpm.

Lathes such as shown in Fig. 46-4 are equipped with a lock pin in the **bull gear,** which is located on the headstock spindle. For direct drive, the

Fig. 46-4.   Lathe with flat-belt drive.

Fig. 46-6.   Manual-shift V-belt drive.

pin is pushed in and the back-gear lever is placed in the released position. For back-geared drive, the pin is pulled out to the released position and the back-gear lever is engaged.

Lathes of the type shown in Fig. 46-5 may be equipped with several types of V-belt drive systems for changing spindle speeds. With the **manual-shift** system, as shown in Fig. 46-6, the belt tension is released with a lever provided for this purpose, and the V-belt is shifted manually to the position desired on the step-type V-pulley.

With the **lever shift** V-belt drive system shown in Fig. 46-7, speed changes are made by shifting the two-speed change levers to any of four different positions. With this arrangement, four speeds are available in direct drive and four speeds in back-geared drive.

With a **geared-head drive,** as on the lathe shown in Fig. 46-8, it is not necessary to shift belts for making changes in spindle speeds. Speed changes are made by shifting speed-change levers, Fig. 46-9.

Speed changes on most geared-head lathes may be made only while the machine is stopped. To shift the gears while the machine is running could cause serious damage to the speed-change

Fig. 46-7.   Lever-shift V-belt drive.

(Sheldon Machine Company, Inc.)

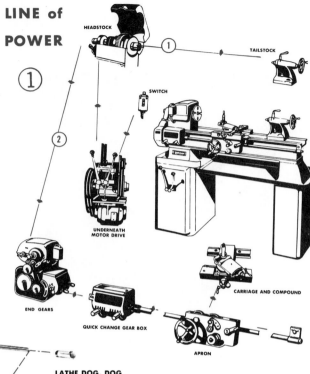

LINE of POWER

HEADSTOCK

TAILSTOCK

SWITCH

UNDERNEATH MOTOR DRIVE

END GEARS

QUICK CHANGE GEAR BOX

APRON

CARRIAGE AND COMPOUND

Fig. 46-5. How a modern lathe operates.

The only purpose of the lathe is to change the shape of metal by rotating a workpiece against a movable cutting tool. All lathe parts, controls, and accessories are designed to perform this one purpose. All lathes have the same basic design. Their components serve one of three primary functions:

a. Driving the lathe — line of power.

b. Holding and rotating work.

c. Holding and moving tool.

Learn these primary functional divisions and then their parts and purposes as well and the lathe operation will be easy to understand.

HOLDING and ROTATING WORK

HEADSTOCK

Headstock supports spindle which rotates on "Zero Precision" tapered roller bearings. Work holders are mounted on spindle nose.

COLLET ATTACHMENT

Collet attachment passes through the hole through entire length of the spindle. Special jaws on collet, release or grip small diameter work at the spindle nose.

LATHE DOG, DOG PLATE AND CENTERS

The lathe dog clamps around the work piece. The dog plate mounts on the spindle. As work piece is placed on the spindle nose center, the tail of the lathe dog is slipped into a slot in the dog plate so that when dog plate revolves it turns the work piece.

TAILSTOCK

Tailstock center supports right end of work held "between centers." It can be offset to cut tapers, locked in any position along lathe bed, and has handwheel feed for tailstock tools.

CHUCK

Chuck mounts on spindle nose. Adjustable jaws permit holding of larger diameter, odd-shaped or stub-end work.

HOLDING and MOVING TOOL

CARRIAGE AND COMPOUND

Carriage provides rigid support for cross-slide and travels — either to the right or left along the bed. Cross slide moves compound in or out with power feed or handwheel. Compound swivels to provide angular feeds.

END GEARS

Outboard gear on spindle drives end gear train which operates lead screw through gear box.

QUICK CHANGE GEAR BOX

Double tumbler levers permit rapid selection of desired ratio (Pitch and feed) between spindle r.p.m. and lead screw r.p.m.

APRON (FRONT VIEW)

Apron controls are centrally grouped with selector lever for power longitudinal and cross feeds, friction clutch for engaging feeds, half-nut lever for thread cutting, and hand wheel for hand traverse of carriage. Built-in safety mechanisms prevent engaging half-nuts and power feeds at the same time.

DETAIL FROM APRON (REAR VIEW)

Lead screw transmits power through apron by (1) spline drive for power feeds and (2) by half nuts for thread cutting. Precision lead screw threads are used only for thread cutting.

(South Bend Lathe, Inc.)

Fig. 46-8.   Modern geared-head lathe.

(South Bend Lathe, Inc.)

Fig. 46-9.   Headstock with all-geared drive.

(Clausing Corporation)

Fig. 46-10. Lathe with a variable-speed drive system.

gears. With a two-speed motor, eighteen spindle speeds ranging from 30 to 1580 rpm are available.

Some lathes are equipped with a **variable-speed drive** system, as in Fig. 46-10. This type of drive is usually found on small- or medium-duty lathes. **With this system, the motor must be running while speed changes are made.** Lathes of this type also are equipped with a back-gear mechanism, thus providing for low- and high-speed ranges.

Feeding and Threading Mechanism

Lathes are equipped so that the cutting tool may be fed manually or automatically along the work which is being turned or faced. The feed is called **longitudinal feed** when the tool travels along the work, parallel to the lathe bed. It is called **cross feed** when the tool travels across the end of the workpiece. Thus the cross feed is used for facing the end of a workpiece.

The amount of **longitudinal feed** is the distance that the tool advances along the workpiece during one revolution of the work. The amount of **cross feed** refers to the distance the tool travels across the end of the workpiece during one revolution.

The entire power feed mechanism includes the following three basic components: the end-gear train (Fig. 46-11), the quick-change gear box

(South Bend Lathe, Inc.)

Fig. 46-11. End-gear train.

(Figs. 46-12 and 67-3), and the carriage and apron assembly (Fig. 46-13).

The **end gears** transmit power from the lathe spindle to the lead screw, through the gear box. The position of the levers on the **quick-change gear box** determine the ratio between the spindle speed and the speed of the lead screw to the carriage.

The controls on the carriage and apron assembly control all carriage and tool movements. The apron **handwheel** is used to move or feed the carriage along the lathe bed manually. The **cross-feed knob** is used to move or feed the tool crosswise (in and out) manually. The **compound-rest knob** is used to feed the compound rest manually.

The compound rest may be set at any angle required. Unless necessary for special operations, it normally is turned at an angle of about 30° from the crosswise position, as shown in Fig. 46-13. The lock screw is used to lock the carriage in position for operations such as facing.

The **feed-change lever** is used to select one of three types of feed: longitudinal feed, cross feed, or threading feed. While threading, the feed-change lever is left in the center position as shown. The feed is then engaged with the **half-nut lever,** which is used only for threading operations.

For longitudinal feed or cross feed, the feed-change lever is located in either the upper or lower position, as desired. The **automatic feed clutch** then is used to engage or disengage the feed for automatic operation.

When it is necessary to reverse the direction of travel of either the longitudinal or the cross feed, the lead screw must be reversed. This is done with the **feed-reverse lever.**

It should be noted that there are large numbers and also small decimal numbers located on the index plate on the quick-change gear box, Figs. 46-12 and 67-3. On machines having metric measurements, the **large** numbers indicate the

(South Bend Lathe, Inc.)

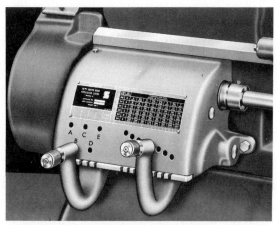

Fig. 46-12. Quick-change gear box with chart.

(South Bend Lathe, Inc.)

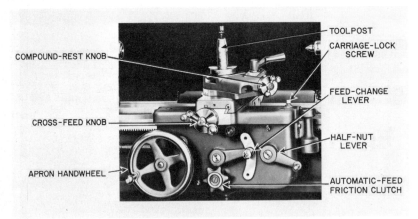

Fig. 46-13. Parts of lathe carriage and apron assembly.

pitch in millimeters of threads obtained when using the half-nut lever for power feeding. On machines having inch measurements, the **large** numbers indicate the number of threads per inch.

The **small** decimal numbers indicate the amount of feed per revolution obtained when using the automatic feed clutch. In metric measurement, this is expressed in hundredths of a millimeter; in inch measurement, it is expressed in thousandths of an inch. However, the cross feed does not always equal the longitudinal feed indicated. On the lathe in Fig. 46-11, the cross feed is equal to 0.375 times the longitudinal feed.

For example, if the feed selected were indicated by the small decimal number 0.135 mm (0.0053″), the longitudinal feed would be 0.135 mm (0.0053″) per revolution of the work. But the cross feed would be only 0.375 × 0.135 mm (0.0053″), or 0.05 mm (0.002″) per revolution of the lathe spindle. The ratio of cross feed to longitudinal feed varies for different lathes, and it usually is indicated on the index plate on the quick-change gear box.

Lathe Accessories

Numerous accessories are required for various machining operations on a lathe. Several standard accessories needed for beginning lathe operations are described in this unit. Other accessories are described in those units dealing with lathe operations for which they are commonly used.

Lathe Centers

Two lathe centers (similar to those shown in Fig. 46-14) are required for turning work between centers on a lathe. Figure 58-2 shows work being turned between centers. The centers have a Morse-taper shank which fits into the tapered hole in the tailstock and the tapered hole in the spindle. Centers which are too small to fit the tapered hole in the headstock spindle can be adapted to fit by the use of a spindle sleeve, Fig. 46-15.

The center in the headstock rotates with the spindle and the work, and, therefore, is called the **live** center; that in the tailstock is stationary and is called the **dead** center. The center in the tailstock always must be made of hardened steel or

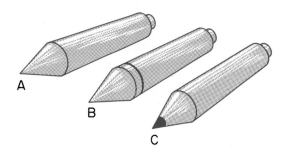

Fig. 46-14. Lathe centers.
- A. Soft
- B. Hardened
- C. Carbide-tipped

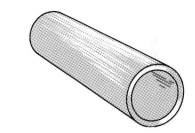

Fig. 46-15. Spindle sleeve.

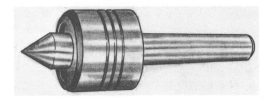

Fig. 46-16. Rotating live center for lathe tailstock.

carbide. That in the headstock may be either hardened or unhardened steel. Hard lathe centers are made of either carbon tool steel or high-speed steel, or they may be equipped with carbide tips which are highly wear- and heat-resistant.

Center holes are drilled into the workpiece, thus permitting the workpiece to be supported between the centers. The center hole in which the dead center is inserted must be lubricated to prevent it from becoming overheated and scored or burned. A mixture of white lead or red lead and oil is a satisfactory lubricant.

Some manufacturers make **rotating live centers** for the tailstock, Fig. 46-16. Such centers are equipped with ball bearings or tapered roller bearings which permit the center to rotate with the work. It is not necessary to lubricate the center holes when mounting work on centers of this type.

For accurate work, lathe centers must be free from nicks or chips. The holes in the tailstock and headstock spindle must be cleaned each time the center is replaced in the hole.

When hard lathe centers become scored, nicked, or damaged, they must be resharpened by grinding, Fig. 46-17. They are ground with a tool post grinder mounted on the compound rest on a lathe. The point angle of the lathe center is 60° and is checked with a **center gage,** Fig. 46-18.

Lathe Dogs

When a workpiece is mounted between centers on a lathe, it is driven with a lathe dog. Several lathe dogs are shown in Fig. 46-19. The clamp dog may be used on work of many diameters within its size range. The dogs with the square-head screw or the recessed safety screw are available in various sizes with work capacities ranging from 9.5 to 152 mm (3/8″ to 6″) diameter.

Other Accessories

Other lathe accessories are described in conjunction with lathe units which present the procedures for performing various lathe operations.

They include toolholders, cutting tools, steady rest, follower rest, knurling tools, chucks, faceplates, types of spindle noses, boring tools and toolholders, mandrels, taper attachments, and milling attachments.

(South Bend Lathe, Inc.)

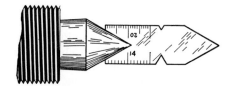

Fig. 46-18. Testing angle of lathe center with center gage.

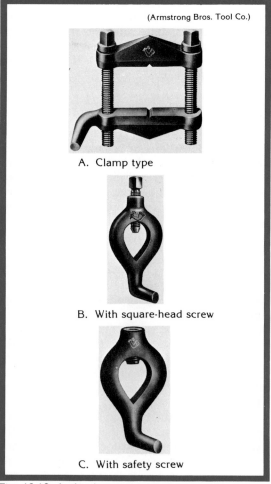

(Armstrong Bros. Tool Co.)

A. Clamp type

B. With square-head screw

C. With safety screw

Fig. 46-19. Lathe dogs.

(South Bend Lathe, Inc.)

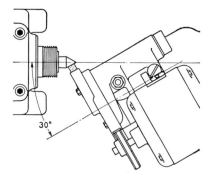

Fig. 46-17. Truing 60° lathe center point.

Safety Precautions for the Lathe

General Turning

1. Wear approved safety goggles.
2. See that all guards are in place.
3. Before starting the lathe, turn the spindle by hand to insure that it turns freely. If the spindle is locked in a stationary position with the back-gear lever as well as the bull-gear pin, release one of these devices for the desired kind of drive and speed.
4. Stop the machine to make all adjustments.
5. Stop the machine for all measurements.
6. Stop the machine to remove chips. Do not remove them with the hands — always use a brush or stick.
7. Stop the machine for oiling.
8. Always stop the machine when adjusting the tool in the tool post.
9. Keep the machine clear of long chips, rags, and unnecessary hand tools.
10. Use the right type of cutting tool for the job.
11. Adjust the feed, speed, and depth of cut according to the size and type of metal.

Turning Between Centers

12. Be sure that the tailstock and the tailstock spindle are locked securely.
13. When available, use a safety dog to drive the workpiece. Select the smallest dog which will do the job and clamp it tightly.
14. Lubricate the tailstock dead center properly and frequently.
15. Always cut toward the headstock whenever possible.
16. Before starting the power feed, make certain that the carriage has sufficient free travel to complete the cut without running into the lathe dog.

Turning Work in a Chuck

17. Place a board under the chuck when mounting it or removing it from the spindle. Keep the fingers clear.
18. Be sure that the chuck is mounted tightly to the spindle.
19. Be sure that the work is mounted tightly in the chuck.
20. Always remove the chuck wrench or key from the chuck immediately after using it.
21. Turn the chuck one complete revolution by hand after the work is mounted to see that it clears the carriage and the ways.
22. Never allow the cutting tool or toolholder to come into contact with the revolving chuck jaws.

Turning Work on a Faceplate

23. Be sure the faceplate is secured tightly to the lathe spindle.
24. Use the shortest bolts and clamps possible for clamping work to the faceplate, and clamp the work securely. The clamps should be supported at the outer ends and should be parallel to the faceplate.
25. Use a counterweight, if necessary, to balance an off-center workpiece.
26. Before starting the machine, turn the work one complete revolution by hand to see that it clears the carriage and the ways.
27. Stand to one side of the revolving faceplate.

Unit
47

Types of Metalworking Production Lathes

(Hardinge Brothers, Inc.)

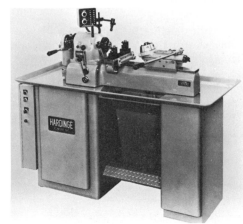

Fig. 47-1.   Hand-screw machine.

(Warner and Swasey Company)

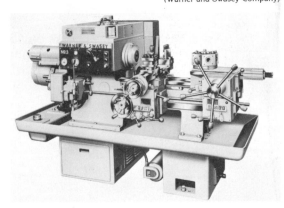

Fig. 47-2.   Ram-type turret lathe.

Most variations of the basic metalworking lathe have been designed to speed production of duplicate parts. **Hand screw** machines, Fig. 47-1, are small, manually operated **turret lathes.** A basic hand screw machine is provided with a six-station bed turret and a cross slide with a front and rear tool post. These machines are not usually provided with power feeds for tool movement. Thus, they require a skillful operator in order to consistently obtain close tolerance parts with fine finishes. Hand screw machines are designed for efficient, low-volume production of fairly simple parts.

Manually operated turret lathes are larger than hand screw machines, and are usually capable of making more complex parts. The smaller sizes of turret lathes are generally **ram-type** machines, Fig. 47-2. In this type of turret lathe, the bed turret is attached to a casting called a **ram,** which is free to slide in and out of a casting that is bolted to the lathe bed. The entire turret assembly can be unbolted and relocated on the lathe bed wherever it is needed to make a particular part. The machine operator controls the movement of the ram, and thus the turret, either manually by turning a hand wheel or by using the power feed. A four-station tool post turret is often provided on the front of the cross slide, with a cutoff blade mounted in a rear tool post. Most turret lathes of this type have power feeds to the carriage, cross slide, and bed turret. They are generally used for production runs of a few dozen to a few hundred parts.

Fig. 47-3.   Saddle-type turret lathe.

Fig. 47-4.   Numerically controlled turret lathe equipped for chucking work.

Fig. 47-5.   A numerically controlled vertical turret lathe, with a close-up of the overhead turret, revolving table, and workpiece.

Saddle-type turret lathes, Fig. 47-3, have the bed turret mounted directly on a second saddle, which is mounted on the bed ways like the saddle which carries the cross slide. Larger turret lathes are usually of the saddle type. The main advantages are that the travel of the turret mounted tools is not limited to the length of travel of a ram, and the turret remains rigidly supported on the saddle rather than being mounted on an overhanging ram. In all other respects, saddle-type turret lathes are similar to ram-type turret lathes.

Numerically controlled (N/C) turret lathes are now available in considerable variety and have proven to be far more productive than manually operated machines. They are made in several sizes and styles. Some have only a bed turret. Others have both a bed turret and a cross slide turret. The machine shown in Fig. 47-4 has a six-station cross slide turret as well as a six-station bed turret. When properly programmed and tooled, these machines can economically produce from a few to a few thousand parts.

Vertical turret lathes are also available as manually operated or numerically controlled machines. See Fig. 47-5. These machines are built for turning large diameter workpieces such as punch press flywheels and gear blanks. The overhead turret is mounted on a cross rail so that it can be used for facing cuts, and the entire cross rail moves vertically for boring and turning

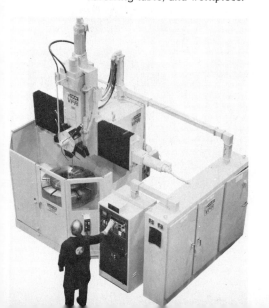

operations. Some machines also have a side turret, enabling the outside diameter to be machined at the same time the overhead turret is being used for facing or boring.

Automatic lathes are designed for efficient high-volume production of quantities that run from several hundred into the tens of thousands and beyond. Figure 47-6 shows a **single-spindle, turret-type** automatic lathe designed for making parts from standard lengths of bar stock. In addition to the six-station turret, two or more cross slides are provided for profile and cutoff operations. All tool movements and stock feeding are controlled either by mechanical devices, such as cams and trip dogs, or by hydraulic or electromechanical systems. Numerically controlled automatic lathes are also now available. Amazingly versatile attachments enable these machines to perform operations not normally performed on lathes. Operations which are routinely performed include broaching, drilling holes crosswise into the workpiece, and transferring the cutoff workpiece to an auxiliary work station for milling screw slots, or other operations that cannot be performed until after the workpiece is

cut off. Cycle times are very fast, often amounting to less than ten seconds per part.

Single-spindle automatic chucking machines are designed for efficient production of parts which must be held in a chuck. These machines are made in many sizes. Although all tool movement, and speed and feed changes are controlled automatically, parts are usually loaded and unloaded manually. These machines are built with numerical controls or with electromechanical or hydraulic control systems. Figure 47-7 shows a conventional machine with a five-station turret and two cross slides. The numerically con-

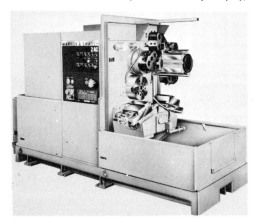

Fig. 47-7.   Single-spindle automatic chucking lathe.

Fig. 47-6.   Single-spindle cam-operated automatic screw machine.

Fig. 47-8.   Numerically controlled single spindle automatic chucking lathe.

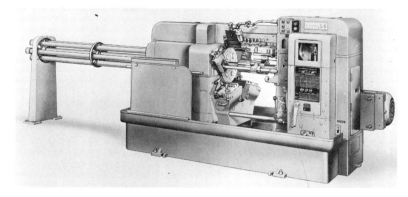

Fig. 47-9. Six-spindle automatic lathe equipped for bar work.

(Warner and Swasey Company)

trolled machine shown in Fig. 47-8 has a twelve-station turret which can move crosswise as well as endwise, thus eliminating the need for cross slides.

Horizontal multiple-spindle automatic lathes are made in four-, six-, and eight-spindle versions. Figure 47-9 shows a six-spindle machine. Instead of the tools being mounted in a revolving turret on this type machine, the spindles are mounted in a turret called a **spindle carrier.** Each spindle carries a bar of material. The end working tools are mounted on a reciprocating slide, one tool station opposite each spindle. Cross slides are also provided for profile work and stock cutoff. At the first station on a six-spindle machine, the bar is automatically unclamped, pushed out against a preset stop, and reclamped. The end working slide automatically comes forward, performing machining operations at each spindle, and the cross slide at station six cuts off a finished workpiece. After the end working slide retracts, the spindle carrier revolves one station, and the cycle is repeated. Each time the spindle carrier indexes one station, a finished workpiece is cut off, and a new piece is started. Cycle times are similar to those obtained with single-spindle automatic lathes.

Vertical multiple-spindle automatic lathes, Fig. 47-10, require a minimum of floor space and are easier to set up because the work stations are more accessible than on horizontal machines. The cutting tools are mounted on vertical slides around a central column and move independently of each other. The method of operation is much the same as for horizontal multiple-spindle machines. A workpiece is finished and a new one

(Bullard Company)

Fig. 47-10. Vertical multiple-spindle automatic turret lathe.

is loaded each time the spindle carrier is indexed. Vertical automatic lathes are made with from six to sixteen spindles, depending on the capacity of the machine. Some machines have two spindles at each work station, so that two workpieces are finished each time the spindle carrier is indexed.

Unit

48

Methods of Holding Workpieces for Machining

Lathe chucks and faceplates are devices for mounting and holding work while it is being machined on a lathe. When work is mounted on a faceplate, it usually is clamped with bolts and metal straps. The plate is equipped with slotted holes to allow for work of varying size and shape. Figures 48-1 and 48-2 show work mounted ready for machining.

Chucks

Chucks hold work by means of jaws which may be adjusted to accommodate work of varying size or shape. The jaws of the four-jaw independent chuck, Fig. 48-3B, move independently. This enables the chuck to hold work which is round, square, rectangular, or irregular in shape. As pictured, the jaws are installed for holding work of fairly large diameter. They should be removed and reversed for holding work of small diameter.

All jaws of the three-jaw universal chuck in Fig. 48-3 operate as a unit, opening or closing together, automatically centering the workpiece. This makes it possible to center workpieces easily and rapidly. Two sets of jaws normally are included with the universal chuck because the jaws are not reversible. Universal chucks also are available with six jaws.

Universal three-jaw and six-jaw chucks are intended primarily for holding round workpieces. They hold work accurately to within 0.05 or 0.8 mm (0.002" or 0.003") and retain this accuracy until either the jaws or the internal threads become worn or damaged.

(South Bend Lathe, Inc.)

Fig. 48-1.   Work mounted on faceplate with metal straps.

(South Bend Lathe, Inc.)

Fig. 48-2.   Work bolted on faceplate.

(South Bend Lathe, Inc.)

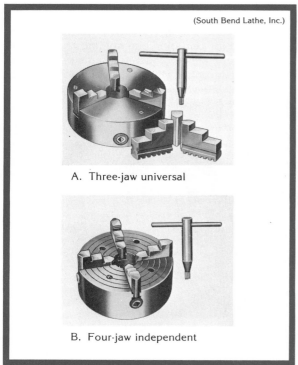

A. Three-jaw universal

B. Four-jaw independent

Fig. 48-3.   Chucks.

(South Bend Lathe, Inc.)

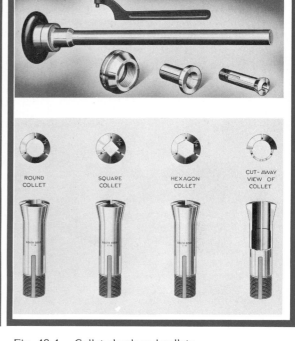

ROUND COLLET   SQUARE COLLET   HEXAGON COLLET   CUT-AWAY VIEW OF COLLET

Fig. 48-4.   Collet chuck and collets.

The four-jaw independent chuck is more accurate than the universal chuck, since the workpiece may be centered exactly with the use of a dial indicator.

Collet Chuck

The collet chuck is used widely in production work because it centers accurately, holds very tightly, and does not mark the workpiece. Collets are made in every standard bar stock size and are designed for holding round, square, or hexagon stock, Fig. 48-4.

A spring collet should be used only for the size of stock it is designed to hold. To avoid danger of breakage, it should not be used for holding stock that is more than 0.08 mm (0.003″) oversize or undersize.

Figure 48-5 shows an assembly of one form of draw-in collet chuck. The chuck is drawn tightly

(South Bend Lathe, Inc.)

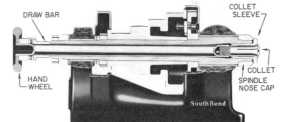

DRAW BAR   COLLET SLEEVE   HAND WHEEL   COLLET   SPINDLE NOSE CAP   SouthBend

Fig. 48-5.   Assembly of draw-in collet chuck.

into the collet sleeve by turning the handwheel at the rear end of the drawbar toward the right, and loosened by turning the wheel toward the left.

Before installing a collet chuck, always clean the hole in the spindle, the collet sleeve, and the collet. Grit or chips left between mating surfaces can spoil the accuracy of the chuck and the parts held in it.

Spindle Chuck

The spindle chuck, Fig. 48-6, resembles an ordinary drill chuck, except that it is designed to screw onto the nose of the lathe spindle. The movable jaws are tightened with a pinion key. A spindle chuck is used primarily for chucking small round work, which it does quickly and accurately within 0.05 to 0.08 mm (0.002″ to 0.003″).

(South Bend Lathe, Inc.)

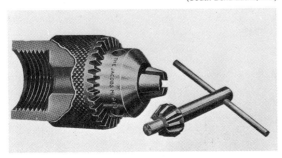

Fig. 48-6.   Spindle chuck.

Step Chuck and Closer

The step chuck and closer unit, Fig. 48-7, is designed for centering and holding small round discs. It operates on much the same principle as the draw-in collet chuck. Small diameter step chucks fit directly into the collet sleeve and do not require a separate closer.

(South Bend Lathe, Inc.)

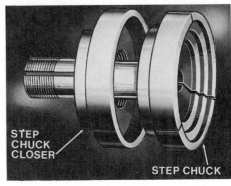

Fig. 48-7.   Step chuck and closer.

Unit 49

Lathe Cutting Tools and Toolholders

For efficient machining on a metalworking lathe, the correct type of cutting tool, called a **tool bit,** must be used. A tool bit is also referred to as a **single-point** cutting tool. A variety of lathe cutting tools are shown in Figs. 49-1, 49-2, and 49-3.

For machining, the tool bit must be mounted in a toolholder, Fig. 49-4. A tool bit must be sharp and the cutting edge must be well supported. It must be the correct type for the job, and

it must be set in the right position in relation to the work being machined. The cutting angles must be ground correctly in order for the tool bit to cut efficiently.

Tool Bits

Tool bits are designed with many different shapes for special purposes. The principles involved in cutting metals with a single-point cutting tool apply to nearly all lathe, shaper, and

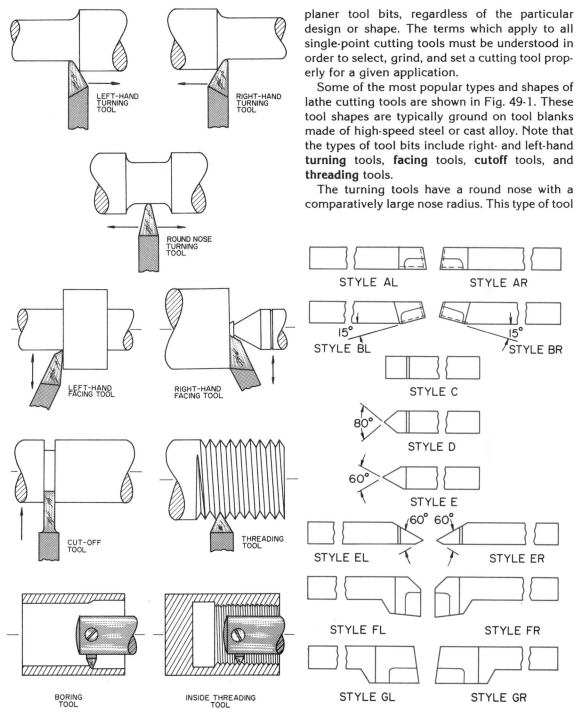

planer tool bits, regardless of the particular design or shape. The terms which apply to all single-point cutting tools must be understood in order to select, grind, and set a cutting tool properly for a given application.

Some of the most popular types and shapes of lathe cutting tools are shown in Fig. 49-1. These tool shapes are typically ground on tool blanks made of high-speed steel or cast alloy. Note that the types of tool bits include right- and left-hand **turning** tools, **facing** tools, **cutoff** tools, and **threading** tools.

The turning tools have a round nose with a comparatively large nose radius. This type of tool

LEFT-HAND TURNING TOOL

RIGHT-HAND TURNING TOOL

ROUND NOSE TURNING TOOL

LEFT-HAND FACING TOOL

RIGHT-HAND FACING TOOL

CUT-OFF TOOL

THREADING TOOL

BORING TOOL

INSIDE THREADING TOOL

Fig. 49-1. Commonly used high-speed steel lathe tool bits and their applications.

STYLE AL

STYLE AR

STYLE BL 15°

STYLE BR 15°

STYLE C

STYLE D 80°

STYLE E 60°

STYLE EL 60°

STYLE ER 60°

STYLE FL

STYLE FR

STYLE GL

STYLE GR

Fig. 49-2. Standard shapes for carbide-tipped single-point tools.

is designed primarily for finish turning with relatively light cuts. For rough turning with heavier cuts, a tool with a very small nose radius is used. A rough-turning tool also is designed with a wider nose angle for additional support of the cutting edge.

Standard shapes for **carbide-tipped** turning and facing tools are shown in Fig. 49-2. Carbide-tipped cutoff and boring tools are also available. These tools are resharpened as needed, using special silicon-carbide or diamond grinding wheels.

Standard shapes for **indexable throwaway inserts** made of carbide, ceramic, cermet, or diamond are shown in Fig. 49-3. Throwaway inserts are made so precisely that after the initial cutting edge has dulled, the tool may be unclamped, rotated to the next cutting edge, reclamped, and cutting continued without any change in the position of the toolholder. Regrinding of these tools is generally more costly than replacing them. Thus, they are discarded after all cutting edges have been dulled.

(Armstrong Bros. Tool Co.)

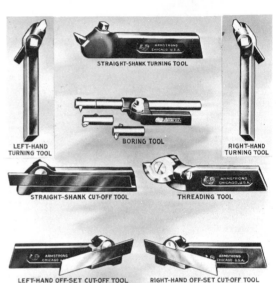

Fig. 49-4.   Commonly used lathe toolholders and cutting tools.

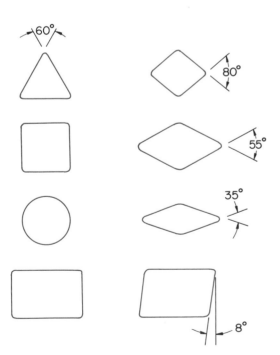

Fig. 49-3.   Standard shapes for indexable throwaway-insert cutting tools.

Toolholders

Various toolholders are available for holding tool bits on lathes, shapers, and planers. Some are standard types which may be used for many standard applications. See Fig. 49-4. Numerous toolholders are designed for special lathe applications. Some of the standard types of toolholders are explained briefly in this unit.

Shank Angle

Toolholders for turning and cutoff tools are available with three standard types of shanks: **straight, right-hand,** and **left-hand,** Fig. 49-4.

Left-hand toolholders permit machining operations close to the lathe chuck or faceplate. Right-hand ones are used for facing and for machining operations which are very close to the tailstock of the lathe. Straight-shank toolholders work best for machining long workpieces.

Types of Toolholders

The angle at which a turning tool bit is held in a toolholder, in relation to the base of the toolholder shank, is important. A standard turning toolholder (of the type shown in Fig. 49-4) holds the tool bit at an angle of about 16½°. This type of toolholder is intended for use with tool bits made of high-speed steel. It is called a **16½° toolholder** in this book.

The straight shank toolholder shown at the top of Fig. 49-4 holds the tool bit parallel to the base of the toolholder shank. This type is intended for holding carbide-tipped tool bits, but it also may be used for holding cast-alloy bits. In this book, it is called a **zero-degree toolholder.**

The toolholder shown at the top right in Fig. 49-4 is a **throwaway insert toolholder.** It is used to hold a carbide or ceramic insert cutting tool and chip breaker. The carbide chip breaker has a beveled edge and is clamped on the top of the cutting tool. The chip breaker may be replaced when badly worn. Many types of throwaway insert toolholders are made to hold different insert shapes in different positions.

Boring toolholders, Fig. 49-4, are used for boring operations on a lathe. Other typical types are shown in Unit 63.

Cutoff toolholders, Fig. 49-4, are used for holding cutoff tools in a lathe. Cutoff tools are used for cutting grooves (Fig. 49-1), cutting to a shoulder, or for cutting off stock.

Threading toolholders are used for holding commercially produced threading tools of the type shown in Fig. 49-4. Standard lathe tool bits also may be ground for thread cutting for use in standard turning toolholders. More information concerning threading tools and how to grind them is included in Unit 50.

Knurling tools, Fig. 49-4, are used for performing knurling operations. Information concerning knurling tools and knurling operations is included in Unit 61.

Many types of **quick-change toolholding systems** are now available. A specially designed toolpost provides for quick clamping and release of the toolholders without the need to handle loose wrenches. With a full assortment of toolholders, any turning operation can be quickly set up, Fig. 49-5.

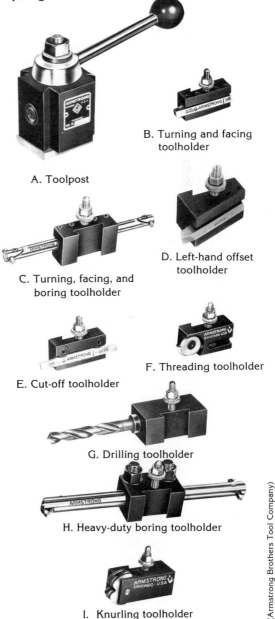

A. Toolpost

B. Turning and facing toolholder

C. Turning, facing, and boring toolholder

D. Left-hand offset toolholder

E. Cut-off toolholder

F. Threading toolholder

G. Drilling toolholder

H. Heavy-duty boring toolholder

I. Knurling toolholder

(Armstrong Brothers Tool Company)

Fig. 49-5. A quick-change lathe toolholder system.

Special toolholders of many kinds are available for use in performing various lathe, turret lathe, and screw machine operations.

Cutting Tool Materials

Cutting tools for metalworking are made of **high-speed steel, cast alloys,** and **cemented carbide.** They also may be made of **ceramics** and **diamonds** for special purposes. Cutting tools must be made of materials which possess special properties:

1. They must have sufficient hardness to cut other materials;
2. They must be capable of retaining hardness at the high temperatures which are produced at the cutting edge;
3. They must rank high in wear resistance; and
4. They must possess sufficient toughness (impact strength) to prevent chipping or fracturing.

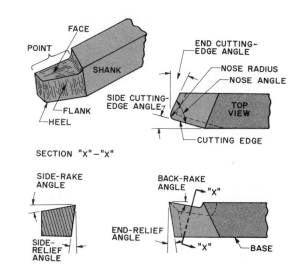

Fig. 49-6.   Cutting-tool angles and terms for lathe, shaper, or planer tools.

High-Speed Steel

High-speed steel tools are most commonly used in schools. They retain their hardness up to a dull-red temperature, about 538° C (1000° F). They generally rank higher in toughness, but lower in wear resistance than cast-alloy or cemented-carbide tools. All basic types of single-point tools, drilling tools, and milling cutters are available in high-speed steel. Suggested cutting speeds for high-speed steel, single-point cutting tools are listed for various materials in Table 49-1.

Cast Alloys

Cast alloys are used in the manufacture of solid tool bits, brazed tips on tool shanks, and as inserts in toolholders and milling cutters. Some common brand names for cast alloys include Stellite, Rexalloy, Armaloy, and Tantung. Cast alloys can withstand higher temperatures and higher cutting speeds than high-speed steel, but generally they are not as tough. They do not lose significant hardness at temperatures below 760° C (1400° F). They may be used at cutting speeds 50 to 70 percent higher than for high-speed steel tools, Table 49-1.

Cemented Carbide

Cemented carbide is used for brazed tips on single-point tools, Fig. 49-2. It is used for insert tools and chip breakers which are clamped in a toolholder, as shown at the top right of Fig. 49-4. It also is used for tips brazed on drills, reamers, or milling cutters. Cemented carbide is so named because the tiny particles used in its manufacture (usually tungsten carbide) are cemented together with cobalt.

Cemented carbide retains its hardness at temperatures higher than those for either high-speed steel or cast alloys. It does not lose significant hardness at temperatures below 927° C (1700° F). Therefore, cutting speeds from two to four times greater than for high-speed steel, may be used, Table 49-1. However, cemented carbides are much more brittle than high-speed steel, and, for this reason, they should be rigidly supported in the toolholder or other holding device. The machine and the work setup should be free from vibration. Zero-degree toolholders, as shown at the top left of Fig. 49-4 should be used. Interrupted cuts should be avoided when possible. Because of the pressure-welding characteristics of carbide tools, they should be used at recommended cutting speeds.

Table 49-1
Cutting Speeds for Metals and Plastics*

Material	Cutting Tool Material	Heavy Cut		Finishing Cut	
		mpm[1]	fpm[2]	mpm	fpm
Free-machining steels	H.S.S.	35.0	115	91.4	300
	cast alloy	76.2	250	144.8	475
	carbide	122.0	400	205.7	675
Low-carbon steels	H.S.S.	30.5	100	79.2	260
	cast alloy	65.5	215	129.5	425
	carbide	106.7	350	190.5	625
Medium-carbon steels	H.S.S.	29.0	95	68.6	225
	cast alloy	58.0	190	106.7	350
	carbide	91.4	300	152.4	500
High-carbon steels	H.S.S.	24.2	80	61.0	200
	cast alloy	53.3	175	91.4	300
	carbide	76.2	250	137.2	450
Cast iron, soft gray	H.S.S.	24.4	80	41.1	135
	cast alloy	42.7	140	76.2	250
	carbide	68.6	225	122	400
Brass and bronze free machining	H.S.S.	53.3	175	106.7	350
	cast alloy	106.7	350	167.6	550
	carbide	175.3	575	274.3	900
Aluminum	H.S.S.	38.1	125	91.4	300
	cast alloy	53.3	175	114.3	375
	carbide	76.2	250	182.9	600
Plastics	H.S.S.	30.5	100	76.2	250
	cast alloy	45.7	150	114.3	375
	carbide	61.0	200	152.4	500

*Speeds should be adjusted ± 10-20% to suit cutting conditions
1. Meters per minute.
2. Feet per minute.

Further information concerning the above cutting-tool materials, as well as ceramic and diamond cutting tool materials, is included in Unit 117, "Cutting Tool Materials."

Single-Point Cutting Tool Terms

The following are some of the most important terms, definitions, and angles which apply to all single-point cutting tools, including those used on lathes, shapers, and planers. Unless otherwise indicated, refer to Fig. 49-6 for illustrations of each term.

The term **tool bit** commonly is applied to relatively small pieces of cutting tool material which are inserted in a toolholder or tool shank in a manner that permits easy removal for regrinding or replacement. (See Fig. 49-4.)

The **cutting edge** is the part of the tool bit that does the actual cutting.

The **face** is the top surface of the tool upon which the chips bear as they are removed from the workpiece and slide away.

The **flank** of the tool is the surface adjacent to and just below the cutting edge.

(South Bend Lathe, Inc.)

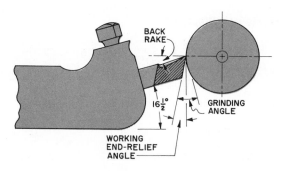

Fig. 49-7. Cutting edge set on center with 16-1/2° toolholder.

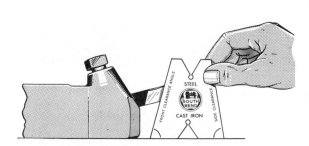

Fig. 49-9. Checking relief (clearance) angles on tool bit.

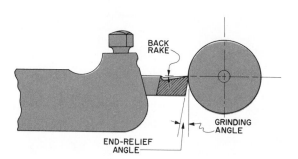

Fig. 49-8. Cutting edge set on center with zero-degree toolholder.

The **nose** is the corner or arc which joins the side cutting edge and the end cutting edge.

The **nose radius** is the dimension of the round arc which forms the nose of the tool bit. For rough turning, a small nose radius — usually about 0.4 mm (1/64″) — is used. For finish turning, a radius from 0.4 mm to 1.6 mm (1/64″ to 1/16″) is used, depending on the size of the tool. A turning tool with a nose radius of 0.8 mm (1/32″) will produce a satisfactory finish for general rough or finish turning.

The **shank** is the body portion of the tool.

The **point** of the tool includes all of that portion of the tool which is shaped to produce the face and the cutting edges.

The **base** of the tool is that portion of the tool which bears against the supporting toolholder or tool block.

A lathe **tool bit** is designated right-hand or left-hand, depending on the direction in which it cuts;

see Fig. 49-1. A **right-hand** tool has its cutting edge on the left, and it cuts from right toward left. A **left-hand** tool has its cutting edge on the right, and it cuts from left toward right.

Tool Angles

Toolholder Angle

The toolholder angle is measured between the bottom of the tool bit slot and the base of the toolholder shank. Toolholders are designed to hold tool bits at either fixed or variable angles. A standard 16½° toolholder commonly used for high-speed steel bits is shown in Figs. 49-7 and 49-9. A zero-degree toolholder commonly used for cast-alloy and cemented-carbide tool bits is shown in Fig. 49-8. Shaper and planer tool-holders usually support the tool at a zero-degree angle, that is, parallel to the base of the tool-holder. The angle at which the tool bit is held in the toolholder must be considered before grinding a tool bit.

Side-Relief Angle

The side-relief angle is ground into the flank below the cutting edge. The angle is measured between the ground flank and a line passing through the cutting edge perpendicular to the base of the tool or the toolholder, Fig. 49-6. The side-relief angle allows the cutting edge to penetrate into the metal and promotes free cutting by preventing the side flank of the tool from rubbing against the work.

End-Relief Angle

The end-relief angle is ground below the nose of the tool. It is measured between this surface and a line passing through the nose cutting edge perpendicular to the base of the tool bit or the toolholder, Figs. 49-6 and 49-8. The end-relief angle permits free cutting by preventing the flank below the nose cutting edge from rubbing against the work.

Working Angles

Working angles are located between the tool and the work. In addition to the shape of the tool, these angles depend upon the location of the tool in relation to the work.

Working Relief Angle

The working relief angle is the angle formed between the ground flank of the tool and a line passing through the active cutting edge tangent to the machined surface, Fig. 49-7.

Comparison of the end-relief angle measured at the tool bit (grinding angle) and the working end-relief angle in Fig. 49-7 makes their differences apparent. The **grinding angle** is the angle which must be ground on the tool bit in order to produce the proper **working** end-relief angle.

End-relief angles may be measured with a tool-angle gage of the type shown in Fig. 49-9. In this figure, the **working** end-relief angle of the tool is being checked, since the tool is mounted at an angle in the toolholder.

When a tool bit is placed in a zero-degree tool-holder in a lathe, and when the tool is mounted on the center line of the work (Fig. 49-8), the end-relief angle, the **working** end-relief angle, and the grinding angle are equal. This also is true for shaper and planer tool bits which are mounted with the base of the tool bit parallel to the base of the toolholder.

Amount of Relief Angle

The purpose of the end- and side-relief angles is to permit free cutting by preventing the flanks below the cutting edge from rubbing against the work. For different metals, the working relief angles vary from about 3° to 15°. The **amount** of

relief angle depends on the following factors:

1. Kind of material being cut.
2. Hardness of the material being cut.
3. Kind of cutting-tool material.
4. Position of the tool in relation to the work.
5. The nature of the cut.

Only the required amount of end or side relief should be used. Excess relief angle reduces the support under the cutting edge and weakens the tool, thus reducing tool life. Hard materials such as high-carbon steel or hard cast iron require smaller relief angles than soft and more ductile materials. Also, smaller relief angles are used with the harder and more brittle cast-alloy or tungsten-carbide tools than with the tougher high-speed steel tools. Since the working relief angle is reduced when a lathe tool is positioned above the center line of the work, an increased end-relief grinding angle is required in order to provide adequate clearance.

The nature of the cut also determines the amount of end and side relief used. For interrupted cuts, such as those required for turning shafts with keyways or irregularly shaped objects, small relief angles are used. Shaper and planer tools, therefore, should be provided with end- and side-relief angles of 3° to 5°.

The suggested working end-relief and side-relief angles for single-point, high-speed steel cutting tools for average uses are given in Table 49-2. End- and side-relief angles from 8° to 10° are fairly standard for turning many common metals with high-speed steel tools. In many shops, it is the practice to grind both the end and side relief at the same angle. Table 49-3 gives recommended angles for carbide tools.

For general-purpose turning applications, the following working end- and side-relief angles may be used: 10° for high-speed steel, 7° for cast alloy, and 7° for tungsten-carbide tools.

Further information concerning cutting-tool angles for single-point cutting tools is included in standard handbooks for machinists.

Back-Rake Angle

The back-rake angle is measured between the face of the tool bit and a line perpendicular to the work at the cutting edge, Figs. 49-7 and 49-8. This angle depends largely on the position at which the tool is held. In the case of a tool bit held

Table 49-2
Average Tool Angles for Single-Point High-Speed Steel Tools

Material	Side[1] Relief	End[2] Relief	Side[3,4] Rake	True Back[4] Rake
Free-machining steel	10°	10°	10°-22°	16°
Low-carbon steel (.05%-.30%)	10°	10°	10°-14°	16°
Medium-carbon steel (.30%-.60%)	10°	10°	10°-14°	12°
High-carbon tool steel (.60-1.70%)	8°	8°	8°-12°	8°
Tough alloy steel	8°	8°	8°-12°	8°
Stainless steel	8°	8°	5°-10°	8°
Stainless steel, free-machining	10°	10°	5°-10°	16°
Cast iron, soft	8°	8°	10°	8°
Cast iron, hard	8°	8°	8°	5°
Cast iron, malleable	8°	8°	10°	8°
Aluminum	10°	10°	10°-20°	35°
Copper	10°	10°	10°-20°	16°
Brass	10°	8°	0°	0°
Bronze	10°	8°	0°	0°
Molded plastics	10°	12°	0°	0°
Plastics, acrylics	15°	15°	0°	0°
Fiber	15°	15°	0°	0°

[1]End- and side-relief angles from 3° to 5° generally are recommended for shaper and planer tools.
[2]End-relief and side-relief angles averaging 8° to 10° are fairly standard for turning most metals. For general machining operations, both side- and end-relief angles often are equal.
[3]Use the lower angle when no chip breaker is used. Use the higher angle with a chip breaker.
[4]Rake angles are true angles measured from horizontal and vertical planes.

in a zero-degree toolholder, with the tool parallel to the base of the holder, the back-rake angle is measured between the face of the tool and a line parallel to the top of the tool.

When the tool is in an angular toolholder, such as a 16½° toolholder, the back rake is established largely by the toolholder. The back rake then is measured between the face of the tool and a line perpendicular to the work at the cutting edge.

An increase in rake angle increases the shear angle at the chip, thereby reducing the cutting force and power required. However, the increase in rake also reduces the cutting angle of the tool and thereby reduces the amount of material which supports the cutting edge.

Generally, small rake angles are used for machining hard materials, while steeper rake angles are used for more ductile materials. Exceptions to this rule include tools for brass, bronze, certain plastics and nonmetals.

Back-rake angles may vary from 0° to 35° for various applications. Suggested rake angles for single-point, high-speed steel tool bits are included in Table 49-2.

For a majority of the general machining applications encountered in the school shop, the back-rake angle established by a 16½° toolholder will produce satisfactory results with high-speed steel tools. When little or no back-rake angle is desired with high-speed steel tools, a zero-degree toolholder should be used.

Table 49-3
Recommended Angles for Single-Point Carbide Tools

Material	Normal end-relief, degrees	Normal side-relief, degrees	Normal back-rake, degrees	Normal side-rake, degrees
Aluminum and magnesium alloys	6 to 10	6 to 10	0 to 10	10 to 20
Copper	6 to 8	6 to 8	0 to 4	15 to 20
Brass and bronze	6 to 8	6 to 8	0 to −5	+8 to −5
Cast iron	5 to 8	5 to 8	0 to −7	+6 to −7
Low-carbon steels up to SAE 1020	5 to 10	5 to 10	0 to −7	+6 to −7
Carbon steels SAE 1025 and above	5 to 8	5 to 8	0 to −7	+6 to −7
Alloy steels	5 to 8	5 to 8	0 to −7	+6 to −7
Free-machining steels SAE 1100 and 1300 series	5 to 10	5 to 10	0 to −7	+6 to −7
Stainless steels, austenitic	5 to 10	5 to 10	0 to −7	+6 to −7
Stainless steels, hardenable	5 to 8	5 to 8	0 to −7	+6 to −7
High-nickel alloys (Monel, Inconel, etc.)	5 to 10	5 to 10	0 to −3	+6 to +10
Titanium alloys	5 to 8	5 to 8	0 to −5	+6 to −5

Side-Rake Angle

Side rake is the angle ground **across** the top of the tool face and is measured between the tool face and a line which represents the top of the unground tool as it is viewed from the end, Fig. 49-6. By providing a shearing action for chip removal, this angle enables the tool to cut more freely.

For side-cutting tools, the side-rake angle is much more important than the back-rake angle. It weakens the tool less than a steep back-rake angle. The side-rake angle largely determines the type of chip produced and the direction at which the chip leaves the tool face.

A steep side-rake angle causes long wire-like chips on ductile materials. **This type of chip is a safety hazard.** A decreased side-rake angle causes the chip to curl up and break off more readily.

Side-rake angles vary from 0° to 22° or more for different applications. Suggested averages for machining various materials with high-speed tools are shown in Table 49-2.

These side-rake angles are listed within a range, such as 10° to 22° for free-machining steel. For steels of highest machinability, the ideal angle in this case would approach 22°. However, with the steeper angle, a chip breaker generally is required to cause the chips to curl up and break off readily. (Chip breakers are described later in this unit.) If a chip breaker is not used, the lower angle (in this case 10°) generally should be used.

Angle of Keenness

The angle of keenness is the included angle of the tool between the face of the tool and the ground flank adjacent to the side-cutting edge, Fig. 49-10.

Side Cutting-Edge Angle

The side cutting-edge angle is formed by the straight side cutting edge and a line representing

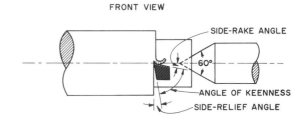

Fig. 49-10. Application of side-relief and side-rake angles on lathe tool bit.

the side of the tool shank before grinding, Fig. 49-6. This angle may vary from 0° to 30° for machining various materials. Angles of more than 30° tend to cause tool chatter. An angle of 15° generally is used for rough turning, while one of 20° produces good results for general machining applications.

End Cutting-Edge Angle

The end cutting-edge angle, Fig. 49-6, is formed by the end cutting edge of the tool and a line at right angles to the straight side of the tool shank. This angle may vary from 7° to 30°.

An average angle of 15° is used for rough turning. An angle of 30° produces good results for general turning applications. The 30° angle also permits the toolholder to be swiveled when turning close to shoulders or when turning close to the lathe dog or faceplate. Increased end cutting-edge angles decrease the material which supports the nose and the end cutting edge of the tool, thus reducing tool life when heavy cuts are made.

Nose Angle

The nose angle is the included angle between the side cutting edge and the end cutting edge of the tool bit. See Fig. 49-6.

Chip Breakers

When relatively steep side-rake angles are used on single-point cutting tools, ductile materials are cut more freely. However, as previously mentioned, with the steep angles, long continuous chips are formed. This type of chip is a **safety hazard** to the machine operator, and the chips are more difficult to remove from the machining area.

A **chip breaker** causes the chip to coil up tightly and break off readily, thus removing the safety hazard. The short, broken chips occupy less space, and also permit better flow of cutting fluid to the tool point.

Three common types of chip breakers used on cutting tools are shown in Fig. 49-11. In order to grind chip breakers of this type accurately, a tool and cutter grinder is used. A surface grinder with a special compound-angle tool-holding fixture also may be used for this purpose.

With the **angular-shoulder chip breaker** shown at **A** in Fig. 49-12, the angle **Y** may vary from about 5° to 15°, with 7° or 8° being the average. The width **W** and depth **D** depend on the feed, speed, depth of cut, and kind of material. The width **W** at the end of the tool usually varies from 1.6 mm (1/16″) to 4.8 mm (3/16″), and the depth, from 0.4 mm (1/64″) to 1.6 mm (1/16″). Tools with a large nose radius have a secondary angle as shown at **B**. The width **Z** should be about one-and-one-half times the nose radius.

The **groove chip breaker** (C, Fig. 49-11) has a groove ground parallel to the side-cutting edge of the tool. For average applications, the following dimensions may be used: **E,** 0.8 mm (1/32″); **F,** 1.6 mm (1/16″); and depth of groove, 0.8 mm (1/32″).

A removable **tungsten-carbide chip breaker** with a beveled edge is shown at the top right of Fig. 49-4. The chip breaker is clamped above the cutting face of the tungsten-carbide throwaway insert.

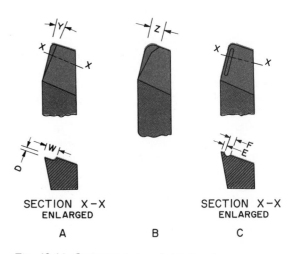

SECTION X–X ENLARGED
A

B

SECTION X–X ENLARGED
C

Fig. 49-11. Common types of chip breakers.

Unit

50

Sharpening Lathe and Shaper Tools

Accurate work with a good finish cannot be done on the lathe or shaper unless the cutters are sharp and are ground at the correct angle. A soft or medium-soft aluminum oxide grinding wheel is recommended for grinding high-speed steel cutting tools. See Table 99-2. A standard bench- or floor-model tool grinder of the type shown in Fig. 28-1 may be used, or a tool grinder with a tilting table may be used, Fig. 28-2.

For grinding cast-alloy tool bits, a tool grinder with a tilting table should be used. The grinder should be equipped with an aluminum-oxide grinding wheel. The tool may be ground wet or dry, **but it should not be quenched after grinding dry.**

For grinding tools with cemented-carbide tips, a tool grinder with a tilting table, Fig. 50-1, should be used; however, it should be equipped with either a silicon-carbide or diamond grinding wheel. Aluminum-oxide grinding wheels are too soft for grinding carbide tools. More information concerning the selection of grinding wheels for grinding carbide tools is included in Units 99 and 117.

(Union Carbide Corporation — Stellite Div.)

Fig. 50-1.   Table of tool grinder tilted to correct grinding angle.

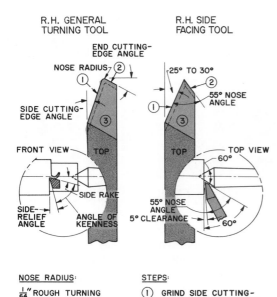

R.H. GENERAL TURNING TOOL

R.H. SIDE FACING TOOL

END CUTTING-EDGE ANGLE

NOSE RADIUS ②

25° TO 30°

②

55° NOSE ANGLE

SIDE CUTTING-EDGE ANGLE

①

③

①

③

FRONT VIEW

TOP

TOP

TOP VIEW

60°

SIDE RAKE

SIDE-RELIEF ANGLE

ANGLE OF KEENNESS

55° NOSE ANGLE

5° CLEARANCE

60°

NOSE RADIUS:

$\frac{1}{64}$" ROUGH TURNING

$\frac{1}{32}$" GENERAL TURNING

$\frac{1}{16}$" FINISH TURNING

STEPS:

① GRIND SIDE CUTTING-EDGE ANGLE WITH SIDE RELIEF

② GRIND END CUTTING-EDGE ANGLE AND NOSE RADIUS WITH END RELIEF

③ GRIND SIDE-RAKE AND BACK-RAKE ANGLE

Fig. 50-2. Grinding the two most-used tool bits, using inch measurements.

Turning and Facing Tools

The two most-used high-speed steel tool bits for lathes are the **right-hand turning tool** and the **right-hand side-facing tool.** The suggested shape and the suggested grinding angles for these two tools for general-purpose work are shown in Figs. 50-2 and 50-3. A left-hand tool ground for use in a zero-degree toolholder, as it should be for use in a shaper, is shown in Fig. 50-4. The angles should be altered for machining different metals as suggested in Table 49-2.

Radius Tools

Radius tools are used to cut a **fillet** of a given radius at a shoulder, to cut a radius in the bottom of a groove, or to round the corner on the end of a shaft, Fig. 50-5. During grinding, the accuracy of the tool shape is checked with a radius gage, Fig. 5-15. Radius tools should be ground with less side and end relief for maximum support of the cutting edge, and without back or side rake to prevent chatter. Both the tool and workpiece must be rigidly supported. Low cutting speeds, moderate feed, and lubrication are important to obtaining good results.

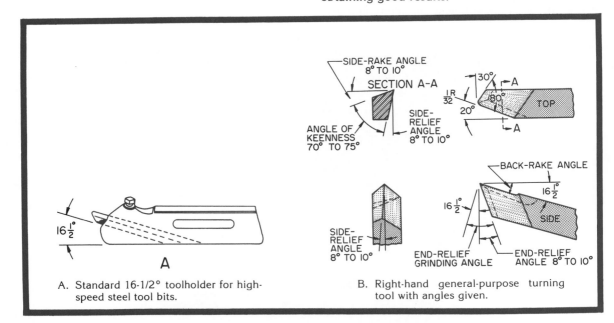

SIDE-RAKE ANGLE 8° TO 10°

SECTION A-A

30°

A

$\frac{1}{32}$ R

80°

TOP

SIDE-RELIEF ANGLE 8° TO 10°

20°

A

ANGLE OF KEENNESS 70° TO 75°

BACK-RAKE ANGLE

$16\frac{1}{2}$°

$16\frac{1}{2}$°

SIDE

SIDE-RELIEF ANGLE 8° TO 10°

END-RELIEF GRINDING ANGLE

END-RELIEF ANGLE 8° TO 10°

$16\frac{1}{2}$°

A

A. Standard 16-1/2° toolholder for high-speed steel tool bits.

B. Right-hand general-purpose turning tool with angles given.

Fig. 50-3. Tool point geometry.

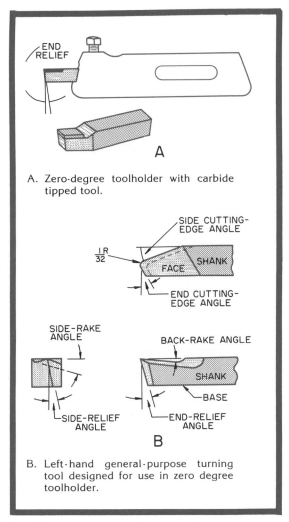

A. Zero-degree toolholder with carbide tipped tool.

B. Left-hand general-purpose turning tool designed for use in zero degree toolholder.

Fig. 50-4.   Terms describing tool point geometry.

Cutoff Tool

The cutoff or **parting** tool, Fig. 50-6, is used for cutting grooves or for cutting off stock. Cutoff tool blades which are mounted in special cutoff toolholders, Fig. 49-4, are widely used.

Some cutoff tool blades are designed with beveled sides which provide several degrees of side relief on each side, Fig. 50-6. Cutoff blades are also made with a "T" cross section. The blade is ground on the front and top only, never on the sides. An end-relief angle from 3° to 10° may be used; an angle of 5° produces good results for general-purpose cutoff work.

No side rake is used for general-purpose cutoff work. No back rake is required, but an angle of 0° to 5° sometimes is used.

A cutoff tool may also be ground from a standard lathe tool bit, Fig. 50-7. The angles X and Y should be very small — just enough to clear the sides of the groove being cut.

Grinding an angle of 10° to 15° across the end of the parting tool will allow parts to be cut off more cleanly than if the end of the blade is ground square, Fig. 50-8.

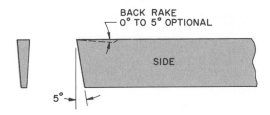

Fig. 50-6.   Cutoff tool blade.

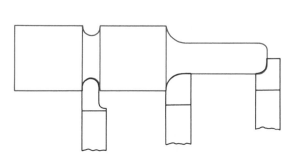

Fig. 50-5.   Applications of radius tools.

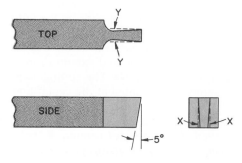

Fig. 50-7.   Cutoff tool ground on tool bit.

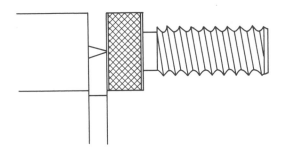

Fig. 50-8.   Angle ground on end of cutoff tool makes clean cutoff.

(Armstrong Bros. Tool Co.)

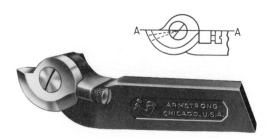

Fig. 50-10. Threading tool.

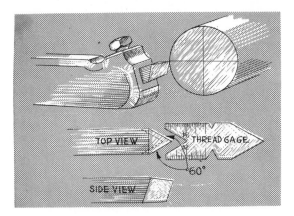

Fig. 50-9.   Cutting angles for threading tools.

(Armstrong Bros. Tool Co.)

Fig. 50-11. Threading tool for internal threading.

Cutoff operations may be performed only on work that is mounted in a chuck. **Never attempt to cut off stock that is mounted between centers.** To do so will cause the work to bend, bind on the sides of the tool, and fly out of the lathe before the piece is cut off.

For cutoff operations, the work and the tool must be mounted rigidly. Locate the cutoff tool as close to the chuck as possible. The tool should extend from the toolholder as little as possible, and the toolholder should extend from the tool post as little as possible. The cutting edge of the tool should be mounted at the center line of the work.

Cutting speeds used should be about one-third less than those used for turning. If tool chatter develops, it may be reduced or eliminated by reducing the cutting speed or increasing the feed rate. The tool should be fed by hand until ex-

perience is acquired. Use of a cutting fluid is particularly helpful for cutting off tough materials or large diameters.

Threading Tools

Threading tools for metric, Unified, or American (National) form threads are ground V-shaped with an included angle of 60°. The tool bit is given from 8° to 10° end relief and from 8° to 10° side relief on each side, Fig. 50-9. The point of the bit generally is provided with a radius or a flat according to the type of thread it is to cut. The cutting edge of the tool is set at the height of the center line of the work, with no working back-rake angle required, Fig. 50-9.

A threading toolholder with a formed threading tool for V-threads is shown in Fig. 50-10. This type of threading tool should only be resharpened across the top of the tool. As necessary, the tool is repositioned on the holder by adjusting the set screw.

For internal threading, a threading tool bit is mounted in a boring bar as shown in Fig. 50-11. This particular boring bar holds the tool at a 30° angle. Boring bars which hold the tool at a 90° angle also are used. With this type, the threading tool is ground in the same manner in which threading tool bits are ground for external thread cutting.

Procedure for Grinding Lathe and Shaper Tool Bits

1. Examine the faces of the grinding wheel for trueness and freedom from grooves. If necessary, dress all of the surface to be used until all ridges or grooves have been removed and the wheel is running true.

CAUTION
Protect the eyes against flying particles of loosened abrasive by using a shield and goggles.

2. If using a wet grinder, be sure there is sufficient fluid in the container. Turn on the grinder first, then turn on the coolant and allow it to run on the face of the wheel for a minute before grinding.
3. If a grinding holder (Fig. 50-12) is available, mount the tool bit in the holder. This procedure is much safer than holding the tool with your fingers.
4. The angles should be ground in the order suggested in Fig. 50-2, as follows:
   a. Grind the side cutting-edge angle with side relief.

(Armstrong Bros. Tool Co.)

Fig. 50-12. Grinding holder for holding tool bits while grinding.

   b. Grind the end cutting-edge angle and nose radius with end relief.
   c. Grind the side-rake and back-rake angles.
5. Grind the side cutting-edge angle, being sure to provide the proper side-relief angle. Check the angle with a tool-angle gage of a type similar to that shown in Fig. 49-8.

When the tool is to be used in a 16½° tool-holder, the actual or working side-relief angle is measured with the tool in the tool-holder.
6. Grind the end cutting-edge angle and nose radius, being sure to provide the proper end-relief angle.

If the tool is to be used in a 16½° tool-holder, the actual or working end-relief angle must be measured with the tool in the tool-holder.
7. When resharpening a tool that already has the correct shape and relief angles, bring the heel of the tool bit into contact with the wheel first (**A,** Fig. 50-13). Raise the rear of the tool until the flank being ground rests its full length on the surface of the wheel (**B,** Fig. 50-13). Holding the cutter at this angle, move it back and forth across the wheel. The nose radius is ground by rotating the tool around its point while maintaining the same angle.
8. A grinder with an adjustable tool rest is recommended for sharpening tools with cast alloy and carbide tips. Set the adjustable rest or table at the relief angle required, and clamp it in position. Place the cutter on the table, and then bring it against the face of the wheel.

Continue grinding until all parts of the cutting edge have been brought into contact with the wheel. Be sure to maintain the correct tool shape.
9. Remove the tool from the grinder, and carefully examine the cutting edge. If any previously worn bright spots are visible, again bring the part or parts of the cutter into contact with the grinding wheel. Continue grinding until all spots have been removed.
10. If necessary, grind the top of the cutter to give it the amount of rake recommended. Ordinarily, grinding on the top of a previously shaped tool bit should be kept at a minimum.

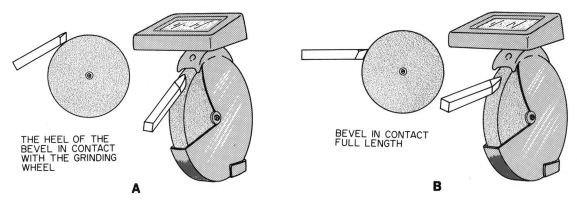

THE HEEL OF THE
BEVEL IN CONTACT
WITH THE GRINDING
WHEEL

BEVEL IN CONTACT
FULL LENGTH

A

B

Fig. 50-13. Grinding cutter bits.

Procedure for Honing Lathe and Shaper Tools

Honing is done either to provide a keener edge on a newly sharpened tool, or to "touch up" a tool just before a finishing cut. Cast alloy and high-speed steel tools can be honed by hand. Carbide tools require special equipment for honing.

CAUTION
Hand honing must be done carefully so as not to destroy the relief angles. Otherwise, more harm than good may be done.

1. Obtain a medium-soft oilstone, and apply a little machine oil to one surface.
2. Hold the oiled side of the oilstone flat against the tool flank, Fig. 50-14. Then move the oilstone up and down until the entire cutting edge has been honed. Bear more heavily on the downward stroke than on the upward stroke.
3. After the flanks of the tool have been honed, lightly hone the top of the tool.
4. Wipe the tool free of abrasive grit.

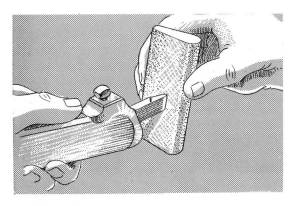

Fig. 50-14. Whetting a lathe tool bit.

Unit

51

Lathe Tool Setup

The height at which a lathe tool should be set depends upon the metal to be turned and, to some extent, upon the operation being performed. As a general rule, the point of a high-speed steel tool bit may be set up to about 5° above center, except when turning brass or copper or when turning a taper, cutting a thread, boring or cutting off stock. In all of these cases, the point of the tool should be at exactly the same level as the axis of the work.

When turning steel or cast iron of small diameter, the point of the tool should be set on or only very slightly above the axis of the work. For turning aluminum, a tool especially ground and sharpened for that purpose is recommended, with the point of the tool set considerably higher above the axis of the work than when turning steel.

The point of cast-alloy tools or cemented-carbide tools should be set exactly at the height of the center line of the work.

Regardless of the operation to be performed or the kind of metal being turned, always set the cutter well back into the toolholder, and position the holder itself so that it projects but slightly beyond the edge of the compound rest. Also, make certain that the toolholder is fastened securely in the tool post and the cutter in the holder. A loose cutter or toolholder may cause an accident or damage the machine or the work.

Procedure

1. Select a sharp tool of the shape and kind desired.
2. Insert the cutter in the toolholder, allowing it to project about 13 mm (½"), as in Fig. 51-1. Then tighten the setscrew that holds the cutter. Do not allow the cutter to project more than the recommended amount, unless conditions **absolutely** necessitate this.

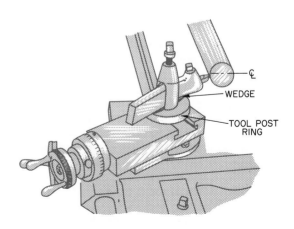

Fig. 51-1.   Adjusting toolholder in toolpost.

3. Insert the toolholder in the tool post, and adjust the point of the cutter to the height desired by sliding the wedge backward or forward in the tool post ring. See Fig. 51-1.
4. Clamp the toolholder in position by tightening the tool post clamp screw.
5. Test the height of the tool by running in the cross slide until the point of the tool is as close as possible to the point of the dead center, Fig. 51-2, or make a trial cut near the center of the end of the workpiece to establish its center.
6. If necessary, partially release the tool post clamp screw and tap the toolholder to move the point of the tool on center. Be sure to retighten the toolholder after making any adjustments.

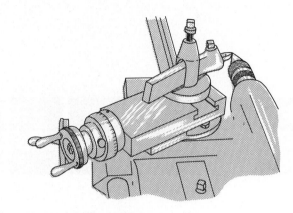

Fig. 51-2.   Testing height of tool.

Unit

52

Cutting Speeds for Lathe Work

Use of correct cutting speeds is important to good tool life and efficient machining. Excessively high cutting speeds will cause overheating of the tool and premature cutting edge failure. Use of cutting speeds that are too slow will reduce productivity and increase manufacturing costs.

For lathe work, cutting speed refers to the rate in meters per minute (mpm) or feet per minute (fpm) at which the surface of the workpiece moves past the cutting tool. Conditions that affect cutting speed include: kind of material being cut, kind of material the cutting tool is made of, shape of the cutting tool being used, rigidity of the workpiece, rigidity of the machine, and kind of cutting fluid being used.

Table 49-1 gives cutting speed recommendations for cutting common metals with high speed steel, cast alloy, and carbide cutting tools. Further data on cutting speeds for specific alloys and for specific machining applications are included in standard handbooks for machinists.

Calculating RPM to Obtain the Desired Cutting Speed

The usual problem facing the machine operator or numerical control programmer is to find the correct rpm to run the tool or workpiece. The tool or workpiece diameter, kind of material to be cut, and the kind of cutting tool material to be used are usually known. Obtaining the recommended cutting speed from a table or chart completes the data needed to calculate the correct rpm.

RPM Formulas:

Metric:

$$rpm = \frac{cutting\ speed\ (c.s.)\ (mpm) \times 1000}{D\ (mm) \times \pi}$$

Inch:

$$rpm = \frac{c.s.\ (fpm) \times 12}{D'' \times \pi}$$

Also, the following formula is easier to remember and use, and gives satisfactory results:

$$rpm = \frac{4 \times c.s.\ (fpm)}{D''}$$

EXAMPLE:

What rpm should be used for a heavy cut of 30.5 mpm (100 fpm) on a piece of low carbon steel of 50.8 mm (2″) diameter?

Metric Solution:

$$rpm = \frac{30.5 \times 1000}{50.8 \times 3.1416} = \frac{30500}{159.59328} = 191$$

Inch Solutions:

$$rpm = \frac{100 \times 12}{2 \times 3.1416} = \frac{1200}{6.2832} = 191$$

$$rpm = \frac{4 \times 100}{2} = \frac{400}{2} = 200$$

In some cases, it is necessary to determine the cutting speed which would be obtained under a given set of conditions. The formulas then become as follows:

Cutting Speed Formulas:

Metric:

$$c.s. = \frac{D\ (mm) \times \pi \times rpm}{1000}$$

Inch:

$$c.s. = \frac{D'' \times \pi \times rpm}{12}$$

or

$$c.s. = \frac{D'' \times rpm}{4}$$

Using the figures from the rpm problem above, cutting speeds are obtained as follows:

Metric Solution:

$$c.s. = \frac{50.8 \times 3.1416 \times 191}{1000} = \frac{30482.316}{1000}$$

$$= 30.5\ mpm$$

Inch Solutions:

$$c.s. = \frac{2 \times 3.1416 \times 191}{12} = \frac{1200.0912}{12}$$

$$= 100\ fpm$$

or

$$c.s. = \frac{2 \times 200}{4} = \frac{400}{4} = 100\ fpm$$

With the diameter of the work known and the cutting speed known, one can look up the approximate rpm from an rpm table such as Table A-10, Appendix. Tables of this type are often posted in machine shops. The competent machinist, however, must know how to calculate the correct cutting speed and rpm, since tables of cutting speeds usually are limited to small diameters.

Unit

53

Locating and Drilling Center Holes for Lathe Work

The approximate center of round and irregular pieces of stock may be found in a number of ways. The most commonly used of these are discussed in the following paragraphs.

Procedure with a Center Head

1. Apply layout dye or chalk to the ends of the workpiece.
2. Place the center head across one end of the piece, as shown in **A**, Fig. 53-1. Then carefully scribe a line across the chalked surface.
3. Give the square a third of a turn, and scribe a second line.
4. Give the square another third turn, and scribe a third line (**B**, Fig. 53-1).
5. Place the point of a sharp center punch at the center of the intersection of the lines and, holding the center punch in a vertical position, strike it a light blow with a hammer.

   Examine the position of the punch mark. If it is located in the center, punch it a little deeper.

   If the punch mark is not in the center of the workpiece, correct it before punching it deeper.

Procedure with Dividers

1. Apply layout dye or chalk to the end of the workpiece.
2. Set a pair of dividers so that the distance between the points is slightly more or less than half the diameter of the workpiece.
3. Lay the workpiece on a flat surface, preferably a surface plate, and hold it in position with one hand.

4. Place one leg of the dividers on the plate at the position where the workpiece lies, Fig. 53-2. Make a short arc near the center of the workpiece.
5. Give the piece a quarter turn, and repeat the operation.
6. Continue this until the short arcs form a four-sided figure (**A**, Fig. 53-2).
7. Place the point of a center punch at the center of this figure and strike it a light blow with a hammer.
8. Examine the position of the punch mark. If it seems to be located in the center of the four-sided figure, punch it to the depth desired.

   To check whether the punch mark is accurately located in the center of the workpiece,

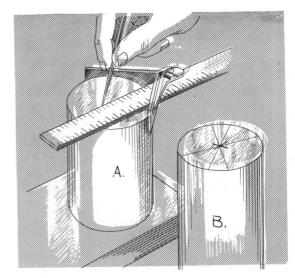

Fig. 53-1.   Finding the center with a center head.

set a pair of dividers to a distance equal to the radius of the stock. Place the point of one leg in the punch mark. Then rotate the point of the other leg around the circumference of the stock, observing how accurately the point follows the circumference.

Correct the position of the punch mark if necessary.

Procedure with Hermaphrodite Caliper

1. Chalk or apply layout dye to the ends of the piece.
2. Set the hermaphrodite caliper so that the distance between the point of the bent leg and that of the straight leg is about 1 mm (1/32″) greater than half the diameter of the stock.
3. Place the bent leg against the side of the stock **near the top edge.** Hold it firmly in this position with the left hand, and, with the straight leg of the caliper, draw a short arc on the chalked end of the stock, Fig. 53-3.
4. Move the bent leg of the caliper half of the distance around the circumference of the stock, and draw a second short arc. Then move the bent leg one-fourth of the distance and draw a third short arc. Continue thus until four short arcs have been drawn, Fig. 53-3.
5. Place the point of a sharp center punch at the center of the small enclosure made by the four arcs, and, holding the center punch in a vertical position, strike it a light blow.
6. With the hermaphrodite caliper, check the location of the center punch mark, as described in the paragraph following step 8 under "Procedure with Dividers."

If necessary, correct the position of the punch mark.

7. When the punch mark has been correctly located, punch it a little deeper.

Procedure with a Bell Centering Cup

1. With a file, remove any burrs or projections on the circumference of the stock at the end or ends to be centered. Then check the end(s) to be center-punched to be sure they are square with the axis of the stock.
2. Place the stock on its end on a flat solid surface, preferably a heavy surface plate.
3. Place the bell centering cup over the end of the work, as in Fig. 53-4. Be sure the barrel of the cup is held in a plane **parallel with the axis of the work.**
4. Strike the center punch or plunger a sharp blow with a hammer, thus locating the center. When necessary to increase the depth of the punch mark, do so with an ordinary center punch.

Fig. 53-3.   Finding the center with an hermaphrodite caliper.

(South Bend Lathe, Inc.)

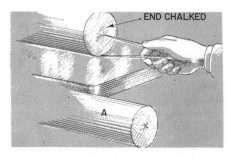

Fig. 53-2.   Finding the center with dividers.

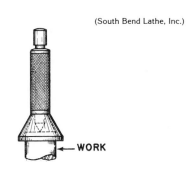

Fig. 53-4.   Bell centering cup.

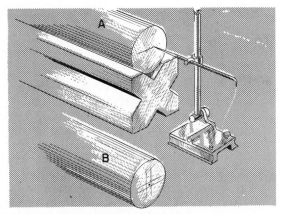

Fig. 53-5.   Finding the center with surface gage.

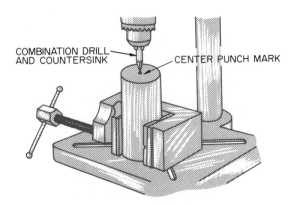

Fig. 53-7.   Drilling centers with combination drill and countersink.

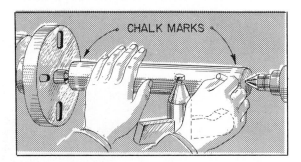

Fig. 53-6.   Testing centers with chalk.

Procedures with a Surface Gage

1. Apply layout dye or chalk to the ends of the workpiece.
2. Mount the stock in a V-block or V-blocks on a surface plate, Fig. 53-5.
3. Set a surface gage, and draw a line near the approximate center of the piece (**A**, Fig. 53-5).
4. Turn the workpiece a quarter of a turn or as far as its shape will permit without raising center. Draw a second line.
5. Repeat step 4 until a four-sided figure has been described (**B**, Fig. 53-5).
6. Place the point of a center punch in the center of this figure, and strike the punch a light blow with a hammer. Inspect to see that the punch mark is on center. Correct if necessary, then sink the punch mark to the depth desired with repeated blows.

Procedure for Testing Centers

1. Place the work between the lathe centers, Fig. 53-6.
2. Hold a piece of chalk close to the work at one end. The hand holding the chalk should be supported by resting the arm on the compound rest or other suitable support.
3. With one hand, revolve the workpiece, and, at the same time, advance the chalk until it just touches the revolving piece. The chalk will make a mark on the high side, or the point farthest from the center.
4. Repeat the operation on the other end of the piece.
   If the stock is accurately centered, the chalk will make a complete line around the workpiece.
5. If greater accuracy is required, use a **dial indicator** in place of the chalk, see Unit 10.

Procedure for Drilling Centers on the Drill Press

1. Locate the centers of the workpiece to be drilled with a center punch mark.
2. Insert a combination drill and countersink of suitable size in the drill press chuck, Fig. 53-7.
   For light work, a No. 2 combination drill and countersink is suitable. One of this size has a body diameter of about 7.9 mm (5/16").
3. Place the work in position on the drill press table and start the machine. The workpiece should be securely held in a V-block or a vise

so that it is as perpendicular as possible to the drill press table, see Fig. 53-7.

4. Draw the point of the drill into the impression made by the center punch, and drill until the tapered part of the drill has entered about **three-quarters of its full length.** If drilling steel, keep the drill well lubricated.

A, Fig. 53-8, shows a correctly drilled and countersunk hole. Notice that the lathe center fits the tapered hole accurately with clearance at the point. **B** is not countersunk at the correct angle, and **C** is drilled too deep to fit the center. When center-drilled and countersunk as at either **B** or **C**, the lathe center will be damaged.

Procedure for Drilling on the Lathe

If the workpiece can be accurately centered and securely held in a chuck or attached to a faceplate, it is not necessary to center-punch the hole location before center drilling. A center drill is installed in a drill chuck and mounted in the lathe tailstock, the workpiece being rotated for drilling.

Workpieces which cannot be accurately centered and securely held in a chuck or on a faceplate may be center drilled in the following manner:

1. Locate the centers of the stock, and mark with a center punch.
2. Insert the shank of the drill chuck in the headstock spindle.
3. Insert a combination drill and countersink in the drill chuck, and fasten it securely with a chuck wrench.
4. Hold the center mark against the point of the drill. Then draw the tailstock up to within 13 mm (½") of the other end of the work.

(South Bend Lathe, Inc.)

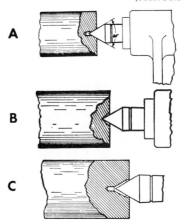

Fig. 53-8.   Good and poor center holes.

5. Tighten the nut that holds the tailstock to the ways of the lathe.
6. Place a little lubricant on the drill, and then start the lathe on medium or slow speed, about 700 rpm.
7. Hold the work against the point of the dead center with one hand, palm up, as in Fig. 53-9.
8. Feed the work forward slowly by turning the handwheel at the rear of the tailstock until the point of the drill enters the marked center.
9. Continue advancing the combination drill slowly and carefully, until the tapered part has entered the work about three-quarters of the length of the taper.
10. When drilled to depth, retract the tailstock spindle and work until the drill clears the hole. Then stop the lathe.

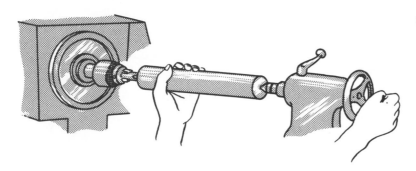

Fig. 53-9. Drilling center hole on the lathe.

Unit
54

Mounting Work between Lathe Centers

Mounting work between centers is a common method of holding a workpiece while it is being machined. To rotate the workpiece, a faceplate having an open slot on one side is mounted on the spindle. A lathe dog is mounted on the stock and engaged in the slot in the faceplate.

Procedure

1. Remove the chuck if one is mounted on the lathe spindle.
2. Clean the threads on the lathe spindle with a shop rag; then apply a few drops of lubricating oil.
3. Test the points of the lathe centers for alignment, as in Fig. 54-1. Both points should be in the same horizontal and vertical planes.
4. Select a faceplate of suitable size. Clean the threads and apply a few drops of lubricating oil.

5. Hold the hub of the faceplate squarely against the nose of the spindle with one hand; then rotate the faceplate clockwise as in Fig. 54-2. Continue rotating the pulley until the faceplate firmly contacts the shoulder of the spindle.

CAUTION
Do not allow the hub of the faceplate to strike hard against the shoulder of the spindle, as this may make it very difficult to remove the faceplate.

6. Thoroughly clean the openings in the headstock and tailstock spindles with a clean shop rag.

(South Bend Lathe, Inc.)

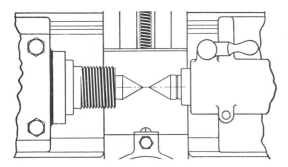

Fig. 54-1.   Testing lathe centers for alignment.

Fig. 54-2.   Mounting the faceplate.

Fig. 54-3.   Engaging the live center.

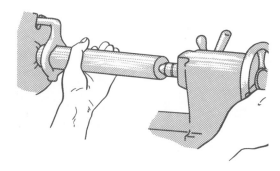

Fig. 54-4.   Engaging the dead center.

CAUTION
Never put your finger in the hole of the lathe spindle while it is revolving.

7. Wipe the centers free of dust, oil, and grit.
8. Insert the soft center in the headstock spindle and the hardened one in the tailstock. The hardened center usually has a groove cut near the cone end, Fig. 54-1.
9. Fasten the lathe dog on one end of the workpiece, with the bent tail pointing outward.

   If the workpiece where the dog is attached is a finished surface, insert a small piece of soft sheet aluminum between the end of the screw in the dog and the workpiece. This will prevent marring.

10. Place center lubricant, such as white lead and oil, in the center to be engaged by the dead (tailstock) center. **Note:** Whenever possible, use a live or revolving tailstock center.

11. Engage the center hole in the work with the point of the live (headstock) center, and with the tail of the lathe dog in the slot in the faceplate, Fig 54-3. Hold the work in this position with one hand.

CAUTION
Be sure the tail of the dog does not rest on the bottom of the slot in the faceplate, as that will prevent the center from entering the countersunk hole in the work correctly.

12. Grasp the tailstock with the right hand, and move it forward until the point of the dead center enters the center hole in the stock.
13. Fasten the tailstock to the lathe bed. **Note: Before clamping the tailstock in position, be sure there is sufficient room for the saddle and tool block to operate.**
14. Advance the tailstock spindle by turning the handwheel at the rear of the tailstock clockwise, Fig. 54-4, until the point of the dead center enters the countersunk hole in the end of the work and all motion endwise is eliminated.
15. Turn the handwheel back very slightly. Then rotate the workpiece by the tail of the dog back and forth, while adjusting the tailstock handwheel until only a slight resistance is felt. Tighten the tailstock spindle binding lever. The workpiece is now properly mounted and ready to run.

Unit

55

How to Mount and Remove Lathe Chucks

There are three types of spindle noses in use for attaching chucks and faceplates. They are (1) the **threaded** spindle nose, Fig. 55-1; (2) the spindle nose with the long taper **key drive,** Fig. 55-2; and (3) the spindle with the **cam-lock drive,** Fig. 55-3.

With the threaded spindle, the chuck (or the faceplate) is threaded directly to the spindle.

With the key-drive spindle, the chuck is mounted with the internal key slot properly aligned with the key on the spindle nose. The threaded collar is then threaded to the shoulder of the chuck and tightened with a spanner wrench.

With the cam-lock spindle, the notched holding pins on the chuck are inserted into the holes in the spindle flange. The chuck then is locked in position by turning the cam-locking screws in the flange with a T-handle chuck key.

When mounting or removing lathe chucks or faceplates, take care to avoid damage to the threads in the chuck or on the nose of the lathe spindle. With either the tapered key drive or the cam-lock spindles, care must be taken to avoid nicking or damaging the mating surfaces of the chuck and the spindle nose. A small chip, nick, or burr will prevent the chuck from running true.

(South Bend Lathe, Inc.)

Fig. 55-2. Spindle nose with long taper key drive.

(South Bend Lathe, Inc.)

Fig. 55-1.   Threaded spindle nose.

(South Bend Lathe, Inc.)

Fig. 55-3.   Cam-lock spindle.

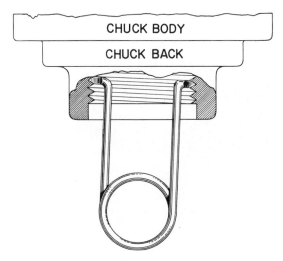

Fig. 55-4.   Cleaning threads in chuck.

Fig. 55-6.   Lathe chuck supported with wooden cradle.

4. In a similar manner, clean the threads and shoulder of the spindle. Apply a few drops of oil.
5. Hold the chuck squarely against the nose of the spindle with the right hand and arm, as in Fig. 55-5. Then with the left hand, rotate the chuck clockwise or, if possible, rotate the spindle forward. Continue until the hub of the chuck rests firmly against the shoulder of the spindle.

CAUTION

Large chucks and faceplates are very heavy and cannot be mounted without assistance. Do not risk injury to yourself or damage to the chuck or machine by attempting to lift a chuck that is too heavy.

Fig. 55-5.   Mounting lathe chuck.

Procedure for Mounting Chuck to Threaded Spindle

1. Remove the live center by tapping it from behind with a rod inserted into the hole in the headstock spindle. Hold the center with one hand to prevent it falling free when it is jarred loose.
2. With the tool shown in Fig. 55-4, clean the threads in the chuck.
3. Wipe the threads with a shop rag, and then apply a few drops of oil to the threads.

Procedure for Removing Chuck from Threaded Spindle

A lathe chuck is a heavy piece of apparatus. Consequently, when it is removed or mounted, the operator can be injured or the chuck or the ways of the lathe damaged unless care is exercised. The ways of the lathe may be protected by supporting the chuck with the right hand. However, this is practicable only when the chuck is small.

Another means is a cleated board. This device is hooked over the lathe bed under the chuck. Still another device is the wooden cradle shown in Fig. 55-6.

(South Bend Lathe, Inc.)

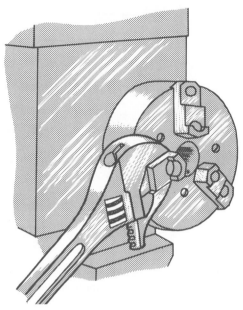

Fig. 55-7.   Loosening chuck with adjustable wrench.

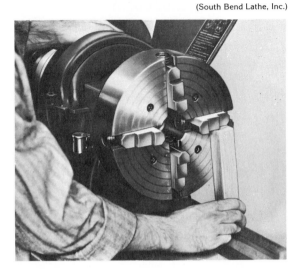

Fig. 55-8.   Loosening chuck with block of wood.

1. To remove the chuck, engage one jaw with an adjustable wrench as in Fig. 55-7 or place a strong bar between the jaws.
2. Engage the back gear lever or otherwise shift the lathe drive system into the slowest speed setting to keep the spindle from revolving. Then give the handle of the wrench or the end of the bar a sharp jerk, which should loosen the chuck.
3. If the chuck will not loosen in this manner, place a block of wood under one jaw, with one end resting on the rear of the lathe bed, as in Fig. 55-8. Then engage the back gears, and give the spindle a quick backward turn by hand.
4. After loosening the chuck, be sure to place a cleated board or a cradle under it to protect the ways should the chuck fall. Unscrew the chuck by rotating it counterclockwise or by rotating the spindle backward. As the chuck approaches the end of the thread on the spindle, press it against the nose of the spindle and hold it firmly to prevent it from falling.
5. Carefully lift the chuck from the board or cradle, and place it in the rack where it is kept when not in use.

Unit

56

How to Center Work
in a Four-Jaw Independent Chuck

A four-jaw independent chuck can be used to hold a wider variety of workpiece shapes than any other chuck. It can be used, for example, when chucking an irregularly shaped piece or a cylindrical piece in such a position that a hole or a recess may be bored **off center.** Another feature of this chuck is that the jaws may be taken out and reversed for certain types of work. With a four-jaw independent chuck, work may be centered with any degree of accuracy desired. It is possible to center as accurately as 0.0025 mm (0.0001″) by using a dial indicator, Fig. 56-1.

Procedure for Centering Stock without Mechanical Aids

1. Adjust the jaws of the chuck to approximately the size of the work by withdrawing or advancing the adjusting screw of each jaw with a chuck wrench. Use the concentric circles on the face of the chuck for approximate positioning of the jaws.
2. Place the work in the jaws of the chuck, leaving the desired amount projecting, and fasten it by lightly tightening the adjusting screw of two opposite jaws, for example, jaws 1 and 3. Tighten these just enough to hold the work securely.
3. In a similar manner, tighten jaws 2 and 4. **Remove the chuck wrench.** Make a practice of doing this; otherwise you may forget it and an accident may result.
4. Mount a suitable tool in the toolholder, and set it at the correct height for the work to be

machined. In this case, the tool bit will be used as an aid in centering the work in the chuck.
5. Revolve the spindle slowly by hand, and advance the tool slowly until the point nearly touches the high side of the work, **A,** Fig. 56-2. **Center the work between two opposite jaws, such as jaws 1 and 3, before attempting to center it between the two remaining jaws. The jaws with which you are working must be in a horizontal position.**

Fig. 56-1.   Centering work with dial indicator.

(South Bend Lathe, Inc.)

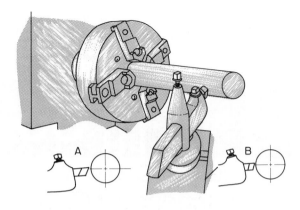

Fig. 56-2.   Truing work in a four-jaw independent
chuck.

6. Give the workpiece a half turn. Note the distance from the point of the tool to the edge of the work, **B,** Fig. 56-2. In order to center the workpiece between these opposite jaws, it will be necessary to move the workpiece toward the low side **half** of this distance.

7. Loosen the jaw on the **low** side of the workpiece and tighten the opposite jaw on the high side of the workpiece. Repeat until the workpiece is centered between the two jaws on which you are working.

8. Test the accuracy of the location of the work between the two jaws by backing the tool off slightly and inserting a slip of paper between the point of the tool and the work. Apply just enough pressure to the tool with the cross-feed screw to cause the paper to drag, but not tear, when it is pulled between the tool and the workpiece. Run the carriage and tool to the right, clear of the work. Then rotate the spindle a half turn. Bring the carriage and tool back, and insert the piece of paper between the tool and the workpiece on the opposite side. If the paper drags, but does not tear, the work is centered between the two opposite jaws to within several thousandths of an inch. If the paper tears or does not drag, an adjustment should be made be-

tween the two jaws, using the procedure described in steps 5 through 8.

9. Give the chuck a quarter turn, and center the work between the remaining two jaws. Use the procedure described in steps 5 through 8 above.

10. Check to see that all jaws are tight against the work and that the work is centered accurately to within the thickness of the paper.

Procedure for Centering Stock with a Dial Indicator

A dial indicator is used to locate a workpiece precisely. Note in Fig. 56-1 that a dial indicator is being used to center the hole in a workpiece in an independent chuck. Either the hole or the outside surface may be centered, depending on which is more important to the proper functioning of the finished workpiece.

1. Install the workpiece and center it approximately, using the point of a cutting tool as a reference as described earlier.

2. Install a dial indicator and bring it lightly into contact with the workpiece at the point where it is farthest off center. Then advance the cross slide until the dial pointer has moved almost a complete revolution.

3. Rotate the chuck until the nearest jaw is horizontal, and note the dial reading. Then rotate the chuck 180° and note the dial reading on the opposite jaw. Moving the workpiece **half the difference** of the two dial readings will center the workpiece between the first pair of jaws. Loosen the jaw at the low reading and tighten the jaw at the high reading until the dial readings are the same.

4. Rotate the chuck 90° and adjust the second pair of jaws until the dial reading is the same as that obtained after the workpiece was centered between the first pair of jaws. **Note: if the workpiece is not perfectly round, there will be a difference in dial readings between the two pairs of jaws.**

5. Tighten each chuck jaw in turn until the workpiece is securely fastened. Make a final check of dial readings, and make the necessary adjustments.

6. If necessary, test the face of the workpiece for wobble, Fig. 56-3. Remove any wobble by tap-

(South Bend Lathe, Inc.)

Fig. 56-3.   Testing face of workpiece with dial indicator.

(L.S. Starrett Company)

Fig. 56-4.   Wiggler with interchangeable points.

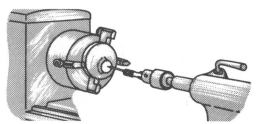

Fig. 56-5.   Centering a workpiece with a wiggler.

ping with a lead hammer. Then recheck to see whether the workpiece must be centered again.

Procedure for Centering Stock with a Wiggler

Wigglers, also called **center finders,** Fig. 56-4, enable a workpiece to be centered from a prick punch mark or a small hole. They are particularly useful for locating holes or workpieces that must be positioned off center in the lathe.

1. Carefully lay out and prick punch the center location on the workpiece.
2. Mount the workpiece in the lathe and center it approximately.
3. Install a wiggler and insert its point in the prick punch mark, Fig. 56-5. When the work-

piece is rotated **by hand,** the point of the wiggler will follow the circular path of the prick punch mark. The farther off center the punch mark is, the greater the amount of "wiggle" made by the wiggler.

4. Depending on the accuracy required, install either a cutting tool or a dial indicator in the tool post. By following the appropriate procedure given above, the wiggler point, and thus the workpiece, will be quickly and accurately centered.

Unit

57

How to Use the Steady Rest and Follower Rest

The steady rest, sometimes called a **center rest,** Fig. 57-1, is a device used to support long shafts or spindles of small diameter while they are being turned, bored, or threaded. When in use, the rest is mounted on the lathe bed and held in position with a clamp. A follower rest, Fig. 57-2, is a supporting device which, when correctly attached to the saddle of the lathe with the supporting jaws adjusted to the work, follows along the finished face of the work and holds it steady against the cutting tool.

When turning very long rods, shafts, or spindles, it sometimes is necessary to use both a steady and a follower rest in combination. Both the steady rest and the follower rest are being used for cutting threads in Fig. 57-3.

Procedure for the Steady Rest

1. Mount the rest on the bed of the lathe, and fasten it loosely.
2. If mounting the workpiece between centers, do not draw the tailstock center too tightly, as there is danger of springing the work.
3. Adjust the steady rest so as to support the work most advantageously. Usually this will be near the point where work is to be performed or near the center. Tighten the clamping bolt.

Fig. 57-1. Using steady rest for boring or internal threading.

(South Bend Lathe, Inc.)

Fig. 57-2. Follower rest in position.

(South Bend Lathe, Inc.)

Fig. 57-3. Using both steady rest and follower rest for cutting threads.

(South Bend Lathe, Inc.)

4. Adjust the jaws so that each one comes lightly into contact with the work. Keep the contact surface lubricated with a high pressure lubricant. The workpiece should rotate quite freely.
5. Fasten the jaws securely, withdraw the tailstock if necessary, and then proceed to machine the work in the usual manner.
6. When the work is completed, remove the steady rest, clean the lathe, and return equipment to the place where kept when not in use.

Procedure for the Follower Rest

1. Attach the follower rest to the saddle of the lathe.

2. Mount the work between centers, and turn a small portion of the work to size at the tailstock end.
3. Adjust the jaws of the follower rest so that they bear directly against the turned part of the work, just back of the cutting tool, and on top of the work, Fig. 57-2.
4. Clamp the jaws in position; then proceed to machine the work in the usual manner. Keep the follower rest contacts well lubricated.
5. When the work is completed, remove the follower rest, clean the lathe, and return equipment to the place where kept when not in use.

Unit

58

How to Machine Precision Diameters

For parts that are machined to their final size in the lathe, one or more **rough cuts** are first made in order to remove the bulk of the excess material. Then the **finish cuts** are made, bringing the part to its final dimensions. Allowance for finish cuts should be on the order of 0.25-0.76 mm (0.010″-0.030″). Parts that must be finished to size by grinding after they are hardened are machined oversize and therefore only require rough cuts on those surfaces. Grinding allowances are about the same as finishing allowances.

Rough cuts should be as deep as the tool, machine, and workpiece permit in order to minimize machining time. Ideally, one rough cut and one finish cut should complete the workpiece, but obviously this is not always possible. When machining cast iron, cast steel, or other metals that have a hard scale on the surface, it is important to make the first cut deep enough to cut under the scale. Otherwise, the hard scale will quickly dull the cutting edge.

Before machining stock between centers, be sure the centers are in line. Test by moving the tailstock center close to the headstock center, Fig. 54-1. This is only a rough check and should not be relied upon when accurate work is required.

An accurate test may be made by turning a section at each end of a workpiece and then measuring with a micrometer or vernier caliper. To make such a test, mount a piece of stock between centers, and feed the tool in as at **A,** Fig. 58-1, until the work has been reduced about 1.6 mm (1/16″) in diameter. Then with the longitudinal feed, advance the tool until a straight cut about 13 mm (1/2″) has been made. Stop the lathe, note the cross slide micrometer dial reading, and withdraw the tool by backing off the cross slide crank one revolution. Then run the carriage to the right until the tool clears the right-hand end of the stock. Advance the cross slide crank until the micrometer dial reads the same as for the first cut. Then make a short longitudinal cut as at **B,** Fig. 58-1. With a micrometer, measure the turned sections at **A** and **B.** If the lathe centers are in line with each other, there should be no difference in the diameters of the two sections.

If the difference in diameters is greater than can be tolerated, the centers should be brought into alignment by setting the tailstock over in the proper direction one-half the difference in diameters. For procedure in setting the tailstock over, see Unit 66.

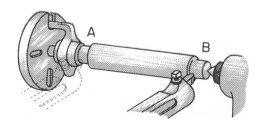

Fig. 58-1.   Testing accuracy of lathe centers.

Procedure for Machining Precision Diameters with Stock Mounted between Centers

1. Mount the work in the lathe. See Unit 54.
2. Obtain or prepare a sharp cutting tool of the shape and size needed. For rough cuts, the tool must be strong enough to take a heavy cut. If the tool is to travel toward the headstock as in Fig. 58-2, a right-hand turning tool should be used, Fig. 49-1 and 50-3. A left-hand tool is used when the tool travels toward the tailstock.
3. Insert the tool in the holder, with the point projecting about 13 mm (1/2″).
4. For turning steel or cast iron, set the toolholder in the tool post so that the point of the tool is between 0° and 5° above the center of the workpiece. Position the toolholder so that its nose projects only slightly beyond the edge of the compound rest.
5. Swing the toolholder so that the cutting edge of the tool is no more than 90°, preferably 80°, from the surface of the workpiece. This angle should be measured from the side toward which the cutting tool will be moving. See Fig. 58-3.
6. With the cross-feed crank, withdraw the tool until it clears the work. Start the lathe, and carefully advance the tool until it just touches the work. If possible, set the micrometer collar on the cross-feed screw at zero, Fig. 58-4. Advance the tool the desired

amount. Dials are usually calibrated either in hundredths of a millimeter or in thousandths of an inch. **Dual reading dials** have both metric and inch graduations, Fig. 58-5. With **direct reading micrometer dials,** move the dial the full amount the diameter is to be reduced, for example, 6.35 mm (1/4″). With **indirect reading micrometer dials,** move the **dial half the amount** the diameter is to be reduced, or 3.175 mm (1/8″) in order to reduce the diameter 6.35 mm (1/4″). Indirect reading dials are calibrated to show the amount the tool is moved. Doubling this amount gives the amount the diameter is

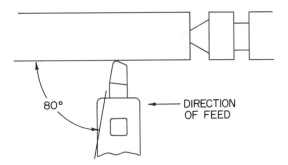

Fig. 58-3.   Position cutting edge at 80° to workpiece for straight turning and facing.

(South Bend Lathe, Inc.)

Fig. 58-4.   Micrometer collar on cross-feed screw.

(South Bend Lathe, Inc.)

Fig. 58-2.   Turning a steel shaft mounted between centers.

Fig. 58-5.   Dual-reading dials have both metric and inch graduations.

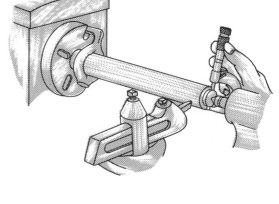

Fig. 58-6.   Testing workpiece diameter with a micrometer.

reduced. Direct reading dials are calibrated to read the amount the diameter is reduced without the need for any calculations.

7. Set the rate and direction of feed desired, advance the tool to the end of the workpiece with the carriage hand wheel, and engage the automatic feed.

8. When about 6.35 mm (1/4″) of the piece has been machined, disengage the power feed, stop the machine, and move the tool back to the end of the workpiece with the carriage hand wheel. Test the piece for size with a micrometer, as in Fig. 58-6.

9. If the piece is undersize or oversize, make the necessary adjustment. Be sure to allow about 0.25-0.76 mm (0.010″-0.030″) for the finish cut.

10. After making adjustments, proceed with the rough cut, using a feed rate of about 0.13-0.25 mm (0.005″-0.010″) per revolution.

11. Continue until the tool or the side of the compound rest is within about 6.35 mm (1/4″) of the dog. Stop the lathe, and run the carriage back until the point of the tool is past the end of the workpiece.

12. If the entire length of the workpiece must be machined, remove the piece from between the centers, remove the dog, and place it on

the turned end. Again mount the workpiece in the lathe and proceed as before, turning the remainder of the workpiece to diameter.

Procedure for Finish Cuts

1. Adjust the feed rate and rpm for the finish cut. A feed rate of 0.04-0.08 mm (0.0015″-0.003″) per revolution should give good results.

2. Advance the tool for the finish cut.

3. Machine about 6.35 mm (1/4″) at this setting. Then stop the machine and test for size.

4. Make adjustments if necessary; then test again for size.

5. When the tool has been set so as to machine the workpiece to the diameter desired, engage the power feed and complete the cut.

Procedure for Machining Precision Diameters with Stock Mounted in a Chuck

1. Mount the stock in a chuck. See Unit 56.

2. Rough turn using the same procedure as when mounted between centers. See Fig. 58-7. Be sure to stop the power feed when the edge of the compound rest or the point of the tool comes within 6.35 mm (1/4″) of the jaws of the chuck. With care, the tool may be fed as close to the chuck jaws as 1.6 mm (1/16″) with the carriage hand wheel.

3. Make the finish cut in the same manner as described above.
4. If the entire surface of the workpiece must be machined, remove the workpiece from the chuck, turn it around, and rechuck it. Protect the workpiece from being marred by placing a soft strip of metal between the chuck jaws and the finished surface. Be sure to seat the workpiece against either the chuck jaws or the face of the chuck by tapping with a soft hammer. If this is not done, the workpiece may wobble in the chuck instead of running true. If the workpiece will not run true enough in a universal chuck, then use an independent chuck and dial indicator to center it. Then proceed with the rough and finish cuts as before.

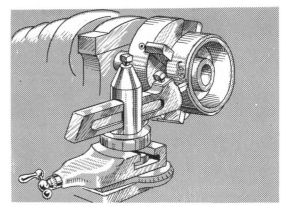

Fig. 58-7.   Turning circumference of stock mounted in a chuck.

Unit

59

How to Machine Precision Lengths

Procedure for Obtaining Accurate Overall Length with Workpiece Mounted between Centers

1. Mount the workpiece between centers. See Unit 54.
2. Position the compound rest 90° to the cross slide, Fig. 59-1. This allows its use for precision adjustment of the tool lengthwise, or parallel to the workpiece axis.
3. Select a facing tool of the kind and shape desired, mount it in the toolholder, and position the point of the tool on center, Fig. 59-1.
4. Start the machine, and advance the tool until the point is almost in contact with the center. Then, with longitudinal feed, advance the tool into the end of the workpiece so that it takes a

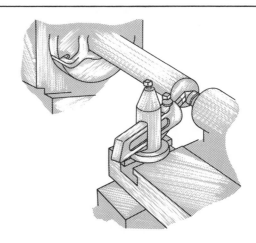

Fig. 59-1.   Facing work mounted between centers.

light cut. Use of a **half-center** permits the point of the tool to enter the center hole. This enables the facing cut to be made without leaving a burr around the center hole. See Fig. 59-2.

5. With the cross-feed crank, or with power-cross feed, feed the tool toward the outer surface. Make additional cuts if necessary until the end is completely machined.

6. Reposition the workpiece and lathe dog so that the other end can be machined.

7. Bring the tool into contact with the end of the workpiece, lock the lathe carriage in place, advance the tool into the workpiece, and make a facing cut.

8. Stop the lathe and measure the overall length of the workpiece. It may be necessary to remove the workpiece from the lathe and the dog from the workpiece. Subtracting the desired length from the measured length gives the distance the tool must be moved to machine the part to the specified overall length.

9. Make one or more additional facing cuts, moving the tool a precise amount with the compound rest crank. When extreme accuracy is required, the piece should be measured after each facing cut.

Procedure for Obtaining Accurate Overall Length with Workpiece Mounted in a Chuck

1. Mount the workpiece in a chuck. Be sure to seat it against the chuck jaws or the face of the chuck.

2. Position the compound rest 90° to the cross slide, Fig. 59-3.

3. Select, install, center, and position a left-hand facing tool.

4. Set the power cross feed to move the tool from the outside diameter towards center. Make one or more rough cuts until the entire end is machined. Next, make all other rough machining operations on the workpiece. Then make all the finish cuts, including a finish facing cut.

5. Remove the workpiece from the chuck, turn it around, and rechuck it. Be sure to use strips of soft metal between the chuck jaws and the machined surface. Seat the workpiece solidly

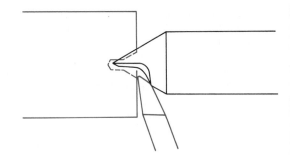

Fig. 59-2. Half-center permits facing end completely flat.

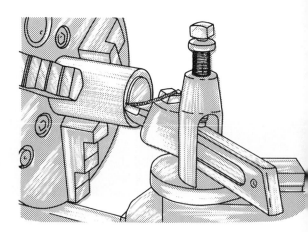

Fig. 59-3. Facing stock mounted in a chuck.

against the chuck face or chuck jaws with a lead hammer.

6. Position the tool to make a facing cut and then lock the carriage in place.

7. Make a facing cut, stop the machine, and measure the length of the workpiece. If the workpiece is seated against the face of the chuck, use a depth micrometer. Otherwise, use a vernier caliper. Subtracting the desired length from the measured length gives the amount the tool must be moved to obtain the specified part length.

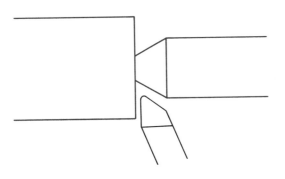

Fig. 59-4.   Cutting edge positioned parallel to end of workpiece.

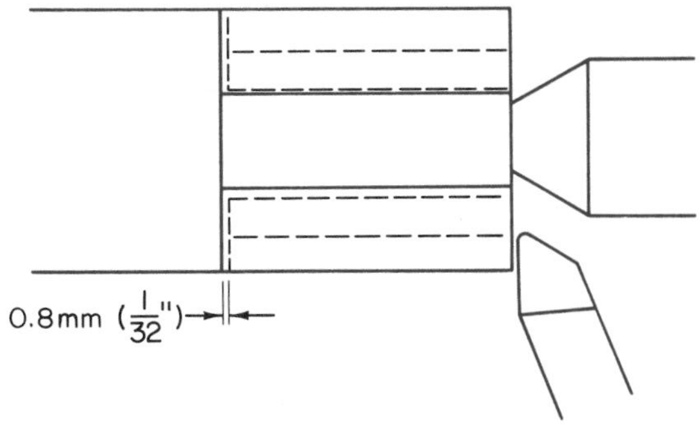

Fig. 59-5.   Dotted lines indicate rough cuts taken prior to the finish cuts.

0.8mm ($\frac{1}{32}$")

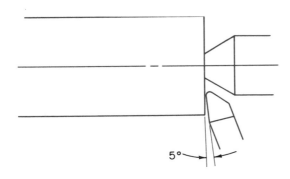

5°

Fig. 59-6.   Tool positioned with cutting edge 5° to end of workpiece.

8. Make additional facing cuts by moving the tool the precise amount with the compound rest crank, until the specified length is obtained. Measure after each cut if extreme accuracy is required.

Procedure for Turning an Accurate Square-Shouldered Length

1. Position the compound rest 90° to the cross slide.
2. Install a facing tool and face the end of the workpiece.
3. Install a right-hand turning tool, positioning the cutting edge parallel to the end of the workpiece, Fig. 59-4.
4. Make the necessary rough cuts, stopping the tool short of the desired length by about 0.8 mm (1/32"), Fig. 59-5. Do not face the shoul-

der until after all the rough cuts are made on the diameter.
5. Reposition the tool so that the cutting edge makes an angle of about 5° to the end of the workpiece, Fig. 59-6.
6. Turn the diameter to its final size, note the cross-slide dial reading, and at the end of the cut, lock the carriage and face off the shoulder.
7. Stop the machine and measure the length of the turned section with a depth micrometer.
8. Start the machine and return the cross slide to the dial reading used for the finish cut on the workpiece diameter. Then, using the compound rest crank, advance the tool into the shoulder the amount needed to make the turned section the required length. Back the tool out making the facing cut. Several cuts may be needed since the shallow tool angle severely limits the depth of each facing cut.

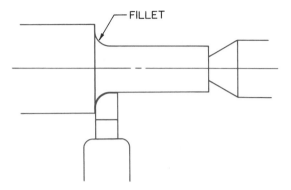

Fig. 59-7.   Radius tool machining a fillet.

Procedure for Filleted Shoulder Turning

1. A **fillet** is a rounded surface at the intersection of the smaller diameter and the shoulder, Fig. 59-7. Proceed as for turning a square-shouldered length, leaving sufficient material in the corner to machine the fillet.
2. After turning the small diameter to size, exchange the right-hand turning tool for a round nose tool of proper radius. This tool should have no back rake and should be set exactly on center.
3. Reduce the lathe rpm to suit the tool radius; the larger the radius, the slower the rpm.
4. Carefully machine away the excess material in the corner, keeping the tool well lubricated.

Procedure for Turning a Diameter between Two Shoulders

1. Install a suitable parting tool.
2. Set an outside caliper slightly larger than the finished diameter to be machined between the shoulders.
3. Make parting tool cuts to locate the two shoulders, Fig. 59-8.
4. Exchange the parting tool for a right-hand turning tool, and turn the diameter to final size between the shoulders, Fig. 59-9.
5. Use a left-hand turning tool to finish the diameter up to the right-hand shoulder.
6. Measure the distance between the shoulders and, if necessary, adjust by facing one of the shoulders.

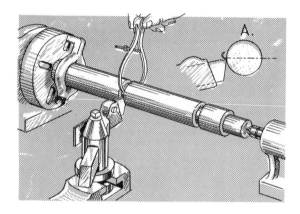

Fig. 59-8.   Testing diameter with calipers.

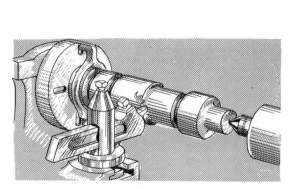

Fig. 59-9.   Removing stock between square shoulders.

Unit

60

How to File and Polish Work in a Lathe

Filing work while it is revolving in a lathe often is resorted to when only a very small amount of stock is to be removed from a diameter, or for removing sharp corners on the workpiece.

Procedure for Filing

1. Obtain a clean, sharp, single-cut mill file of 200 or 250 mm (8" or 10") length.
2. Grasp the handle of the file in the right hand and the opposite end between the thumb and fingers of the left.
3. For filing, set the rpm the same as for turning.
4. Place the file flat on the work near the left end of the part to be filed, with the tip of the file pointed toward the right, as in Fig. 60-1.
5. Press lightly on the file, and, at the same time, move it forward and to the right. At the end of the stroke, release the pressure, raise the file from the work slightly, and draw it back for the next stroke. Take long strokes, and con-

tinue in this manner until the piece is filed smooth. Clean the file frequently with a file card.

Procedure for Polishing

1. Secure a strip of abrasive cloth, usually aluminum oxide. A strip about 25 to 30 mm (1" to 1-1/4") wide by about 250 mm (10") long will suffice.
2. For polishing, set the rpm two or three times as fast as for turning.
3. Grasp the ends of the abrasive between the fingers. Then place the cloth across the work, and draw it across the rotating surface as when filing.
4. Greater force may be applied if the abrasive is held under a file and used as in Fig. 60-2.
5. To produce a finer finish, use a finer grade of abrasive and, for ferrous metals (iron or steel), apply a few drops of machine oil on the abrasive.

(South Bend Lathe, Inc.)

Fig. 60-1.   Filing work revolving in lathe.

(South Bend Lathe, Inc.)

Fig. 60-2.   Polishing work revolving in lathe.

Unit

61

How to Knurl in a Lathe

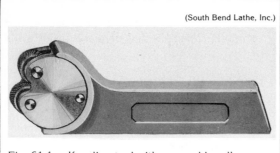

(South Bend Lathe, Inc.)

Fig. 61-1.   Knurling tool with removable rollers.

(Armstrong Bros. Tool Co.)

Fig. 61-2.   Removable knurling rolls. Diamond pattern is shown on left and straight-line pattern on right.

(South Bend Lathe, Inc.)

COARSE   MEDIUM   FINE

Fig. 61-3.   Diamond pattern knurls.

Certain machine parts such as knobs, cylindrical handles, round nuts, and collars are knurled to improve their appearance and to provide a surface which can be gripped more effectively. Knurling tools, Fig. 61-1, are made with removable rollers, Fig. 61-2. To produce a fine, medium, or coarse knurl, Fig. 61-3, merely select and insert the appropriate roller.

On the knurling tool shown in Fig. 61-4, the revolving head has three pairs of removable knurling rolls — coarse, medium, and fine. Knurling rolls are available which produce a knurled surface with either a diamond pattern or a straight-line pattern. With each pattern, the surface may be coarse, medium, or fine.

Procedure

1. Finish-turn the work to the dimension desired.
2. Select a knurling tool of the pattern and degree of coarseness desired.
3. Insert the knurling tool in the tool post, with the front of the rollers flat against the side of the work and both rollers in contact with it.

(Armstrong Bros. Tool Co.)

Fig. 61-4.   Knurling tool with turret head.

4. Fasten the tool **firmly** in the tool post.
5. Set the lathe for a longitudinal feed of 0.5 to 0.8 mm (0.020″ to 0.030″) and for back-gear drive. Start the lathe on a slow speed.
6. With the apron handwheel, move the carriage over until the right side of the rollers pass the right end of the piece about 3 mm (1/8″).
7. With the cross-feed crank, force the rollers about 0.5 mm (0.020″) into the work. At the same time, engage the longitudinal feed of the carriage, and allow the knurling tool to travel across the face of the workpiece, Fig. 61-5. Apply oil liberally throughout the operation if knurling steel or steel alloys.
8. When the left end of the knurling roller reaches the left end of the surface to be knurled, stop the power feed.
9. Release most of the pressure on the knurling tool, hand-crank the tool to the right end of the workpiece, and repeat the operation until the knurling is the depth desired.

(South Bend Lathe, Inc.)

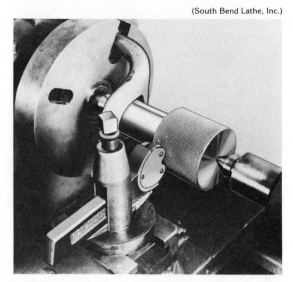

Fig. 61-5.   Using the knurling tool.

Unit
62

How to Drill and Ream in a Lathe

Drilling, reaming, and other drill press operations such as countersinking and counterboring can be done readily in a lathe. In performing these operations, the workpiece is usually mounted in a lathe chuck or attached to a faceplate, and the drill or other cutting tool is mounted in the tailstock.

For center drilling with a combination drill and countersink, Fig. 62-1, the tool is mounted in a drill chuck which is installed in the tailstock. The tailstock is clamped to the lathe bed, and the tool is fed to the desired depth with the tailstock handwheel.

(South Bend Lathe, Inc.)

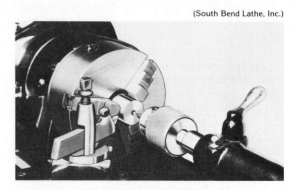

Fig. 62-1.   Center drilling in a lathe.

(South Bend Lathe, Inc.)

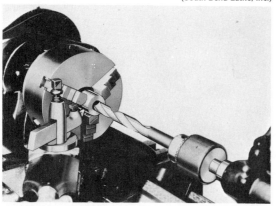

Fig. 62-2.   Straight-shank drill mounted in drill chuck in tailstock.

(South Bend Lathe, Inc.)

Fig. 62-3.   Drilling with taper-shank drill mounted in tailstock.

(South Bend Lathe, Inc.)

Fig. 62-4.   Drilling with crotch center in tailstock.

Straight-shank drills (Fig. 62-2) and reamers also are mounted in drill chucks. Larger drills with taper shanks (Fig. 62-3) are mounted directly into the tailstock for drilling.

Taper-shank reamers, counterbores, and other taper-shank cutting tools also may be mounted in the tailstock for hole-machining operations. Small parts may be drilled with the drill mounted in the headstock chuck and the workpiece mounted against a crotch center, as shown in Fig. 62-4.

A **pad center** is used like a crotch center, but it has a flat surface for supporting flat workpieces for drilling operations.

Procedure

1. Check that the lathe centers are properly aligned. If they are not, the hole will not be machined straight.
2. Mount the workpiece in the lathe; be sure that it is centered properly.
3. Install a drill chuck in the tailstock if required, Fig. 62-1. Be sure that the hole in the tailstock spindle is clean.
4. Install the drilling tool in the drill chuck, and tighten the chuck. Tap the end of taper-shanked tools **lightly** with a lead hammer to insure a snug fit in the tailstock.
5. Determine the proper cutting speed (Table 49-1) and rpm (Table A-10, Appendix). Then start the lathe.
6. Feed the drill or reamer with the tailstock handwheel, Fig. 62-3. Feed at a steady rate for the desired chip formation. Small drills are fed at a slower feed rate than larger drills. Apply cutting fluid as recommended in Table A-11, Appendix. When drilling or reaming deep holes, withdraw the tool periodically for chip removal. This practice will prevent tool breakage in deep holes.
7. When the drilling operation has been completed, remove the cutting tool and proceed with the next machining operation.

Unit
63

How to Bore Holes in a Lathe

Lathes are often used for boring holes and recesses. Boring can produce rounder, straighter, smoother, and more accurately sized holes than by drilling and reaming. Also, boring is the only way to produce holes with internal grooves or other shapes and holes with diameters larger than the largest available drills. For boring, the workpiece may be mounted on a faceplate, Fig. 63-1, or in a chuck, Fig. 63-2.

Many different styles of boring bars and toolholders are used in lathes. The boring tool in Fig. 63-3 includes the toolholder, boring bar, and insert tool bits which are ground in the same manner that standard lathe tool bits are ground. The boring bar is equipped with end caps which hold

(South Bend Lathe, Inc.)

Fig. 63-2.   Boring work mounted in a chuck.

(South Bend Lathe, Inc.)

Fig. 63-1.   Boring work mounted on faceplate.

(Armstrong Bros. Tool Co.)

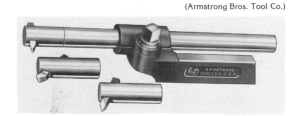

Fig. 63-3.   Boring tool with 90°, 45°, and 30° end caps.

(Armstrong Bros. Tool Co.)

Fig. 63-4.   Boring tool with forged boring bar.

(Carboloy Systems)

Fig. 63-6.   Boring bar for holding throwaway insert tools.

(Armstrong Bros. Tool Co.)

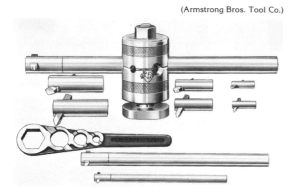

Fig. 63-5.   Three-bar boring tool.

(Armstrong Bros. Tool Co.)

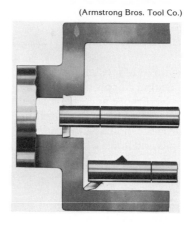

Fig. 63-7.   Boring tools in relation to workpiece.

the tool bit at 90°, 45°, and 30° angles. This type boring bar is available in a variety of sizes for boring holes about 16 mm (5/8") in diameter or larger.

The boring tool in Fig. 63-4 has reversible bar clamps which allow it to hold boring bars from 6.35 to 12.7 mm (1/4" to 1/2") in diameter. The boring tool in Fig. 63-5 is made to hold any of three boring bars of different sizes. Boring bars are also available for holding throwaway inserts of different shapes, Fig. 63-6.

Procedure for Boring a Hole

1. Mount the workpiece in a chuck, Fig. 63-2, or on a faceplate, Fig. 63-1, and position it accurately.
2. If the workpiece is solid, center drill and drill a hole large enough to admit the boring bar.
3. Obtain a suitable boring bar with a sharp cutting tool of proper shape.

4. Set the tool so that the point is on center and fasten both the tool and toolholder securely.
5. Set the lathe for proper rpm and feed rate.
6. Start the lathe and bring the point of the tool into contact with the surface of the hole. Then move the tool out of the hole and adjust it for the first cut.
7. With the apron handwheel, advance the tool into the work about 6.35 mm (1/4").
8. Stop the machine, back the tool away with the apron handwheel, and measure the diameter of the hole with a rule, vernier caliper, or inside micrometer, Fig. 63-2.
9. Adjust the tool if necessary and then complete the cut. When machining a shoulder or a recess with a closed bottom, be sure to stop the power feed just before the point of the tool reaches the shoulder or bottom, Fig. 63-7. Feed the tool the rest of the way by hand with the carriage handwheel.

10. If other cuts are necessary to enlarge the hole, proceed as in steps 7, 8, and 9, leaving enough material for a finish cut.

11. Adjust the rpm and feed rate for a finish cut, and finish the hole to size.

Procedure for Machining the Bottom of a Hole

1. Select a boring tool which will enable you to machine to the center and to the outside edge of the bottom surface. A tool mounted in a 45° cap, as shown in Fig. 63-7 is satisfactory for large diameter holes. For smaller diameter holes, use a forged boring bar as in Fig. 63-4.

2. Starting a short distance from the center of the workpiece, feed the tool slowly into the bottom of the hole with the apron handwheel. Moving the tool slightly back and forth with the cross-slide crank while this is done helps prevent tool chatter.

3. When the tool has entered the metal the depth desired, feed it toward center with the cross-feed crank until the center is reached.

4. Reverse the direction of travel, and feed toward the inside wall.

5. When the cut almost reaches the wall, stop the machine, and set the toolholder at an angle which permits machining into the corner. A round nose tool will leave a small fillet in the corner, which may or may not be permissable. If a square corner is required, remove the round nose tool and insert a right-hand corner or side tool.

6. Advance the corner tool with the apron handwheel until it just strikes the bottom. Then with the cross-feed crank, feed the tool slowly toward the wall of the recess.

7. Stop the feed when the side of the tool just touches the wall of the recess.

8. Finish both the bottom and the hole diameter with light cuts if necessary.

Unit

64

How to Machine a Workpiece Mounted on a Lathe Mandrel

Sometimes it is necessary to machine the exterior of a cylindrical workpiece accurately in relation to a hole that previously has been bored in the center of the piece. In such cases, the work often is mounted on a mandrel (Fig. 64-1) and turned between centers, Fig. 64-2. A mandrel usually is a cylindrical piece of hardened steel turned with a slight taper. For an emergency, a mandrel may be turned from a piece of soft steel. Since a mandrel of given diameter has only limited application, an expanding bushing (Fig.

(South Bend Lathe, Inc.)

Fig. 64-1. Mandrel.

64-3) often is used. The illustration shows a mandrel equipped with a slotted bushing. The bushing is bored with a taper corresponding to the taper of the mandrel. A series of such bushings greatly extends the usefulness of a mandrel of a given size.

Procedure

1. Secure or turn a mandrel of suitable size.
2. Apply a little oil to the surface of the mandrel and to the side walls of the hole.

   Unless a lubricant is applied, the mandrel may stick, in which case it cannot be removed without damaging the workpiece, the mandrel, or both.
3. Insert the small end of the mandrel in the hole in the object, and force it in tightly with a press or a **lead** hammer. The mandrel must be tight enough to prevent slipping while the workpiece is being machined.
4. Mount the workpiece between centers, apply center lubricant on the dead center, and turn in the usual manner.
5. To remove the mandrel, apply pressure to the small end with a press or strike it with a lead hammer.

(South Bend Lathe, Inc.)

Fig. 64-2.   Workpiece mounted on a mandrel.

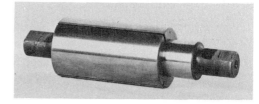

(South Bend Lathe, Inc.)

Fig. 64-3.   Mandrel with slotted expanding bushing.

Unit

65

Calculations Required for Taper Turning

Tapers are specified either as an angle of a given number of degrees, or as a given amount of change in diameter for a standard unit of length. Metric tapers are specified in millimeters per meter (mm/m). Inch tapers are normally specified in inches per foot (in/ft).

Calculating Angle of Taper

When the angle of taper is not given, it can be found by using the mathematics of trigonometry.

Example: Find the angle of taper for the part shown in Fig. 65-1.

Solution:

1. Construct the line A-B parallel to the center line, and intersecting the corner of the workpiece, forming the right triangle A-B-C.
2. The length of the base of the right triangle, line A-B, is 25 mm (0.984") as read from the drawing.
3. The height of the right triangle is found by subtracting the end diameter (d) from the workpiece diameter (D) and dividing by 2:

$$\frac{D\text{-}d}{2} = \frac{25\,mm\,(0.984") - 12\,mm\,(0.472")}{2}$$

$$= \frac{13\,mm\,(0.512")}{2} = 6.5\,mm\,(0.256").$$

4. Using the tangent function of trigonometry:

$$\text{Tan} \angle x = \frac{\text{side opposite}}{\text{side adjacent}}$$

$$= \frac{6.5\,mm\,(0.256")}{25.0\,mm\,(0.984")}$$

$$= 0.26\,(0.26)$$

5. The table of trigonometric functions, Table A-15, Appendix, shows that 0.26 in the tangent column corresponds to an angle of approximately 14°35'.

Note: If the compound rest is used for turning the taper, it should be set for the angle obtained above. If a taper attachment is used, however, it should be set for double this angle. Taper attachments are calibrated for the **included** angle of taper, not the angle from center line to one side as was calculated above.

Calculating Tailstock Setover

The amount of tailstock setover may be calculated with the formula:

$$\text{Setover} = \frac{T \times L}{2}, \text{ where } T = \text{rate of taper}$$

(mm/m or in/ft), and L = the **total length** of the workpiece **expressed in meters or feet.**

Example: Calculate the setover required to machine the taper on the workpiece shown in Fig. 65-2.

Solution: $\text{Setover} = \dfrac{T \times L}{2}$

$$= \frac{66.67\,mm/m\,(0.801"/ft) \times .380\,m\,(1.247')}{2}$$

$$= \frac{25.33\,mm\,(0.999")}{2} = 12.67\,mm\,(0.499"),$$

the amount of setover required.

When the rate of taper is not given, it may be calculated by subtracting the small diameter (d) from the large diameter (D), and dividing by the length of the **tapered section, L_t,** expressed in meters or feet.

Example: Using the workpiece in Fig. 65-2 as an example, find the rate of taper.

Solution:

$$\frac{D - d}{L_t} = \frac{40\,mm\,(1.575") - 30\,mm\,(1.181")}{0.150\,m\,(0.492')}$$

$$= \frac{10\,mm\,(0.394")}{0.150\,m\,(0.492')} = \begin{array}{l} 66.67\,mm/m \\ (0.801\,in/ft) \\ \text{rate of taper} \end{array}$$

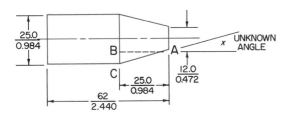

Fig. 65-1.   Problem in calculating angle of taper.

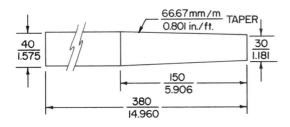

Fig. 65-2.   Problem in calculating tailstock setover required for turning a taper.

Unit

66

Procedure for Taper Turning

The most common ways of turning tapers are: (1) the tailstock setover method, (2) using a taper attachment, and (3) with the compound rest. The first two methods are used for long, shallow tapers. The third method is used for short, steep tapers.

(South Bend Lathe, Inc.)

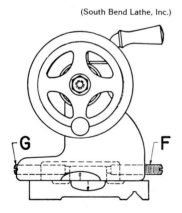

Fig. 66-1.   Tailstock setover screws.

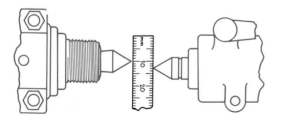

Fig. 66-2.   Checking amount of tailstock setover.

Procedure for Tailstock Setover Method

1. Mount the workpiece between centers.
2. Turn the stock to the size desired; then remove it from the lathe.
3. Calculate the amount the tailstock should be set over.
4. Unclamp the tailstock from the lathe bed, and loosen the tailstock setover screw on the far side of the tailstock, **F,** Fig. 66-1. This screw should be backed out several turns farther than the offset required. Next, screw in the tailstock setover screw on the front side of the tailstock until it is set off center the distance desired. Measure the amount of offset between the two witness marks on the base of the tailstock, Fig. 66-1. For lathes without witness marks, the amount of offset can be measured between the lathe centers, as in Fig. 66-2. Tighten the rear setover screw just enough to prevent sidewise movement of the tailstock.
5. Fasten the tailstock in position by tightening the clamping nut.
6. Mount the stock between centers. Be sure to apply center lubricant on the dead center.
7. Install a tool of suitable size and shape, and position the point of the tool exactly on center.
8. With the apron handwheel, draw the tool forward until it is in position to make a cut. Then with the cross-feed crank, advance the tool so that it will take a cut.

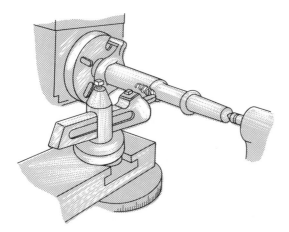

Fig. 66-3. Cutting a taper with work mounted between centers.

(South Bend Lathe, Inc.)

Fig. 66-4. Cutting a taper with a taper attachment.

(South Bend Lathe, Inc.)

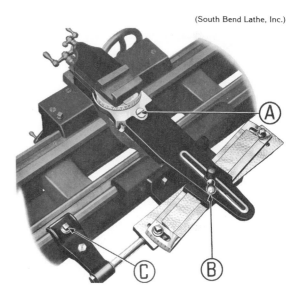

Fig. 66-5. Plain taper attachment.

9. Set the lathe for the desired rpm and feed rate. Start the lathe, and feed the tool toward the left in the usual manner, Fig. 66-3.

10. At the end of the cut, move the carriage back to the starting point. Take a second cut in a similar manner.

11. Stop the machine, and mark a unit of length on the taper. Measure both ends of the taper.

12. Calculate the amount of taper to determine whether the rate of taper is correct. If a taper gage is available, draw two chalk lines lengthwise on the tapered surface 90° apart. Remove the workpiece, insert the tapered end into the gage as far as it will go, then twist the gage about half a turn and remove it. If the taper is too shallow, more of the chalk will be rubbed off on the small end. If the taper is too steep, more of the chalk will be rubbed off on the large end.

13. Adjust the taper if necessary.

14. With repeated cuts, reduce the stock to the size desired.

15. When the job has been completed, set the tailstock back in normal position.

Objections to this method of turning tapers are: (1) the center holes wear unevenly, (2) calculation is necessary for each different length of stock, and (3) the procedure cannot be used to bore tapered holes, except in special cases.

Procedure for Taper Attachment Method

Figure 66-4 shows a standard No. 5 Morse taper being cut on a lathe with a taper attachment. Two types of taper attachments are commonly available. The taper attachment in Fig. 66-5 is called a plain taper attachment; that in Fig. 66-6 is a telescopic taper attachment. When the latter is used, it is not necessary to disconnect the cross-feed screw, as described in step 3.

Fig. 66-6.   Telescopic taper attachment.

Fig. 66-7.   Cutting an internal taper.

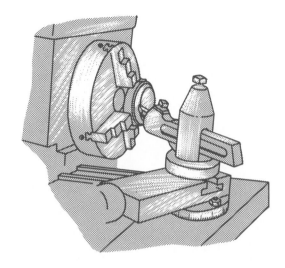

Fig. 66-8.   Cutting a taper by setting compound rest.

1. Set the taper bar at the angle desired, and fasten it in position with its clamp screws.
2. If a plain taper attachment is used, position the compound rest parallel with the cross slide, as in Fig. 66-5.
3. Disengage the cross-feed screw by removing the screw that holds the cross-feed control nut to the cross slide, **A** in Fig. 66-5.
4. Install the toolholder, tool, and workpiece.
5. Position the taper attachment on the ways, and fasten it with the clamp screw, **C.**
6. Position the cross slide so that the tool is close to the workpiece, then lock it to the taper bar by tightening handle **B.**
7. Set the lathe rpm and feed rate, advance the tool with the compound rest crank, and commence cutting.
8. Check the taper for accuracy after the first two or three cuts. Follow the procedure described

for the tailstock setover method, steps 11 and 12.
9. After the taper has been machined, unclamp the cross slide from the taper bar. Then re-attach the cross-slide nut to the cross slide by replacing the screw removed in step 3. Finally, release the clamp holding the taper attachment to the rear machine way.

The advantages of this method of turning tapers are: (1) the centers are always in line, (2) a taper of a given amount may be cut independent of workpiece length, and (3) both external and internal tapers can be cut with this method, Figs. 66-4 and 66-7.

Procedure for Compound Rest Method

1. Mount the stock in the lathe. Do whatever machining must be done before the taper can be machined.
2. Set the compound rest to the angle desired, and lock it in place. See Fig. 66-8.
3. Install a suitable toolholder and tool, setting the point of the tool exactly on center.
4. Position the tool for the first cut with the apron hand wheel and the cross-feed crank.
5. Lock the carriage to the lathe ways.
6. Make the first cut, feeding the tool with the compound-rest crank. At the end of the cut,

return the tool to the end of the workpiece and adjust it with the cross-feed crank for a second cut.

7. Make repeated cuts until the taper is the size desired, finishing with a light cut.

The advantages of this method of turning a taper are: (1) the centers are always in line, (2) only one calculation is required since the length of the workpiece does not affect the angle of taper, and (3) both external and internal tapers may be cut.

Unit
67

Cutting Threads in a Lathe

Cutting threads in a lathe by making multiple cuts with a single-point tool is called **thread chasing.** While the method is slow, and therefore expensive, it is often necessary for cutting large diameter threads or threads calling for accurate pitch or lead. External and internal threads may be cut in a lathe, either straight or tapered, and right hand or left hand.

Procedure for Cutting 60° External, Right-Hand Threads

1. Install the workpiece in the lathe, and do all the machining necessary prior to the thread chasing. Always chamfer the end of the workpiece, and make an undercut where the thread will end, as in Fig. 67-1.

2. Set the levers on the quick-change gearbox, Fig. 67-2, to obtain the thread pitch desired. The chart on the gearbox, Fig. 67-3, lists the thread pitches available. For example, to set the gearbox shown in Fig. 67-2 to cut 20 threads per inch, find the 20 on the gearbox chart, and place the right-hand lever directly below the column in which the 20 is listed.

(South Bend Lathe, Inc.)

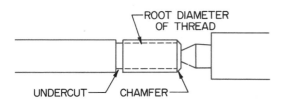

Fig. 67-1.   Workpiece chamfered and undercut in preparation for threading.

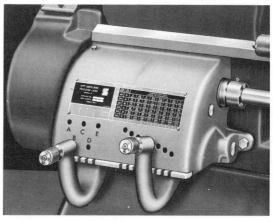

Fig. 67-2.   Quick change gearbox.

(South Bend Lathe, Inc.)

POWER CROSS FEED .375 TIMES LONGITUDINAL FEED	STUD GEAR	LEFT HAND TUMBLER	THREADS PER INCH FEEDS IN THOUSANDTHS						
	48	A	4 .0841	4½ .0748	5 .0673	5½ .0612	5¾ .0585	6 .0561	6½ .0518	7 .0481
	24	A	8 .0421	9 .0374	10 .0337	11 .0306	11½ .0293	12 .0280	13 .0259	14 .0240
	24	B	16 .0210	18 .0187	20 .0168	22 .0153	23 .0146	24 .0140	26 .0129	28 .0120
	24	C	32 .0105	36 .0093	40 .0084	44 .0076	46 .0073	48 .0070	52 .0065	56 .0060
	24	D	64 .0053	72 .0047	80 .0042	88 .0038	92 .0037	96 .0035	104 .0032	112 .0030
	24	E	128 .0026	144 .0023	160 .0021	176 .0019	184 .0018	192 .0017	208 .0016	224 .0015

Fig. 67-3.   Index chart for quick change gearbox.

(South Bend Lathe, Inc.)

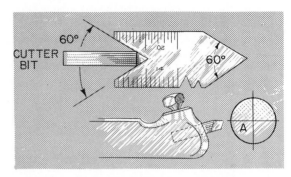

Fig. 67-5.   Testing a threading tool for correct angle.

Fig. 67-4.   Chasing an external thread.

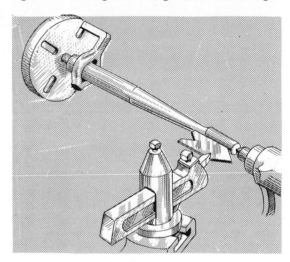

Fig. 67-6.   Setting threading tool square with work.

To find where to place the left-hand lever, find the capital letter assigned to the row in which the 20 is listed. The letter is listed in the second column from the left marked "left-hand tumbler" and, in this case, the letter is "B". Place the left-hand lever in the "B" position.

3. Set the feed selector lever on the carriage apron in neutral, which unlocks the half-nut lever for use.

4. For cutting a 60°, right-hand, external thread, set the compound rest at 29° or 29-1/2° to the right as in Fig. 67-4.

5. Sharpen a threading tool, testing the angle for accuracy with a center gage, Fig. 67-5. The tool should have no back rake, but may have several degrees of side rake. If available, a formed threading tool, Fig. 49-4, may be used.

6. Install the threading tool so that it is square with the work, Fig. 67-6, and with the point exactly on center.

7. Set the lathe rpm for about one-third the speed normally used for turning material of that kind and diameter.

8. Advance the tool until the point almost touches the work, then set the micrometer collar on the cross-feed crank to zero. **The cross slide must always be reset to this zero before each cut.**

9. With the carriage handwheel, position the tool just off the end of the workpiece. Then start the lathe and engage the half-nut lever when a line on the thread dial is even with the reference mark on the thread dial housing, Fig. 67-7.

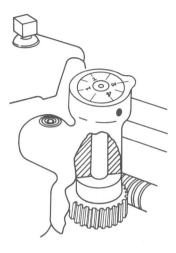

Fig. 67-7.   Thread dial attached to lathe carriage.

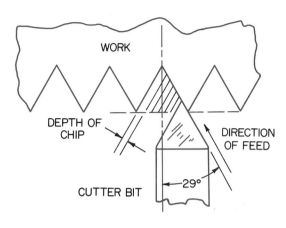

Fig. 67-8.   Action of thread-cutting tool when compound rest is set at 29° angle.

When cutting an even number of threads per inch, the half-nut may be closed at any line on the dial. When cutting an odd number of threads per inch, only numbered lines may be used. For half-threads, use only even numbered lines; for quarter threads, use only odd numbered lines.

10. As the tool moves across the surface of the workpiece, advance it with the **compound-rest crank** until the tool has entered the surface slightly. At the end of the cut, stop the carriage by disengaging the half-nut lever, and withdraw the tool with the **cross-slide crank** far enough for it to clear the work. Then crank the carriage to the right to position the tool for the next cut, and return the cross-slide crank to its zero dial reading.

11. Advance the tool for the next cut with the **compound-rest crank.** For the first several cuts, the tool should be moved about 0.08 mm (0.003"). This amount should be reduced to about 0.05 mm (0.002") when the thread is about half depth, and to 0.025 mm (0.001") when the thread nears full depth.

Threads are normally chased by advancing the tool from the compound rest, causing the tool to cut primarily on one side of the thread groove, see Fig. 67-8. This technique is preferred because it allows tools to cut cleanly and produce threads with good finishes.

Plunge cutting of threads by advancing the tool with the cross-slide crank, causes the tool to cut on both sides at once, making use of a side rake impossible. As a result, thread finishes are generally inferior. Under certain conditions, however, plunge cutting can produce acceptable results.

12. Continue cutting the thread as explained above. When threading steel, keep the tool lubricated with a good cutting oil, Fig. 67-4.

13. When the thread nears completion, stop and check it for size, using whatever thread measuring system is appropriate for the degree of accuracy required. See Unit 26.

14. Machine the thread to final size, measuring after each cut to avoid machining the thread undersize.

Procedure for Cutting 60° Internal Threads

For cutting threads on the inside of a hole, a boring bar fitted with a threading tool is required, Fig. 67-9. The tool must be provided with more clearance than a tool for cutting external threads in order to prevent the heel of the tool from contacting the hole wall before the point.

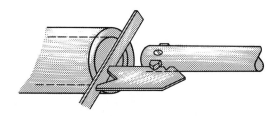

Fig. 67-9.   Setting the boring bar for cutting internal threads.

1. Drill and bore a hole of the size required. Be sure to provide an undercut at the end of the section to be threaded.
2. Set the compound rest at 29° or 29-1/2° **to the left,** which positions the compound-rest crank in front, or use the **reciprocal** setting (180° from this position), positioning the crank to the rear.
3. Install a correctly sharpened tool for cutting internal threads. Align the tool with a center gage as shown in Fig. 67-9.
4. Securely fasten the tool in the boring bar and the boring bar toolholder in the tool post.
5. When cutting internal threads where the end of the threaded section cannot be seen, measure the distance from the end of the workpiece to the center of the undercut. Mark this length on the boring bar with layout fluid or a marking pen, measuring from the point of the tool. Then, when the mark on the boring bar is even with the end of the workpiece, the half-nut can be disengaged with confidence, knowing that the point of the tool has reached the center of the undercut.
6. With the apron handwheel, advance the tool into the hole about 3 mm (1/8″). Then with the cross-feed crank, adjust the tool until the point is close to the surface of the hole. Set the cross-feed dial for zero.
7. Set the lathe for proper rpm and thread pitch. Start the lathe, then engage the half-nut when the appropriate line on the thread dial is even with the line on the rim of the thread dial housing. As the tool enters the hole, advance it with the **compound-rest crank** until it enters the surface of the hole slightly.
8. When the threading tool reaches the undercut, stop the carriage by disengaging the half-nut. Then move the tool away from the hole surface by turning the **cross-feed crank** one revolution **clockwise.**
9. Crank the carriage back to the starting point, return the cross slide to the zero position, and advance the tool with the **compound-rest crank** for the next cut.
10. Close the half-nut at the appropriate line to start the cut. Make repeated cuts as above until the thread is finished to size.

Some operators prefer to cut internal threads by **starting at the back of the hole.** To cut threads in this manner, mount the point of the tool to the rear of the boring bar, cutting edge up. Set the lathe spindle to run in reverse, and the carriage to move from left to right. Follow the same general procedure for cutting the thread as described above.

Resetting the Tool After Its Removal

When it is necessary to remove the threading tool for sharpening or other reasons before the thread has been cut to the depth desired, the tool may be reset in the partially cut thread groove as described below.

First, the tool must be set square with the work. Then, with the tool clear of the workpiece, start the lathe, and engage the half-nut lever. Stop the lathe, leaving the half-nut engaged. Then feed the tool into the original thread groove by manipulating the cross-feed crank and the compound-rest feed crank as necessary.

When the tool is fitted to the original thread groove, set the micrometer collar on the cross-feed crank to **zero.** Then proceed with cutting the thread in the usual manner.

Cutting Left-Hand Threads

A left-hand thread is cut in the same manner as for a right-hand thread, except that the direction in which the feed screw rotates is reversed, thus causing the direction of travel of the tool to change. On most lathes, this is accomplished by means of a lever at the left-hand end of the headstock. On some, however, it is accomplished by means of a lever on the apron.

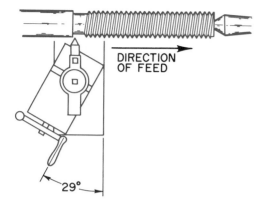

Fig. 67-10. Setup for cutting left-hand screw threads.

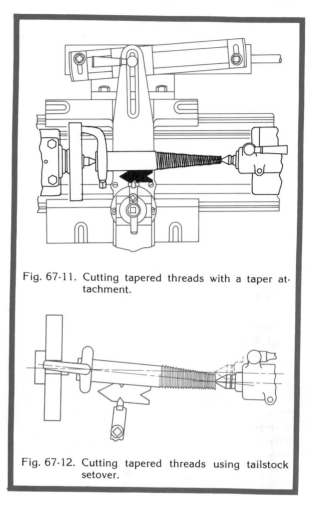

Fig. 67-11. Cutting tapered threads with a taper attachment.

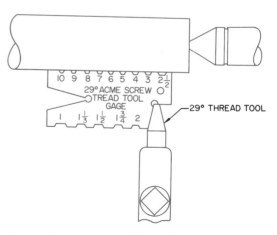

Fig. 67-12. Cutting tapered threads using tailstock setover.

In cutting left-hand threads, the compound rest is swiveled 29° to the left, instead of to the right. The thread is cut with the carriage feeding from left to right (toward the tailstock). See Fig. 67-10.

Cutting Tapered Threads

For cutting tapered threads, such as pipe threads, the workpiece is first machined to the required taper. Either the taper attachment or the tailstock setover method is used.

The threading tool must be aligned **with the straight portion of the workpiece,** regardless of which method is used. See Figs. 67-11 and 67-12. The thread is then cut in the usual manner.

Cutting Acme Threads

The procedure for cutting Acme threads differs only in the shape of the threading tool required and the angle the compound rest must be set.

The threading tool is ground to an included angle of 29°, and the width of the flat on the end of the tool is made to fit the notch in the gage corresponding to the pitch of the thread to be cut. See Fig. 67-13. The compound rest is set at 14°, the threading tool is aligned with the workpiece as in Fig. 67-13, and the thread is cut in the usual manner.

Using a Thread-Cutting Stop

The **thread-cutting stop** eliminates the need to use a zero micrometer dial reading to relocate the threading tool for each cut.

Fig. 67-13. Alignment of threading tool for cutting Acme thread.

When all preparations have been made for threading to begin, advance the tool until it almost touches the workpiece. Then attach the thread-cutting stop to the saddle dovetail, leaving a gap of about 6.35 mm (1/4″) between it and the back of the cross slide, Fig. 67-14. Then, insert the **thread-cutting stop screw,** turning it clockwise until its shoulder is tight against the stop. When the tool is backed away from the workpiece at the end of a cut, the head of the screw will move away from the thread-cutting stop exactly the same amount. When the tool is repositioned for the next cut, the cross-slide crank is simply turned clockwise until the shoulder of the thread-cutting stop screw strikes the thread-cutting stop. This method is faster and more accurate than using a dial reading.

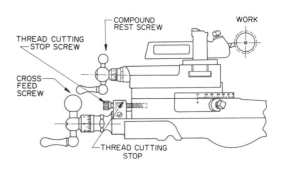

Fig. 67-14. Thread-cutting stop attached to dovetail of saddle.

Unit

68

Milling and Grinding in a Lathe

Some manufacturers offer milling and grinding attachments which extend the usefulness of the lathe. These are particularly useful to small shops which have only occasional need to do milling and grinding.

Milling Attachments

The attachment shown in Fig. 68-1 is a vise for holding the work in the position desired. It may be swiveled horizontally or pivoted vertically for angular cuts. The cutter may be mounted on a taper shank which fits the tapered hole in the lathe spindle, Fig. 68-2, or, if a cutter with a straight shank is used, it may be mounted in a collet or spindle chuck. Cutters also may be

(South Bend Lathe, Inc.)

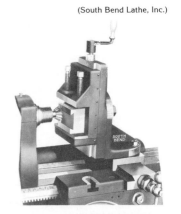

Fig. 68-1. Milling a dovetail on a lathe.

(South Bend Lathe, Inc.)

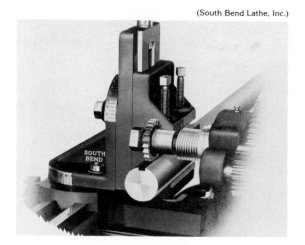

Fig. 68-2. Milling a keyway on a lathe.

(South Bend Lathe, Inc.)

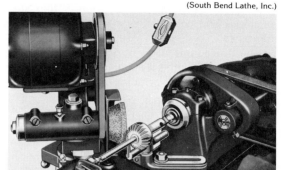

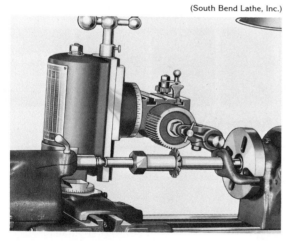

Fig. 68-3. Milling gear teeth on a lathe.

mounted on an arbor held between centers, Fig. 68-3. The workpiece is brought into contact with the cutter by means of the handwheel on the carriage, the cross-slide crank, and the vertical adjusting screw at the top of the milling vise. The cut is controlled through the power cross-feed system.

The attachment shown in Fig. 68-3 is designed for cutting teeth on spur and bevel gears, splines, and other forms of milling requiring the use of a dividing head.

The procedures in setting up a lathe to perform a milling operation after the attachment has been

(South Bend Lathe, Inc.)

Fig. 68-4. Grinding a milling cutter on a lathe.

(South Bend Lathe, Inc.)

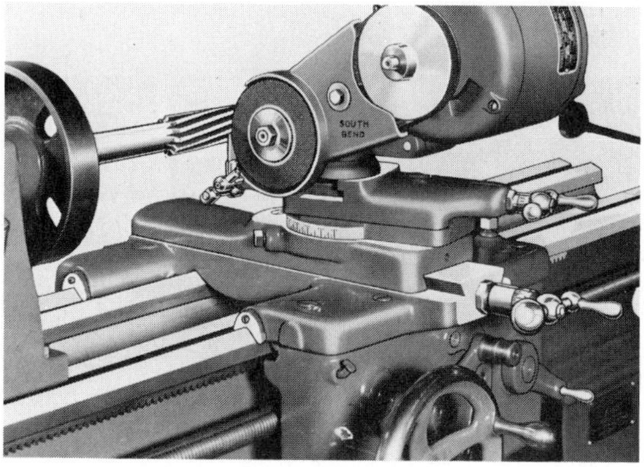

Fig. 68-5. Grinding a reamer on a lathe.

mounted on the carriage are very similar to those described in setting up a standard milling machine. See Section 11.

Grinding Attachments

Lathe-grinding attachments are mounted on the compound rest in place of the tool post, and thus are also known as tool post grinders. An **external** grinding attachment enables the lathe to do precision cylindricai grinding of hardened steel bushings and shafts. With appropriate accessories, it is also capable of sharpening milling cutters, Fig. 68-4, reamers, Fig. 68-5, and other precision cutting tools.

An **internal** grinding attachment, Fig. 68-6, is used to finish round holes in parts like steel bushings and cutting dies after they have been hardened.

The procedures used in performing grinding operations in a lathe are much the same as for a standard cylindrical grinding machine. See Section 13.

(South Bend Lathe, Inc.)

Fig. 68-6.   Internal grinding attachment.

Unit
69

Using a Hydraulic Tracing Attachment

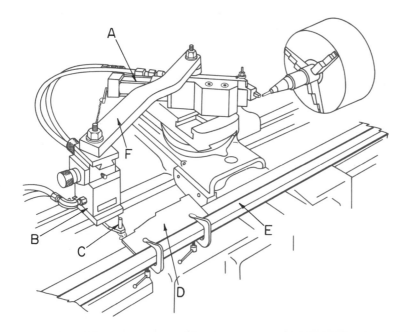

Fig. 69-1. A lathe hydraulic tracer attachment.
  A. Hydraulic cylinder
  B. Valve body
  C. Tracing stylus
  D. Template
  E. Template holder
  F. Arm connecting valve body with tool slide

A **hydraulic tracer attachment,** Fig. 69-1, converts an ordinary lathe into a copying lathe. The tracing attachment consists mainly of (1) a hydraulic cylinder to which a toolholder is attached, (2) a sensitive valve which detects the movement of the **tracing stylus,** (3) a rigid arm connecting the valve body to the toolholder, (4) a template holder, and (5) a hydraulic pump. Tracer attachments are capable of turning parts of almost any shape, but are valued most for their ability to efficiently turn parts with curved contours. Figure 69-2 shows several parts made on a tracer lathe.

The shape of the object to be machined is either turned on a round model or cut into a flat template. As the tracing stylus is drawn across the round model or template, it signals its movements through the valve to the hydraulic cylinder, causing a like amount of tool movement. The profile of the template or round model must be made to the same size and shape as that desired on the workpiece. The surface which is traced should be finished smoothly, since any roughness will be transmitted to the workpiece. It is important to provide about 25.4 mm (1") of land area on each end of the template, Fig. 69-3. The surplus land area at the starting end of the template is necessary to provide a location for the stylus to contact the template without the tool touching the workpiece. At the trailing end, the surplus land area ensures that the tool traverses over the entire length of the workpiece. If only a portion of the workpiece is involved, the template may be shaped to automatically withdraw the tool from the workpiece surface at the desired location.

Setup and Operation Procedure

1. Install the tracer attachment on the compound rest of the lathe. Position the body of the tracer so that its center line bisects the widest angle between opposite or adjacent surfaces on the template. Figure 69-4 shows three setup positions used for machining commonly encountered shapes.
2. Turn on the hydraulic pump.
3. Sharpen and install a cutting tool with a point of the same size and shape as the tracing stylus. A 60° stylus with a small nose radius is quite versatile.
4. Loosen both ends of the arm connecting the tracer slide to the valve body.

(South Bend Lathe, Inc.)

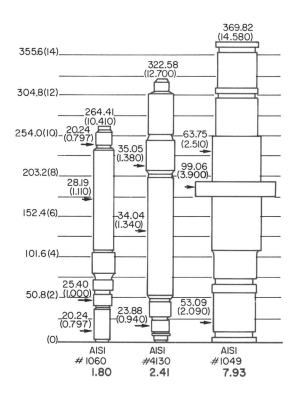

Fig. 69-2. Typical parts made by hydraulic tracing in a lathe. Estimated turning times (numbers in color) are for South Bend Gallicop lathes.

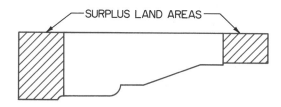

Fig. 69-3. Templates require surplus land areas.

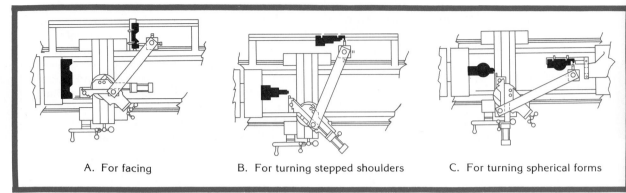

A. For facing   B. For turning stepped shoulders   C. For turning spherical forms

Fig. 69-4.   Tracer body positions most commonly used.

5. Position the point of the cutting tool in line with the end of the workpiece.

6. Place the template on the template bar, and advance the valve body, keeping it square with the template bar. Slide the template to the point where the tracing stylus point is even with the end of the profile to be traced. Clamp the template to the template bar.

7. Tighten the arm connecting the valve body and the tracer slide.

8. Using the cross-slide crank and the carriage handwheel, move the tracing stylus until it just contacts the template at its "highest" point. Note the gap between the tool and the workpiece. Before cutting can begin, this gap should be about 1.6 mm (1/16″). If this gap is more than about 12.7 mm (1/2″), readjust the connecting arm. To make minor adjustments, back the stylus away from the template by adjusting the slide on the valve body. **This moves only the tracing stylus.** Then advance the cross-slide crank until the tracing stylus again just touches the template at its highest point. **This moves both the tool and the stylus.** If the gap between the tool and the workpiece is still too large, continue this procedure until the correct gap is obtained.

9. Reposition the tool at the end of the workpiece, and check to be sure the point of the tracing stylus is aligned with the template at its starting point. If necessary, adjust the stylus by moving the slide on the valve body. Cutting may now begin.

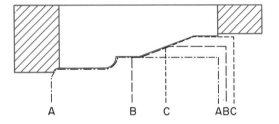

A   B   C   ABC

Fig. 69-5.   General pattern of roughing cuts for hydraulic tracer machining.

10. Move the tool off the end of the workpiece slightly, advance the tool to take the first cut, turn on the lathe, and engage the power feed. Cuts will follow the general pattern shown in Fig. 69-5. At the end of each cut, retract the tracer slide so that the tool is well clear of the workpiece, and crank the carriage back into position for the next cut. When the entire surface has been machined, the workpiece will be oversize by 1.6 mm (1/16″), the size of the gap left between the tool and the workpiece before cutting began.

11. Finishing cuts are made by backing away the tracing stylus from the template the same amount as the amount of tool advancement desired. The adjusting screw on the valve body has micrometer calibrations to allow very precise adjustment of the tracing stylus. Finishing cuts should traverse the entire workpiece surface.

Unit

70

Turret Lathe Tooling and Setup Procedure

Manually operated turret lathes are designed for low and moderate volume production of duplicate parts. Turret lathes appear much more complex than a standard lathe, Figs. 47-2 and 47-3. However, all lathes are operated by the same basic principles. Anyone familiar with the operation of standard metal lathes can quickly learn how turret lathes work.

Turret lathes are fitted with all of the tools needed to machine a particular part. Some parts can be completely machined in a single setup. Often, however, a second setup is required in order to machine surfaces on the back of the piece which were inaccessible during the first setup. Lengthwise movement of tools is controlled by setting adjustable stops built into the machine. Diameters are usually determined by the machine operator who sets the tool location according to a micrometer dial reading. In the hands of skilled machine operators, these machines are capable of making duplicate parts of almost any shape with remarkably consistent accuracy.

Turret Lathe Tooling

A great variety of toolholders and cutting tools have been designed especially for use on the bed turret. Only a basic few can be shown here. Some commonly used turret lathe tools are shown and described in Fig. 70-1.

Tooling may be either **shank-mounted** or **flange-mounted.** Shank-mounted tooling fits into a hole in the turret face where it is held in place by a built-in clamp, Fig. 70-1B. Flange-mounted tooling is bolted directly to the face of the turret, Fig. 70-1D. Shank-mounted tooling can also be held in flange-mounted toolholders. **Multiple**

toolholders are also available, permitting several operations to be carried out from the same position, Fig. 70-2.

The square turret mounted on the cross slide uses conventional turning tools. Large high-speed steel or cast alloy tool bits are mounted directly in the turret. Throwaway carbide inserts are mounted in their special toolholders which in turn are mounted in the turret. The square turret is tooled for any of the usual facing and turning operations. The rear tool post normally mounts an inverted cutoff blade; but chamfering, grooving, and other cuts may be made from this position as well. When fully tooled, the turret lathe may carry from 11 to 16 or more cutting and forming tools.

Turrets are often provided with **permanent setups** of universal toolholders capable of accepting a wide variety of tools. Figure 70-3 shows a permanent setup recommended for making parts from bar stock. With a permanent setup, job

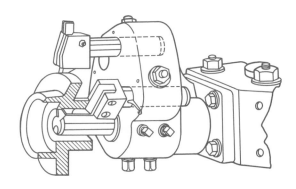

Fig. 70-2.   Multiple toolholder for turret lathe.

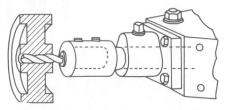

A. Adjustable toolholder suitable for holding drills, reamers, counterbores, etc.

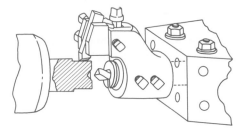

B. Adjustable knee tool for turning and drilling short lengths at the same time if necessary.

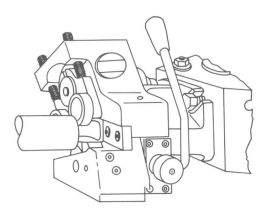

C. Roller turner which cradles the workpiece between ball bearing rollers while it is being turned, thereby preventing the workpiece from springing away from the tool.

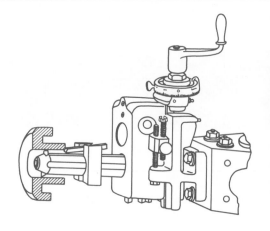

D. Boring bar held in a flange-mounted slide tool which is adjustable to micrometer accuracy.

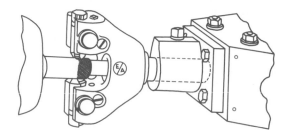

E. Knurling tool capable of making straight, diamond, or square knurling patterns.

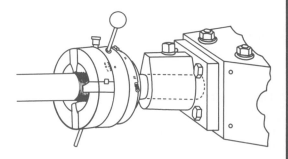

F. Self-opening die head for cutting external threads.

Fig. 70-1.   Turret lathe tools.

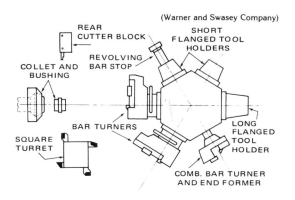

(Warner and Swasey Company)

REAR
CUTTER BLOCK

SHORT
FLANGED TOOL
HOLDERS

REVOLVING
BAR STOP

COLLET AND
BUSHING

LONG
FLANGED
TOOL
HOLDER

SQUARE
TURRET

BAR TURNERS

COMB. BAR TURNER
AND END FORMER

Fig. 70-3.   Permanent bed turret setup recommended for bar machines.

(Warner and Swasey Company)

Fig. 70-4.   Permanent setups quickly adapt to machine a variety of parts.

changeover times are sharply reduced. Figure 70-4 shows an assortment of parts which were turned by installing appropriate tools in the permanent setup shown in Fig. 70-3.

Setup Procedure

Following is an explanation of the setup procedure for machining the male end of a two-piece key chain holder, Fig. 70-5. First, the sequence of machining operations should be determined and recorded on a programming sheet, Fig. 70-6. Then, the tooling arrangement should be sketched out as in Fig. 70-7. The setup would proceed as follows:

1. Sharpen the cutoff and turning tools.
2. Install the correct size collet in the spindle, obtain a bar of the specified material, and adjust the collet closing mechanism for the desired holding power.

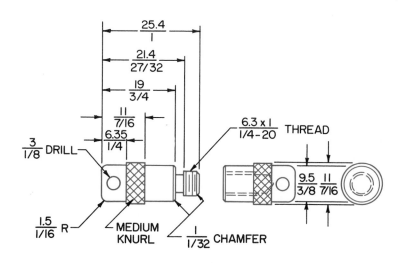

Fig. 70-5.   Two-piece key chain holder.

Turret Lathe Program

Operation	Cross Slide		Carriage Stop # (1 - 4)	Bed Turret Position (5 - 10)
	Tool # (1 - 4)	Dial Reading		
Position stock				5 — stock stop
Knurl				7 — knurling tool
Turn body diameter	1	10	1	8
Turn thread diameter	2	137	2	8
File chamfers				8
Thread				9 — die head
Undercut thread	3	197	2	10
Form rear of body	4	175	3	10
Cut off				4	10

Fig. 70-6.   Program sheet for the male half of a key chain holder to be tooled on a turret lathe.

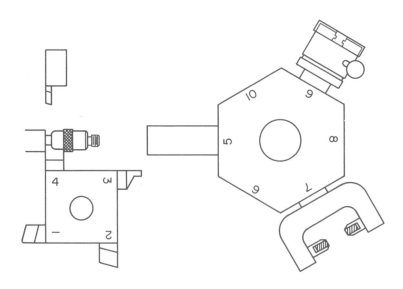

Fig. 70-7.   Turret lathe tooling arrangement for machining male half of key chain holder.

Be sure to support stock projecting from the outboard side of the machine spindle. Unsupported rods may bend and flail about, and may cause serious injury should they strike anyone.

3. Set the lathe for the desired rpm and feed rate.
4. Install the cutoff blade in the rear tool post and adjust it to make a clean cutoff on both the end of the workpiece and the end of the bar. An angle of about 15° on the end of the tool helps obtain a clean cutoff on the workpiece.
5. Install the square turret tools according to the arrangement shown in Fig. 70-7.
6. Mount the stock stop, knurling tool, and self-opening die head in the bed turret as shown in Fig. 70-7.
7. Pull or push about 50.8 mm (2″) of stock out of the collet, and tighten the collet.
8. Rotate the knurling tool into operating position, start the lathe, and adjust the knurling tool to the workpiece diameter. Then advance the knurling tool to within 6.35 mm (1/4″) of the spindle nose. Stop the lathe. Set the turret stop screw which controls the knurling tool travel.
9. Start the lathe, back off the knurling tool, and rotate the turret to bring the stock stop into operating position. Stop the lathe.
10. Measuring from the end of the knurl next to the spindle, mark where the end of the workpiece should be. Cut off the excess stock with the cutoff tool.
11. Bring the stock stop into contact with the end of the workpiece, and set the corresponding turret stop screw.
12. Rotate the square turret to bring tool #1 into operating position. Turn the large diameter to size and length, and note the micrometer dial reading on the program sheet.
13. Set #1 carriage stop to locate square turret tool #1.
14. Rotate the square turret to bring the second turning tool into operating position. Turn the

thread diameter to size and length, and note the micrometer dial reading on the program sheet.
15. Set #2 carriage stop to locate square turret tool #2.
16. File chamfers on the end of the thread diameter and on the body diameter.
17. Back the square turret away and bring the die head into operating position. With the lathe **off** and the die head tripped **open**, advance the die head until it is about 0.4 mm (1/64″) from the shoulder at the end of the thread diameter. Set the turret stop screw controlling the die head travel.
18. With the machine running at slow speed, cut the thread. Then stop the machine and test the thread for correct pitch diameter. Adjust the die head if necessary. Back the turret away until it automatically indexes.
19. Rotate square turret tool #3 into operating position and undercut the thread. Run a thread gage or a nut on the thread to be sure it will go all the way to the shoulder. Note the dial reading on the programming sheet.
20. Set the carriage stop to locate turret tool #3.
21. Rotate square turret tool #4 into position, and advance it to turn the end radius. Note the dial reading on the programming sheet.
22. Set carriage stop #4 to locate the radius tool.
23. Cut off the workpiece. If carriage stop #4 cannot be used to locate the cutoff tool, use any of the other stops together with a suitable spacer if necessary.

   To begin the next cycle:
   a. Open the collet chuck.
   b. Advance the bar of metal.
   c. Rotate the bed turret to bring the stock stop into position.
   d. Close the collet chuck just enough to grip the bar loosely.
   e. Push the bar into the collet by advancing the bed turret as far as it will go.
   f. Retighten the collet.

Machine each successive part by following the procedure on the programming sheet. Check each of the first few parts made to be sure that all measurements are correct. Make any adjustments necessary. Thereafter, check one out of every ten parts or so to be sure that measurements stay within the specified tolerance.

Test Your Knowledge of Section 9

Unit 46: The Metalworking Lathe

1. List seven external machining operations which may be performed on a metalworking lathe.
2. List six hole-machining operations which may be performed on a lathe.
3. To what degree of accuracy can work be produced on a metalworking lathe?
4. How are the size and capacity of a lathe designated?
5. Describe how changes in spindle speed are accomplished on lathes equipped with flat belt drive, geared drive, and variable speed drive.
6. What is the function of the back-gear mechanism on a lathe?
7. How is the amount of longitudinal feed on a lathe expressed?
8. How is the amount of cross feed on a lathe expressed?
9. List the three basic components of the feeding and threading mechanism on a lathe.
10. What control is used to select either power cross feed or power longitudinal feed?
11. What control is used to engage the automatic feed on a lathe?
12. What control is used to reverse the direction of power feed on a lathe?
13. What do the large numbers on the index plate on the quick-change gear box indicate?
14. What do the small decimal numbers on the index plate on the quick-change gear box indicate?
15. How does one determine the amount of cross feed in comparison to longitudinal feed on a lathe, at a given setting?
16. Describe the difference between a live center and a dead center.
17. Why must the center hole in which the dead center is located be lubricated?
18. Why are rotating live centers for tailstocks gaining widespread use?
19. What is the included angle of the point of a lathe center?
20. For what purpose are lathe dogs used?
21. List several important, general safety precautions which should be observed in setting up and operating a metalworking lathe.

Unit 47: Types of Metalworking Production Lathes

1. In what ways do hand screw machines differ from manually operated turret lathes?
2. Explain the main difference in design between ram type and saddle type turret lathes.
3. What is the principle of control used in operating N/C turret lathes?
4. Describe the operation of a vertical turret lathe.
5. List two types of single-spindle automatic lathes and explain the kind of work each is designed to do.
6. Name two kinds of multiple-spindle automatic lathes and describe briefly how their operation differs from single-spindle automatics.
7. For what production volume are hand screw machines suited? Manually operated turret lathes? N/C turret lathes? Automatic lathes?

Unit 48: Methods of Holding Workpieces for Machining

1. Describe two ways that workpieces are held on lathe faceplates for machining.
2. List five different kinds of lathe chucks and the workpiece shapes each can hold.

Unit 49: Lathe Cutting Tools and Toolholders

1. List four types or shapes of single-point cutting tools commonly used on a lathe.
2. How does the size of the nose radius compare on rough- and finish-turning tool bits?
3. Explain how to determine whether a tool is right-hand or left-hand.
4. What kinds of cutting tools are available as indexable throwaway inserts?
5. For what kinds of tools are 16-1/2° toolholders intended?
6. For what kinds of tools are zero-degree toolholders intended?
7. Describe the properties of high-speed steel as a cutting-tool material.
8. List the properties of cast-alloy cutting tools in comparison with high-speed steel tools.
9. What are the properties of cemented-carbide cutting tools in comparison with high-speed steel and cast-alloy tools?
10. Define the meaning of the following terms as they apply to single-point cutting tools:

(a) cutting edge, (b) face, (c) flank, (d) nose, (e) nose radius, (f) shank, (g) point, (h) base.
11. Define the side-relief angle on a tool bit.
12. Define the normal end-relief angle on a tool bit.
13. What end- and side-working-relief angles may be used for general-purpose turning applications with the following kinds of turning tools: (a) high-speed steel tools, (b) cast-alloy tools, (c) tungsten-carbide tools?
14. Define the meaning of back-rake angle.
15. What is meant by side-rake angle?
16. How does a decrease in the side-rake angle affect the type of chip produced by a tool bit?
17. Explain how a chip breaker functions on a tool bit.

Unit 50: Sharpening Lathe and Shaper Tools

1. What kind of grinding wheel should be used for sharpening high-speed steel tools? Cast-alloy tools? Carbide-tipped tools?
2. List the following approximate angles for a general-purpose turning tool: (a) working side-relief, (b) working end-relief, (c) side-rake, (d) back-rake.
3. Describe the recommended rake and relief angles for radius tools.
4. What end-relief angle generally is recommended for cutoff tools?
5. List several precautions to be observed in making setups for cutoff operations.
6. Describe a formed threading tool and tell how it should be resharpened.
7. List the recommended procedure for grinding a high-speed steel tool bit.
8. Why must hand honing of tool bits be done with considerable care?

Unit 52: Cutting Speeds for Lathe Work

1. Find the correct rpm for a heavy cut on a 19 mm (3/4") diameter medium carbon-steel workpiece using a high-speed steel tool.
2. Calculate the correct rpm for a finishing cut on a 50.8 mm (2") diameter cast-iron workpiece using a carbide-tipped tool.
3. Calculate the correct rpm for a heavy cut on a 152.4 mm (6") diameter aluminum workpiece using a high-speed steel tool.

4. Find the cutting speed which results when turning a workpiece of 25.4 mm (1") diameter at 1400 rpm.

Unit 65: Calculations Required for Taper Turning

1. Calculate the angle to set the compound rest for turning a center punch whose point diameter is 3.175 mm (1/8"), body diameter is 9.525 mm (3/8"), and length of taper is 50.8 mm (2").
2. Calculate the angle to set the compound rest to turn a taper 50.8 mm (2") long, with the large diameter 19.05 mm (3/4"), and the small diameter 6.35 mm (1/4").
3. Calculate the tailstock setover required to turn a 50 mm/m (0.600 in/ft) taper on a workpiece of 254 mm (10") length.
4. Calculate the tailstock setover required for a workpiece of 304.8 mm (12") overall length, with a tapered section of 152.4 mm (6") measuring 19.05 mm (3/4") in diameter at the large end, and 12.7 mm (1/2") in diameter at the small end.

Unit 66: Procedures for Taper Turning

1. Explain how to set a tailstock off-center a given amount for taper turning.
2. Describe how a taper attachment is set up for taper turning.
3. List three advantages of using taper attachments for taper turning.
4. Explain how to use the compound rest for taper turning.
5. List three advantages of using the compound-rest method of taper turning.

Unit 67: Cutting Threads in a Lathe

1. What kinds of threads may be cut by chasing in a lathe?
2. Explain briefly how to set up the lathe to cut a 60° thread.
3. Why should plunge cutting of threads be avoided?
4. Explain how the thread dial is used for cutting threads of different pitches.
5. What is the purpose of a thread-cutting stop?

Unit 68: Milling and Grinding in a Lathe

1. Describe briefly how a lathe may be adapted for milling.
2. Name two kinds of grinding attachments made for lathes and give several examples of the kind of work each can do.
3. How should the lathe be protected when using grinding attachments? Why?

Unit 69: Using a Hydraulic Tracing Attachment

1. List the main parts of a lathe hydraulic tracer system.
2. Why should templates have surplus land areas at each end?
3. What is the rule for determining the angle to set the tracer body?
4. At what angle should the valve body be set?

5. Explain how the tool is advanced to make finishing cuts.

Unit 70: Turret Lathe Tooling and Setup Procedure

1. How is lengthwise movement of turret lathe tools controlled?
2. How is cross-slide movement of turret lathe tools controlled so as to machine precision diameters?
3. Explain the difference between shank-mounted and flange-mounted turret lathe tools.
4. What type of machining operations are usually carried out from the bed turret position? Tool post turret? Rear tool post?
5. What is meant by a permanent setup for a turret lathe? Why is it recommended?

Section 10

Milling Machine Operations

Unit

71

The Milling Machine

(Cincinnati Milacron)

A milling machine uses one or more revolving cutters to shape the workpiece, Figs. 71-1 and 71-2. The workpiece is usually held in a vise or fixture attached to a movable table. Cutting takes place by feeding the workpiece against the revolving cutter. On some very large machines, however, the revolving cutter is fed past the stationary workpiece.

Fig. 71-1. A single heavy-duty plain milling cutter mounted in a vertical milling machine spindle.

(Cincinnati Milacron)

Fig. 71-2. Milling a casting with several cutters mounted on the arbor of a horizontal milling machine. This is known as gang milling.

Milling Operations

Milling machines are widely used for machining flat surfaces. Shoulders, grooves, keyways, T-slots, and dovetails are also common milling operations. See Fig. 74-3. Curved and irregular surfaces can also be made with special milling techniques and milling cutters.

Milling machines may also be used for all of the common hole-machining operations normally done on a drill press. With the milling machine, holes may be precisely located without the use of drill jigs. The **micrometer collars** that are attached to each table feed screw are graduated in hundredths of a millimeter (thousandths of an inch) or less. This feature greatly simplifies the production of holes that must have accurate location and depth.

The milling machine is one of the most important machine tools. It is capable of a wide variety of machining operations, and its high metal removal rates make it very efficient. For these reasons, it ranks equal in importance with the metalworking lathe.

Types of Milling Machines

Milling machines may be classified in a general way according to two types: the **column and knee** type and the **bed** type. Machines of the column and knee type, Figs. 71-3 and 71-4, are the most versatile. The bed type generally is a larger

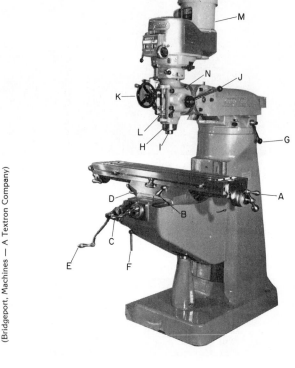

(Bridgeport, Machines — A Textron Company)

Fig. 71-3. A vertical milling machine of the turret type.
A. Longitudinal traverse hand crank
B. Longitudinal traverse lock
C. Cross traverse hand crank
D. Cross traverse lock
E. Vertical traverse hand crank
F. Vertical traverse lock
G. Ram positioning lever
H. Quill
I. Spindle
J. Quill feed lever for coarse feed
K. Quill feed handwheel for fine feed
L. Quill depth stop
M. Motor
N. Adjusting bolt for swiveling the head

machine used for more specialized production milling operations, Fig. 72-3. Milling machines are also classified by whether the spindle is normally vertical, Fig. 71-3, or horizontal, Fig. 71-5.

Vertical Milling Machines

The vertical milling machine shown in Fig. 71-3 is known as a **turret** milling machine. The turret feature allows the head to be rotated in a

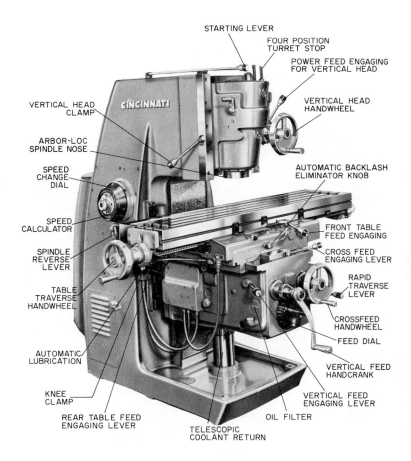

STARTING LEVER

FOUR POSITION
TURRET STOP

POWER FEED ENGAGING
FOR VERTICAL HEAD

VERTICAL HEAD
HANDWHEEL

VERTICAL HEAD
CLAMP

ARBOR-LOC
SPINDLE NOSE

AUTOMATIC BACKLASH
ELIMINATOR KNOB

SPEED
CHANGE
DIAL

SPEED
CALCULATOR

FRONT TABLE
FEED ENGAGING

CROSS FEED
ENGAGING LEVER

SPINDLE
REVERSE
LEVER

RAPID
TRAVERSE
LEVER

TABLE
TRAVERSE
HANDWHEEL

CROSSFEED
HANDWHEEL

FEED DIAL

AUTOMATIC
LUBRICATION

VERTICAL FEED
HANDCRANK

KNEE
CLAMP

VERTICAL FEED
ENGAGING LEVER

REAR TABLE FEED
ENGAGING LEVER

OIL FILTER

TELESCOPIC
COOLANT RETURN

Fig. 71-4.   A vertical milling machine with a fixed head. The head can be raised and lowered but cannot be swiveled or tilted.

horizontal plane wherever needed to be in the best operating position to machine a workpiece. With this type of mill, attachments mounted on the rear of the ram may be brought into operating position over the table. The spindle is carried in a **quill** that can move in and out of the head like a drill press quill. The universal head may be both swiveled and tilted for angular milling operations or for drilling holes at any angle. This type of machine is popular in schools and in tool and die, maintenance, and job shops. Some vertical milling machines have heads that cannot be swiveled or tilted, Fig. 71-4. They are built to withstand the stress of heavy milling operations.

Vertical milling machines use end-milling cutters of various types and sizes, Fig. 74-2. Vertical mills can machine horizontal surfaces, vertical surfaces, angular surfaces, shoulders, grooves, fillets, keyways, T-slots, and dovetails.

Vertical milling machines often are used for hole-machining operations that require extreme accuracy in hole location, such as jig boring operations.

Horizontal Milling Machines

Horizontal milling machines are made in two types, **plain**, Fig. 71-5, and **universal**, Fig. 71-6.

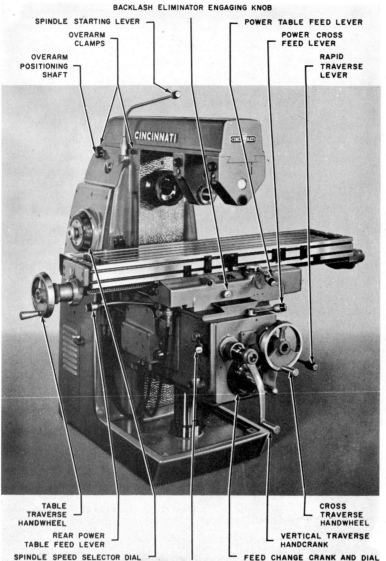

BACKLASH ELIMINATOR ENGAGING KNOB

SPINDLE STARTING LEVER

OVERARM CLAMPS

OVERARM POSITIONING SHAFT

POWER TABLE FEED LEVER

POWER CROSS FEED LEVER

RAPID TRAVERSE LEVER

TABLE TRAVERSE HANDWHEEL

REAR POWER TABLE FEED LEVER

SPINDLE SPEED SELECTOR DIAL

POWER VERTICAL FEED LEVER

CROSS TRAVERSE HANDWHEEL

VERTICAL TRAVERSE HANDCRANK

FEED CHANGE CRANK AND DIAL

Fig. 71-5. A plain horizontal milling machine. The close-up view shows the cutter finishing a groove.

(Cincinnati Milacron)

Plain milling machines have the table fixed at a right angle to the knee. Universal milling machines have a table that can be pivoted in a horizontal plane. This feature allows the universal milling machine table to be swiveled to whatever angle is needed for milling a helical groove.

Attachments, described in Unit 73, extend the range of work that can be done on a plain horizontal milling machine. They enable it to do many of the operations normally done on a universal horizontal or vertical milling machine.

The universal horizontal milling machine is used widely by toolmakers and maintenance machinists. With its full range of attachments, it is the most versatile of all types of milling machines.

Column and Knee Milling Machines

Milling machines of the column and knee type have a table that travels longitudinally on the **saddle**. The saddle, which travels transversely (crosswise), is mounted on the **knee**. The knee is mounted on the **column** in a way that permits it, with the table, to be lowered or raised to the desired height. Thus, the three directions in which the table may be moved on a column and knee machine are **longitudinal, transverse,** and **vertical,** Fig. 71-7.

Bed Milling Machines

Bed-type milling machines, Fig. 72-3, are usually manufacturing or production machines used for producing duplicate parts rapidly. They are not as versatile as standard column and knee machines. The table on a bed machine rests on a stationary bed. Therefore, it is very sturdy and rigid. On many bed-type machines, the table travels longitudinally only.

Bed-type milling machines are often equipped with more than one spindle. These are located in spindle heads that may be advanced or withdrawn to adjust the depth of cut. The heads also may be adjusted so that the cutter can be properly located over the workpiece.

Size

Four basic size factors are used in identifying milling machines: horsepower, capacity, model, and type.

The **power rating** of a machine is based on the power of the spindle drive motor. On machines of standard size, this rating may vary from 2.238 to 37.3 kW (3 to 50 hp). On smaller machines, it may be as low as 0.373 kW (1/2 hp).

The **capacity** or **size** of the machine is based on the amount of longitudinal table travel. The overall capacity of the machine is also related to the amount of cross and vertical table travel. However, only the longitudinal travel is used in identifying the machine size. The six standard sizes that apply to column and knee milling machines follow:

Amount of Table Travel

No. 1	559 mm (22")
No. 2	711 mm (28")
No. 3	864 mm (34")
No. 4	1067 mm (42")
No. 5	1270 mm (50")
No. 6	1524 mm (60")

The column and knee machine shown in Fig. 71-5 is No. 2 size.

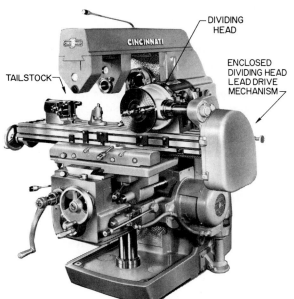

(Cincinnati Milacron)

Fig. 71-6. A universal horizontal milling machine equipped for helical milling. The close-up shows the workpiece mounted between the dividing head centers. A fluting cutter is mounted on the arbor.

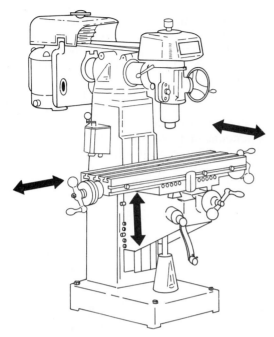

Fig. 71-7. A column and knee design permits three axes of table movement. The spindle drive system guard has been cut away to show the V-belt drive.

The **model** designation is determined by the manufacturer. Features vary with different brands.

The **type** of milling machine is either plain or universal, horizontal or vertical, and column and knee or bed type. Machines may also have other special designations of type.

Part Names and Functions

As Figs. 71-3, 71-4, and 71-5 show, milling machines have many parts. The names and functions of these parts can be learned by studying the illustrations and reading the accompanying descriptions.

Spindle Speed Adjustment

Change in spindle speed is made by operating a **speed-change dial** (or lever), Fig. 71-4. Sixteen different speeds are available on standard machines. Machines designed to operate at high speeds have an even wider range of speeds. On newer machines, speed and feed changes can be made from either the front or side of the machine.

Light-duty machines are usually powered with V-belt drives, Fig. 71-7. With standard V-belt drives, speeds are changed by shifting the V-belt to another pulley combination. Some V-belt drives provide four direct speeds and four back-geared speeds.

Variable speed V-belt drives provide an infinite range of speeds within two speed ranges. See Fig. 71-8.

CAUTION
Variable speed drive systems normally should be changed only when the machine is running.

Power Feed Systems

Many milling machines have power longitudinal and transverse feeds. Some column and knee milling machines also have power vertical feed. Most machines with power feed have adjustable **trip dogs**. These devices can be set to automatically stop the power feed at the end of the cut.

Feed rate is selected by operating a dial or lever at the front or side of the machine, Fig. 71-5. Change in the direction of feed is made with a reversing lever at the front of the machine. Levers are provided at the front of the machine to start and stop each of the power feeds.

Some milling machines are equipped with a **power rapid feed**. This enables the operator to move the knee or table rapidly in any direction with power. Skilled operators are able to speed up production through careful use of power rapid feed.

CAUTION:
Serious damage can result if the workpiece strikes the cutter or arbor at a high rate of feed. Beginners should not use power rapid feeds until they have received proper instruction in their use.

Direction of Feed

The direction of feed in relation to the direction of cutter rotation is one of the most important factors in milling operations. The direction of feed may be set for either of two methods of milling: down milling and up milling. When the direction of feed opposes the direction of cutter rotation, as in Fig. 71-9, the method is called **up milling**. When the direction of feed is in the same direction as the cutter rotation, Fig. 71-10, the method is called **down milling**.

Until recent years, up milling was used almost exclusively. Today, however, down milling is being used more because it results in smoother finishes.

CAUTION

Down milling is recommended only on machines provided with anti-backlash devices or on machines equipped with preloaded recirculating ball bearing table screws.

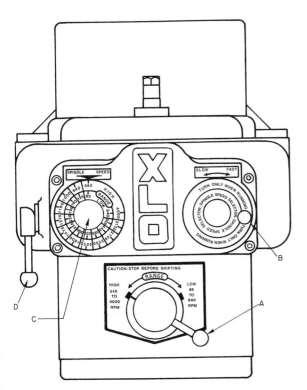

Fig. 71-8. A variable speed drive system on a vertical milling machine.
A. Speed range selector lever
B. Speed adjusting handwheel
C. Speed indicator dial
D. Spindle brake lever

Most light-duty machines and older machines do not have anti-backlash devices. Modern machines of the type shown in Fig. 71-5 have this feature.

Backlash

Backlash is caused by normal wear between a conventional table screw and the nut in which it turns. The wear produces a certain amount of clearance between the screw and the nut. The clearance allows the table screw to revolve a certain amount without a corresponding amount of table movement. This can result in table positioning errors great enough to scrap workpieces. It can also cause cutter breakage during down milling.

To manually compensate for backlash, the operator must make all table moves by revolving the table screw in the same direction. For example, two holes are to be drilled 25.4 mm (1″) apart. The first hole position is located by first turning the table screw clockwise far enough to eliminate backlash. (The table must actually be moving as a result of the screw rotation.) The micrometer dial on the table screw is then set for zero. The first hole is drilled, then the table screw is moved clockwise just far enough to move the necessary 25.4 mm (1″). The second hole is then drilled.

To return the table to the starting point, the operator must turn the table screw counterclockwise. If the operator revolves the screw only enough to return to the zero location used for the first hole, the table will have moved **less** than 25.4 mm (1″). The difference will be equal to the amount of backlash in the table screw.

To accurately reposition the table, the operator must rotate the table screw about a half-turn counterclockwise past the desired dial reading. Then by rotating the table screw clockwise to the desired dial reading, the operator will eliminate the backlash and the table will be accurately positioned.

Up Milling

In up milling, the cutter tooth starts with a chip of zero thickness and ends with a thick chip. The cut starts in clean metal and ends by lifting off the rough surface scale. This procedure increases cutting tool life where a hard or dirty surface exists on the workpiece. The forces caused by the cutter on the workpiece act in a direction that tends to pull the workpiece out of the vise or fixture. Hence, the workpiece must be fastened very securely for up milling. Up milling also tends to push the workpiece away from the cutter, thus eliminating backlash.

Up milling should, therefore, be used on all milling machines not equipped with anti-backlash devices or ball bearing table screws. Up milling is also recommended for cutting the softer steels and other ductile metals. Because of its hard surface, cast iron should be machined this way.

Down Milling

With down milling, the cutter tooth starts with a chip of maximum thickness and ends with a thin chip of zero thickness. A scraping action results as the thinned edge of the chip is removed. Hence, down milling generally produces a good-quality surface finish.

The direction of cutter rotation in down milling tends to push the work down against the table, thus resulting in more rigid setups. This factor is an advantage when cutting thin workpieces held in a vise or workpieces held on a magnetic chuck.

Metal sawing with thin cutters tends to cause the cutter to **walk** while up milling. This tendency is reduced and straighter cuts are made by using the down-milling technique. Down milling often is recommended where carbide cutters are used. Since the cut starts with a chip of maximum thickness, there is less rubbing action and, therefore, less wear on the cutting tool.

Methods of Holding Work

There are many ways of holding a workpiece while it is being machined on a milling machine. The most common are: (1) using a special jig to accurately position the work for machining and at the same time hold the work securely with clamping bolts, studs, and screws; (2) fastening the work securely to an angle plate with clamps or bolts; (3) using a vise that is bolted to the machine table; (4) supporting the work between the centers of a dividing head; and (5) using a chuck that screws onto the spindle of a dividing head. (The dividing head is shown and explained in Units 73 and 80.)

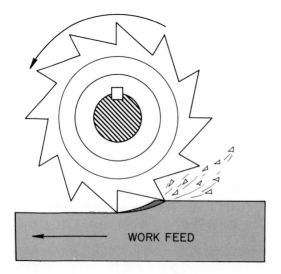

Fig. 71-9. Up or conventional milling

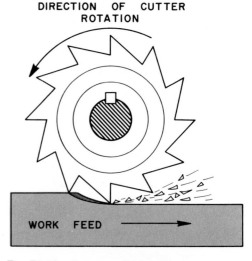

Fig. 71-10. Down or climb milling

Safety Precautions for Milling Machines

1. Wear approved safety goggles.
2. See that the table is clean and dry before mounting holding devices for the work.
3. Be sure that the bases of work-holding accessories or workpieces are wiped clean before they are fastened to the table.
4. Check to see that the workpiece is mounted securely.
5. Select and mount the proper cutter, and see that it revolves in the proper direction.
6. Be sure that the arbor, cutter, and collars are clean before mounting them on the machine. Use a rag to handle sharp cutters.
7. Keep your fingers clear of the arbor hole when replacing the overarm support. Fasten the overarm and overarm support securely.
8. Use only correctly fitting wrenches on the machine.
9. **Tighten or loosen the arbor nut only with the overarm support in place. Doing otherwise may result in a bent arbor.**
10. Use the proper spindle speed, feed, and depth of cut. With constant speed machines driven by gears or belts, make these adjustments only while the machine is stopped. However, change variable speed drive systems only while the machine is running.
11. Use a lead or soft-head hammer to seat workpieces in the vise for setups.
12. Make certain that the table, holding device, and workpiece will clear the arbor and arbor support during the cut.
13. Disengage the manual feed hand wheels when automatic feed is to be used.
14. Loosen the table, knee, and saddle clamps when making setup adjustments. After completing the setup, keep one clamp loose to allow the desired direction of table movement. However, tighten all other clamps to prevent unwanted movement in the other directions.
15. Make sure that unauthorized persons are outside the safety zone of the machine when it is started.
16. Keep clear of the revolving cutter. Do not reach over the cutter.
17. Stop the cutter to remove chips, and use a brush for chip removal.
18. Release all automatic feed controls when the job is completed.
19. Keep the floor around the machine free of chips and oil.
20. When you are finished, remove and return all tools and accessories. Brush the machine free of chips and wipe up all excess oil.

Unit

72

Milling Machines for Special Uses

In addition to the basic horizontal and vertical milling machines discussed in Unit 71, other milling machines have been developed to satisfy one or more of the following needs: (1) to increase production rates, (2) to perform specialized operations, (3) to accommodate very large workpieces.

Hand millers, Fig. 72-1, are small, light-duty horizontal or vertical machines equipped only for hand feeding. They are normally tooled to quickly perform a single operation, such as milling screw slots, milling setscrew or wrench flats, or milling keyways.

Jig boring machines, Fig. 72-2, are ultra-precise vertical milling machines. They are made in several sizes and styles and are used mainly for close-tolerance tool and die and jig-and-fixture work. The table-positioning accuracy of jig boring machines is 0.00254 mm (0.0001") or less. On older machines, a system of measuring rods and dial indicators was used to make precise table movements. These have been largely replaced with **digital readout** systems that provide a continuous numerical readout of the table position. See Fig. 72-2 and Units 90 and 91. Jig borers may also be provided with numerical control systems.

Turret drilling and milling machines, Fig. 37-6, are bed-type vertical machines usually equipped for operation by numerical control. The turret is programmed to rotate the desired tool into operating position as needed. Vertical move-

ment of the turret, spindle speeds, table movements, and other machine functions are all programmed to produce parts automatically.

(U.S. Burke Machine Tool Co.)

Fig. 72-1.   A hand milling machine.

Horizontal boring mills, Fig. 72-3, have a horizontal spindle in a quill that can be moved in and out of the machine head like a drill press quill. The machine head can be rapidly positioned anywhere on its vertical column. Some horizontal boring mills have a worktable that can be traversed at right angles to the tool-carrying spindle. The worktable can also be moved toward or away from the spindle to accommodate workpieces of widely different size. Often the workpiece is mounted on a built-in rotary table, allowing the workpiece to be positioned at any angle to the spindle, Fig. 72-4. Many horizontal boring mills are designed

Fig. 72-2.   A jig boring machine equipped with a digital readout system.

Fig. 72-3.   A horizontal boring mill.

for numerical control operation. These mills have automatic tool-changing systems with storage for 30 or more tools, Fig. 72-4.

Some very large horizontal boring mills are made as **traveling column** machines, Fig. 72-5. With this type of machine, the workpiece is anchored to a stationary table and cannot move. Horizontal positioning of the cutter is accomplished by moving the entire machine column along horizontal ways. Multiple-spindle horizontal boring mills are also made, Fig. 72-6.

Planer milling machines are large bed-type machines constructed and operated in much the same manner as the older single-point cutting tool planers. Figure 72-7 shows a large planer mill of the double-housing type with two milling heads on the overhead crossrail, and two side heads. These machines can accommodate very large workpieces such as machine bedframes. They also can efficiently gang-mill quantities of smaller workpieces.

(Giddings and Lewis Machine Tool Company)

Fig. 72-5. A traveling column horizontal boring mill.

(Cincinnati Milacron)

Fig. 72-4. A numerically controlled horizontal boring mill with an automatic tool-changing system and a built-in rotary table.

(Cincinnati Milacron)

Fig. 72-6. A multiple-spindle horizontal boring mill. The operation is automatic by numerical control, machining three parts at a time.

Fig. 72-7. A planer-type milling machine with four milling heads. The close-up shows the overhead cutters and one of the side heads.

Duplex and **triplex milling machines** are planer mills designed for rapid machining of fairly large parts. See Fig. 72-8. The duplex mill has its worktable mounted between two horizontal spindles opposite each other. This permits simultaneous milling of two opposite surfaces on large workpieces, or two-at-a-time machining of smaller workpieces. The triplex machine is like the duplex but has a third milling head mounted vertically over the bed, enabling three surfaces to be milled at the same time.

Milling machines equipped with hydraulic tracers are called **profiling machines**. They shape the workpiece to the exact contour of a master template or three-dimensional pattern. In Fig. 72-9, the tracer and the master pattern are at the right; the tool and workpiece are at the left. Any shape can be quickly copied. These machines are most useful for milling punches and dies used in stamping and forging presses. They are also used for milling mold cavities for plastic, rubber, and metal molding.

(Cincinnati Milacron)

Fig. 72-8.   A triplex milling machine, with a close-up of the three milling heads.

Fig. 72.9.   A vertical milling machine with a hydraulic tracing (copying) attachment.

(Cincinnati Milacron)

Pantograph engraving machines, Fig. 72-10, called **copy millers,** also use rotating cutters. The machine is designed so that the cutting tool is connected to a tracing stylus by pantograph arm linkage. When the tracing stylus is moved to trace the profile of a template or model, the cutter exactly duplicates the path of movement. The size of the engraving is determined by the position of the pantograph arms. Some machines offer reduction ratios between 1:1 and 50:1. Models are often made oversize for convenience and accuracy, and then reduced by pantograph milling to the size required. Copy millers are widely used for precision line engraving, such as for steel rule and micrometer collar graduations. Other uses include sign engraving and mold making with three-dimensional models. Figure 72-11 shows examples of work done on an engraving machine.

Gear hobbing machines cut gear teeth by milling with a cutter called a **hob.** Gear hobbers may have either horizontal or vertical spindles for the gear-shaping cutter, Fig. 72-12. When possible, several gear blanks are fitted on the work holding arbor, so that each machining cycle can produce several identical gears.

(Lars Machine Inc.)

(Kuhlmann American, Inc.; Photo Courtesy of Illinois State University)

A. Precision graduations on machine dial

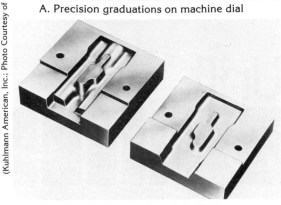

B. Sheet metal stamping dies

Fig. 72-11.   Examples of engraving machine work

Fig. 72-10.   A pantograph engraving machine.

(Barber-Colman company)

Fig. 72-12.   A close-up of gear hobbing on a horizontal gear hobber.

Thread milling machines cut threads by using formed milling cutters that usually cut the thread to full depth in one pass. Universal thread millers may be set up to cut threads of any profile and pitch, and normally use a cutter that cuts a single-thread groove at a time. Production thread millers use cutters wide enough to cut the entire length of thread simultaneously with little more than one revolution of the workpiece, Fig. 72-13.

Fig. 72-13.   A close-up of production thread milling.

Unit
73

Milling Machine Attachments

(Colex International)

This unit presents the most commonly used milling machine attachments. Their use greatly extends the range of work that can be efficiently done on milling machines.

Vises

Vises are often used to hold the workpiece. They are accurately made and have replaceable hardened steel jaws. **Plain vises**, Fig. 73-1, have a flanged base with slots that allow the vise to be bolted to the milling table in one of two positions: (1) with the jaw faces parallel to longitudinal table travel and (2) with the jaw faces at 90 degrees to longitudinal table travel.

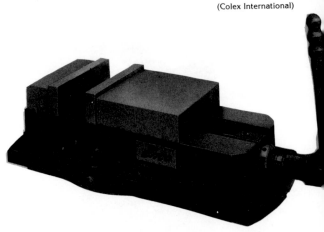

Fig. 73-1.   A plain vise with a lock-down movable jaw. The lock-down feature reduces or eliminates the need to seat the workpiece with a dead-blow hammer, as is required with a conventional vise.

Swivel vises, Fig. 73-2, have a swivel base that is bolted to the milling table. The swivel base is marked with degree graduations. This enables the jaws of the vise to be positioned at any angle desired without moving the swivel base.

Universal vises, Fig. 73-3, can be tilted up, tilted sideways, or swiveled. They are used for holding work that must be machined at a **double** or **compound** angle. **Tilting vises** are similar to universal vises, but they cannot be tilted sideways.

(Colex International)

Fig. 73-2.   A swivel vise with a lock-down movable jaw.

(Universal Vise & Tool Co.)

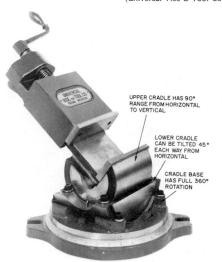

Fig. 73-3.   A universal vise.

The bases of most vises are fitted with **keys**, which are small blocks of steel that fit into the milling table T-slot to provide quick alignment of the vise. See Fig. 73-4.

Vise sizes are designated by the width of the vise jaw in millimeters or inches. Plain and swivel vises range in size from 76 mm (3″) to 254 mm (10″). Tilting and universal vises have jaw widths of 76 mm (3″), 102 mm (4″), or 127 mm (5″).

The Dividing Head

The dividing head, Fig. 73-5, is also called an **indexing head**. It is a very versatile attachment that is mounted on the milling machine table. It is

(Photo Courtesy of Illinois State University)

Fig. 73-4.   Keys attached to the bottom of a vise provide quick alignment on the milling table.

(Cincinnati Milacron)

Fig. 73-5.   A universal dividing head.

(Cincinnati Milacron)

Fig. 73-6.   A chuck being used with a dividing head to hold a workpiece for angular milling.

(Troyke Manufacturing Co.)

Fig. 73-7.   A rotary table with hand feed.

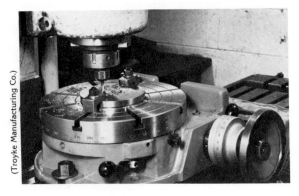

(Troyke Manufacturing Co.)

Fig. 73-8.   Using the rotary table attachment to mill a radial slot.

designed mainly for holding work between centers for machining surfaces, grooves, or gear teeth at precise angular distances apart, Fig. 73-16. A chuck that screws onto the nose of the dividing head further extends its usefulness, Fig. 73-6. The operation of the universal dividing head is explained in Unit 80.

The Rotary Table Attachment

With the rotary table, Fig. 73-7, the operator can perform work such as milling curved slots and spacing holes at given angles apart. Rotary tables can be equipped for accurate indexing just like dividing heads. They are also available with power feeds. A setup with the rotary table in use is shown in Fig. 73-8.

Boring Heads

Boring heads provide the best way to machine round, accurately sized, straight holes with fine finishes. The end of the boring head that carries the boring tool is adjustable radially so that holes of different sizes may be bored. A micrometer dial provides precision adjustment of the tool. Some boring heads are capable of limited facing as well. See Fig. 73-9.

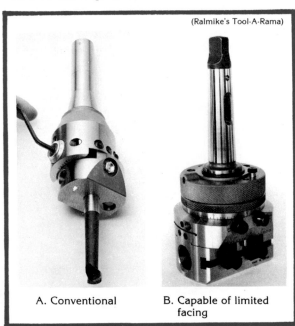

(Ralmike's Tool-A-Rama)

A. Conventional   B. Capable of limited facing

Fig. 73-9.   Boring heads used in milling machines.

The Slotting Attachment

The slotting attachment in Fig. 73-10 will perform vertical shaping operations. The attachment changes the rotating motion of the drive motor to a back-and-forth (straight-line) motion of the tool. Figure 73-11 shows a set of slotting tools.

Angle-Milling Attachments

Ninety-degree attachments, Fig. 73-12, allow a vertical mill to position cutting tools in a horizontal plane. **Universal-angle attachments** enable small cutting tools to be positioned at any angle between vertical and horizontal, Fig. 73-13.

(Bridgeport Machines — A Textron Company)

Fig. 73-10.   A slotting attachment for a vertical milling machine.

(Bridgeport Machines — A Textron Company)

Fig. 73-11.   A set of slotting tools.

(Bridgeport Machines — A Textron Company)

Fig. 73-12.   A ninety-degree attachment for a vertical milling machine.

(Bridgeport Machines — A Textron Company)

Fig. 73-13.   A universal-angle attachment on a vertical mill. Settings at many different angles are possible.

The Vertical Milling Attachment

The vertical milling attachment, Fig. 73-14, is attached to the column and driven by the spindle of horizontal milling machines. The speed of the vertical spindle is usually the same as the speed of the horizontal spindle. The head may be swiveled between 0° and 45° to the right or left of the vertical position. See Fig. 73-6.

(Cincinnati Milacron)

Fig. 73-14.   A vertical milling attachment.

(Cincinnati Milacron)

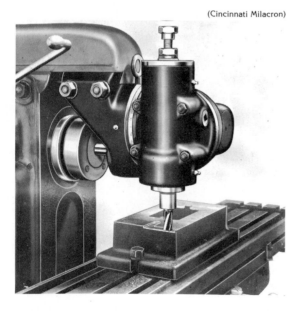

Fig. 73-15.   A high-speed universal milling attachment.

The High-Speed Attachment

The high-speed universal milling attachment, Fig. 73-15, is used on horizontal milling machines. It is designed to adapt these machines to a wide range of end-milling operations. Because of its compound swivel arrangement, the head may be swiveled either parallel or at right angles to the face of the column. The cutter spindle runs at a higher speed than the machine spindle.

The Universal Spiral Attachment

A universal spiral milling attachment makes it possible to mill spirals with a plain horizontal milling machine, Fig. 73-16. This attachment is mounted to the face of the column and is driven by the machine spindle. Its compound swivel arrangement enables the head to be swiveled in both the vertical and horizontal planes. The cutter spindle speed is the same as the machine spindle speed.

This attachment also may be used on universal horizontal milling machines. When used on these machines, spirals with angles greater than 45° may be machined to produce items such as screw threads, gears, and worms. This attachment also may be used to mill rack teeth, Fig. 73-17.

(Cincinnati Milacron)

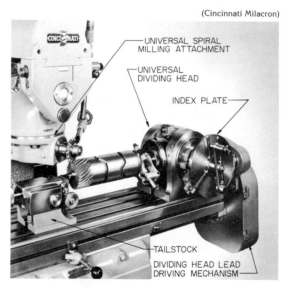

Fig. 73-16.   A universal spiral milling attachment installed on a plain horizontal mill to cut helical gear teeth.

The Rack Milling Attachment

The rack milling attachment, Fig. 73-18, may be used on either plain or universal horizontal

milling machines. For cutting long racks, a long **rack vise,** also shown in Fig. 73-18, is used to hold the workpiece. A special **rack indexing attachment** often is used with the rack milling attachment.

(Cincinnati Milacron)

(Cincinnati Milacron)

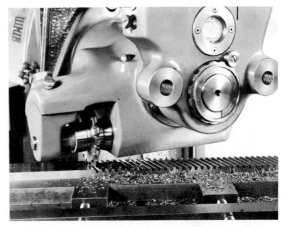

Fig. 73-17. Using a universal spiral attachment to mill gear rack teeth.

Fig. 73-18. A rack milling attachment.

Unit
74

Milling Cutters and Holders

Milling cutters are multiple-toothed cutting tools made especially for use on milling machines. They are made in many standard shapes and sizes, Fig. 74-1.

Cutter Design and Composition

Milling cutters are designed three ways: (1) as solid cutters made of one piece of high-speed steel, cast alloy, or cemented carbide; (2) as solid cutters made of cemented carbide teeth brazed to an alloy steel body; and (3) as inserted-tooth cutters, which have an alloy steel body with replaceable cutters made of high-speed steel, cast alloy, or cemented carbide.

Increasingly, milling cutters with replaceable carbide inserts are being used for production

(Photo Courtesy of Illinois State University)

Fig. 74-1. An assortment of milling cutters. Most standard types are shown.

milling. Their ability to withstand high cutting speeds and feed rates makes them the most economical choice for many applications.

Most cutters used in schools, however, are made of solid high-speed steel or cast alloy. They have good impact and wear resistance. They also are able to withstand the vibration that often occurs on lightweight milling machines most often used in schools.

Rigidity of the machine and the workpiece is important when machining with carbide cutters. Vibration or chatter will often cause the cutters to chip or shatter because of their extreme hardness.

The properties and characteristics of high-speed steel, cast alloy, and cemented-carbide cutting-tool materials are explained in Unit 117. The cutting speeds recommended for milling with cutters made of these materials are included in Unit 75.

Types of Milling Cutters

Milling cutters fall into two major classifications — standard and special. **Standard** types are made according to dimensional standards established by the American Standards Association. The dimensions apply to diameter, width, hole size, size of keyway, etc. **Special** types may or may not be made to standard dimensions. Sometimes they are designed to combine several different milling operations. Some of the most common types of milling cutters and their uses are explained below.

End Milling Cutters

Solid end-milling cutters are made of one piece of cutting tool material, Fig. 74-2. Some end mills are of the **shell type**, Fig. 74-7, in which the cutter body and its shank are separate. End-milling cutters have teeth on the periphery (circumference) and on the end. The teeth on the periphery may be either straight or helical. Solid end mills may have either a **straight shank** or a **tapered shank**. Shell end mill adapters normally have tapered shanks.

End mills may be used for machining horizontal, vertical, angular, or irregular surfaces. Common operations include the milling of slots, keyways, pockets, shoulders, and flat surfaces, Fig. 74-3.

Two-flute end mills have only two teeth. The end teeth are designed so that they can cut to the center of the mill. Therefore, two-flute end mills may be fed into the work like a drill; they then may be fed lengthwise to form a slot. These mills may be either the single-end type with teeth on one end only, or they may be the double-end type, Fig. 74-2.

Multiple-flute end mills have three, four, six, or eight flutes, and normally are available in diameters up to 50 mm (2"). They also may be either the single- or double-end type.

Ball end mills, Fig. 74-2, are used for milling fillets or slots with a radius bottom, for rounding pockets and the bottoms of holes, and for all-round die-sinking and die-making work. Four-fluted ball end mills with center cutting lips also are available.

Roughing end mills, Fig. 74-2, are made to rapidly remove excess metal with minimum horsepower. They have from three to eight flutes depending on the tool diameter.

Insert-type end mills, Fig. 74-4, use replaceable inserts of high-speed steel or carbide. Sizes as small as 25 mm (1") diameter use two inserts. Larger sizes use three or four inserts.

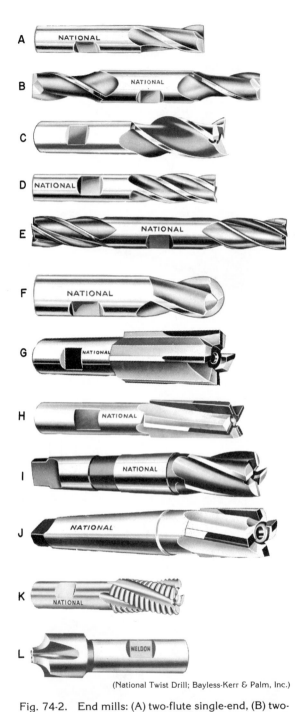

(National Twist Drill; Bayless-Kerr & Palm, Inc.)

Fig. 74-2. End mills: (A) two-flute single-end, (B) two-flute double-end, (C) three-flute single-end, (D) four-flute single-end, (E) four-flute double-end, (F) two-flute ball-end, (G) carbide-tipped straight-flute, (H) carbide-tipped right-hand helical-flute, (I) four-flute with taper shank, (J) carbide-tipped with taper shank, (K) roughing, (L) corner-rounding.

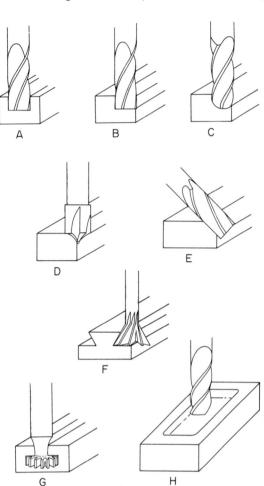

Fig. 74-3. Commonly performed end-milling operations: (A) slot or groove, (B) step, (C) concave radius or fillet, (D) convex radius, (E) angled surfaces, (F) dovetail, (G) T-slot, (H) pocket.

(Kennametal, Inc.)

Fig. 74-4. An end mill designed to use three replaceable insert cutters.

T-Slot Milling Cutters

The T-slot cutter, Fig. 74-5, is a special type of end mill designed for cutting T-slots. In producing a T-slot, a groove for the narrow portion of the T-slot is first machined with an end mill or side mill. The T-slot cutter is then used to complete the slot.

Key Seat Cutters

These cutters are of special design for cutting key seats for Woodruff keys (which have the shape of a half circle). An end mill Woodruff key seat cutter is shown at the bottom of Fig. 74-6. An arbor type also is shown in the same illustration.

Shell end mills, Fig. 74-7, have a hole for mounting the cutter on a short (stub) arbor. The center of the shell is recessed to provide space for the screw or nut that fastens the cutter to the arbor. The teeth usually are helical. These mills are made in larger sizes than solid end mills; normally they are available in diameters from 32 mm (1-1/4″) to 152 mm (6″). Cutters of this type are intended for milling wide, flat surfaces.

Face Milling Cutters

A face milling cutter, Fig. 74-8, is a special form of a large end mill. They are made in sizes 152 mm (6″) in diameter or over. Similar cutters under 152 mm in diameter are called **shell end mills.** Facing cutters usually have inserted teeth that cut on the periphery and the face. Most of the cutting takes place on the periphery, but some finishing also is done by the face teeth. Face milling cutters are used for milling of large flat surfaces.

(Brown and Sharpe Manufacturing Company; Cincinnati Milacron)

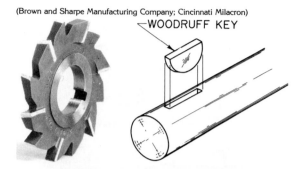

Fig. 74-6.   Woodruff key seat cutters and assembly drawing of a shaft, key seat, and Woodruff key.

(National Twist Drill; Kennametal, Inc.)

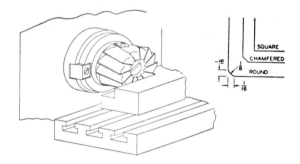

(Cincinnati Milacron)

Fig. 74-5.   A T-slot cutter.

A. One-piece type   B. Replaceable-insert type

Fig. 74-7.   Shell end mills.

(Valenite)

Fig. 74-8. A face milling cutter, replaceable-insert type.

(National Twist Drill)

Fig. 74-11. A heavy-duty plain milling cutter.

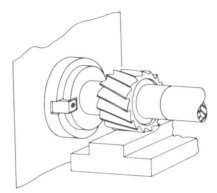

Fig. 74-9. Light-duty plain milling cutters.

(National Twist Drill)

Plain Milling Cutters

Plain milling cutters are cylindrical, with teeth on the periphery only, Fig. 74-9. They have an accurately ground hole for mounting on the milling machine arbor. They are used primarily for milling plain flat surfaces. However, they may be combined with cutters of other types to produce surfaces with various shapes. Plain milling cutters are available in many widths and diameters. The names of the different parts of milling cutters are shown in Fig. 74-10 on the next page. Refer to this figure as you read the following discussion of the three groups of plain milling cutters:

1. **Light-duty plain milling cutters** are available in two different forms. Those that are narrower than 19 mm (3/4″) generally have straight teeth parallel to the axis of the cutter (left, Fig. 74-9). Cutters wider than 19 mm generally have helical teeth at an angle of about 25° (right, Fig. 74-9). The helical teeth cut with a shearing action that makes starting the cut easier than with straight-toothed cutters. In addition, they cut with less chatter and produce a better surface finish. Since the teeth on these cutters are relatively small and have a fine pitch, they are best for light cuts with fine feeds.

2. **Heavy-duty plain milling cutters** are called **coarse-tooth milling cutters,** Fig. 74-11. They are made in the larger widths only and

MILLING CUTTER PARTS

RELIEF ANGLE

CLEARANCE ANGLE

LAND

HEEL

FLUTE OR GASH

TOOTH

RADIAL RAKE ANGLE
POSITIVE SHOWN

FILLET

OFFSET
BACK OF CENTER SHOWN

{PERIPHERAL
CUTTING EDGE}

TOOTH FACE

CHIP BREAKERS

SIDE CLEARANCE

CONCAVITY
OR DISH

RADIAL RAKE ANGLE
POSITIVE SHOWN

RELIEVED SURFACE

TOOTH

FLUTE OR
GASH

FILLET

OFFSET
BACK OF CENTER SHOWN

TOOTH FACE

SIDE RELIEF

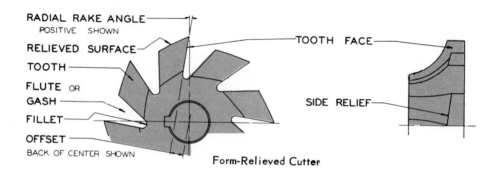

Form-Relieved Cutter

FACE WIDTH

HELICAL TEETH

HELIX ANGLE,
L.H. SHOWN
(AXIAL RAKE ANGLE)

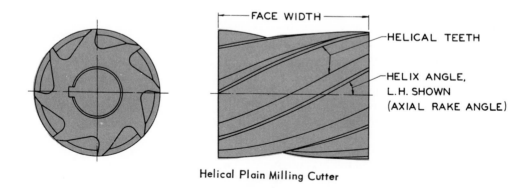

Helical Plain Milling Cutter

Fig. 74-10.   Parts of milling cutters.*

*Extracted from American Standard Milling Cutters, (ASA B5.3 — 1960), with the permission of the publisher, The American Society of Mechanical Engineers.

have larger and fewer teeth than light-duty cutters. The teeth have a helix angle between 25° and 45°. These cutters are designed for heavy plain milling cuts. The strongly supported cutting edges and the wide flutes provide strength and adequate space for heavy chip removal. Wide cutters often are called **slab mills.**

3. **Helical plain milling cutters,** Fig. 74-12, have fewer and coarser teeth than the heavy-duty type. The helix angle of the teeth is between 45° and 60°, and it may be even greater. The high helix angle tends to absorb the load in end thrust. This type of cutter is efficient for taking wide, shallow profiling cuts on brass or soft steel. However, it is not as efficient as the heavy-duty type for heavy feeds and deep cuts on wide, flat surfaces.

Side Milling Cutters

Side milling cutters, Fig. 74-13, are similar to plain milling cutters, but they also have teeth on one or both sides. The teeth on the periphery do most of the actual cutting, while those on the sides finish the side of the cut to size. The teeth may be either straight or helical. These cutters are used for side milling, slotting (Fig. 74-13), and straddle milling. In **straddle milling,** Fig.

74-14, two side milling cutters are mounted on a milling arbor, with the desired distance between the cutters established with spacers. Thus, both sides of the part are machined parallel to each other simultaneously. Several types of side milling cutters are in common use:

1. **Plain side milling cutters,** Fig. 74-13, have straight teeth on the periphery and on both sides. The teeth on the side are provided with **concavity** or taper toward the center of the cutter, thus giving side relief or clearance, Fig. 74-10.

(National Twist Drill)

Fig. 74-12.   A helical plain milling cutter.

(National Twist Drill; Standard Tool Co.)

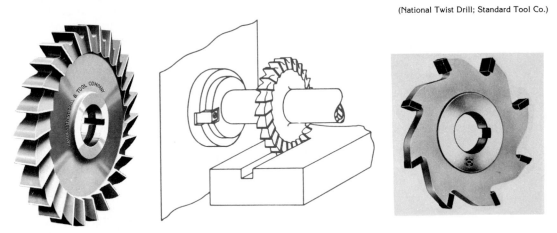

Fig. 74-13.   Plain side milling cutters. A solid high-speed steel cutter is shown at the left, and a carbide-tipped cutter is shown at the right.

 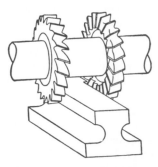 

A. Left-hand cutter   B. Straddle milling setup   C. Right-hand cutter

Fig. 74-14.   Half side milling cutters.

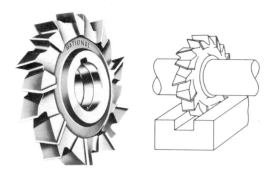

Fig. 74-15.   A staggered-tooth side milling cutter.

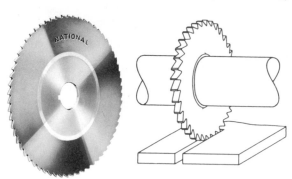

Fig. 74-16.   A plain slitting saw.

2. **Half side milling cutters,** Fig. 74-14, have helical teeth on the periphery and on one side only. Cutters of this type are recommended for heavy-duty face milling and straddle-milling operations where teeth are needed only on one side of the cutter. The teeth are deeper and longer on the side, and thus provide more chip clearance.

3. **Staggered-tooth side milling cutters,** Figs. 74-10 and 74-15, are narrow cutters with teeth alternating on opposite sides. This tooth arrangement reduces dragging and scoring, thus providing a free cutting action. It also provides more space for chip removal. This cutter is recommended for heavy-duty keyway and slotting operations.

Slitting Saws

Slitting saws are cutters designed for cutoff operations and for cutting narrow slots. For deep cuts, saws with side chip clearance should be used.

Plain slitting saws, Fig. 74-16, are essentially thin plain milling cutters. However, the sides are slightly tapered toward the hole, thus providing side relief, which prevents binding on the sides of the cutter. The teeth are much finer than on plain milling cutters. Hence, the feed rate must be much less, usually from one-eighth to one-quarter of that used with plain milling cutters. Plain slitting saws are available in widths from 0.8 mm (1/32") to 4.8 mm (3/16") and in diameters from 63.5 mm (2-1/2") to 203 mm (8").

Slitting saws with side teeth (Fig. 74-17, left) are similar to side milling cutters. The side teeth provide clearance for chips and prevent the cutter from binding on the sides. These cutters are available in widths from 1.5 mm (1/16") to 4.8 mm (3/16"), and in diameters from 63.5 mm (2-1/2") to 203 mm (8"). Cutters of this type are designed for deeper slotting and cutoff operations than those normally done with plain slitting saws.

Staggered-tooth slitting saws (Fig. 74-17, right) are similar to staggered-tooth side milling cutters. They are recommended for cuts of 4.8 mm (3/16") and wider. They may be used for deeper cuts and with standard feeds.

Screw-slotting cutters are plain slitting saws that are designed for cutting slots in screw heads. These cutters have fine-pitch teeth, and therefore are designed for fine feeds. The sides of the cutter are ground straight and parallel. Hence, no side relief is provided. They are available in widths from 0.5 mm (0.020") to 4.6 mm (0.182"), and with a maximum diameter of 70 mm (2-3/4").

Angle Milling Cutters

Angle milling cutters, Figs. 74-18 and 74-19, are used for angle milling operations such as cutting V-grooves, dovetails, serrations, and reamer teeth. Two basic types of angle cutters are available.

Single-angle cutters, Fig. 74-18, have one angular surface, and they have teeth on the angular surface as well as the straight side. These cutters are normally made with 45° or 60° angles.

Double-angle cutters, Fig. 74-19, are used for machining V-grooves. Double-angle cutters with equal angles on both faces normally are available with an included angle of 45°, 60°, or 90°. The angles, however, need not be the same on both sides of the cutter.

(National Twist Drill)

A. With side teeth   B. With staggered teeth

Fig. 74-17.   Slitting saws.

(National Twist Drill)

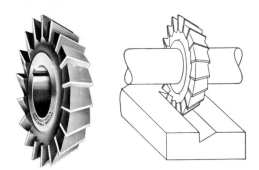

Fig. 74-18.   A single-angle milling cutter.

(National Twist Drill)

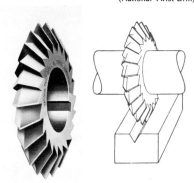

Fig. 74-19.   A double-angle milling cutter.

(National Twist Drill)

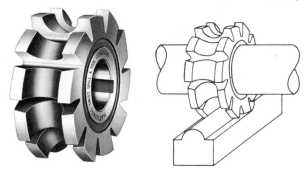

Fig. 74-20.   A concave milling cutter.

Form-Relieved Cutters

Formed-tooth milling cutters are used for machining surfaces with a curved outline. **Concave** milling cutters are used to mill convex half-circles, Fig. 74-20. **Convex** milling cutters are used to cut concave surfaces, Fig. 74-21.

Corner-rounding cutters are used for rounding outside corners, Fig. 74-22. They are available in either a right-hand or left-hand style. **Gear cutters** (**A**, Fig. 74-23) are used for cutting gear teeth. **Fluting cutters** (**B**, Fig. 74-23) are used for cutting flutes in reamers and milling cutters. Formed-tooth cutters also are available in special shapes, such as those shown in Fig. 74-24.

(National Twist Drill)

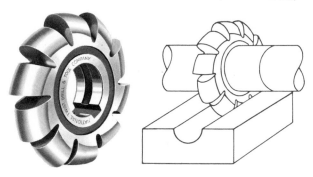

Fig. 74-21.   A convex milling cutter.

(National Twist Drill)

A. Gear milling cutter   B. Fluting cutter

Fig. 74-23.   Form-relieved milling cutters.

(National Twist Drill)

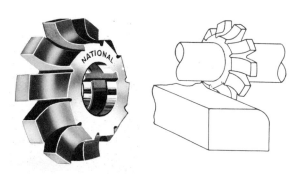

Fig. 74-22.   A corner-rounding milling cutter.

(Brown and Sharpe Manufacturing Company)

Fig. 74-24.   Special form-relieved cutters.

(DeVlieg Machine Co.)

Fly Cutters

A fly cutter consists of one or more single-point tool bits or cutters mounted in a bar or cylinder. There are three types of fly cutters.

One kind of fly cutter is used for boring holes. This type consists of an arbor into which a single-point cutting tool is fastened, Fig. 74-25. Special multiple cutting-tool fly cutters are custom-made for boring several holes of different diameters simultaneously, Fig. 74-26. Each cutting tool is individually adjustable.

A second kind of fly cutter is used primarily for planing flat surfaces, Fig. 74-27.

A third kind of fly cutter is used for cutting grooves. It is also used to cut holes by plunging a single-point cutting tool into the workpiece, Fig. 74-28. This method of cutting is known as **trepanning.**

The "Hand" of a Milling Cutter

Helical fluted cutters are either left-handed or right-handed. **Looking at the cutting end of the cutter, if the flutes twist to the right, the cutter is right-handed.** The **hand** of the cutter determines the direction it must rotate in order to cut. **A right-handed cutter requires counterclockwise rotation when viewed from the cutting end.** A left-handed cutter requires clockwise rotation.

Care of Milling Cutters

Milling cutters should not be nicked by bumping against tools, machines, or accessories. When not in use, they should be stored in a way that prevents damage. When cutters show evidence of

Fig. 74-26. A special fly cutter for boring several diameters simultaneously.

(Colex International)

Fig. 74-27. 5/16″ fly cutter tool bits are used in the above holders to plane flat surfaces.

Fig. 74-25. A fly-cutter adapter used with a single-point cutting tool for cutting holes.

(Courtesy of Clark Tool Co., San Fernando, California)

Fig. 74-28. A fly cutter used for cutting grooves or holes. It is adjustable for cutting different diameters.

A. Spring collet

B. Holder for straight-shank end mill

C. Adapter for taper-shank end mill

Fig. 74-29. Devices for holding end mills.

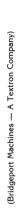

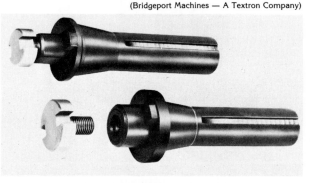

Fig. 74-30. Each of these shell end mill adapters accepts a different size shell end mill.

Fig. 74-31. This quick-change tool system for milling machines uses a ball bearing locking mechanism.

becoming dull, they should be sharpened. If they are allowed to become very dull, extreme forces build up at the cutting edge of the teeth, thus causing possible chipping or fracture. Dull cutters also cause extreme forces on the milling arbor and other parts of the machine.

Milling cutters never should be forced onto the arbor. Doing so may damage the cutter or the arbor.

Methods of Holding Cutters

Several different methods are used to hold milling cutters on milling machines. These methods include (1) collets and special holders, (2) quick-change toolholding systems, (3) arbors, and (4) adapters.

Collets and Holders

Straight-shank end mills are held either in **spring collets** or in **end mill holders**, Fig. 74-29. When a spring collet is tightened, its hole is reduced in size and the collet grips the end mill shank evenly around its circumference. **If the end mill is not tightened securely in a spring collet, it will slip under the pressure of cutting. It may then pull out of the collet and damage the workpiece, the cutter, or both.**

End mill holders for straight-shank end mills use a setscrew to lock the end mill in place, Fig. 74-29B. **Take care to ensure that the setscrew is tightened against the flat surface provided on the end mill**, see Fig. 74-2.

Taper-shank end mills are held in adapters that have holes with matching tapers, Fig. 74-29C. If the taper shank on the tool is smaller than the tapered hole in the adapter, a reducing sleeve is used in the adapter.

Shell end mill adapters are made in several standard sizes to accept the different size shell end mills, Fig. 74-30.

Quick-Change Toolholding Systems

Many quick-change toolholding systems are now available, Figs. 74-31 and 74-32. They speed tool-changing and thereby increase valuable machining time. A full range of adapters is available for holding end mills as well as drill chucks, drills, reamers, and other common drilling tools.

(Cincinnati Milacron)

Fig. 74-32. A quick-change tool system using a flange-and-ring locking mechanism.

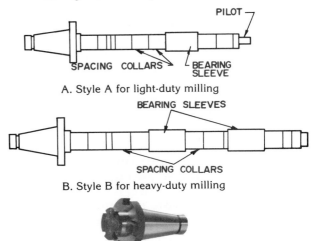

A. Style A for light-duty milling

B. Style B for heavy-duty milling

C. Style C for shell end mills

Fig. 74-33. Milling machine arbors.

Fig. 74-34. A style A arbor used in straddle milling.

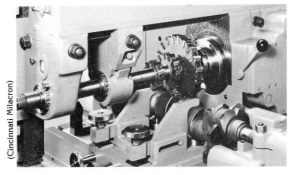

Fig. 74-35. Milling a crankshaft in a special fixture. In this heavy-duty milling operation, a style B arbor is used with two arbor supports.

Arbors

Plain, side, angle, and form-relieved cutters are used mostly on horizontal milling machines. They are usually held on either a style A or style B arbor, Fig. 74-33.

Style A arbors have a pilot at the outer end. The pilot fits in a bearing in the style A arbor support, which is suspended from the overarm, Fig. 74-34. This type of arbor is used chiefly in smaller milling machines and primarily for light-duty milling applications. It provides the distinct advantage of allowing the work to be brought up close to the arbor. Hence, in many setups, small-diameter milling cutters may be used more readily than with the style B arbor. It also is possible to use an inner arbor support with the style A arbor. With this setup, a bearing sleeve is keyed to the arbor and runs in the bearing of the inner arbor support.

Style B arbors do not have a pilot. Rather, they are provided with one or more bearing sleeves which are keyed to the arbor, Fig. 74-33. The bearing sleeves run in the bearings of style B arbor supports, as illustrated in Fig. 74-35. Style B arbors are used on both large and small machines for heavy-duty milling operations where maximum clearance is not required under the arbor supports. Care must be taken to see that the bushings in the support fit the bearing sleeves properly. If they are too loose, chatter will develop; if they are too tight, heat will develop.

Spacing collars are provided on both style A and style B arbors for spacing the cutters and the bearing sleeves. They also keep the arbor straight and rigid. The collars are precision-ground and are lapped on the faces to hold the arbor straight.

CAUTION

Care should be taken to avoid nicks or scratches on the face of the collar. A tiny nick is sufficient to bend the arbor when the arbor nut is turned up tightly.

Before the collar and cutter are installed, they always should be wiped clean. The cutter also should be keyed to the arbor with a properly fitted key. If not, it may slip and score the faces of the collars, thus damaging them seriously.

Style C arbors, Fig. 74-33, are also called shell end mill arbors. They are used for holding shell end mills and face mills that are too small to be bolted directly to the spindle nose of the machine.

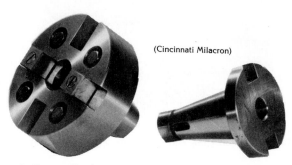

(Cincinnati Milacron)

A. Face mill adapter   B. Collet adapter

Fig. 74-36. Milling machine spindle adapters.

Adapters

Adapters are devices that are used to mount cutters of various types and sizes on a milling machine spindle. The **arbor adapter (A,** Fig. 74-36) is used to mount face mills on the spindle.

The **collet adapter (B,** Fig. 74-36) is used for mounting end mills on the spindle. The tapered hole in this adapter is a **self-holding type,** the most common being the **Morse** taper and the **Brown and Sharpe** taper.

Arbor Shanks

Most manufacturers have adopted the **national milling machine taper,** which is available in four sizes for use on milling machine arbors. These sizes are designated by the numbers 30, 40, 50, and 60. The No. 50 taper is the most common on production machines. No. 30 and No. 40 tapers are used on smaller machines.

Standard milling machine tapers are steep tapers of the **self-releasing type.** The amount of taper is 291.67 mm/m (3-1/2″ per foot) (16° 36′ included angle). Arbors or adapters with this type of taper must be retained in the spindle socket with a locking device such as a **collar** or a **draw-in bolt.** The draw-in bolt is also called a **drawbar.** See Fig. 74-37. Positive drive is provided by two keys bolted to the face of the spindle. The keys engage the slots in the backs of arbors, adapters, and face mills.

Removing Self-Holding Arbors

Some older milling machines and certain small vertical milling machines are equipped with a spindle nose that has a self-holding taper. It is usually a Brown and Sharpe taper. Arbors and arbor adapters usually are installed and held in spindles

Fig. 74-37. The drawbar system for holding milling machine arbors with self-releasing tapers.

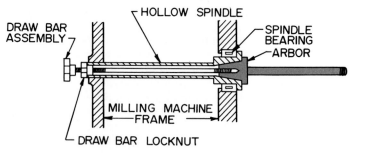

of this type with a drawbar. **The drawbar should not be pulled up too tightly when these devices are installed. Otherwise, they are very difficult to remove.**

The arbors or adapters are removed by turning the drawbar opposite to the direction used for installation. This may be to either the right or the left, depending on the type of machine or adapter. As the drawbar is turned, a shoulder on the drawbar is pushed against a retaining collar; this forces the arbor out of the spindle.

If the arbor does not release when a normal force is applied to the arbor bolt with a wrench, request further instructions from your instructor.

Removing Standard Arbors

To remove a self-releasing arbor or adapter which is held with a drawbar:

1. Loosen the lock nut on the drawbar about two turns.
2. Strike the end of the drawbar with a lead hammer. This will release the arbor or adapter from the spindle hole.
3. Hold the arbor or adapter so that it cannot fall out of the spindle when the drawbar is removed.
4. Unscrew the drawbar and remove the arbor or adapter from the spindle.

Use of Cutting Fluids

The use of a cutting fluid is recommended for all milling operations on steel, aluminum, and copper alloys. Gray cast iron may be machined dry, or it may be cooled with compressed air. For recommended cutting fluids for various metals, see Table A-11, Appendix. Detailed information concerning the nature and purpose of various cutting fluids is included in Unit 116.

Unit
75

Calculating Milling Cutter RPM and Feed Rates

Milling machine operators and numerical control programers must know how to calculate correct milling cutter rpm and feed rates. The cutter rpm rate varies according to the diameter of the cutter, the kind of tool material the cutter is made of, and the cutting speed for the kind of material to be cut. The feed rate varies according to the type of cutter, the size of the cutter teeth, the rigidity of the machine and work setup, and whether a rough or finish cut is to be made.

Cutting Speed

Cutting speed for milling is the circumferential speed of the milling cutter, expressed in meters per minute (mpm) or feet per minute (fpm). It is the distance that the cutting edge of a milling cutter tooth travels in one minute. If the cutting speed is too high during milling, the cutter becomes overheated and dulls rapidly. If the cutting speed is too low, the production rate is low and inefficient.

Rpm refers to the number of revolutions of the milling cutter during one minute. To cut at a given cutting speed, a small milling cutter must rotate at a higher rpm than a larger cutter. To cut at 15.24 mpm (50 fpm), a 25.4 mm (1″) diameter cutter must rotate at 191 rpm, while a 76.2 mm (3″) diameter cutter must rotate at only 64 rpm.

Therefore, to prevent cutter damage, the milling machine operator must know how to calculate or select the correct rpm for a cutter of any size to run at any cutting speed.

Cutting Speed Selection

There is no exact cutting speed for milling any single type of material. It is common practice to start with an average cutting speed, which is then increased or decreased according to the results produced.

Machinability

One of the most important factors determining cutting speeds is the machinability of the metal. **Machinability** refers to the ease with which the metal may be machined.

Machinability ratings are normally compared with that of 1112 steel, which is rated at 100 percent. 1015 steel (a low-carbon steel) is rated at 50 percent. A satisfactory cutting speed for milling this steel is about 24.38 mpm (80 fpm). As the machinability rating increases, the cutting speed also may be increased. Since the rating for 1112 steel is double that for 1015 steel, the cutting speed may be doubled also. Hence a cutting speed of 48.76 mpm (160 fpm) generally will produce satisfactory results with 1112 steel. Some leaded free-machining steels have ratings as high as 300 percent. Steel of this type, therefore, may be machined at about six times the cutting speed for 1015 steel.

The following factors affect cutting speeds (and also feeds) selected for milling operations:

1. Machinability rating of the material being machined.
2. Kind of cutting-tool material to be used (carbon steel, high-speed steel, cast alloy, or cemented carbide).
3. Whether cutting fluid is used, and, if so, the kind of fluid.
4. Type of cutter, its size, and the coarseness of the teeth.
5. Amount of metal being removed (rough cut or finish cut).

Suggested Cutting Speeds

The range of cutting speeds recommended for milling different metals with high-speed steel cutters and with carbide cutters is given in Table 75-1.

For heavy roughing cuts, use the lower end of the recommended speed range. For light finishing cuts, use the upper end of the recommended range. **For carbon-steel cutters, use half the cutting speeds recommended for high-speed steel cutters.** Cutters made of cast alloys may be run at cutting speeds 50 to 75% higher than those recommended for high-speed steel cutters.

When cutting fluids are not used, the lower range of the suggested cutting speeds should be selected. Cast iron is an exception, since it generally is machined dry. Further specific cutting speed recommendations are included in standard handbooks for machinists.

Calculating Revolutions Per Minute

With the cutting speed selected, the problem is to determine the rpm of the cutter, so that the machine spindle speed can be set accordingly. The rpm for a given cutting speed may be calculated with the following formulas:

Metric: $\text{rpm} = \dfrac{\text{c.s. (mpm)} \times 1000}{D \text{ (mm)} \times \pi}$

Inch: $\text{rpm} = \dfrac{\text{c.s. (fpm)} \times 12}{D'' \times \pi}$

Also, the following inch formula is easier to remember and use, and gives satisfactory results:

$$\text{rpm} = \frac{4 \times \text{c.s. (fpm)}}{D''}$$

Example: Calculate the cutter rpm for a 76.2 mm (3″) diameter cutter which is to mill steel at 24.38 mpm (80 fpm).

Metric Solution:

$$\text{rpm} = \frac{24.38 \times 1000}{76.2 \times 3.1416} = \frac{24380}{239.390}$$
$$= 101.84 \text{ or } 102$$

Table 75-1

Cutting Speeds Recommended for Milling in MPM (FPM)*

Material	Brinell Hardness	High-Speed Steel Cutters	Carbide Cutters
Free-machining low-carbon steel, resulphurized, 1111, 1112	100-150 150-200	36.6-48.8 (120-160) 36.6-54.9 (120-180)	121.9-182.9 (400-600) 121.9-274.3 (400-900)
Free-machining low-carbon steel, leaded, 10L18, 12L14	100-150 150-220	30.5-68.6 (100-225) 33.5-76.2 (110-250)	76.2-152.4 (250-500) 76.2-182.9 (250-600)
Plain low-carbon steels, 1006-1026	100-125 125-175	24.4-45.7 (80-150) 24.4-42.7 (80-140)	91.4-182.9 (300-600) 76.2-152.4 (250-500)
Plain medium-carbon steels, 1030-1052	125-175 175-225	24.4-42.7 (80-140) 21.3-39.6 (70-130)	76.2-152.4 (250-500) 68.6-121.9 (225-400)
Plain high-carbon steels, 1060-1095	125-175 175-225	21.3-36.6 (70-120) 18.3-33.5 (60-110)	76.2-137.2 (250-450) 68.6-121.9 (225-400)
Tool steels, W1-W7	150-200	24.4-36.6 (80-120)	91.4-106.7 (300-350)
Tool steels, H20-H43	200-250	12.2-25.9 (40-85)	53.3-91.4 (175-300)
Tool steels, D1-D7	200-250	9.1-18.3 (30-60)	30.5-61.0 (100-200)
Stainless steel, 302	135-185	21.3-30.5 (70-100)	68.6-106.7 (225-350)
Stainless steel, 430F	135-185	30.5-42.7 (100-140)	106.7-137.2 (350-450)
Gray cast iron, ASTM Class 20 Through scale: Under scale:	100-140	24.4-36.6 (80-120) 33.5-36.6 (110-150)	99.1-152.4 (325-500) 121.9-182.9 (400-600)
Malleable iron, ferritic, 32510 Through scale: Under scale:	110-160	42.7-61.0 (140-200) 39.6-68.6 (130-225)	106.7-213.4 (350-700) 121.9-243.8 (400-800)
Aluminum, cold-drawn wrought alloys	. . . .	152.4-243.8 (500-800)	304.8-548.6 (1000-1800)
Aluminum, casting alloys (as cast)	. . . .	182.9-304.8 (600-1000)	365.8-609.6 (1200-2000)
Brass, 360 free-cutting, cold-drawn	. . . .	91.4-152.4 (300-500)	182.9-548.6 (600-1000)
Bronze, 220 commercial, annealed	. . . .	24.4-42.7 (80-140)	54.9-83.8 (180-275)

*Figures are for annealed or cold-drawn materials. Source: Erik Oberg, Franklin D. Jones, and Holbrook L. Horton, **Machinery's Handbook** (New York: Industrial Press Inc., 1979).

Inch Solutions:

$$rpm = \frac{80 \times 12}{3 \times 3.1416} = \frac{960}{9.4248}$$
$$= 101.86 \text{ or } 102$$

$$rpm = \frac{4 \times 80}{3} = \frac{320}{3} = 106.67 \text{ or } 107$$

The need to calculate cutter rpm, however, can be practically eliminated by the use of an rpm chart. Table A-10 in the Appendix gives the correct rpm for selected diameters at different cutting speeds.

Calculating Cutting Speed

Occasionally, the machinist must determine the cutting speed of a cutter of a given diameter when the machine is set at a given rpm. In this case, the following formulas would be used:

Metric: $\text{c.s.} = \dfrac{D\,(\text{mm}) \times \pi \times \text{rpm}}{1000}$

Inch: $\text{c.s.} = \dfrac{D'' \times \pi \times \text{rpm}}{12}$

or $\text{c.s.} = \dfrac{D'' \times \text{rpm}}{4}$

The figures from the rpm problem above can be used to obtain cutting speeds as follows:

Metric Solution:

$\text{c.s.} = \dfrac{76.2 \times 3.1416 \times 101.84}{1000}$

$= \dfrac{24379.469}{1000} = 24.38 \text{ mpm}$

Inch Solutions:

$\text{c.s.} = \dfrac{3 \times 3.1416 \times 101.86}{12} = \dfrac{960.01}{12}$

$= 80 \text{ fpm}$

$\text{c.s.} = \dfrac{3 \times 106.67}{4} = \dfrac{320.01}{4} = 80 \text{ fpm}$

Rate of Feed

Feed is the rate at which the work is moved past the cutter. It is the most important factor in determining the rate of metal removal and overall machining efficiency. The feed rate, in conjunction with the width and depth of cut, determines the number of cubic millimeters (cubic inches) of metal removed per minute.

The tendency with inexperienced milling machine operators is to use a cutting speed that is too high and a feed rate that is too low. The feed rate should be as great as the machine, the cutter, the workholding method, and the workpiece will

Table 75-2

Feeds per Tooth for Milling with High-Speed Steel Cutters

Material	End Mills[1]	Shell end and Face Mills	Plain Mills Heavy-Duty	Plain Mills Light-Duty	Slitting Saws	Form-Relieved Mills
Low-carbon steel, free-machining	0.08 (0.003)*	0.20 (0.008)	0.25 (0.010)	0.13 (0.005)	0.08 (0.003)	0.10 (0.004)
Low-carbon steel	0.08 (0.003)	0.15 (0.006)	0.20 (0.008)	0.13 (0.005)	0.08 (0.003)	0.08 (0.003)
Medium-carbon steel	0.08 (0.003)	0.15 (0.006)	0.20 (0.008)	0.13 (0.005)	0.05 (0.002)	0.08 (0.003)
High-carbon steel	0.05 (0.002)	0.10 (0.004)	0.10 (0.004)	0.05 (0.002)	0.05 (0.002)	0.05 (0.002)
Stainless steel, free-machining	0.08 (0.003)	0.20 (0.008)	0.20 (0.008)	0.15 (0.006)	0.05 (0.002)	0.05 (0.002)
Stainless steel	0.05 (0.002)	0.10 (0.004)	0.15 (0.006)	0.10 (0.004)	0.05 (0.002)	0.05 (0.002)
Cast iron, soft	0.10 (0.004)	0.20 (0.008)	0.31 (0.012)	0.20 (0.008)	0.10 (0.004)	0.10 (0.004)
Cast iron, medium	0.08 (0.003)	0.15 (0.006)	0.25 (0.010)	0.15 (0.006)	0.08 (0.003)	0.08 (0.003)
Malleable iron	0.10 (0.004)	0.20 (0.008)	0.25 (0.010)	0.15 (0.006)	0.08 (0.003)	0.08 (0.003)
Brass and bronze, medium	0.08 (0.003)	0.20 (0.008)	0.25 (0.010)	0.15 (0.006)	0.08 (0.003)	0.08 (0.003)
Aluminum, wrought, cold-drawn	0.10 (0.004)	0.25 (0.010)	0.36 (0.014)	0.20 (0.008)	0.10 (0.004)	0.13 (0.005)

[1]Feeds cited are for 12.7 mm (1/2″) diameter end mills taking a 6.35 mm (1/4″) depth of cut. For finishing cuts, reduce the above feed rates by about 50%.
*millimeters (inches)

Table 75-3

Feeds Per Tooth for Milling with Cemented Carbide Cutters[1]

Material	Brinell Hardness	Shell End and Face Mills	Slotting and Side Mills
Low-carbon steel, free-machining	140-180	0.20-0.50 (0.008-0.020)*	0.15-0.31 (0.006-0.012)
Low-carbon steel	150-200	0.20-0.50 (0.008-0.020)	0.15-0.31 (0.006-0.012)
Medium- and high- carbon steel	120-180	0.20-0.50 (0.008-0.020)	0.15-0.31 (0.006-0.012)
	180-220	0.15-0.50 (0.006-0.020)	0.13-0.25 (0.005-0.010)
Alloy steels with less than 0.3% carbon	220-300	0.13-0.25 (0.005-0.010)	0.08-0.20 (0.003-0.008)
	125-220	0.15-0.50 (0.006-0.020)	0.13-0.31 (0.005-0.012)
	220-280	0.10-0.31 (0.004-0.012)	0.08-0.25 (0.003-0.010)
	280-320	0.08-0.20 (0.003-0.008)	0.05-0.15 (0.002-0.006)
Alloy steels with more than 0.3% carbon	170-220	0.13-0.50 (0.005-0.020)	0.13-0.31 (0.005-0.012)
	220-280	0.10-0.31 (0.004-0.012)	0.08-0.20 (0.003-0.008)
	280-320	0.08-0.20 (0.003-0.008)	0.05-0.15 (0.002-0.006)
Tool steel, annealed	. . . .	0.15-0.50 (0.006-0.020)	0.13-0.31 (0.005-0.012)
Stainless steel	135-185	0.20-0.38 (0.008-0.015)	0.15-0.31 (0.006-0.012)
Gray cast iron	150-220	0.20-0.50 (0.008-0.020)	0.13-0.31 (0.005-0.012)
	220-300	0.15-0.31 (0.006-0.012)	0.10-0.20 (0.004-0.008)
Malleable iron	110-160	0.20-0.50 (0.008-0.020)	0.10-0.31 (0.004-0.015)
Zinc die-casting alloys	. . . . . .	0.13-0.50 (0.005-0.020)	0.10-0.38 (0.004-0.015)
Brass and bronze	100-150	0.20-0.50 (0.008-0.020)	0.15-0.31 (0.006-0.012)
	150-250	0.15-0.36 (0.006-0.014)	0.10-0.25 (0.004-0.010)
Cast aluminum, as cast	. . . .	0.25-0.50 (0.010-0.020)	0.20-0.41 (0.008-0.016)
Wrought aluminum, cold-drawn	. . . .	0.25-0.50 (0.010-0.020)	0.15-0.41 (0.006-0.016)

[1]For finishing cuts, these feed rates should be reduced by about 50%.
*Millimeters (inches)

safely stand, and, at the same time, produce a satisfactory finish. Of course, if the feed rate is too great, the cutter may fracture or the machine or the workpiece may be damaged. Suggested feed rates per tooth per revolution are included in Tables 75-2 and 75-3. The feed rate on a milling machine is controlled in the three ways described below.

Manual Feed

Manual feed is used on some small milling machines that are not provided with a power feed mechanism. This type of feed is accomplished by manually turning the table hand crank to obtain the desired direction of travel. The rate of feed is a matter of operator judgement. With small cutters or with fine-tooth cutters, the rate of feed per cutter tooth should be less than for larger and coarser-toothed cutters.

If the feed rate is too rapid, a poor-quality finish will result, or the cutter may fracture. With a feed rate that is too slow, the cutter will become dull more rapidly, and the rate of metal removal will be inefficient. With hand-feed machines, it usually is advisable to start with a slow feed rate. This may then be increased as the operator acquires the "feel" of the cutting operation. A slower feed rate is better for heavy roughing cuts or for cuts requiring a smooth finish. A faster feed rate may be used for lighter cuts.

Millimeters (Inches) Per Revolution of the Spindle or Cutter

This type of feed mechanism is found on a few small milling machines and on some older machines. As the rpm of the spindle or cutter increases, the feed rate also increases. For example: with the feed rate controls set at 0.127 mm

(0.005") per revolution, a spindle speed of 600 rpm produces a feed rate of 76.2 mm (3") per minute [600 × 0.127 mm (0.005")]. Doubling the rpm to 1200 without resetting the feed controls results in a doubling of the feed rate to 152.4 mm (6") per minute [1200 × 0.127 mm (0.005")].

If a feed of 0.599 mm (0.022") were selected for a 4-tooth end mill, the feed per tooth would be 0.1397 mm (0.0055") regardless of the spindle rpm. If the same feed were used with a 10-tooth plain cutter, the feed per tooth would be 0.0559 mm (0.0022").

Millimeters (Inches) Per Minute

The majority of large milling machines have a feeding mechanism that is set in terms of millimeters (inches) per minute. This type of feed is independent of spindle rpm. Thus, if a feed of 102 mm (4") per minute were selected, this feed rate would remain constant regardless of the spindle rpm selected. **A change in spindle speed would not affect the feed.**

The following is the usual procedure in determining the feed rate setting in millimeters (inches) per minute:

1. Select the desired cutting speed, for example, 24.38 mpm (80 fpm) for low-carbon steel, Table 75-1.
2. Determine the rpm of the cutter.
3. Count the number of teeth on the cutter.
4. Select the amount of feed per tooth, Tables 75-2 and 75-3.
5. Calculate the feed rate, using the following formula: $F = R \times T \times rpm$
   Where: F = Feed rate in millimeters (inches) per minute
   R = Feed per tooth per revolution in millimeters (inches)
   T = Number of teeth
   rpm = Revolutions per minute of cutter

Example: Determine the feed rate for machining low-carbon steel at 24.38 mpm (80 fpm), 122 rpm using a heavy-duty plain milling cutter which is 63.5 mm (2-1/2") in diameter, with 8 teeth, and with 0.127 mm (0.005") feed per tooth.

Metric solution:
   $F = 0.127 \times 8 \times 122 = 123.95$
   or 124 mm/min.

Inch solution:
   $F = 0.005" \times 8 \times 122 = 4.88"/min.$

With the feed-selector dial or feed-selector levers, set the feed rate that is nearest to 124 mm (4.88") per minute for the machine.

Suggested Feed Rates

The feed rates for different milling operations vary considerably. Factors determining the rate of feed include depth of cut, width of cut, diameter of the cutter, number of teeth in the cutter, speed at which the cutter revolves, manner in which the cutter is held, power of the machine, and rigidity of the machine and work. Accuracy cannot be achieved if a heavy cut is taken on a machine lacking in rigidity or on work whose nature makes it impracticable to eliminate vibration.

When all of the above factors have been considered, the feed rate may be determined. It is good practice first to select a feed rate per tooth per revolution from Tables 75-2 and 75-3, and then to modify this rate according to the operating conditions. The rates given are production feed rates for use on sturdy machines. For light duty machines, it may be necessary to reduce these rates 50 percent or more. For finishing cuts, they generally should be reduced 50 percent. Finally, the feed desired must be calculated in millimeters (inches) per minute and the machine must be set accordingly.

Normally at least two cuts are required to finish work satisfactorily — a **rough cut** followed by a

Fig. 75-1.   Undercutting results from the feed being stopped while the cutter continues to revolve.

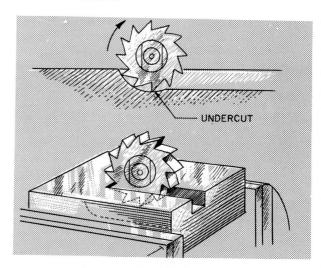

finish cut. Standard practice is to take a heavy rough cut, using as rapid a feed as the cutter will stand without causing excessive vibration. The finishing cut usually is a light one, ordinarily from 0.254 to 0.813 mm (0.010" to 0.032") depth, with the use of a finer feed and, generally, a higher cutting speed.

The feed never should be stopped before the cut is finished. If it is, a slight groove will be milled into the surface of the work, Fig. 75-1, which can be removed only by taking another cut.

Since it is practically impossible to mount a cutter on a milling machine so that it will run perfectly true, some teeth cut slightly deeper than others. This is one cause of a groove being cut into the metal when the feed is stopped. Another cause is that under normal cutting conditions, the cutter tends to spring away from the work because of the pressure being exerted against it. When the feed is stopped, the pressure is released, so the teeth dig a little deeper into the metal, leaving a groove at that point.

Unit
76

Procedure for Milling Flat Surfaces, Bevels, and Chamfers

Broad, flat surfaces are milled in one of two ways: by **plain milling** or by **face milling**. In **plain milling**, the flat surface is machined with a plain milling cutter mounted on a horizontal milling machine arbor. The flat surface is cut parallel to the axis of cutter rotation, Fig. 76-1. Milling wide, flat surfaces this way is called **slab milling.**

Face milling is the process of machining a flat surface parallel to the face of the cutter, Fig. 76-2. Small surfaces are face-milled with end mills, shell end mills, or fly cutters. Large surfaces are face-milled with inserted-tooth face-milling cutters, see Fig. 72-8.

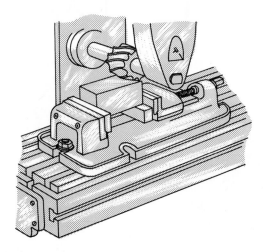

Fig. 76-1. Plain milling a flat surface.

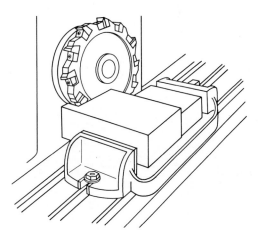

Fig. 76-2. Face-milling a flat surface with a horizontal milling machine.

313

Bevels and chamfers are flat surfaces cut at an angle to the main workpiece surfaces. A **bevel** is an angular cut extending from side to side; it completely removes the perpendicular edge. A **chamfer** is an angular cut that removes only a part of the perpendicular edge. Bevels and chamfers may be cut in two ways: (1) moving the workpiece into an angular cutter, Fig. 76-3, and (2) holding the workpiece at the desired angle while moving it into a plain cutter or end mill. Angular positioning of the workpiece may be with a vise or by using a fixture in a vise as in Fig. 76-4.

Procedure

1. Remove all burrs and surface irregularities from the workpiece by filing or grinding.

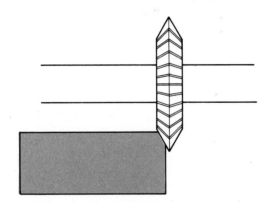

Fig. 76-3.   An angle cutter being used to cut a chamfer.

2. Position the vise on the milling table so that the direction of cut will be parallel to the longest workpiece dimension.

   If the entire workpiece surface is to be milled with a plain milling cutter or with a face-milling cutter in a vertical milling machine, the vise need not be precisely aligned with the worktable. However, for face milling the end of a workpiece square in a horizontal mill, use a dial indicator to align the solid vise jaw perpendicular to the column of the machine. See Fig. 76-5. Swivel and tilting vises should be set at zero degrees, and the solid jaw should be aligned parallel with one axis of the machine table before being fastened in place. Then, for machining a bevel or chamfer, tilt or swivel the vise to the required angle.

3. Obtain a **sharp** cutter of suitable size and shape, and mount it securely in proper position. Be sure to wipe clean all mating surfaces before installing the cutter.

4. Wipe the vise, workpiece, and parallels clean of dirt, chips, and oil. Then mount the workpiece in the vise, seating it firmly with a lead hammer. When the workpiece is properly seated on parallels, the parallels cannot be moved by hand. Be sure the workpiece is

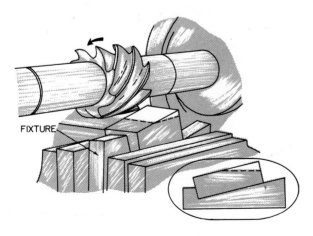

Fig. 76-4.   A fixture being used to position a workpiece at an angle for milling.

Fig. 76-5.   Aligning the vise square with the milling machine column using a dial indicator.

positioned so that the vise jaws will not interfere with cutting. **The workpiece should normally be centered in the vise so that when the vise is tightened the jaws remain parallel. Small workpieces that must be held to one side can only be held securely when balanced with a piece of the same size on the opposite side of the vise.** See Fig. 76-6.

Workpieces too large to hold in a vise may be fastened directly to the machine table with clamping straps and bolts.

5. Position the workpiece so that the surface to be milled is centered on the cutter.

6. Determine the correct rpm for the cutter and set the speed accordingly.

7. Determine the correct feed rate and set the machine controls accordingly.

8. Start the machine and check the direction of rotation of the cutter and the direction of feed. The cutter teeth must point in the same direction as the desired rotation of the spindle, Fig. 76-4. Stop the machine and correct if necessary.

9. With the cutter running, slowly advance the workpiece until it just touches the cutter. Since this results in removing an unknown amount of material, an alternate method is to lightly hold a strip of ordinary notebook paper between the revolving cutter and the workpiece surface. The paper should be about 12.7 mm (1/2″) wide and long enough to keep the hand well away from the cutter. When the cutter is the thickness of the paper away from the surface, a distance of about 0.08 mm (0.003″), a distinct tug will be felt.

Stop the cutter rotation and back the workpiece away so as to position the cutter at the starting side of the cut. Position the table so that the cutter will remove the desired amount from the workpiece for the rough cut. Lock all directions of table movement except the direction in which the cut will be made.

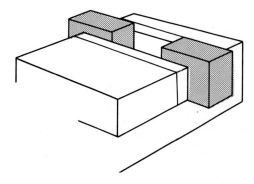

Fig. 76-6. A method of keeping the vise jaws parallel when small workpieces cannot be centered.

10. Start the cutter rotation, advance the workpiece to the cutter, engage the power feed control, and allow the cut to proceed about 6.35 mm (1/4″).

11. Stop the feed. Then stop the cutter rotation. Back the workpiece away from the cutter far enough to measure the workpiece. If the cutter has made the rough cut the desired amount, start the machine and complete the cut. Otherwise, make the necessary adjustment before completing the cut.

12. At the end of the rough cut, adjust the rpm and feed rate for the finish cut.

13. If the rough cut was a heavy cut, start the machine and feed the workpiece past the cutter about 6.35 mm (1/4″) at the same setting used for the rough cut. A small amount of additional material will be removed, indicating the amount the cutter had sprung away from its normal centerline during the rough cut.

14. Measure the workpiece, adjust its position for the finish cut, and again allow the cut to proceed about 6.35 mm (1/4″). Stop the machine, re-check the workpiece size, and make any necessary adjustments before completing the cut.

Unit

77

Procedure for Squaring Stock with a Milling Machine

Sometimes it is necessary to machine all six surfaces of a piece of rectangular stock to a specific thickness, width, and length. The sequence of operations generally followed in performing this task is given below. See also Fig. 77-1. This procedure may be varied, and often is, to meet particular conditions.

Procedure for First Broad Surface

1. Prepare the stock and mount a vise on the table of the machine. Follow steps 1-8 in Unit 76.
2. Proceed as in steps 9-14, in Unit 76. Remove only enough material to fully machine the surface.

Procedure for First Edge

1. Lay a parallel at the bottom of the vise, if needed; then place the machined surface against the stationary vise jaw. Protect it if necessary with soft metal or a heavy piece of paper.
2. Place a rod of about 6.4 mm (1/4") diameter between the movable jaw of the vise and the workpiece, **A,** Fig. 77-2. Be sure to place the rod about halfway between the bottom of the workpiece and the top of the vise jaw. Tighten the vise screw and seat the workpiece with a lead hammer. This procedure assures that the machined surface will align itself parallel with the stationary vise jaw.
3. Proceed as in step 2 under "Procedure for First Broad Surface." Remove only enough stock to fully machine the edge.

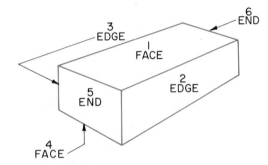

Fig. 77-1.   The machining sequence for squaring a block by milling.

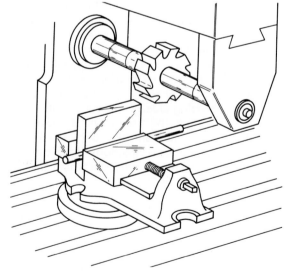

Fig. 77-2.   The stock in position for machining an edge.

Procedure for Second Edge

1. Loosen the vise and turn the workpiece edge-for-edge. Make sure the machined edge lies flat on the bottom of the vise or on a parallel.
2. Proceed as for machining the first edge, but finish the width to size.

Procedure for Second Broad Surface

1. Place the workpiece in the vise with the finished face down. Support it on parallels if necessary, and tighten it securely.
2. Proceed as under "Procedure for First Broad Surface," step 2, but finish the thickness to size.

Procedure for First End

1. The workpiece may be held vertically for machining the ends square. It must not, however, project so far out of the vise as to flex under the pressure of cutting.

   Long, relatively thin workpieces should be positioned horizontally, broad surface down, for vertical milling, **A**, Fig. 77-3. For horizontal milling, place the piece on edge, as in **B**, Fig. 77-3. Square the end with a vertical cut.
2. If the workpiece can be held securely in a vertical position, place one end on the bottom of the vise or on a parallel. Then with a try square held against one edge, test for perpendicularity, Fig. 77-4.

   If the piece does not stand perpendicular, place shims of paper or metal under the low corner; then test again. It is important that the workpiece be positioned precisely vertical. Otherwise, the ends will not be square with the sides of the workpiece.
3. Proceed as under "Procedure for First Broad Surface," step 2, removing only enough material to fully machine the end.

Procedure for Second End

1. Turn the stock end-for-end, and clamp it securely.
2. Proceed as for machining the first end, but finish the workpiece to length.

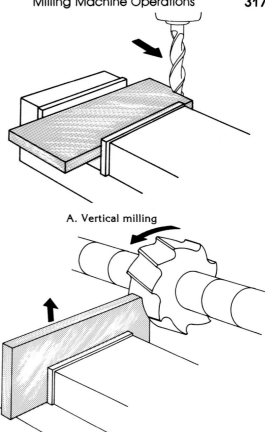

A. Vertical milling

B. Horizontal milling

Fig. 77-3.   Methods of squaring the ends of long, thin workpieces.

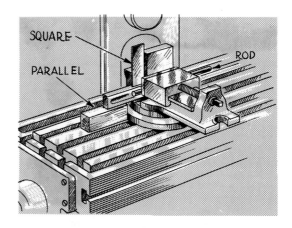

Fig. 77-4.   Checking for squareness before machining an end.

Unit
78

Procedure for Milling Grooves and Keyways

A groove or keyway may be cut with a plain milling cutter, an end mill, or a Woodruff key cutter. Any of these cutters can be used in a horizontal mill. The vertical mill can use only the last two cutters listed.

Procedure for Cutting a Groove in a Rectangular Block

1. Obtain the workpiece, deburr it, and mount it in the vise. Be sure the vise is properly aligned with the machine table.

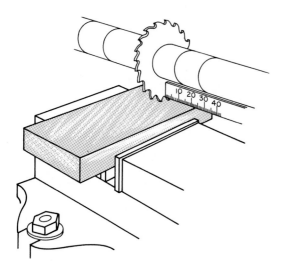

Fig. 78-1. Locating the cutter for machining a slot. Use a depth micrometer for more accurate positioning.

2. If using a horizontal mill, install a plain or side milling cutter of appropriate diameter and width. Install an end mill of appropriate diameter if using a vertical mill.
3. Calculate the correct cutter rpm and feed rate, and set the machine controls accordingly. Check for proper direction of cutter rotation.
4. Move the workpiece under the cutter with the table handwheels. Raise the workpiece until it either touches the revolving cutter or is the thickness of a piece of paper from it (see step 9, Unit 76). Stop the cutter.
5. Locate the cutter where the groove is to be milled by measuring the distance from the side of the workpiece to the side of the cutter with a steel scale or a depth micrometer, Fig. 78-1. Set the transverse table lock.
6. Back the workpiece away so that it is clear of the cutter. Set the vertical feed micrometer dial to zero, and raise the table the amount desired for the first cut. Set the vertical table lock.
7. Start the cutter and feed the workpiece into the cutter by hand. Allow the cut to proceed only far enough to measure the location of the groove. Stop the cutter, back the workpiece away, and check the position of the cut. Correct the position of the cut if necessary, then finish the cut, Fig. 78-2.
8. Make any additional cuts as needed to machine the groove to final width and depth.

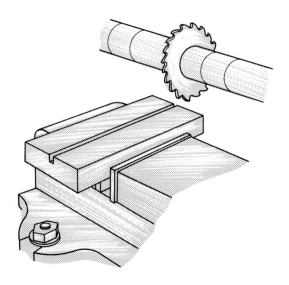

Fig. 78-2.   A groove machined with a horizontal mill.

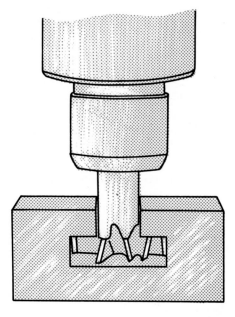

Fig. 78-3.   Machining a T-slot with a vertical mill.

Procedure for Milling a T-Slot or a Dovetail

1. Proceed as in steps 1 through 8 above for cutting a plain groove. Machine the groove to the dimensions of the vertical part of the T-slot or the center portion of the dovetail.
2. Install a T-slot cutter or a dovetail cutter of correct size.
3. Set the correct rpm and feed rate.
4. Carefully align the T-slot or dovetail cutter, and set all table locks except for the direction in which the cut will be made.
5. Start the cutter and feed the workpiece into it by hand, stopping to check its position when the full cutter diameter has entered the workpiece.
6. Make any necessary corrections in the position of the cutter, then finish the cut. See Figs. 78-3 and 78-4.

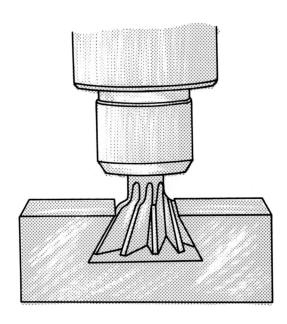

Fig. 78-4.   Machining a dovetail with a vertical mill.

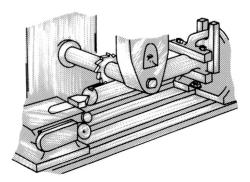

Fig. 78-5. Milling a groove with the workpiece held between centers in a dividing head.

Fig. 78-6. Using a vise to hold a shaft for milling a keyway.

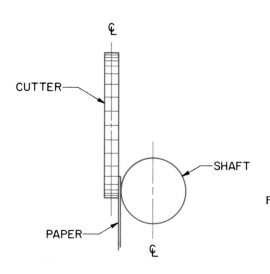

Fig. 78-7. A method of locating a cutter prior to centering it for cutting a keyway. To center the cutter, move the workpiece half its diameter plus half the cutter width plus the paper thickness.

Procedure for Milling a Plain Keyway in a Shaft

1. Obtain the workpiece and fasten it securely in one of the following ways:
   a. held between dividing head centers, Fig. 78-5,
   b. held in a vise, Fig. 78-6, or,
   c. if the diameter is large enough, held cradled in a T-slot and secured with clamps.
2. Install a plain or side milling cutter of appropriate width and diameter if using a horizontal mill. Install an end mill of appropriate diameter if using a vertical mill.
3. Raise the table until the diameter of the workpiece has passed the edge of the cutter. Then, using a strip of paper as a feeler gage between the cutter and the workpiece, move the workpiece towards the cutter until you feel a pull on the paper, Fig. 78-7.
4. Lower the table until the cutter is slightly above the top of the workpiece.
5. Center the workpiece under the cutter. Move the workpiece toward the cutter a distance equal to half the workpiece diameter **plus** half the cutter diameter **plus** the thickness of the paper, Fig. 78-7.
6. Start the cutter and raise the workpiece until it touches the cutter, or is the thickness of a piece of paper from it. Stop the cutter.
7. Back the workpiece away from the cutter in the appropriate horizontal axis. Set the vertical feed micrometer dial at zero, then raise the table the amount required for the size of keyway being cut. Consult a handbook for machinists for standard keyway dimensions.
8. Start the cutter and proceed to cut the keyway. At the end of the cut, stop the cutter, then back the workpiece away.

Procedure for Milling an Elongated Slot for a Captive Key

1. Obtain the workpiece and mount it securely.
2. Install a **center cutting** end mill of the required size.
3. Center the end mill over the shaft as described in "Procedure for Milling a Plain Keyway," steps 3 through 6.
4. Mark the location of each end of the keyway slot.
5. Locate the cutter at one end of the keyway slot, and set the transverse and longitudinal table locks.
6. Start the cutter, and raise the workpiece the amount required for the depth of the slot. Release the appropriate table lock and cut the slot to required length, Fig. 78-8.
7. Stop the cutter, lower the table, and unclamp the workpiece.

Fig. 78-8.   Milling a slot for a captive key with a vertical mill.

Procedure for Cutting a Woodruff Key Slot

1. Obtain the workpiece and mount it securely.
2. Mark the location of the Woodruff slot on the workpiece.
3. Install a Woodruff key cutter of required size.
4. Center the cutter on the shaft as described in "Procedure for Milling a Plain Keyway," steps 3 through 6.
5. Position the cutter opposite the location marked for the Woodruff key slot. Set the elevation and longitudinal table locks.
6. Set the transverse feed micrometer dial to zero, start the cutter, and move the workpiece into the cutter the required amount, Fig. 78-9. Watch the cut carefully as it proceeds. Adjust the longitudinal position of the workpiece as needed in order to keep the cutter centered between the layout lines.
7. Back the workpiece away from the cutter, stop the cutter, and unclamp the workpiece.

Fig. 78-9.   Cutting a Woodruff key slot with a vertical mill.

Unit
79

Procedure for Drilling and Boring Holes

Vertical milling machines are commonly used for all drilling and boring operations. The micrometer dials on the table handwheels make it possible to locate hole centers very precisely.

Procedure for Locating and Drilling Holes

1. Install a drill chuck in the spindle.
2. Install a dial indicator in the drill chuck.
3. Check the spindle to see if it is square with the table. This is done by "sweeping the table" with the dial indicator, Fig. 79-1. The dial indicator should read the same on both sides of the spindle parallel to the longitudinal and the transverse axes of the table.
4. Mount a vise on the table. Use a dial indicator to check the solid vise jaw, Fig. 76-5. This jaw should be made parallel with either the longitudinal or the transverse axis of the table.
5. Mount the workpiece in the vise so that the holes will not be drilled into the vise bottom or supporting parallels.
6. Install an **edge finder** in the drill chuck. Position the end of the edge finder about 6.35 mm (0.250″) below the top of the workpiece.
7. Set the spindle rpm at about 600 rpm and loosen the micrometer dials on the hand wheels.
8. Move the table screws so as to position the workpiece near the edge finder in the longitudinal axis, Fig. 79-2.
9. Start the machine. Move the table very slowly in the longitudinal axis towards the edge finder. Stop when the workpiece contacts the edge finder and makes it jump off-center, Fig. 79-3. Stop the machine.

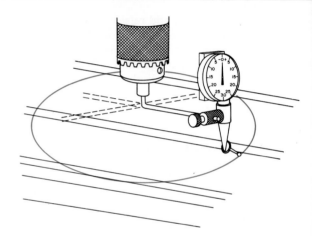

Fig. 79-1. "Sweeping the table" of a vertical milling machine with a dial indicator.

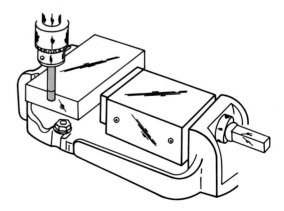

Fig. 79-2. Using an edge finder to locate a hole center a precise distance from the end of a workpiece.

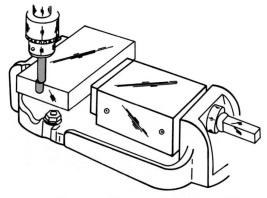

Fig. 79-3. The end of the edge finder jumps off-center when it contacts the workpiece.

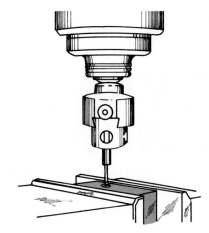

Fig. 79-4. Boring a hole with a vertical milling machine.

10. Raise the edge finder above the top surface of the workpiece. Set the longitudinal micrometer dial to zero. Then move the longitudinal table hand wheel half the edge finder diameter — usually 6.35 mm (0.250″) — plus the distance the hole center is located from the end of the workpiece. Set the longitudinal table lock.

11. Move the workpiece in the transverse axis far enough to position the end of the edge finder 6.35 mm (0.250″) below the top of the workpiece.

12. Repeat steps 9 and 10 to locate the center of the hole in the transverse direction. Set the transverse table lock.

13. Remove the edge finder from the drill chuck.

14. Install a center drill in the drill chuck and drill a center hole.

15. Follow the center drill with drills and other drilling tools as required. **To provide room for tool-changing, raise the quill or lower the table.** DO NOT MOVE THE WORKPIECE UNTIL THE HOLE IS FINISHED.

16. After finishing the first hole, release the table locks. Use the micrometer dials to position the workpiece accurately for the next hole. Remember to reset the table locks before center-drilling and drilling each hole.

Procedure for Boring a Hole

1. Use the procedure described above to locate and drill a hole about 1.6 mm (0.0625″) smaller than the diameter of the finished hole desired.

2. Install a boring head in the spindle, Figs. 73-9 and 79-4.

3. Install a boring tool of suitable size in the boring head.

4. Adjust the boring head until the boring tool just touches the inside of the drilled hole. Withdraw the boring tool from the hole and adjust it to enlarge the hole by about 0.5 mm (0.020″).

5. Set the machine to provide a power feed of about 0.08 mm (0.003″) per revolution of the spindle. Set the spindle rpm also.

6. Start the machine, engage the power feed, and bore the hole about 3.2 mm (0.125″) deep. Stop the machine, withdraw the boring tool, and measure the hole diameter.

7. If the hole is not oversize, start the machine and complete the rough cut.

8. Stop the machine, withdraw the boring tool, and measure the hole. Adjust the boring head to make a second rough cut if necessary. Be sure to leave about 0.25 mm (0.010″) for a finish cut.

9. Change the feed rate to about 0.04 mm (0.0015″) per revolution for the finish cut.

10. Measure the hole and adjust the boring head for the finish cut. Start the machine and bore the hole about 3.2 mm (0.125″) deep. Then stop the machine and measure the hole.

11. Adjust the boring head, if necessary, to make the hole the correct diameter. Start the machine and make another trial cut as above. Continue making trial cuts until the diameter is correct, then complete the finish cut.

Unit 80

The Dividing or Indexing Head and Its Operation

The dividing head is one of the most important milling machine accessories. The complete dividing head has two parts, the head and the tailstock, Fig. 80-1. The **head** of a universal dividing head is complex. In contrast, the tailstock is a simple piece used for supporting the outer end of a workpiece. The tailstock center can be moved longitudinally with a handwheel. Also, the center can be elevated while remaining in a horizontal plane or adjusted vertically in an inclined plane.

Typical milling operations performed with the use of the dividing head are shown in Figs. 81-2 and 83-1.

Direct Indexing

Direct indexing is the simplest form of indexing. Every dividing head has a **direct index plate** attached to its spindle so that the ratio of its movement is 1:1 with the spindle. Direct index plates normally have 24 equally spaced holes. A lever-operated **plunger pin** is provided for positioning the plate.

The 24-hole plate can be used for dividing a circle into 24, 12, 8, 6, 4, 3, and 2 equal divisions. Dividing the number of spaces desired into 24 yields the number of holes that must be indexed.

Fig. 80-1.   A dividing head with its principal operating parts labeled. A close-up of the side index plate is shown on the right.

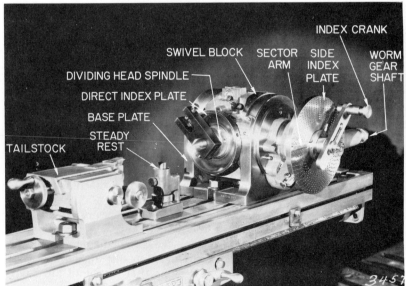

(Cincinnati Milacron; Photo Courtesy of Illinois State University)

For example, to cut a hexagon on the end of a shaft: 24 ÷ 6 = 4. Indexing 4 holes for each cut will produce the desired hexagon. Angles in multiples of 15° can also be indexed. Dividing 15° into the angle desired yields the number of holes to be indexed for moving the angle desired. For example: 60° ÷ 15 = 4 holes.

For direct indexing, retract the plunger pin, rotate the index plate through the number of spaces desired, and re-insert the plunger pin in the index plate hole. On some dividing heads, it is possible to disengage the worm gear attached to the index head crank. This allows the operator to move the direct index plate without turning the index head crank, thus providing faster indexing.

Simple Indexing

Simple indexing makes it possible to index many divisions not possible with direct indexing. Divisions of 5, 7, 9, 11, 13, and many other numbers of equally spaced divisions can be indexed by this method.

For simple indexing, the worm shaft that connects the index crank with the work spindle must be engaged. The work spindle is then turned by withdrawing the **index plunger pin** and turning the index crank the required amount. The plunger pin is then set again before the workpiece is machined.

The side index plates have several circles of holes. A different number of holes is equally spaced around each hole circle. Some side index plates have circles of holes on each side. The plate may be reversed to obtain the desired hole circle for a particular indexing operation. An index plate of the type in Fig. 80-1 has the following circles of equally spaced holes:

One side —
24-25-28-30-34-37-38-39-41-42-43
Other side —
46-47-49-51-53-54-57-58-59-62-66

A **sector,** consisting of **sector arms,** is located on the front of the index plate, Fig. 80-2. A **sector clamp screw** is used to lock the two arms apart for a space representing a fraction of a circle of holes. This makes it convenient to index a fractional part of a revolution with the index crank.

The following formula is used to determine the number of revolutions of the index crank for a desired number of equally spaced holes:

$$T = \frac{40}{N}$$

Where:
  T = Number of turns (including fractional parts of a turn) of the index crank
  40 = Number of teeth on the worm wheel
  N = Number of divisions desired

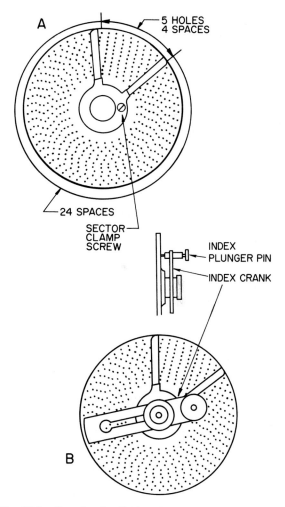

Fig. 80-2. Details of a dividing head side index plate. The sector arms are set to enclose 4/28 or 1/7 of an index crank revolution.

Example 1:

Determine the indexing required to machine a reamer with eight equally spaced teeth.

Solution: $T = \dfrac{40}{8}$

$T = 5$ turns

To make the eight-tooth reamer, revolve the index crank five full turns after cutting each tooth.

Example 2:

Determine the indexing required to cut a gear with 30 teeth.

Solution: $T = \dfrac{40}{30}$

$T = 1\ 1/3$

The fractional part of a revolution of the index crank is established between the two sector arms. To determine how far the arms must be set apart, select a circle with a number of holes that is divisible by the denominator of the fraction. For the example above, a 30-hole circle may be used. This circle is divisible by three (30 ÷ 3 = 10). Thus, one-third of the circle is equal to 10/30. Space the sector arms 10 spaces apart on the 30-hole circle. (**Note:** This is the space between 11 holes. It is the number of spaces between holes that is important.)

To make the 30-tooth gear in the example above, revolve the index crank 1-1/3 turns after cutting each tooth (one complete turn of the index crank, plus the 10/30 revolution between the sector arms). After each indexing, remember to move the sector immediately so it will be properly positioned for the next indexing.

Angular Indexing

Simple indexing also can be used for indexing angles. Forty turns of the index crank will revolve the spindle one full turn. Therefore, for each turn of the index crank, a point on the circumference of the work will rotate through 1/40 of a circle or

(Cincinnati Milacron)

Fig. 80-3.   A special wide-range dividing head.

9° (360° ÷ 40 = 9°). Thus, to find the number of index crank turns needed to index for a given angle, simply divide the angle desired by 9.

Example 3:

Index for two holes to be drilled 145° apart.

Solution: T = 145 ÷ 9 = 16-1/9

The 1/9 fraction would be expanded to 2/18 or 3/27, since no plate is available with only 9 holes. After drilling the first hole, turn the index crank 16 full turns, plus 2 spaces on an 18-hole circle or 3 spaces on a 27-hole circle.

Fine Spacing

For indexing of very fine spacing not possible with standard index plates, high number plates are available. These plates are interchangeable with standard index plates.

Special wide-range dividing heads are also available. The dividing head shown in Fig. 80-3 can index from 2 to 400,000 divisions. Angles can be indexed as fine as six seconds of a degree. Additional index plates or change gears are not required for these operations.

Unit 81

Procedure for Milling a Square, Hexagon, or Similar Shape

Milling squares or other polygons with flat surfaces may be done in several ways. A long workpiece should be held between the centers of a dividing head as in Fig. 81-1. A short workpiece may be held either horizontally or vertically in a dividing head chuck, Fig. 81-2.

Procedure for Milling a Square with an End Mill in a Vertical Milling Machine

1. Place the dividing head on the table of the milling machine and fasten it in position.
2. Withdraw the index plunger pin if necessary. Then disengage the worm gear so that rapid indexing can be done.

3. Remove the dividing head center, then tilt the dividing head spindle to the vertical position, Fig. 81-2.
4. Remove the nose guard, if any, and mount a chuck on the dividing head spindle.
5. Insert the cap that closes the hole in the dividing head spindle. This prevents chips from falling through it and becoming lodged between the swivel block and base plate.
6. Obtain a suitable sharp end mill. Select an end mill adapter of the correct size. Clean the shank of the end mill adapter and the

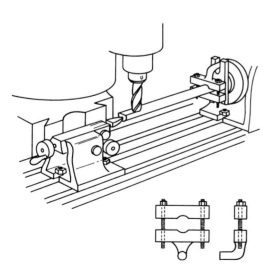

Fig. 81-1. Cutting a square on stock held between centers.

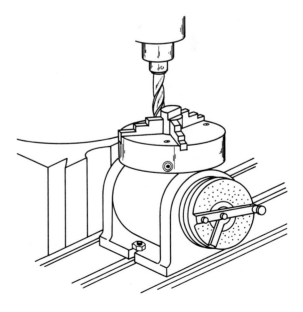

Fig. 81-2. Milling a square with an end mill.

hole in the milling machine spindle. Insert the end mill adapter into the spindle and tighten it in place.

7. Clean the shank of the end mill and the hole in the end mill adapter. Insert the end mill into the end mill adapter. Align the flat on the end mill shank so that it is opposite the adapter setscrew. Tighten the setscrew to lock the end mill in place.

8. Obtain the workpiece and fasten it securely in the chuck.

Assume that the stock has been turned to 38.1 mm (1-1/2") diameter and that a 25.4 mm (1") square section is to be milled on the end. Subtracting 25.4 mm (1") from 38.1 mm (1-1/2") leaves 12.7 mm (1/2") of stock to be removed, or 6.35 mm (1/4") from each side.

9. With the table handwheels, advance the workpiece until it almost touches the cutter. Then, with the vertical adjustment, raise the work so as to take a cut about 3 mm (1/8") in depth.

10. With the cross feed, move the table until the workpiece just touches the cutter, or is the thickness of a piece of paper from it. With the longitudinal feed, back the workpiece away from the cutter. Then, with the cross feed, advance the workpiece toward the cutter the 6.35 mm (1/4") plus the thickness of the paper, if used. Set the column clamp screw and the saddle clamp screw.

11. Start the machine, and with the longitudinal feed, advance the cutter and make the cut.

12. Stop the machine. Then return the table until the cutter clears the work.

13. Revolve the direct index plate on the front of the spindle through one-half turn; for example, 12 spaces on a 24-hole plate. Lock the plate in position with the plunger pin located in the head.

14. Start the machine, and advance the workpiece to cut a flat opposite the first cut.

15. When the cut is completed, stop the machine. Then accurately measure the distance across the flats (from one flat side to the opposite side). If the work measures more than 25.4 mm (1") across the flats, feed the table **toward** the cutter with the cross feed half the amount the stock is oversize. If undersize, feed the table half the amount **away** from the column.

16. When the machine has been set correctly, raise the table so as to make a cut of the depth desired. Start the machine and proceed with the cut.

17. When the first flat has been completely machined, withdraw the plunger and revolve the index plate through one-quarter turn. Lock it in position with the plunger pin.

18. Machine the second flat in a manner similar to the first.

19. Continue as in steps 16 through 18 until all four sides have been machined.

Procedure for Milling a Hexagon with an End Mill in a Vertical Milling Machine

1. Begin as in steps 1 through 7 above.
2. Obtain the workpiece and fasten it securely in the chuck.
3. Assume the stock has been turned to 34.93 mm (1-3/8") diameter and that one end is to be milled in the form of a hexagon that measures 25.4 mm (1") across the flats. Subtracting 25.4 mm (1") from 34.93 mm (1-3/8") leaves 9.53 mm (3/8") of stock to be removed, or 4.77 mm (3/16") from each side.
4. With the cross feed, advance the table until the cutter is in position to remove 4.77 mm (3/16") from one side. Raise the workpiece so that a cut of about 3 mm (1/8") depth will be made.
5. Proceed as in steps 11 through 16 above.
6. When the first flat has been completely machined, withdraw the plunger and revolve the direct index plate through one-sixth turn, for example, 6 spaces in a 24-hole circle. Lock it in position with the plunger pin.
7. Repeat step 6 until all 6 flats have been machined.

Procedure for Milling a Square by Straddle Milling

1. Begin as in "Procedure for Milling a Square With an End Mill," steps 1 through 5 inclusive.
2. Secure two sharp side milling cutters of the same size, as well as an arbor and the necessary collars.

3. Insert the arbor and mount the first cutter on the arbor as close to the column as the work will permit. Next, select collars that will give the **desired width** between cutters, as for example, 25.4 mm (1″). Place these collars in position on the arbor. Then place the second cutter and other necessary collars and, finally, the arbor nut. Position the over-arm support, clamp it in place, and then tighten the arbor nut.

4. Center the cutters over the work and make a trial cut in the usual manner, Fig. 81-3. Then measure for size.

   If necessary, adjust the size by inserting a wider or narrower collar between the cutters. Sometimes very thin discs of metal or even paper are used to adjust the space between the cutters.

5. When the machine has been set correctly, raise the table so that a cut of the depth desired will be made.

6. Start the machine and make the cut.

7. Stop the machine and return the workpiece to the starting side of the cut.

8. Revolve the direct index plate one-quarter turn, and make the second cut, thus completing the square.

9. Stop the machine, back the workpiece to the starting side of the cut, and remove the workpiece.

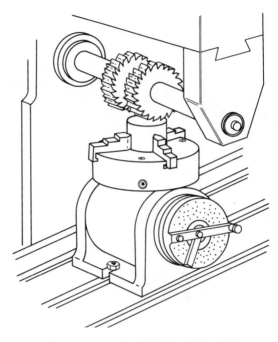

Fig. 81-3. Milling a square by straddle milling.

Procedure for Milling a Hexagon by Straddle Milling

1. Begin as in "Procedure for Milling a Square By Straddle Milling," steps 1 through 7 inclusive.

2. Revolve the direct index plate one-sixth turn and make the second cut.

3. Revolve the index plate another one-sixth turn, and make the third cut, completing the hexagon.

4. Stop the machine, back the workpiece to the starting side of the cut, and remove the workpiece.

Unit

82

Spur Gears and Gearing

(Illinois Gear and Machine Co.)

Fig. 82-1. Spur gears. A large spur gear is shown above and a rack and pinion below.

A gear is a wheel into which teeth have been cut, see Fig. 82-1. The action of gears can be understood more readily by comparing them with **friction wheels,** Fig. 82-2. Although friction wheels will transmit motion and power, they are inefficient because slippage occurs even under heavy loads.

The transmission of power can be made more effective by connecting wheels with belts instead of relying on friction contact. However, some loss of efficiency still results due to belt slippage or stretching during sudden increases in speed. The addition of gear teeth to wheels makes it possible to transmit motion and power at a completely uniform rate.

Gears are used on nearly all types of machines, ranging from wristwatches and egg beaters to machine tools, automobiles, and heavy manufacturing machinery. Gears may be used to transmit motion between shafts that are parallel, intersecting, or neither parallel nor intersecting. The gear connected to the source of power is called the **driver,** and the one to which motion is transmitted is called the **driven.** These terms have no relation to the size of the gears. The speed ratios of the shafts may be increased or decreased as desired. If the shaft to which the driven gear is attached must revolve at three times the speed of the drive shaft, then the driver must have three times as many teeth as the driven gear. If the driven shaft is to revolve at only one-third the rate of the driver, then the driver must have only one-third as many teeth as the driven gear.

Gears normally are mass-produced on special machines. These machines are of several types,

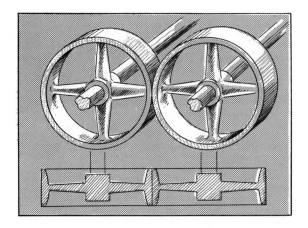

Fig. 82-2.   Friction wheels transmit motion inefficiently.

(Illinois Gear and Machine Co.)

Fig. 82-3.   This gear-generating machine produces large spiral bevel gears.

including **form-cutting** machines and **gear-generating** machines, Fig. 82-3. Gear-generating machines usually are further classified according to two types — **gear-shaping** machines and **gear-hobbing** machines.

Machinists in maintenance, job, and experimental machine shops frequently are required to produce gears of various types on a milling machine. The machinist, therefore, should be familiar with some of the basic kinds of gears and the terminology that applies to them.

When two gears of unequal size are mated, the smaller one is called a **pinion**. A **gear rack** (Fig. 82-1) is a gear that has teeth spaced along a straight line. This permits rotary motion to be changed to straight-line motion, or vice versa. A **basic rack** is one that is the basis for a system of interchangeable gears, see Figs. 82-13 and 82-14.

Types of Gears

Many types of gears are used in industrial machines. The most common are **spur, bevel** and **miter, internal, helical,** and **worm** gears.

Spur Gears

Spur gears are cylindrical and have teeth that are parallel with the axis of the gear, Fig. 82-1. They are designed for use on shafts whose axes are parallel with each other. A **spur rack** has teeth at a right angle to the axis of motion of the rack. Spur gears are the most commonly used type of gear on industrial machines. Normally, they are used under conditions of moderate speed and moderate load.

Bevel Gears

Bevel gears are conical and are used on shafts with intersecting axes, Fig. 82-4. They are used

(Illinois Gear and Machine Co.)

Fig. 82-4.   A bevel gear and pinion.

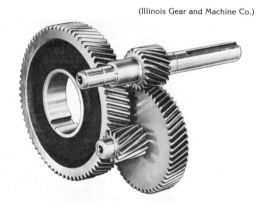

Fig. 82-5. An internal spur gear and pinion.

Fig. 82-6. A helical gear train with parallel shafts.

for transmitting power around corners. The shafts generally are at right angles, but they may be at various angles. Bevel gears on shafts at angles other than 90° are called **angular gears.**

Two bevel gears of equal size, with equal numbers of teeth, and designed for use on shafts at right angles are called **miter gears.** Bevel gears with straight-tooth elements, that pass through the intersection of their axes when extended are called **straight bevel gears,** Fig. 82-4. Those with teeth that are curved or oblique are called **spiral bevel gears;** see Fig. 82-3.

Internal Gears

Internal gears have teeth on the inner surface of a cone or cylinder, Fig. 82-5. This type may also have spur, bevel, or helical teeth. An advantage in combining internal and external gears is their compactness. Since the centers of the gears are closer together, less space is required. They also possess increased operational efficiency: more teeth are in mesh, the tooth lines curve in the same direction, and friction is reduced.

Helical Gears

Helical gears are similar to spur gears, except that the teeth form a helix twisting around the body of the gear, Fig. 82-6. The helical teeth provide greater strength and smoother operation at high speeds. The teeth do not hit each other as in the case of spur gears. Instead, they slide across each other, thus reducing noise and vibration. Since several teeth are in contact at the same time, their strength is greater than for spur teeth of the same size.

A disadvantage of helical gear teeth is that increased friction and heat are caused by the sliding action of the teeth. This disadvantage may be overcome with good lubrication. Hence, helical gears that operate at high speeds generally are run in an oil bath.

Helical gears may be used for connecting shafts that are parallel (Fig. 82-6) or at an angle with each other, provided their axial lines do not intersect. When used on parallel shafts, they are called **parallel helical gears** and are of opposite hand. This means a right-hand gear is meshed with a left-hand gear. Gears that operate on shafts with crossed axes are called **crossed helical gears.** They may have gears with teeth of the same hand or of the opposite hand.

Herringbone Gears

Herringbone gears have the appearance of a pair of right- and left-hand helical gears located side by side, Fig. 82-7. Since they have both right- and left-hand teeth, they frequently are called **double-helical gears.** They are designed for operation on parallel shafts. This tooth design eliminates the side thrust that results from the sliding action that single-helical gears exert

Fig. 82-7. Herringbone gears.

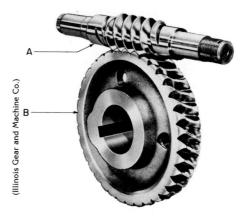

Fig. 82-8. A worm gear mechanism: (A) worm, (B) worm gear.

against each other. Thrust bearings often must be used with single-helical gears to compensate for side thrust.

Herringbone gears have greater surface contact than other gears of similar size. They therefore have greater load-carrying capacity. They generally withstand continuous, heavy-duty, high-speed operation better than other types of gears.

Worm Gears

Worm gears are meshed with a **worm;** the gear and the worm constitute a **worm gear mechanism,** Fig. 82-8. The teeth on the worm gear are helical and conform with the helix angle of the tooth on the worm. The helical tooth on the worm is a form of thread, similar to an acme thread, and it often is called a **worm thread.**

Worms may have single, double, or triple threads. With the single thread, one revolution of the worm revolves the worm gear a distance equal to that between a point on one tooth and a corresponding point on the next worm gear tooth, or one **circular pitch.** One revolution of a double-thread worm revolves the gear an amount equal to two teeth on the gear, and so on.

Worm gearing is used largely for speed reduction. The worm gear cannot turn the worm when a single-thread worm is used. This type of gear mechanism is **self-locking.** Engineers take advantage of this feature when they employ worm gears in steering mechanisms, in hoisting equipment, and in other devices.

Knowing About Gears

Modern gears generally have **involute teeth.** This means that the shape of the tooth is generated or drawn with an involute curve. Such a curve may be drawn with a pencil inserted in the loop of a string wound about a cylinder and held taut as the string is unwound, **A** in Fig. 82-9. This

Fig. 82-9. Methods of generating gear tooth shapes.

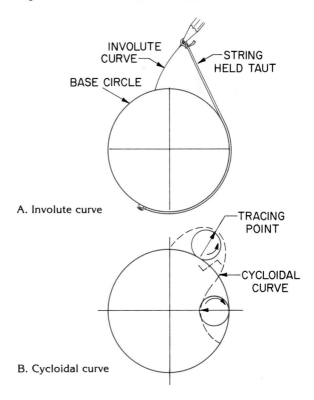

A. Involute curve

B. Cycloidal curve

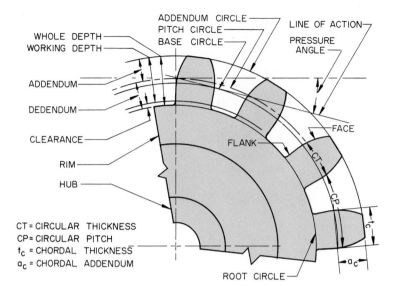

Fig. 82-10.   Gear nomenclature.

form of tooth has been found to give excellent results in terms of quietness and smoothness of operation.

The size of a gear is given in terms of its **diameter at the pitch circle,** which is called the **pitch diameter.** See Fig. 82-10.

Metric gears are always made according to the **module** (m) system, with measurements in millimeters. The module represents **the amount of**

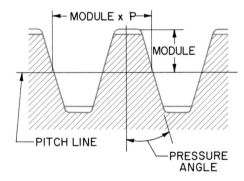

Fig. 82-11.   A profile of gear teeth, showing the module.

pitch diameter per gear tooth, and also corresponds to the **addendum,** Fig. 82-11. Therefore, the higher the module number, the larger the size of the gear tooth. To say that a metric gear has a module of 2 means that it has 2 mm of pitch diameter for each gear tooth. Thus, a 2m, 40-tooth gear would have a pitch diameter of $2 \times 40 = 80$ mm; a 3m, 40-tooth gear would have a pitch diameter of $3 \times 40 = 120$ mm, etc.

Inch gears are usually made according to the **diametral pitch** (P) system, with measurements in inches. Diametral pitch represents **the number of gear teeth for each inch of pitch diameter.** Therefore, the higher the diametral pitch, the smaller the size of the gear tooth. To say that a gear has a diametral pitch of 10 means that it has 10 teeth for each inch of pitch diameter. Thus, a 10P, 40-tooth gear has a pitch diameter of $40 \div 10 = 4''$; a 16P, 40-tooth gear has a pitch diameter of $40 \div 16 = 2.5''$, etc. Inch gears can also be made to **circular pitch** (p) and **inch module** (Mod) standards, but this is rarely done.

The term **circular pitch** refers to the distance along the pitch circle on a gear or along the pitch line on a rack, from a point on one tooth to a corresponding point on the next tooth, Fig. 82-10. Circular pitch corresponds to **linear pitch** on a gear rack, Fig. 82-13, on p. 336.

Diametral pitch, circular pitch, and inch module are all related as shown below:

$$\text{Diametral Pitch} = \frac{\pi}{\text{Circular Pitch}}$$

$$\text{Circular Pitch} = \frac{\pi}{\text{Diametral Pitch}}$$

$$\text{Module} = \frac{\text{Pitch Diameter}}{\text{no. of teeth}}$$

Gear Terms and Definitions

Gear nomenclature, including spur gear tooth forms, terms, definitions, and letter symbols, has become fairly well standardized. The following definitions are extracted from American Standard Gear Nomenclature (ASA B6.10-1954), with the permission of the publisher:

Involute teeth of spur gears, helical gears, and worms are those in which the profile in a transverse plane (exclusive of the fillet curve) is the involute of a circle. (See Fig. 82-9.)

The **base circle** is the circle from which involute tooth profiles are derived. (See Fig. 82-10.)

A **pitch circle** is the curve of intersection of a pitch surface of revolution and a plane of rotation. According to theory, it is the imaginary circle that rolls without slipping with a pitch circle of a mating gear. (See Fig. 82-10.)

A **pitch line** corresponds in the cross section of a rack to the pitch circle in the cross section of a gear. (See Fig. 82-13.)

The **addendum circle** coincides with the tops of the teeth in a cross section. (See Fig. 82-10. Continue to refer to this figure throughout these definitions.)

The **root circle** is tangent to the bottoms of the tooth spaces in a cross section.

The **line of action** is the path of contact in involute gears. It is the straight line passing through the pitch point and tangent to the base circles.

Pressure angle is the angle between a tooth profile and the line normal to a pitch surface, usually at the pitch point of profile. This definition is applicable to every type of gear. The term **pressure angle** originally meant an angle between the line of pressure and the pitch circle. In involute teeth, **pressure angle** is often described as the angle between the **line of action** and the line tangent to the **pitch circles.**

Center distance (C) is the distance between parallel axes of spur gears and parallel helical gears or between the crossed axes of crossed helical gears and worm gears. Also, it is the distance between the centers of pitch circles.

Addendum (a) is the height by which a tooth projects beyond the pitch circle or pitch line; also, it is the radial distance between the pitch circle and the addendum circle.

Dedendum (b) is the depth of a tooth space below the pitch circle or pitch line; also, it is the radial distance between the pitch circle and the root circle.

Clearance (c) is the amount by which the dedendum in a given gear exceeds the addendum of its mating gear.

Working depth (h_k) is the depth of engagement of two gears, that is, the sum of their addendums.

Whole depth (h_t) is the total depth of a tooth space, equal to addendum plus dedendum, also equal to working depth plus clearance.

Pitch diameter (D, d) is the diameter of the pitch circle.

Outside diameter (D_o, d_o) is the diameter of the addendum (outside) circle. In a bevel gear, it is the diameter of the crown circle. In a throated worm gear, it is the maximum diameter of the blank. The term applies to external gears.

Root diameter (D_R, d_R) is the diameter of the root circle.

Circular thickness (t_G, t_P) is the length of arc between the two sides of a gear tooth, on the pitch circle unless otherwise specified.

Chordal thickness (t_c) is the length of the chord subtending a circular-thickness arc.

Chordal addendum (a_c) is the height from the top of the tooth to the chord subtending the circular-thickness arc.

Number of teeth or threads (N) is the number of teeth contained in the whole circumference of the pitch circle.

Gear ratio (m_G) is the ratio of the larger to the smaller number of teeth in a pair of gears.

Full-depth teeth are those in which the working depth equals 2 × the metric module (2.000 divided by normal diametral pitch).

Stub teeth are those in which the working depth is less than 2 × the metric module (2.000 divided by normal diametral pitch).

Measuring Tooth Thickness

The chordal thickness (t_c) and the chordal addendum (a_c) of spur gear teeth, Fig. 82-10, may be accurately checked for size with a gear-tooth vernier caliper, Fig. 85-2. The values for these parts may be secured in standard handbooks for machinists. The vertical scale on the vernier caliper is first set at the handbook value for the chordal addendum (sometimes called corrected addendum). The caliper is then fit onto the tooth, and the chordal thickness is measured by using the sliding vernier scale.

Symbols

The following symbols and abbreviations are used in conjunction with spur gear formulas:

Symbol	Term	Abbreviation
M	Metric module	
	Inch module	Mod
P	Diametral pitch	DP
p	Circular pitch	CP
D	Pitch diameter	PD
D_o	Outside Diameter	OD
N	Number of teeth	N
t	Circular thickness	CT
a	Addendum	A
b	Dedendum	D
h_k	Working depth	WkD
h_t	Whole depth	WD
c	Clearance	C
C	Center distance	
L	Length of rack	

Gear-Tooth Forms

In order for mating gears of the same module (or diametral pitch) to mesh with a smooth, quiet, rolling action, they must have the proper gear-tooth form. Two basic forms (or curves) for gear teeth are used — one with an **involute curve, A** in Fig. 82-9, and a second that is a composite of the involute curve and **cycloidal curves. B** in Fig. 82-9 shows the shape of a cycloidal curve.

Mating gears with either of these forms will roll together smoothly and quietly without interference when operated at the prescribed pressure angle. Several systems of gear-tooth form have been standardized, and each system is designed to operate at a specified pressure angle, usually at 14-1/2°, 20°, or 25°.

Gears with the involute curve tooth design are the most widely used. The teeth of gears with the composite tooth form are very similar to those of the involute form, except for the design of basic rack teeth. This rack has an involute curve in the area of the pitch line, but it is modified slightly with cycloidal curves in areas above and below the pitch line. The modification prevents interference between mating gears that have a small number of teeth.

Gears with the composite tooth form normally are produced on milling machines, with form-type rotary milling cutters, Fig. 82-12. The use of this tooth form largely is limited to the production of gears in small job shops, maintenance shops, and shops where small numbers of gears are produced. This form of gear also may be produced by hobbing or with other gear-generating machines.

SI Metric 20° Full-Depth Involute

The approved SI metric gear tooth form is shown in Fig. 82-13. Rack teeth with this form have straight sides and are full depth, providing the basis for a complete system of interchangeable gears. The pressure angle of 20° is in conformance with international agreement that 20° is the most versatile pressure angle. The addendum is equal to the module, m, which corresponds to the reciprocal of the American diametral pitch, 1 ÷ DP. The dedendum is 1.250 m, corresponding to recent American practice. Root radius is some-

(Cincinnati Milacron)

Fig. 82-12. A form-type of milling cutter for cutting gear teeth.

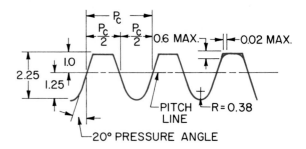

Fig. 82-13. The SI metric 20° gear tooth form.

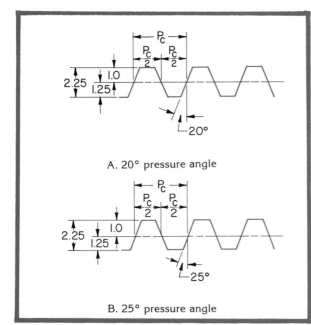

A. 20° pressure angle

B. 25° pressure angle

Fig. 82-14.   American National Standard gear tooth forms.

what greater than American practice, and the tip radius deviates from American practice, which does not specify rounding. This does not, however, prohibit the American practice of specifying a tip radius as near zero as practicable.

In spite of the nearly identical design of the SI metric 20° full-depth tooth form and the American Standard 20° full-depth tooth form, the gears **are not interchangeable**, due to differences in circular pitch.

American National Standard 20° and 25° Full-Depth Involute

The present American National Standard (ANSI B6.1-1968) provides specifications for two full-depth involute spur gear forms. The 20° tooth form may be used for gears with 18 or more teeth and the 25° tooth form for gears with 12 or more teeth. Otherwise, the tooth forms are identical, see Fig. 82-14. The 20° tooth form provides a gear with good strength and is used almost universally. The 25° tooth form is used for applications that require greater tooth strength or fewer teeth than the 20° tooth form can provide.

Rack teeth with these forms have straight sides. The addendum is 1 ÷ DP. The dedendum is 1.250 ÷ DP, although 1.350 or more is permitted for shaved or ground teeth.

Other American Standard Tooth Forms

The former American Standard (ASA B6.1-1932) included the 14-1/2° composite form, 14-1/2° full-depth involute form, and the 20° stub involute form. These forms are no longer recommended for use in new equipment, but are described briefly here because of their wide use in the recent past.

14-1/2° Composite

Gears within the 14-1/2° composite system have full-depth teeth with a pressure angle of 14-1/2°. The tooth form on the basic rack is a composite of involute and cycloidal curves, as described previously. The sides of the basic rack tooth are curved slightly at the top and bottom. This form of gear tooth is used for gears that are milled on milling machines with form-type rotary cutters.

14-1/2° Full-Depth Involute

Gears in the 14-1/2° full-depth involute system have full-depth teeth with a 14-1/2° pressure angle. This type is satisfactory when the numbers of teeth are large enough to avoid excess undercutting of the tooth form. Undercutting occurs when the pinion has less than 32 teeth, and it may become excessive when there are less than 22 teeth. In this system, the teeth on the basic rack have straight sides.

20° Stub Involute

This tooth form differs from the 20° full-depth system in that the teeth are shorter. The whole depth of the tooth is equal to 1.8 divided by the diametral pitch. The shorter tooth, together with the 20° pressure angle, provides for strengthened teeth on pinions with only 12 or 13 teeth without significant undercutting.

Table 82-1
Spur Gear Formulas for SI Metric 20° Full-Depth Tooth Form

To Find	Given	Formula (dimensions in millimeters)
1. Number of teeth	Module and pitch diameter	$N = \dfrac{PD}{m}$
2. Pitch diameter	Module and number of teeth	$PD = mN$
3. Outside diameter	Module and pitch diameter	$OD = PD + 2\,m$
4. Outside diameter	Module and number of teeth	$OD = m\,(N + 2)$
5. Whole depth of tooth	Module	$WD = 2.250\,m$
6. Clearance	Module	$C = 0.250\,m$
7. Thickness of tooth	Module	$CT = 1.5708\,m$
8. Module	Number of teeth and pitch diameter	$m = \dfrac{PD}{N}$
9. Module	Number of teeth and outside diameter	$m = \dfrac{OD}{N + 2}$
10. Circular pitch	Module	$CP = 3.1416\,m$
11. Circular pitch	Pitch diameter and number of teeth	$CP = \dfrac{PD}{N} \times 3.1416$
12. Center distance	Module and number of teeth	$CD = \dfrac{m\,(N + n)}{2}$
13. Center distance	Pitch diameters	$CD = \dfrac{PD + pd}{2}$

Spur Gear Formulas

The formulas in Table 82-1 are used in making calculations for SI metric 20° full-depth involute gears. Table 82-2 gives formulas used in making calculations for the American National Standard (ANSI B6.1-1968) 20° and 25° full-depth involute tooth forms.

The rules and formulas for spur gears with 20° involute stub teeth, 20° involute fine-pitch teeth, 14-1/2° composite full-depth, and 14-1/2° involute full-depth teeth are available in standard handbooks for machinists.

Sample Spur Gear Problems

Metric:

Given a 2-module gear with pitch diameter of 60 mm, find the following:
1. Number of teeth
2. Outside diameter
3. Addendum
4. Dedendum
5. Clearance
6. Whole depth
7. Circular tooth thickness
8. Equivalent diametral pitch

Solution:

1. $N = \dfrac{PD}{m} = \dfrac{60}{2} = 30$ teeth
2. $OD = (N + 2) \times m = (30 + 2) \times 2$
   $= 32 \times 2 = 64$ mm
3. $a = m = 2$ mm
4. $b = m \times 1.250 = 2 \times 1.250 = 2.5$ mm
5. $c = m \times 0.250 = 2 \times 0.250 = 0.5$ mm
6. $WD = m \times 2.250 = 2 \times 2.250 = 4.5$ mm
7. $CT = m \times 1.5708 = 2 \times 1.5708$
   $= 3.1416$ mm
8. $DP = \dfrac{25.4}{m} = 12.7$

Table 82-2
Spur Gear Formulas for American National Standard 20⁰ and 25⁰ Full-Depth Tooth Forms

To find	Given	Formula (dimensions in inches)
1. Number of teeth	Diametral pitch and pitch diameter	$N = DP \times PD$
2. Number of teeth	Diametral pitch and outside diameter	$N = (DP \times OD) - 2$
3. Pitch diameter	Diametral pitch and number of teeth	$PD = \dfrac{N}{DP}$
4. Pitch diameter	Outside diameter and addendum	$PD = OD - 2A$
5. Outside diameter	Diametral pitch and number of teeth	$OD = \dfrac{N + 2}{DP}$
6. Outside diameter	Pitch diameter and number of teeth	$OD = \dfrac{N + 2}{\dfrac{N}{PD}}$
7. Whole depth of tooth	Diametral pitch	$WD = \dfrac{2.250}{DP}$
8. Clearance	Diametral pitch	$C = \dfrac{0.250}{DP}$
9. Thickness of tooth	Diametral pitch	$CT = \dfrac{1.5708}{DP}$
10. Diametral pitch	Circular pitch	$DP = \dfrac{3.1416}{CP}$
11. Diametral pitch	Pitch diameter and number of teeth	$DP = \dfrac{N}{PD}$
12. Diametral pitch	Outside diameter and number of teeth	$DP = \dfrac{N + 2}{OD}$
13. Circular pitch	Diametral pitch	$CP = \dfrac{3.1416}{DP}$
14. Circular pitch	Number of teeth and pitch diameter	$CP = \dfrac{PD}{0.3183N}$
15. Center distance	Pitch diameters	$CD = \dfrac{PD + pd}{2}$
16. Center distance	Diametral pitch and number of teeth	$CD = \dfrac{½ (N + n)}{DP}$

Inch:
Given a 16 diametral pitch gear with a pitch diameter of 3″, find the following:

1. Number of teeth
2. Outside diameter
3. Addendum
4. Dedendum
5. Clearance
6. Whole depth
7. Circular tooth thickness
8. Equivalent metric module

Solution:

1. N = PD × DP = 3 × 16 = 48 teeth

2. OD = $\dfrac{N + 2}{DP}$ = $\dfrac{48 + 2}{16}$ = $\dfrac{50}{16}$ = 3.125″

3. a = $\dfrac{1.0}{DP}$ = $\dfrac{1.0}{16}$ = 1/16 or 0.0625″

4. b = $\dfrac{1.250}{DP}$ = $\dfrac{1.250}{16}$ = 0.078″

5. c = $\dfrac{0.250}{DP}$ = $\dfrac{0.250}{16}$ = 0.016″

6. WD = $\dfrac{2.250}{DP}$ = $\dfrac{2.250}{16}$ = 0.141″

7. CT = $\dfrac{1.5708}{DP}$ = $\dfrac{1.5708}{16}$ = 0.098″

8. m = $\dfrac{25.4}{DP}$ = $\dfrac{25.4}{16}$ = 1.5875

Gear Cutters

Form-tooth gear cutters for milling machines, Fig. 82-12, are made in standard sets of eight cutters for each module and diametral pitch. The profile of the tooth on a 13-tooth gear of a given module or pitch is different from that on a 40-tooth gear. Therefore, a cutter with a different profile form must be used to cut gears with different numbers of teeth. With a standard set of eight cutters, a satisfactory tooth profile can be produced on a gear with any number of teeth, ranging from 12 teeth or more, up to and including a rack.

The individual cutters in a standard set are numbered as follows:

Cutter Number	Cutting Capacity
1	135 or more teeth, including rack teeth
2	55 to 134 teeth
3	35 to 54 teeth
4	26 to 34 teeth
5	21 to 25 teeth
6	17 to 20 teeth
7	14 to 16 teeth
8	12 to 13 teeth

For example, a No. 6 cutter is used for cutting a gear with 18 teeth. In each instance, the cutter is designed for the lower number of teeth. In the case of the No. 6 cutter, a 17-tooth gear would have a profile that would operate more efficiently than a 19-tooth gear.

Cutters are available for producing more accurate tooth profiles on gears requiring more efficient operation, such as those used in automotive transmissions. Cutters of this type are available with half numbers such as 1-1/2, 2-1/2, through 7-1/2. Thus a No. 5-1/2 cutter is designed to cut 19 to 20 teeth. A 19-tooth gear cut with this cutter would operate more smoothly than if it were cut with the No. 6 cutter.

Unit
83

How to Mill Spur Gears

Gear cutting is one of the most interesting operations that can be performed on a milling machine. The dividing head, formed tooth cutters, and other features of modern milling machines provide a system of spur gear cutting that is beautiful in its simplicity, Fig. 83-1. This system is not efficient, however, for mass production of gears.

Procedure

1. Select a fairly short mandrel of a diameter that will fit the hole in the hub of the gear blank.

(Cincinnati Milacron)

Fig. 83-1.   Cutting spur gear teeth.

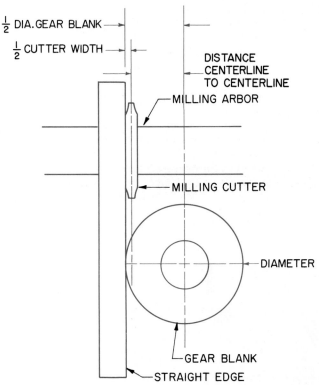

Fig. 83-2.   Aligning the side of the gear cutter with the side of the gear blank.

2. Wipe the mandrel clean and lightly oil the surface.
3. Force the mandrel into the hub with an arbor press or a lead hammer.
4. With the cross feed, run the table as close as practicable to the column.
5. Secure the dividing head and tailstock. Place them on the table of the milling machine as close to the column as the nature of the work and the dividing head will permit.
6. Fasten the dividing head securely in position. Then slide the tailstock forward until the centers just touch.

    Examine the position of the points of the centers. Both must be in the same horizontal plane. If not, adjust the tailstock center vertically until it is in the same plane as the point of the center in the head.
7. Move the tailstock back to the position desired, and fasten it securely.
8. Secure the proper gear cutter, milling machine arbor and collars, and a dog for driving the mandrel.
9. Wipe the shank of the arbor clean and dry. Insert it into the spindle of the milling machine and tighten in place. Then place the cutter on the arbor so that the direction of cut is towards the dividing head.
10. Slide the rest of the arbor collars and the bearing on the arbor, and turn the arbor nut on hand tight. Position the overarm support and tighten it in place. Then tighten the arbor nut.

11. Set the machine controls for proper spindle rotation, spindle rpm, and table feed rate and direction.
12. Install the mandrel and the workpiece between the dividing head centers. **Note:** The large end of the mandrel should be at the dividing head end. Then fasten the tail of the dog in the driver so that it cannot move.
13. Now align the side of the cutter over the side of the gear blank, Fig. 83-2. Move the cross feed screw until the workpiece edge just touches a straightedge held against the far side of the gear cutter.
14. Measure the width of the cutter and divide it by two. Measure the diameter of the gear blank and divide it by two. **Subtract** half the cutter width from half the gear blank diameter.

This is the amount the workpiece must be moved to place its center line directly below the center line of the cutter.

15. Lower the table, if necessary, so that the gear blank will not collide with the cutter. Then move the cross feed screw the required amount to position the center of the gear blank directly under the center of the cutter. Start the cutter and bring the table up until the gear blank just touches the cutter or is the thickness of a piece of paper away. Then stop the cutter and back the workpiece away. Set the vertical feed micrometer dial to zero. Raise the table the full depth of the gear tooth, unless a finish cut is necessary.

16. Adjust the dividing head crank and sector to index the required number of divisions. See Unit 80 if necessary.

17. Start the machine, engage the power feed, and cut the tooth space. Use a cutting fluid if called for.

18. At the end of the cut, disengage the power feed and stop the cutter rotation. Then return the table to the starting position for the next cut.

19. Rotate the gear blank to the position for the next tooth space by turning the dividing head crank the required number of turns.

20. Repeat steps 17 through 19 until all teeth have been cut. If a finish cut is required, raise the table accordingly and recut all tooth spaces.

Unit 84

Bevel Gears

Bevel gears are used to transfer motion from one revolving shaft to another shaft that intersects it at an angle, Fig. 84-1. Usually one gear, the pinion, is much smaller than the other. When both gears are the same size, they generally have a face angle of 45° and are called **miter gears**. Figure 84-2 shows other forms of bevel

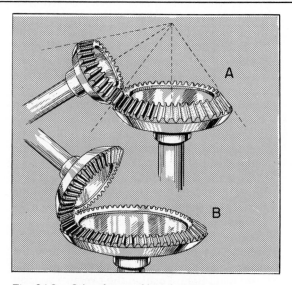

Fig. 84-2.   Other forms of bevel gears.
A. With shafts less than 90° apart
B. With shafts more than 90° apart

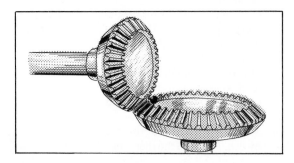

Fig. 84-1.   Bevel gears.

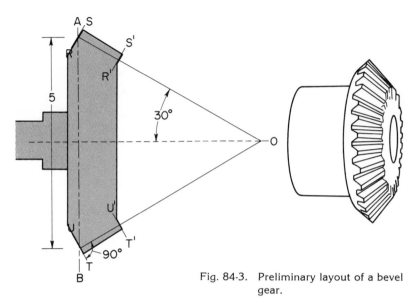

Fig. 84-3. Preliminary layout of a bevel
gear.

gears. In **A**, the two shafts are in planes less than 90° apart, while in **B**, they are in planes more than 90° apart.

Bevel gears usually are produced on special gear-generating machines. However, a machinist occasionally has to make a bevel gear with a milling machine. An explanation of the terms used and the method employed in making a layout for bevel gears follows.

Laying Out a Bevel Gear

The shape and size of the large end of the teeth on a bevel gear are the same as those on a spur gear. The first step in laying out a bevel gear is to determine the center angle, for example, 30° in Fig. 84-3. Then the pitch diameter must be determined, followed by the module or diametral pitch, the number of teeth, and the width of the face. After making the necessary calculations, draw the angle AOB, Fig. 84-3. Then draw line AB at a point where it will be equal to the pitch diameter. Through points A and B respectively, draw lines RS and TU perpendicular respectively to AO and BO. Next, on lines AO and BO, at a distance equal to the width of the face, draw lines R^1S^1 and T^1U^1 parallel to RS and TU. Calculate the outside diameter of the large end of the teeth and draw lines CO and DO, Fig. 84-4. (The necessary formulas are available in standard handbooks for machinists.) Calculate the dedendum of the tooth

(the total height of the tooth below line AO). Then draw lines EO and FO, Fig. 84-4. Next, calculate the thickness of the tooth.

As can be seen in Fig. 84-4, all lines converge toward the apex of the cone. This means that each tooth is thinner and shorter at the apex end than at the base end. Bevel gears cannot be cut as accurately on a milling machine with a standard gear cutter as on special gear-cutting machines.

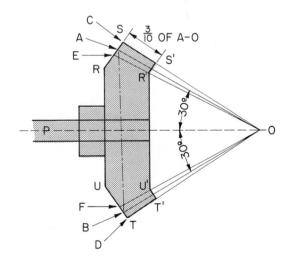

Fig. 84-4. Layout of a bevel gear.

Unit
85

How to Mill Bevel Gears

Normally, bevel gears are cut on special bevel gear-cutting machines, Fig. 82-3. However, bevel gears can be cut with sufficient accuracy for many applications by using a standard milling machine. The machining setup is made by mounting the bevel gear blank in the dividing head chuck or spindle, Fig. 85-1.

Cutters used for cutting bevel gears on a milling machine are similar to those used for cutting spur gears. However, they are thinner, in order to conform to the profile of the teeth at the small end of the gear. Bevel gear cutters are made in sets, and are stamped with the word **bevel**. Instructions for the selection of bevel gear cutters generally are included in cutter manufacturer's catalogs or in standard handbooks for machinists.

Procedure

1. Secure a dividing head with a chuck. Fasten the head on the table of the milling machine, and mount the chuck on the dividing-head spindle.
2. Set the dividing head at the angle required.
3. Obtain the correct gear cutter, milling machine arbor, and collars. Be sure all chips are removed from the arbor and collars.
4. Mount the cutter on the arbor as near the nose of the spindle as practicable. Position the overarm and arbor bearing and fasten securely.

(L.S. Starrett Company)

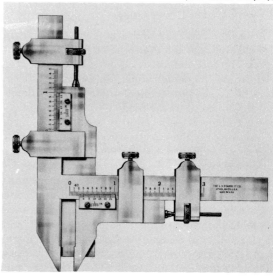

Fig. 85-2.   A gear-tooth vernier caliper.

(Cincinnati Milacron)

Fig. 85-1.   Cutting bevel gear teeth on a gear blank mounted on a dividing head.

5. Mount the gear blank in the dividing head chuck. Make sure it is securely fastened.
6. Adjust the table until the center line of the cutter is in line with the center line of the gear blank.
7. Start the machine; then raise and advance the table until the cutter just touches the surface of the gear blank or is within the thickness of a piece of paper from it.
8. Back the workpiece away from the cutter. Set the micrometer dial for vertical adjustment at zero. Then raise the table the full depth of cut.
9. Make the cut in the usual manner, using a cutting fluid if required.
10. Index for the next tooth, and make the cut.
11. Accurately caliper the thickness of the tooth at the large end. Use a gear-tooth vernier caliper, Fig. 85-2. Assume the large end measures 4.93 mm (0.194″). Subtract from this measurement the finished thickness of the tooth, for example, 3.99 mm (0.157″), and divide the remainder by 2.

$$\frac{4.93 \text{ mm } (0.194″) - 3.99 \text{ mm } (0.157″)}{2} =$$

$$\frac{0.94 \text{ mm } (0.037″)}{2} = 0.47 \text{ mm } (0.0185″)$$

This figure represents the amount of stock to be trimmed from each side of the large end of the tooth, **A,** Fig. 85-3.
12. In a similar manner, calculate the amount to be trimmed from the small end. Ordinarily, only a very small amount, if any, will need to be trimmed.
13. Paint the cut surfaces of the tooth with layout fluid.

    The correct setting of the machine for trimming the gear teeth cannot be achieved by merely moving the index pin a hole or two to the right. This would move the small end of the gear proportionately with the large end and result in too much stock being removed from the small end. The amount of setover and the amount of roll for trimming teeth of various sizes will be found in the handbook of the particular machine being used or in a standard handbook for machinists.
14. Loosen the clamping screws, and offset the table of the milling machine about one-eighth of the thickness of the tooth at the

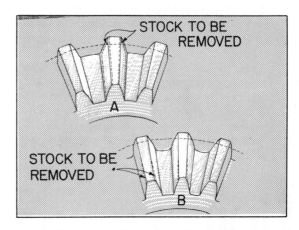

Fig. 85-3. Layout of stock to be removed on the finishing cut.

large end, for example, about 0.51 mm (0.020″). Then clamp the table securely. Be sure to adjust for backlash.
15. Withdraw the index crank pin, and rotate the gear blank until the edge of the cutter is in position to take a cut of the amount desired, for example, 0.47 mm (0.0185″).

    If the pin does not fit in a hole in the index plate when the gear blank has been revolved to the position desired, loosen the index plate lock and move the plate until one of the holes lines up with the pin.
16. Start the machine, and carefully trim the side of the tooth. This should remove all of the dye except at the extreme of the small end.

    Again measure the ends of the tooth. If the tooth is still too thick at either end, offset the blank (as in step 14) in the direction desired and take another light cut.

CAUTION

For this second adjustment, the amount of offset will be the difference between the required thickness of the tooth plus one-half the difference found when the tooth was first measured, for example, 3.99 mm (0.157″) + 0.47 mm (0.0185″) = 4.46 mm (0.1755″), and the magnitude of the last measurement, for example, 4.58 mm (0.1805″). Then 4.58 mm (0.1805″) − 4.46 mm (0.1755″) = 0.127 mm (0.005″), the amount to be removed.

17. When the tooth measures correctly at both ends, index to the next tooth space and start the machine. Proceed in the usual manner, observing the normal precautions. Continue until one of the sides of all the teeth has been cut.
18. Offset the table in the opposite direction an amount corresponding with the offset for the first cut. Be sure to adjust for backlash.
19. Rotate the index crank the same amount as before, but in the opposite direction.
20. Start the machine, and trim the second side of the first tooth.

21. Carefully measure the finished tooth at both ends. Make adjustments, if necessary, and again trim the tooth.
22. When the first tooth has been cut to the size required, proceed to trim the second side of the remaining teeth.
23. Because the cutter does not have the proper curvature for cutting the top part of the small end of the tooth, it will be necessary to remove a little stock with a file. File the top of the teeth until the desired curvature is achieved, **B**, Fig. 85-3.

Unit
86

Helical Milling

Helical milling requires the use of a mechanism that will cause the workpiece to rotate at a constant rate as the cut advances. Such a mechanism is an integral part of a universal horizontal milling machine. This mechanism includes a dividing head, a tailstock, and a lead driving mechanism for the dividing head, Fig. 71-6.

A standard universal horizontal milling machine has a dividing head driving mechanism with a lead adjustable from 63.5 to 2540 mm (2-1/2″ to 100″). **Lead** is the longitudinal distance the helix travels during one complete revolution of the workpiece, Fig. 86-1. Leads from 6.35 to 2540 mm (1/4″ to 100″) can be achieved by using special change gears or, in some cases, an auxiliary reducing-gear bracket. With special long- and short-lead driving attachments, leads from 0.254 to 25 400 mm (0.010″ to 1000″) may be achieved.

Steep helices, such as the helix for a worm thread, may be produced on either a plain or

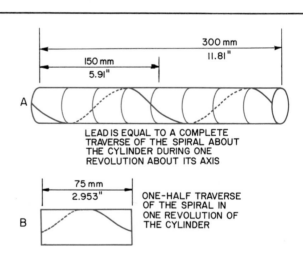

LEAD IS EQUAL TO A COMPLETE TRAVERSE OF THE SPIRAL ABOUT THE CYLINDER DURING ONE REVOLUTION ABOUT ITS AXIS

ONE-HALF TRAVERSE OF THE SPIRAL IN ONE REVOLUTION OF THE CYLINDER

TWO TRAVERSES OF THE SPIRAL IN TWO REVOLUTIONS OF THE CYLINDER

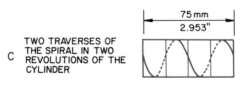

Fig. 86-1. Illustrations of right-handed helices or spirals with different leads.

universal milling machine, Figs. 86-2 and 73-16. To do this, either machine has to be equipped with a universal spiral milling attachment and a dividing head with a short-lead attachment. The universal spiral milling attachment must be used on universal milling machines when the angle of the helix exceeds the swivel angle of the machine table.

A **helix** is a line or a groove which advances longitudinally on a cylinder or cone at a constant rate as the object is rotated about its axis. A helix may be either right- or left-handed. To determine the hand, hold the object in a vertical plane. If the helix travels upward towards the left, it is a left-hand helix. If the helix travels upward towards the right, it is a right-hand helix.

The cylinder in Fig. 86-1**A** is 300 mm (11.811″) in length, and the helix has traveled twice around it in traversing its complete length. Thus, the helix has a lead of 150 mm (5.0955″). If the same cylinder were only 75 mm (2.953″) long, as indicated at **B,** the helix would still have a 150 mm (5.0955″) lead, because (as the illustration shows) the helix has traveled half way around the cylinder in one complete revolution. On the other hand, if in two revolutions of a cylinder 75 mm (2.953″) long, the helix had traveled twice around the cylinder (as in **C**), then the helix would have a lead of 37.5 mm (1.476″).

Change Gears

Included with universal milling machines and with dividing-head driving mechanisms is a set of instructions entitled "Table of Change Gears, Angles and Leads." These instructions tell what gears to use and where they are to be placed in the driving mechanism in order to produce a given lead for each revolution of the stock. In the event of the loss of the instructions or of the need of computing a gear ratio not given, refer to the mathematical formula given in Unit 87.

Ordinarily, there are 12 gears in a set of change gears. Four gears are chosen from the set to place in the lead driving mechanism. In a compound train of four gears, two are driving gears and two are driven, Fig. 86-3. To achieve a given lead, one must select gears having a ratio that will cause

(Cincinnati Milacron)

Fig. 86-2. Milling a worm thread. A universal spiral milling attachment and a dividing head with a short-lead attachment are used.

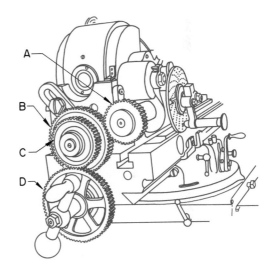

Fig. 86-3. A dividing head geared to a milling machine table screw for helical milling.
A. Gear on worm (driven)
B. First gear on stud (driving)
C. Second gear on stud (driven)
D. Gear on screw (driving)

the work to rotate a given amount while it is moved past the cutter a given distance.

Change in the direction of rotation of the stock is accomplished by introducing an idler gear into the train of gears. Since the sole function of the idler is to change the direction of rotation, it does not affect the gear ratio.

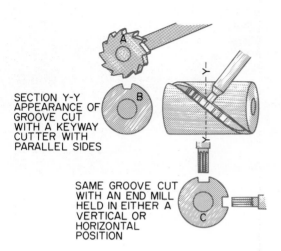

SECTION Y-Y
APPEARANCE OF
GROOVE CUT
WITH A KEYWAY
CUTTER WITH
PARALLEL SIDES

SAME GROOVE CUT
WITH AN END MILL
HELD IN EITHER A
VERTICAL OR
HORIZONTAL
POSITION

Fig. 86-4.   Helices cut with a keyway cutter and with an end mill.

(Cincinnati Milacron)

Fig. 86-5.   Cutting a fairly steep left-hand helix on a milling cutter blank.

Positioning the Work

When a helix is milled, the work must be set at an angle to the axis of the cutter by swiveling the table of the machine, except when using an end mill, **C,** Fig. 86-4. Even when the work is set at an angle, a helix with parallel sides 90° to the bottom of the groove cannot be milled true with a

(Cincinnati Milacron)

Fig. 86-6.   Helical milling with an end mill.

circular cutter having parallel sides. See **A** and **B,** Fig. 86-4. This is because a circular mill with straight sides cannot fit in a curved groove without striking the sides of the groove as it revolves. This makes the groove wider at the top than at the bottom. However, formed circular cutters have teeth whose sides tend to converge to a point. These cutters can be used to mill a helix having sides that conform to the slope of the teeth. See Fig. 86-5. This is possible because the sides of the teeth are in full contact with the sides of the groove only at the moment a given tooth is removing the metal at the bottom of the groove. After passing this point, the sides of the tooth swing clear of the finished part of the groove.

The method of computing the angle at which the table should be swiveled for different helix leads is explained in Unit 87.

For a right-hand helix, the table is swiveled toward the left (counterclockwise) from the zero line on the saddle through the desired number of degrees. For a left-hand spiral, the table is swiveled toward the right from the zero line. Figure 86-5 illustrates the table swiveled to the right for cutting a rather steep helix.

Helical milling also may be done with an end mill on a vertical milling machine equipped with a dividing head and a dividing-head lead driving mechanism, Fig. 86-6.

Unit
87

Calculations for Helical Milling

When a helix is milled on a workpiece, the workpiece must be made to rotate at the same time it is fed into the revolving cutter. This is accomplished by gearing the dividing head to the milling machine table screw, see Figs. 86-3 and 87-1. The lead of the helix, that is, the distance required for the workpiece to make one revolution, is determined by the size and placement of the change gears, labeled A, B, C, and D in Figs. 86-3 and 87-1.

Charts are usually available with the machine or in machinist's handbooks, which list the gears

(Cincinnati Milacron)

Fig. 87-1. A dividing head lead-driving mechanism, with change gears labeled.
A. Gear on worm (driven)
B. First gear on stud (driving)
C. Second gear on stud (driven)
D. Gear on screw (driving)

that will produce various leads. Leads may also be calculated as described below.

Calculation of Change Gears

Before the change gears necessary to obtain a given lead can be calculated, the **lead of the machine** must first be determined. The lead of the machine is the distance the milling table must be moved in order to rotate the spindle of the dividing head one revolution. This lead must be determined when gears of equal size are mounted on both the milling table screw and on the dividing head worm shaft. When so provided, a gear ratio of 1:1 exists. Since most dividing heads have a built-in 40:1 worm-to-spindle ratio, 40 turns of the table screw will result in one turn of the dividing head spindle. And since most milling machines have a table screw with a pitch of 6.35 mm (1/4″), the lead of the milling machine will be 6.35 mm (1/4″) × 40, or 254 mm (10″).

The length of the lead is changed by changing the gear ratio between the table feed screw and the dividing head worm shaft. The formula for calculating the change gears necessary to produce a given lead is as follows:

$$\frac{\text{lead of machine}}{\text{lead of helix desired}} = \frac{\text{driving gears}}{\text{driven gears}}$$

Example 1:

Determine the change gears required for a lead of 381 mm (15″). Assume the milling machine has a lead of 254 mm (10″).

$$\frac{\text{lead of machine}}{\text{lead of helix desired}} = \frac{254\,(10)}{381\,(15)}$$
$$= \frac{\text{driving gears}}{\text{driven gears}}$$

If a simple gear train could be used (one driving and one driven gear), a 10-tooth gear on the table screw meshed with a 15-tooth gear on the dividing-head worm shaft would produce the 381 mm (15″) lead required. However, gears of 10 and 15 teeth are not available, and the drive system is designed for a compound gear train of four gears. Therefore, the fraction 10/15 must be split into two fractions whose product equals 10/15. This is done by factoring as follows:

$$\frac{10}{15} = \frac{5 \times 2}{5 \times 3} = \frac{\text{driving gears}}{\text{driven gears}}$$

If gears with 5 and 2 teeth were possible, they would be the driving gears, and gears with 5 and 3 teeth would be the driven gears. But since this is not possible, then each of the fractions must be expanded by multiplying both the numerator and the denominator by a number which will result in a product that corresponds to the number of teeth on available gears:

$$\frac{5}{5} \times \frac{8}{8} = \frac{40}{40} \quad \text{and} \quad \frac{2}{3} \times \frac{12}{12} = \frac{24}{36}$$

$$\text{or} \quad \frac{5 \times 2}{5 \times 3} = \frac{40 \times 24}{40 \times 36} = \frac{\text{driving gears}}{\text{driven gears}}$$

Thus, gears with 40 and 24 teeth become the driving gears, and gears with 40 and 36 teeth become the driven gears.

These gears would be arranged in the gear train as follows:

Gear A (on dividing-head worm shaft)
      40 teeth (driven)
Gear B (first gear on idler stud)
      24 teeth (driving)
Gear C (second gear on idler stud)
      36 teeth (driven)
Gear D (gear on the table screw)
      40 teeth (driving)

The positions of the driving gears may be interchanged without changing their products; likewise, the driven gears. Thus, several different combinations of driving and driven gears will produce a helix with the same lead.

Example 2:

Determine the change gears required for a lead of 609.6 mm (24″). Assume the machine has a lead of 254 mm (10″).

$$\frac{\text{lead of machine}}{\text{lead of helix desired}} = \frac{254\,(10)}{609.6\,(24)} = \frac{5 \times 2}{4 \times 6}$$

$$= \frac{\text{driving gears}}{\text{driven gears}}$$

$$\frac{5}{4} \times \frac{8}{8} = \frac{40}{32} \quad \text{and} \quad \frac{2}{6} \times \frac{12}{12} = \frac{24}{72}$$

$$\text{Thus,} \quad \frac{10}{24} = \frac{5 \times 2}{4 \times 6} = \frac{40 \times 24}{32 \times 72}$$

$$= \frac{\text{driving gears}}{\text{driven gears}}$$

Gear on worm (A) 32 teeth (driven)
First gear on stud (B) 24 teeth (driving)
Second gear on stud (C) 72 teeth (driven)
Gear on table screw (D) 40 teeth (driving)

Idler Gear

On most machines, the setup for cutting a left-hand helix requires an idler gear in the train of gears. This in no way affects the gear ratio, but merely changes the direction of rotation of the stock. Usually, the idler gear is mounted on an adjustable idler bracket. On some machines, an idler is used when cutting a right-hand rather than a left-hand helix.

The Angle of the Helix

In order to cut a helix, except when using an end mill, it is necessary to set either the table or the cutter at the angle of the helix to be cut.

Swiveling the table at an angle is not feasible when milling very short leads, for example, 127 mm (5″). In such cases, the use of a universal attachment or the use of an end mill is recommended.

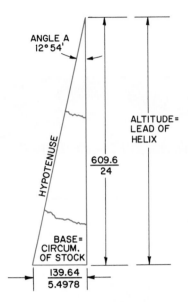

Fig. 87-2.   The graphical method of finding the angle of a helix.

Two methods of determining the angle of a helix are discussed here. These are the graphical and the mathematical methods.

Graphical Method

The angle of the helix may be determined graphically by laying out a right triangle using the circumference of the object or stock as a base and the lead of the helix as the altitude.

For example, assume the stock is a cylinder 44.45 mm (1-3/4") in diameter, and a lead of 609.6 mm (24") is planned. Hence, the base = 44.45 mm (1-3/4") × 3.1416 = 139.64 mm (5.4978") and the altitude = 609.6 mm (24"). Then the angle of the helix equals Angle **A,** Fig. 87-2.

Determine the number of degrees in Angle A by measuring it with a protractor. This is found to be approximately 12°54' or 12-9/10°. The line marked "hypotenuse" in Fig. 87-2 represents the line of the helix as it would traverse the cylinder if the triangle were cut out and wrapped around the cylinder.

Mathematical Method

The angle of a helix may be found by using the mathematics of trigonometry as follows:

Using the figures from the example in the graphical method:

$$\text{Tangent} = \frac{\text{side opposite}}{\text{side adjacent}}$$

$$= \frac{139.64 \text{ mm } (5.4978")}{609.6 \text{ mm } (24")}$$

$$= 0.2291$$

Examination of a table of trigonometric functions reveals that .2291 is approximately the tangent of an angle of 12°54'.

The Lead of the Helix

When the diameter of the workpiece and the angle of the helix are known, the lead of the helix may be calculated by dividing the circumference of the workpiece by the tangent of the angle.

Example:

Find the lead of the helix on a workpiece 57.15 mm (2-1/4") in diameter, having a helix angle of 15°. From a table of trigonometric functions, it is found that the tangent of 15° is 0.2679.

$$\textbf{Solution: } \text{lead} = \frac{\text{circumference of workpiece}}{\text{tangent of helix angle}}$$

$$= \frac{57.15 \text{ mm } (2\text{-}1/4") \times 3.1416}{0.2679}$$

$$= \frac{179.54 \text{ mm } (7.0686")}{0.2679}$$

$$= 670.13 \text{ mm } (26.39")$$

Note: When calculating the helix angle of a helical gear, use the pitch diameter instead of the outside diameter as the base of the triangle.

Setting the Table for the Helix Angle

Swivel the table counterclockwise for a right-hand helix. Set it clockwise for a left-hand helix.

Unit
88 ▌▌▌ ▐ █ █ █ █ █

How to Mill a Helix

The following is an explanation of how to mill the flutes of a right-hand helical milling cutter. The procedure described is for a milling machine that has the dividing head mounted on the right-hand end of the table. For a machine that has the dividing head mounted on the left-hand end, the procedure is reversed. Compare **C** with **D**, Fig. 88-1.

Procedure for Milling with a Double-Angle Cutter

Problem: Machine a cutter 57.15 mm (2-1/4″) in diameter, having 15 teeth with radial faces, 0.8 mm (1/32″) land, helix angle 12°. Use a double-angle cutter with the angles 48° and 12°.

1. Determine the number of grooves or teeth to be cut. The number of teeth and the width of the land will determine the amount of radial offset. See **A,** Fig. 88-3.
2. Determine the lead.

$$\text{Lead} = \frac{\text{Circumference of workpiece}}{\text{Tangent of angle}}$$

Tangent of 12° = .2126 or .213
Applying data gives:

$$\text{Lead} = \frac{57.15\ (2.25) \times 3.1416}{.213}$$
$$= 842.92\ \text{mm}\ (33.15''),\ \text{or roughly}$$
838.2 mm (33″), which is sufficiently close.

3. Mount the dividing head and tailstock on the table of the milling machine.
4. Determine the change gears necessary to produce a helix with the lead desired. Mount

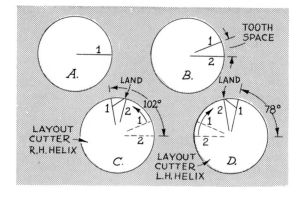

Fig. 88-1.   Layout for right-hand and left-hand helices.

the gears on the respective shafts and studs on the dividing head.

5. Operate the longitudinal hand feed to determine that the dividing head mechanism operates freely.
6. Determine the indexing required to mill 15 teeth equally spaced: The workpiece must be rotated through 24° for each tooth (360° ÷ 15 = 24°). Since one revolution of the dividing head rotates the workpiece through 9°, then 24° ÷ 9° = 2-2/3 turns of the index crank for each tooth space; or, since 360° ÷ 9° = 40, then 40 ÷ 15 = 2-2/3 turns of the index crank. This requires 2 full turns of the crank and 2/3 of any circle of holes divisible by 3, for example, 16 spaces on a 24 circle, 32 spaces on a 48 circle, or 26 spaces on a 39 circle. See Unit 80.
7. Adjust the location of the crank pin and the dividing head sector to index 2-2/3 turns.

8. Secure a suitable mandrel, arbor, cutter, dog, and surface gage.

9. After applying a little oil to the surface of the mandrel, press it firmly into the workpiece.

10. Place the dog on the large end of the mandrel; then place the piece between the centers of the dividing head.

11. Loosen the table swivel clamping bolts, and then temporarily swivel the table to the angle desired, for example, 12°. Move the table crosswise with the transverse hand crank until it is about 25 mm (1″) from the machine column.

12. Apply layout fluid to the tailstock end of the workpiece.

13. Set the point of the surface gage at the exact height of the axis of the dividing head center, and scribe a radial line on the end of the workpiece, as illustrated in Fig. 88-2 and in **A**, Fig. 88-1.

14. Index the work for one tooth space, and draw a second radial line, **B**, Fig. 88-1. Then index the work back one tooth space so that the first radial line will be in its original position **A**, Fig. 88-1.

15. Install the milling machine arbor and the cutter on the arbor as nearly over the axis of the work as possible. Position the overarm and bearing, and then draw the clamping nut tight. In this position, the axis of the cutter and the pivot center of the table should be approximately in the same vertical plane.

16. Move the table longitudinally until the end of the workpiece at which the cut is to start is under the center of the milling machine arbor.

17. Swivel the table back to its normal position (zero on the graduated base). Adjust the table laterally until the axis of the workpiece is exactly under the vertical axis of the cutter.

18. Index the workpiece counterclockwise through 102° from the horizontal, 11-1/3 turns of the index crank. See **A** and **B**, Fig. 88-3. See also **C**, Fig. 88-1.

19. Raise the table until the work just touches the cutter. Then adjust the table transversely until the 12° side of the cutter is aligned with the radial line (in the same plane) on the end of the workpiece. Test by holding a rule against the 12° side of the cutter, as in Fig. 88-4. This relationship will be changed when

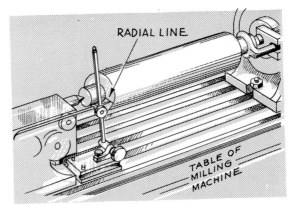

Fig. 88-2.   Scribing a radial line on the tailstock end of a workpiece.

the table is raised to make the cut. To compensate for the change, the table must be adjusted in a manner that will again align the 12° side of the cutter with the radial line. When correctly positioned, the side of the cutter with the 12° angle should split the radial line.

C, Fig. 88-3, shows the work rotated through 102° clockwise and the short side of the cutter aligned with the radial line. Rotated clockwise, the work is in position for milling a left-hand helix.

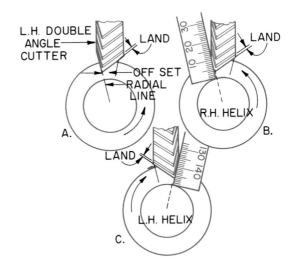

Fig. 88-3.   The method of aligning the workpiece and the cutter.

20. Swivel the table to the angle desired, and tighten the clamping bolts.
21. Determine the depth of the groove.
22. Set the graduated collar on the vertical feed at zero.
23. Set the graduated collar on the cross feed at zero. This facilitates making slight lateral (crosswise) adjustments, if and when necessary. Be sure to adjust for backlash, if any.
24. Start the machine; then engage the reverse traverse. When the cutter has cleared the workpiece, disengage the power feed and raise the table an amount equal to the depth of the helix groove, less about 1.5 mm (1/16").
25. Engage the power feed. When the cutter enters the work and is cutting its full depth, stop the machine. Examine the cut to make certain that the 12° side of the cutter is splitting the radial line on the end of the workpiece.

    If the cutter is not correctly positioned, lower the table, reverse the feed, and move the table back until the work just clears the cutter. Then make the necessary adjustment by moving the table laterally in the direction desired. Be sure to adjust for backlash.
26. If adjustment was found necessary and has been made, again start the machine and continue the cut until the cutter is cutting its full depth. Stop the table feed, and examine the cut. If further adjustment is necessary, adjust as in step 25.
27. When the cutter has been positioned correctly, advance the table until the cutter is cutting full depth. Then reverse the table travel, and index for one tooth space.
28. Proceed until the cutter is cutting its full depth. Then stop the machine and examine the width of the land (the flat spot at the end of the tooth). See **A,** Fig. 88-3. If the land is too narrow, the table must be lowered an amount sufficient to increase the land the amount desired; if the land is too wide, the table must be raised. **In either event, the table must be adjusted laterally to realign the radial line with the 12° side of the cutter.**

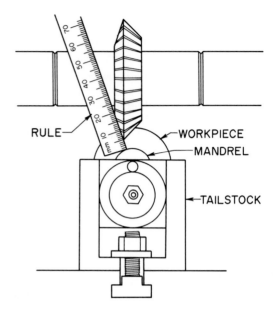

Fig. 88-4. Checking the position of the workpiece before making a trial cut.

29. When the cutter has been properly aligned and set to depth, start the machine and make the cut. Use a cutting fluid freely if required.

    After the cutter has been set for depth of cut, it is good practice to set the graduated collar on the vertical feed at zero. This facilitates resetting for subsequent cuts.
30. At the end of the cut, lower the table. Then reverse the feed, and allow the cutter to travel back to the starting point.
31. Index for the next groove. Raise the table, and proceed as in steps 29 and 30.
32. Proceed as in steps 30 and 31 until all of the grooves have been rough-cut.
33. Adjust the vertical feed for the finishing cut, if one is necessary. Then proceed as in steps 29 through 32 until all of the grooves have been milled.

Unit
89

Cam Milling

Cams are mechanical devices that are widely used for timing and controlling mechanical motions in machinery. There are three kinds of cams: (1) plate or disc, (2) cylindrical or drum, and (3) translating or sliding. See Fig. 89-1.

Translating cams may be accurately machined with the aid of a rotary table to obtain the required angles and curves, Fig. 89-2. Numerically controlled milling machines with contouring capability are ideally suited for this kind of machining problem. See Section 16. Drum cams are machined using standard helical milling practices. See Units 86-88.

Plate or disc cams having leads within the normal range of gearing may also be cut using standard helical milling practice. The cam blank is mounted in the spindle of the dividing head, which is swiveled to a vertical position. A vertical milling attachment is also positioned with its spindle vertical, Fig. 89-3. As the table advances towards the cutter, it rotates the cam blank, continually shortening the distance between the center of the cam blank and the cutter. This results in a spiral curve with a lead equal to the lead for which the machine was geared, Fig. 89-4.

Plate cams such as used on automatic screw machines, Fig. 89-5, frequently have leads that are less than can be provided with standard gearing. For example, the shortest lead normally obtainable on Brown and Sharpe milling machines is 17.02 mm (0.670"). Shorter leads may be obtained by tilting the dividing head to an angle between vertical (90°) and horizontal (0°). As may be seen from Fig. 89-6, zero lead is obtained when the dividing head spindle and the cutter are set at zero degrees. Under these conditions, the

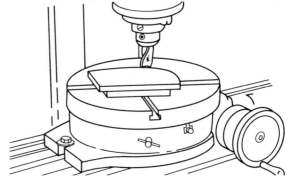

Fig. 89-2.   Milling a translating or sliding cam on a vertical mill with the aid of a rotary table. (Workpiece clamps are removed for clarity.)

Fig. 89-1.   Types of cams.

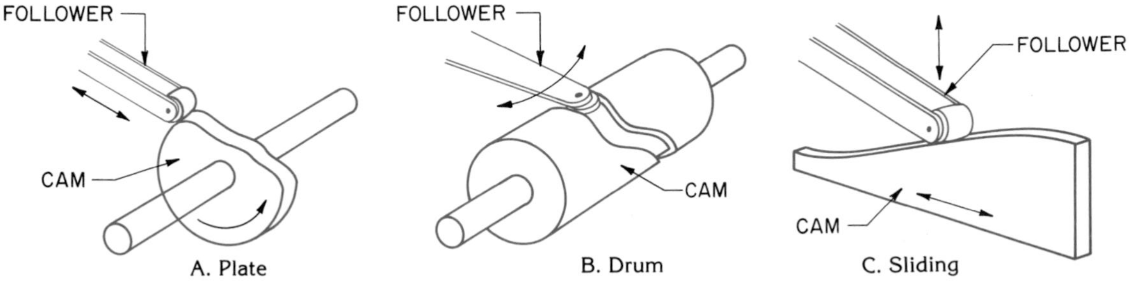

FOLLOWER

CAM

A. Plate

FOLLOWER

CAM

B. Drum

FOLLOWER

CAM

C. Sliding

movement of the cam blank towards the cutter while it is being rotated results only in the cutting of a circle. It follows then, that by setting the dividing head at an angle between zero and 90°,

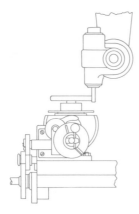

Fig. 89-3. For milling a plate cam or a translating cam, the dividing head and the milling machine spindle are both set at 90°.

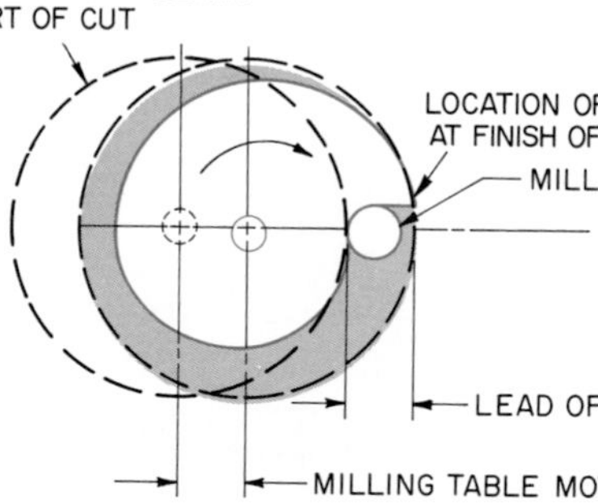

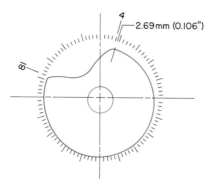

Fig. 89-4. When the dividing head is set at 90°, milling table movement and cam lead are equal.

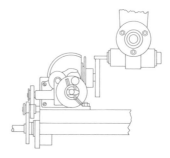

Fig. 89-5. A circle divided into 100 divisions for the layout of an automatic screw machine cam.

any amount of lead may be obtained between the lead for which the machine is geared and zero, Fig. 89-7. **In all cases, it is important that the dividing head spindle and the cutter spindle be set parallel to each other.**

Example: To illustrate how to find the correct angle to set the dividing head spindle when a lead of less than 17.02 mm (0.670″) is required, the plate cam shown in Fig. 89-5 will serve as the machining problem.

1. Calculate the lead of the cam lobe. This is the amount of **rise** (increase in radius) of the cam lobe if it were continued a full 360°. For cams figured in degrees, divide the rise by the number of degrees it occupies and multiply by 360. For cams figured in hundredths of a circle, Fig. 89-5, divide the rise by the number of hundredths the cam occupies and multiply by 100.

   The cam lobe in Fig. 89-5 rises 2.69 mm (0.106″) and occupies 77/100 of a circle.

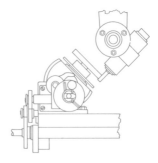

Fig. 89-6. When both the dividing head and the milling machine spindle are set at zero degrees, a zero lead results.

Fig. 89-7. The dividing head and milling machine spindle are set at matching angles between zero and 90° to cut a disc cam with a short lead.

Therefore, its lead is

$$\frac{2.69 \, (0.106)}{77} \times 100 = 3.50 \text{ mm } (0.138'')$$

2. The required lead of 3.50 mm (0.138") is much less than the shortest lead that can normally be obtained, that is, 17.02 mm (0.670"). Therefore, the machine is geared for a lead of 17.02 mm (0.670") and the dividing head is tilted so that a lead of 3.50 mm (0.138") will be obtained.

   Dividing the lead required by the lead for which the machine is geared results in the sine of the angle for which the dividing head must be set. The problem is shown graphically in Fig. 89-8.

   Solution:

   $$\frac{3.50 \text{ mm } (0.138'')}{17.02 \text{ mm } (0.670'')} = 0.20597$$

By consulting a trigonometry table, it will be found that the figure 0.20597 is very close to the sine of 11°52'. Since the dividing head graduations are in quarter degrees, setting the dividing

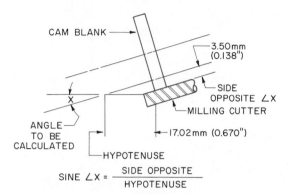

SINE ∠X = $\dfrac{\text{SIDE OPPOSITE}}{\text{HYPOTENUSE}}$

Fig. 89-8.   A diagram explaining the method of determining the dividing head angle required to obtain a short lead.

head to approximately 11-7/8° will be suitably accurate.

The rest of the cam profile from 81/100 to 4/100 may be removed by milling or band sawing and filing to the layout line.

Unit
90

Digital Readouts

Digital readouts (DROs) are electro-mechanical devices that provide a constant numerical display of any condition being monitored. In metalworking, DROs are used on machine tools to accurately position the cutting tool or workpiece. DROs are becoming increasingly popular on manually operated lathes, precision grinders, and milling machines, Fig. 90-1. They are also used on **coordinate measuring machines** to speed inspection of finished parts, Fig. 90-2.

DROs are usually millimeter/inch switchable, which permits parts to be made or inspected in either millimeters or inches on the same machine. Millimeter readouts are available with accuracies of either one hundredth or two thousandths (0.01

or 0.002) of a millimeter. Inch readouts are available with accuracies of either one half-thousandth or one ten-thousandth (0.0005 or 0.0001) of an inch.

Advantages of Using DROs

There are several advantages to using DROs, as follows:

1. Backlash compensation errors can occur when using micrometer dials to measure movement. These errors are eliminated through the use of a DRO. The DRO shows the exact location of the moving machine part, **not** the location of the screw that moves it.

Fig. 90-1. A vertical milling machine equipped with a two-axis digital readout.

Fig. 90-2. A coordinate measuring machine equipped with a digital readout aids in the rapid inspection of parts.

This results in:
  a. improved part accuracy
  b. fewer scrapped parts
  c. faster positioning, and
  d. higher productivity per operator per machine.
2. There is little or no arithmetic for the operator to do. The addition and subtraction required when using micrometer dials is eliminated.
3. Operators need not be as highly skilled.

Kinds of DRO Systems

Three kinds of DRO systems are available: (1) a rack and a pinion-driven encoder, (2) magnetic tape and a tape reader, and (3) an engraved glass scale and an electro-optical reader.

In each case, a measuring device is attached to the moving part of the machine, for example, a milling table. A reader is located in a fixed position next to the measuring device. Any movement of the machine part is sensed by the reader, which converts the movement into electrical signals. The electrical signals are sent to the DRO, which converts them into the numerical display that shows the position of the movable machine part.

In a rack-and pinion encoding system, the rack is the measuring device. Its movements are sensed by the pinion, which in turn drives an encoder. The encoder converts the mechanical movements into an electrical signal that activates the DRO.

Types of DROs

DROs are made either as nonprogrammable or programmable units. **Nonprogrammable units** provide a continuous readout of position in relation to a fixed reference point. The reference point can usually be repositioned by the operator whenever it is necessary. However, no keyboard is provided to allow programming positions in advance of movement.

Part drawings for use with nonprogrammable DROs should have all dimensions taken from a **part reference point.** (This method is known as **absolute dimensioning.**) Then, when the DRO is reset to read zero at the part reference point, its numerical display will always correspond with dimensions on the part drawing. This eliminates errors that result from having to add or subtract dimensions between moves (incremental dimensioning) in order to arrive at the correct DRO number for a given location.

Programmable DROs are equipped with a keyboard that allows entry of either incremental or absolute dimensions, Fig. 90-3. One type can remember only the original **datum** or reference point and the distance entered for one move. If an incremental move is to be made, the operator keys in the distance to the next location. The move is then made and is complete when the DRO display is again zero. If an absolute dimension is keyed in, the DRO calculates the difference, if any, between the absolute dimension and the incremental move needed to get there. The incremental amount is then displayed and the move is made until the DRO display is once again zero.

A second type of programmable DRO has enough memory to allow the operator to enter the program for an entire part before the part is made. The DRO shown in Fig. 90-3 has a memory that can retain up to 64 locations per axis. This memory can be expanded to 90 locations per axis if desired. Entering the program before making several duplicate parts eliminates the need to key in the point locations each time a part is made. Instead, the operator simply recalls the program from memory one step at a time. This causes the point locations to be displayed on the DRO. Duplicate parts can be made more quickly this way because no further programming time is required.

(Dynamics Research Corp.)

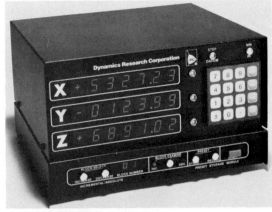

Fig. 90-3. A programmable three-axis digital readout. One-axis and two-axis readouts are also available.

Unit
91

How to Use a Digital Readout

To understand how to use a DRO, you must first study its controls, Fig. 91-1. When you have become familiar with the controls, read the following explanation of how a DRO is used.

Determining the Machining Sequence for the Part

Select the **datum** or part reference point to be used for machining the part and assign it the title **Block Number 1**. See Fig. 91-2. Block numbers assigned to each location are important to the machining of the part. They generally are assigned in the same order as the sequence in which the machining operations will be performed.

Before assigning all the block numbers, **examine the part drawing for operations that are the same.** (For example, identical size holes or slots to be milled.) Plan the machining sequence so that the identical operations will have sequential block numbers. This will save time when machining the part because it will eliminate unnecessary tool changing.

Preparing the DRO for Use

1. Turn the power switch on the back of the DRO to the ON position.
2. Push each axis switch to the ON position. **Each axis must be activated even if it will not be used.**
3. Place the Step/Datum switch in the STEP mode and push the CLEAR button. This will activate the readouts and clear the STEP memory.
4. Place the Step/Datum switch in the DATUM mode, and push the CLEAR button to clear the DATUM memory.
5. Place the Millimeter/Inch switch in the IN mode.

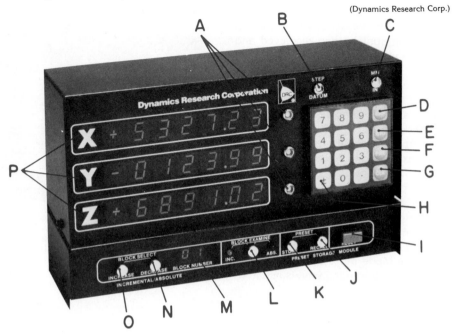

(Dynamics Research Corp.)

Fig. 91-1. A programmable digital readout with a preset storage module. An explanation of the controls is provided below.

A. Axis switches. These switches turn each axis on and off independently to allow use of one keyboard for all three axes. A switch must be depressed to enter or clear dimensions.

B. Step/Datum switch. Dimensions entered with this switch in the STEP mode (depressed position) will be recorded as incremental moves. Dimensions entered in the DATUM mode will be recorded as absolute dimensions. **When a job is run, this switch must be in the STEP postion.**

C. Millimeter/Inch (MM/IN) switch. This switch changes modes from inches to millimeters and vice versa. The number displayed is instantly calculated upon switching. However, numbers in memory are not converted.

D. Examine (EXM) key. When pressed, this key displays a single preset point location that can be stored on readouts not equipped with a preset storage module.

E. Store (STO) key. When pressed, this key stores one number in the memory for each axis after that number has been entered into the display. The Store key is used on readouts not equipped with a preset storage module.

F. Preset (PRE) key. This key recalls a number previously stored. It is used on readouts not equipped with a preset storage module.

G. Clear (CLR) key. This key clears the display in all axes that are turned on.

H. Plus/Minus (±) key. This key changes the sign of a dimension. Numbers must be entered into the display before they key is used.

I. RESET Switch. This switch instantly returns the Block Number Display to Block Number 1.

J. RECALL switch. This switch calls up the dimensions stored in the memory block indicated in the Block Number Display.

K. STORE switch. This switch stores dimensions in the memory block indicated in the Block Number Display.

L. BLOCK EXAMINE switch with incremental (INC.) and absolute (ABS.) pilot lights. Holding this switch in displays the dimensions that correspond to the block number in the Block Number Display. The pilot lights indicate whether the dimensions entered are incremental or absolute.

M. Block Number Display. This display shows the current memory location as a two-digit number.

N. Block Select DECREASE switch. This switch changes the Block Number Display to the next lowest Block Number. Holding the switch in causes a rapid sequencing in this direction.

O. Block Select INCREASE switch. This switch changes the Block Number Display to the next highest Block Number. Holding the switch in causes a rapid sequencing in this direction.

P. Digital displays for X, Y, and Z axes. Together, these displays show the location of a given point on the X, Y, and Z axes.

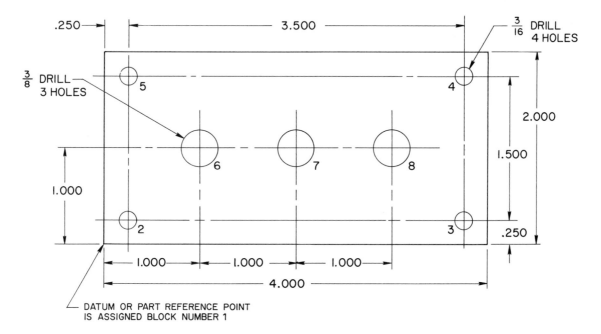

Fig. 91-2. A mounting plate used to explain the operation of a digital readout.

DATUM OR PART REFERENCE POINT IS ASSIGNED BLOCK NUMBER 1

Programming the DRO

The following is the procedure for programming the DRO to make the part shown in Fig. 91-2.

1. Mount the workpiece in a milling machine vise.
2. Using an edge finder, position the center line of the spindle over the **datum** or part reference point. See Unit 79 if necessary.
3. Using the BLOCK SELECT buttons on the memory module, set the Block Number display to Block Number 1.
4. Push the Clear (CLR) key to zero out both axis displays.
5. Push the STORE button on the Preset Storage Module (not the STO key on the keyboard). **This programs Block Number 1 as the datum or part reference point.**
6. Set the Block Number Display to Block Number 2 by using the Block Select INCREASE button.
7. Move the Step/Datum switch to the STEP position.
8. Since the first move is the same in both axes, and both axis displays have been turned on, enter the dimension of +.250.

9. Store the X and Y dimensions for Block Number 2 by pressing the STORE button on the Preset Storage Module.
10. Set the Block Number Display to number 3.
11. Press CLR to zero out both axes, then turn off the Y axis, since the part is already correctly positioned for that axis.
12. Enter the X dimension of +3.500 to reach position 3.
13. Store the X and Y dimensions for Block Number 3 by pressing the STORE button on the Preset Storage Module.
14. Set the Block Number Display to 4.
15. Press CLR to zero out the X axis, then turn the X axis off.
16. Turn the Y axis on. Press CLR, then enter the Y axis dimension of +1.500 to reach position 4.
17. Store the X and Y dimensions for Block Number 4.
18. Set the Block Number Display to number 5.
19. Press CLR to zero out the Y axis, then turn the Y axis off.
20. Turn the X axis on. Press CLR, then enter the X dimension of −3.500.
21. Store the X and Y dimensions for Block Number 5.

22. Set the Block Number Display to 6.
23. Hole 6 is dimensioned from the original datum point. Therefore, set the Step/Datum switch to DATUM.
24. Press CLR, then enter the X dimension of + 1.000. Turn the X axis off.
25. Turn the Y axis on, press CLR, and enter the Y dimension of + 1.000.
26. Store the X and Y dimensions for Block Number 6.
27. Set the Block Number to 7.
28. Return the Step/Datum switch to STEP.
29. Press CLR to zero out the Y axis, then turn the Y axis off.
30. Turn the X axis on, press CLR, then enter the X dimension of + 1.000.
31. Store the X and Y dimensions for Block Number 7.
32. Set the Block Number to 8.
33. Press CLR, enter the X dimension of + 1.000, then turn the X axis off.
34. Turn the Y axis on, then press CLR to zero out the Y axis.
35. Store the X and Y dimensions for Block Number 8.

This ends the programming for the mounting plate in Fig. 91-2. It is advisable to reset the Block Number Display to Block Number 1 (using the RESET switch), and then check each block of the program for errors using the BLOCK EXAMINE switch on the Preset Storage Module.

Using the Preset Program to Machine the Mounting Plate

1. Install a drill chuck and center drill in the spindle.
2. Set the spindle speed desired.
3. Set the Block Number Display to Block Number 1. Turn on both axes and push the CLR button to zero out both displays.
4. Set the Block Number Display to Block Number 2 and press the RECALL button on the Preset Storage Module. The DRO display will flash for an instant. When the flashing stops, move the table until both the X axis and Y axis read all zeros. The spindle is now over position 2. Center-drill the position and set a depth stop if desired.

NOTE: To check where the spindle is at any time while running the program, move the Step/Datum switch to DATUM. The read-out displays will show the X and Y coordinates for that position. **(The table must be stopped while you make this check.)** IMPORTANT: ALWAYS MOVE THE STEP/DATUM SWITCH BACK TO THE STEP MODE BEFORE CONTINUING. **The display will not recall unless it is in the STEP mode.**

5. Press the INCREASE button to set the Block Number to 3. Press the RECALL button and move the table until the X and Y axes both read all zeros. Center-drill position 3.
6. Repeat this procedure for positions 4, 5, 6, 7, and 8.
7. Set the Block Number to Block Number 2. Press the RECALL button, then zero out the X and Y displays by moving the table. This returns the spindle to position 2 on the workpiece.
8. Remove the center drill and install the 3/16″ drill. Reset the spindle rpm if desired, then drill the hole. Set a depth stop if desired.
9. Set the Block Number Display to Block Number 3. Press the RECALL button, then zero out the X and Y displays by moving the table. Drill the hole at position 3.
10. Repeat the above procedure for positions 4, 5, 6, 7, and 8.
11. Set the Block Number Display to Block Number 6. Press the RECALL button, then zero out the X and Y displays by moving the table. Change to the 3/8″ drill. Adjust the spindle speed and drill the hole at position 6. Set a depth stop if desired.
12. Set the Block Number Display to Block Number 7. Press the RECALL button, zero out the X and Y displays by moving the table, and drill position 7.
13. Repeat the above procedure for position 8.

To run the program again for making a second part, reset the Block Number Display to Block Number 1. Move the table to zero out the X and Y axes. Remove the finished workpiece and replace it with a blank of the same size. Use an edge finder, if necessary, to locate the blank in the correct position. The program may now be repeated to make the second part.

Test Your Knowledge of Section 10

Unit 71: The Milling Machine

1. Describe the process of removing metal by milling.
2. Milling machines are widely used for machining what kinds of surfaces?
3. What is the advantage in performing hole-machining operations on a milling machine?
4. For what basic purpose are bed milling machines used?
5. How does a turret-type vertical mill differ from one with a fixed head?
6. List ten common vertical milling machine operations.
7. What types of operations are more readily performed on universal milling machines than on plain machines?
8. List the three directions a work table may be moved on a column and knee milling machine.
9. List four basic factors used in identifying milling machine size.
10. List three methods of changing spindle speeds on milling machines.
11. Why should a person be thoroughly instructed in use of the power rapid feed before using it?
12. Explain the difference between up milling and down milling.
13. What feature must be provided on machines that are used for down milling in order to prevent cutter damage?
14. What are the advantages of down milling?
15. List five methods that may be used for holding workpieces on a milling machine.
16. Why should the arbor nut be tightened or loosened only when the overarm support is in place?
17. What kind of hammer should be used for seating workpieces in a vise?
18. Explain how chips can be safely removed from around milling cutters.

Unit 72: Milling Machines for Special Uses

1. For what kind of work is a hand milling machine used?
2. What kind of machine is a jig borer? For what is it used?
3. What is the unique feature of a turret drilling and milling machine?

4. Describe how a conventional boring mill operates. How does a traveling column type differ from the conventional type?
5. For what kind of work are planer-type milling machines used?
6. Of what advantage are duplex and triplex milling machines?
7. Explain how the path of the cutting tool is controlled on profiling machines.
8. Describe a pantograph engraving machine and the kind of work for which it is used.
9. Describe the process of cutting gears by hobbing.
10. Describe the process of cutting threads by thread milling.

Unit 73: Milling Machine Attachments

1. Name four kinds of vises used on milling machines.
2. How is the size of a milling machine vise measured?
3. What is the range of sizes available in plain and swivel base vises?
4. For what purposes is a dividing head used on a milling machine?
5. Describe the rotary table attachment and its use.
6. Describe how boring heads are able to machine precision holes.
7. Describe the slotting attachment and the kind of operation for which it is used.
8. Name two kinds of angle-milling attachments for vertical milling machines.
9. How does the vertical milling attachment differ from the high-speed universal milling attachment?
10. What milling operations are possible with a universal spiral milling attachment used on a plain horizontal milling machine?
11. For what milling operations is a rack milling attachment used?

Unit 74: Milling Cutters and Holders

1. List three ways milling cutters are designed.
2. Which milling cutter design is becoming more widely used for production milling?
3. Milling cutters used in schools are usually made of what two cutting tool materials?

Why are cutters made of these materials preferred for school use?

4. Why is rigidity of the machine and the workpiece important when using carbide cutters?
5. Of what advantage is it to have center-cutting teeth on an end mill?
6. List six kinds of end mills.
7. How do insert-type end mills differ from solid end mills?
8. Explain how a T-slot is milled.
9. Key seat cutters are used to cut slots for what kind of keys?
10. How do shell end mills differ from solid end mills?
11. How is a face milling cutter different from a shell end mill?
12. Describe three general types of plain milling cutters and give a principal use for each.
13. Describe three kinds of side milling cutters and give an application for each.
14. Describe how straddle milling is done.
15. List three types of slitting saws and give the type of sawing recommended for each type.
16. Name two kinds of angle milling cutters. What angles are normally available?
17. Name five kinds of formed-tooth milling cutters.
18. Describe three kinds of fly cutters.
19. Explain how to determine whether a milling cutter is right- or left-handed.
20. Name two ways of holding straight-shank end mills in the milling machine spindle.
21. List three types of milling machine arbors and indicate a principal use for each.
22. How will nicks on the face of arbor spacing collars affect the accuracy of a milling machine arbor?
23. Name two ways that milling arbors with standard (national milling machine taper) shanks are held in the machine spindle.

Unit 75: Calculating Milling Cutter RPM and Feed Rates

1. List five factors that affect the cutting speeds and feeds used for milling.
2. Explain how the cutting speed for a given milling application is determined or selected.
3. How do the recommended cutting speeds for carbide milling cutters compare with those for high-speed steel cutters?

4. Calculate the rpm for a 69.85 mm (2-3/4″) diameter cutter that is to cut at 27.4 mpm (90 fpm).
5. Calculate the rpm for a 12.7 mm (1/2″) diameter end mill that is to cut at 21.4 mpm (70 fpm).
6. Calculate the approximate cutting speed for a 44.45 mm (1-3/4″) diameter end mill that is operating at 165 rpm.
7. Determine the feed rate in millimeters (inches) per minute for machining annealed high-carbon steel at 18.3 mpm (60 fpm). A heavy-duty plain milling cutter of 63.5 mm (2-1/2″) diameter with 8 teeth will be used, revolving at 90 rpm. The feed per tooth desired is 0.10 mm (0.004″).
8. Explain the basic differences in depth of cut, rate of feed, and rpm for roughing cuts and finishing cuts in milling.
9. Why is it advisable not to stop the feed on a milling machine before the cut is finished?

Unit 80: The Dividing or Indexing Head and Its Operation

1. List the principal operating parts of the dividing head.
2. List several kinds of milling operations for which a dividing head is used.
3. Explain how direct indexing is carried out.
4. State the formula used for simple indexing that determines the number of turns of the index crank required to rotate the spindle a given amount.
5. Explain how the sector is used on a dividing head.
6. Explain how a hexagon is milled using direct indexing.
7. For simple-indexing a 10-flute reamer, how many index crank turns are required for each flute?
8. For simple-indexing a 56-tooth gear, how many index crank turns are required for each tooth?
9. Solve for simple-indexing a circle of holes 30° apart.

Unit 82: Spur Gears and Gearing

1. What are two types of special machine tools that are generally used for mass-producing gears?

2. Define **gear, pinion,** and **gear rack.**
3. List five commonly used types of gears.
4. Explain the difference in the meaning of the terms **pitch diameter** and **diametral pitch.**
5. Explain the meaning of **circular pitch.**
6. Explain the module system of sizing metric gears.
7. Explain how the chordal thickness of a gear tooth is measured.
8. Describe the SI metric full-depth gear tooth form.
9. How is the module related to diametral pitch?
10. Describe the current American National Standard gear tooth forms.
11. Are any American National Standard gear tooth forms interchangeable with the SI metric gear tooth form? Explain.
12. Find the outside diameter, number of teeth, and whole depth for a 1.5 module gear with a pitch diameter of 48 mm.
13. Find the outside diameter, number of teeth, and whole depth for a 16 diametral pitch gear with a pitch diameter of 2 inches.
14. What gear cutter should be used for cutting (a) a gear with 18 teeth, (b) a gear with 40 teeth, (c) a gear rack?

Unit 86: Helical Milling

1. Define the meaning of **lead** as it applies to helical milling.
2. How are variations in the length of the lead accomplished in helical milling?
3. Describe a right-hand helix.
4. How is a change in the direction of rotation of the dividing-head spindle accomplished?

Unit 87: Calculations for Helical Milling

1. Explain how the lead of the machine is determined.
2. Write the formula used for determining the change gears required to produce a desired lead with a dividing head mechanism.
3. Give the formula that may be used for determining the angle of a helix mathematically.
4. Determine the change gears required for a lead of 457.2 mm (18").
5. Determine the helix angle for item 4 above, given a workpiece diameter of 25.4 mm (1").

Unit 89: Cam Milling

1. Name the three kinds of cams.
2. Describe the method of machining each type of cam.
3. Calculate the angle for setting the dividing head to obtain a lead of 6.35 mm (0.250") when the machine is geared for a lead of 17.02 mm (0.670").

Unit 90: Digital Readouts

1. Name two ways digital readouts are used in metalworking.
2. What are the standard digital readout accuracies available in millimeters? Inches?
3. List six advantages of using digital readouts.
4. Name three kinds of digital readout measuring systems.
5. Name two types of digital readouts.
6. What dimensioning system should be used when using a nonprogrammable digital readout? Why?
7. Explain the difference between the two types of programmable digital readouts.

Unit

92

Shaping and Planing Machines

Shaping and planing machines cut by a reciprocating (back and forth) movement of either the tool or the workpiece. Neither the tool nor the workpiece rotates. The basic principles of shaping and planing are discussed in Unit 2, which you may review if necessary.

Shapers cut by moving the workpiece across the path of a reciprocating single-point cutting tool. **Horizontal shapers,** Fig. 92-1, are used mostly for machining flat surfaces, which may be horizontal, vertical, or angular. **Vertical shapers,** Fig. 92-2, are also known as **slotters.** They are used more for machining slots, keyways, splines, and other shapes in large gears, pulleys, and flywheels that are difficult to machine. Vertical shaping attachments allow vertical shaping to be done on milling machines, Fig. 73-10. A highly specialized type of shaper is the **gear shaper,** Fig. 92-3.

Planers cut by moving a single-point cutting tool across the path of a reciprocating workpiece. **Double-housing planers,** Fig. 92-4, have two columns to support a **cross rail** on which **overhead tool heads** ride. A **side tool head** is also provided on the column on the operator's side of the machine. Cutting tools in all three tool heads often cut simultaneously to speed the machining of parts. **Open-side planers,** Fig. 92-5, support the cross rail from a single column. This allows wide workpieces to overhang the table on the open side if necessary.

Planers are used mostly for machining flat surfaces on workpieces too large for shapers. However, both planers and shapers may be fitted with hydraulic tracing attachments that enable them to cut curved surfaces.

Shapers and planers require many strokes of the tool or workpiece to complete a cutting operation. Because milling machines can remove metal at a more rapid rate, horizontal and vertical mills have almost completely replaced horizontal shapers and planers for production work.

(Cincinnati Incorporated; Photo Courtesy of Illinois State University)

(Rockford Machine Tool Co.)

Fig. 92-1. A plain heavy-duty horizontal shaper, with a close-up of the clapper box, tool post, and cutting tool.

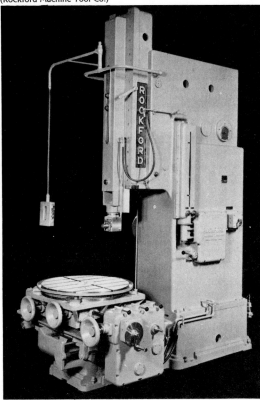

Fig. 92-2. A vertical shaper or slotter. The built-in rotary table is a standard feature.

Fig. 92-3. A gear shaper. The close-up on the left shows how a spur gear is shaped.

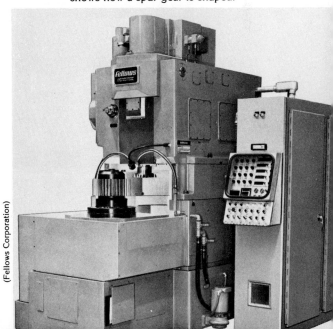

(Fellows Corporation)

(Rockford Machine Tool Co.)

Fig. 92-4.   A medium-size double-housing planer.

(Rockford Machine Tool Co.)

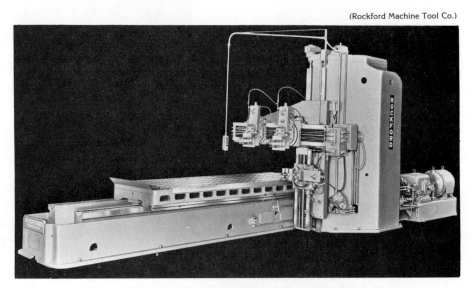

Fig. 92-5.   An open-side planer of medium size.

Unit
93

Horizontal Shapers

Horizontal shapers are inexpensive to maintain, and they use inexpensive single-point cutting tools. For these reasons and because of their ability to machine internal shapes, horizontal shapers continue in limited use.

Shaper Size

The size of a shaper is designated by the maximum length of stroke (cut) it can take. Shapers are also designed for light-duty, medium-duty, or heavy-duty work. The shaper in Fig. 93-1 is sold as a 305 mm (12″) shaper, but has a maximum stroke of 343 mm (13-1/2″).

(Sheldon Machine Company, Inc.)

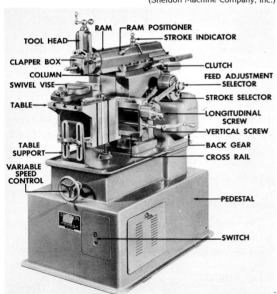

Fig. 93-1.   Principal parts of a 305 mm (12″) horizontal shaper.

The Driving Mechanism

Shaper driving mechanisms use either belts, gears, or hydraulic systems. The **ram** is driven by a slotted **rocker arm**. The rocker arm is connected to the **driving wheel** by a **crankpin** and a **sliding block**, Fig. 93-2. As the driving wheel revolves, the crankpin and the sliding block are carried through a circular path. This causes the rocker arm to move back and forth, driving the ram in a straight line.

Speed Change Mechanisms

When set for any particular **operating speed**, a shaper will make a constant number of strokes, **regardless of the length of the stroke.** However,

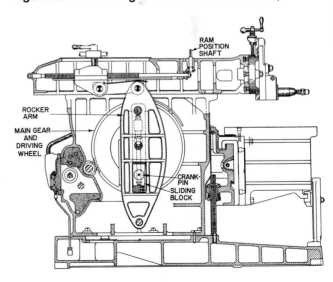

Fig. 93-2.   A sectional side view of a shaper.

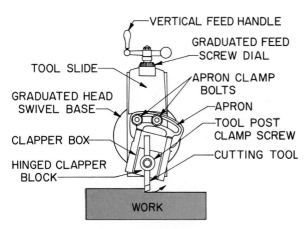

Fig. 93-3. A shaper tool head assembly. The arrow indicates the direction in which the tool swings on the return stroke.

to maintain a constant **cutting speed**, the shaper must make twice as many strokes when making a cut 76 mm (3″) long as when making one 152 mm (6″) long. On larger machines, speed change is accomplished through a system of gears. An index plate shows where to set the control levers for each speed. Some smaller shapers are equipped with a variable speed drive that is continuously variable in the range of 12 to 180 strokes per minute.

The Work Table

The table of a shaper is bolted to a **saddle** that slides in a horizontal plane along the **cross rail**. The saddle is moved by a **longitudinal feed screw**. The cross rail may be raised or lowered with a **vertical adjusting screw**. Before vertical adjustment of the cross rail is made, the clamping bolts holding it to the column should be loosened.

CAUTION

If the table is to be lowered, the table support bolts must also be loosened.

After the cross rail has been adjusted, retighten the table support bolts and cross rail clamp bolts.

The top and sides of the table are provided with T-slots for clamping the work to the table, or for attaching vises or fixtures.

The Tool Head

The tool head, Fig. 93-3, consists of a **tool slide**, a **swivel base**, an **apron**, a **clapper box**, a **clapper block**, and a **tool post**. A **feed screw** on the tool head has a collar graduated in hundredths of a millimeter or thousandths of an inch. The swivel base allows vertical or angular cuts. The tool is held in a tool post that is attached to the hinged clapper block. During the cut, the clapper block is forced against the clapper box and is thus solidly supported. On the return stroke, the clapper block swings free. This allows the tool to be drawn back across the work with only a slight rubbing effect, minimizing wear on the cutting edge.

The apron may be swiveled to the left or right to cause the cutting tool to swing away from the direction of cut. However, for cutting grooves, the apron should be kept vertical.

The Table Feed Mechanism

Shapers are provided with an automatic longitudinal feed. This is controlled with either a **ratchet** or **cam mechanism**. With a ratchet mechanism, the feed rate is determined by the number of notches the ratchet pawl moves on the feed wheel for each stroke of the ram. The number of notches, in turn, is determined by the position of the knob in the feed crank wheel. On a cam-actuated feed, a direct-reading dial is set to the desired feed rate, for example 0.15 mm (0.006″) per stroke.

Shaper Cutting Tools

Shaper, planer, and lathe cutting tools are very similar. Two basic kinds are used: forged tools and tool bits. The former are heavy tools, examples of which are shown in Fig. 93-4. Tool bits are more widely used than forged tools because of their lower cost.

Tool Terminology and Shape

The terminology used for shaper, planer, and lathe tools is the same. For more information on single-point cutting tools, see Units 49 and 50.

Relief Angles

A side-relief angle of 3° to 5° is enough for most shaper tools. An end relief of 3° to 5° is also sufficient.

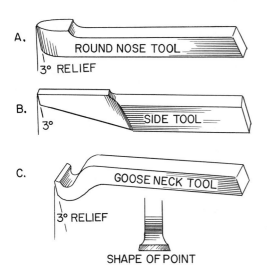

Fig. 93-4. Forged shaper tools.

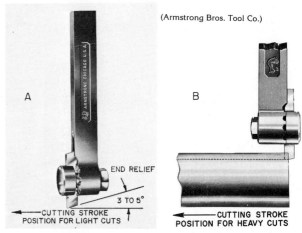

(Armstrong Bros. Tool Co.)

Fig. 93-5. A shaper or planer toolholder for holding tool bits.

Rake Angles

The side rake for shaper tools is usually 8° or 10°. Little or no back rake is needed for most applications.

Grinding Tool Bits

Shaper tool bits are ground in the same way as lathe tool bits. The grinding procedure is included in Unit 50.

Shaper Toolholders

Forged tools and other larger shaper tools are mounted directly in the tool post.

The type of toolholder shown in Fig. 93-5 is used for holding shaper or planer tools bits. With this toolholder, the tool bit may positioned at different angles for various types of cuts, Fig. 93-6.

The position of the tool in a planer and shaper toolholder in relation to the direction of tool travel is important. For light cuts, the tool generally is clamped at the front of the toolholder, as shown in A in Fig. 93-5. **For moderate to heavy cuts, the toolholder and the tool are reversed,** as shown in B in Fig. 93-5. With this procedure, the cutting edge is brought back of the shank of the toolholder, causing an effect similar to a gooseneck tool. This reduces the tendency of the tool to **chatter** or **dig in,** Fig. 93-7.

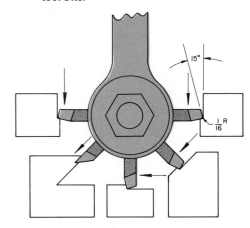

Fig. 93-6. Tool positions and types of cuts possible with a shaper or planer toolholder.

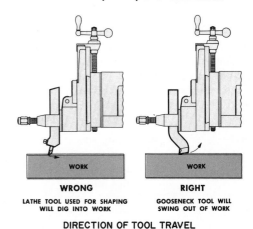

Fig. 93-7. The tool's position and the angle at which the tool is held affect its cutting action.

(Armstrong Bros. Tool Co.)

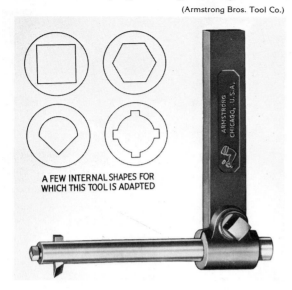

A FEW INTERNAL SHAPES FOR
WHICH THIS TOOL IS ADAPTED

Fig. 93-8.   An extension shaper tool.

An **extension shaper tool** consists of a boring bar mounted in a toolholder, Fig. 93-8. This tool is used for cutting internal keyways and holes of various shapes.

Calculating Shaper Cutting Speeds

On most shapers, the cutting stroke takes approximately one and one-half times as long as the return stroke. The ratio of 1:1-1/2, when each is multiplied by two, may be represented by the products 2 and 3 respectively. Thus three-fifths of each full minute of running time is spent in making cutting strokes and two-fifths in return strokes.

Cutting speed is always given in meters per minute (feet per minute) and is determined as follows:

When: CS = Cutting speed in meters (feet) per minute

N = Number of strokes per minute

L = Length of stroke in millimeters (inches)

3/5 = Portion of time spent actually cutting

12 = Inches per foot

1000 mm = 1 meter

Then: A. The metric formula becomes:

$$CS \ (m/min) = \frac{N \times L \ (mm)}{3/5 \ of \ 1000}$$

$$= \frac{N \times L \ (mm)}{600}$$

B. The inch formula becomes:

$$CS \ (ft/min) = \frac{N \times L \ (inches)}{3/5 \ of \ 12}$$

$$= \frac{N \times L \ (inches)}{7.2}$$

When the shaper is used, the cutting speed desired and the length of the stroke will be known. Since it will be necessary to determine the number of strokes per minute required to operate at the given cutting speed, the above formulas may be converted into the necessary form:

A. Metric: $N = \dfrac{CS \ (m/min) \times 600}{L \ (mm)}$

B. Inch: $N = \dfrac{CS \ (ft/min) \times 7.2}{L \ (inches)}$

Example: A piece of machine steel 254 mm (10″) in length is to be machined with a high-speed steel tool.

The known factors are:

CS = 27.43 m/min (90 ft/min)

L = 254 mm (10″)

Substituting the known values in the formulas:

A. Metric: $N = \dfrac{27.43 \times 600}{254}$

$= \dfrac{16458}{254}$

= 64.79 or 65

B. Inch: $N = \dfrac{90 \times 7.2}{10}$

$= \dfrac{648}{10}$

= 64.8 or 65

Adjusting Stroke Length and Position

The length of the stroke of the ram is determined by the distance the crankpin is from the axis of the driving wheel.

(South Bend Lathe, Inc.)

Fig. 93-9. The rocker arm is graduated for accurate stroke length settings.

Procedure for Setting Stroke Length

1. If necessary, loosen the device that locks the stroke-setting shaft.
2. Turn the stroke-setting shaft until the pointer on the ram or rocker arm registers the length of stroke desired, for example, 102 mm (4"), Fig. 93-9.

CAUTION
The stroke should be at least 19 mm (3/4") longer than the cut. It should extend from at least 12.7 mm (1/2") before the work to at least 6.35 mm (1/4") beyond the work.

3. When the machine has been adjusted for length of stroke, tighten the locking device, if any.

Procedure for Setting Ram Position

1. Loosen the ram by turning the handle on the top of the rear part of the ram, Fig. 93-2.
2. With the clutch lever, engage the clutch lightly, and advance the rocker arm to its extreme forward position. The instant this position is reached, disengage the clutch.

Some machines are equipped with a handwheel to perform this operation.

3. Turn the ram positioning shaft, Fig. 93-1, until the tool is about 6.35 mm (1/4") ahead of the work at the point where the cut ends. **Then tighten the ram clamp.**

Safety Precautions for the Shaper

1. Wear approved goggles or face shield.
2. Be sure that the vise or setup fixture is securely fastened to the table.
3. Make certain that the workpiece is mounted securely.
4. Select the proper tool for the job.
5. Be sure that the tool, tool head, table, and table support are secure before starting the machine.
6. Do not run the ram back into the column with the tool slide set at an angle. Doing so will make the slide strike the column and cause serious damage, Fig. 96-2. On machines equipped with a handwheel, **before starting the machine,** run the ram back by hand to insure that the tool slide clears the column and that the tool clears the work.
7. Remove all wrenches before starting the machine.
8. Select the proper speed, feed, and depth of cut according to the kind of material, type of tool, and method of holding the work.
9. Make certain that no one else is inside the safety zone for the machine.
10. Make sure that chips are prevented from striking anyone in the area.
11. Stand parallel to the direction of the stroke of the machine when it is operating, and never reach across the table between strokes.
12. Do not touch the tool, clapper box, or workpiece while the machine is running.
13. Remove chips only while the machine is stopped. Use a brush to remove chips.
14. Make measurements on the work only when the machine is stopped.
15. Make adjustments on the work setup only when the machine is stopped.
16. Turn off the machine before leaving it.

Unit

94

How to Make Horizontal Cuts with a Shaper

Beginners should read Units 92 and 93 before operating the shaper.

Procedure for Roughing Cut

1. Remove any burrs or bumps from the workpiece by grinding or filing.
2. Remove any burrs or chips from the jaws and bottom of the vise.
3. Open the vise to accommodate the workpiece. Support the workpiece on parallel bars if necessary.
4. Locate the workpiece in the center of the vise whenever possible to avoid twisting the moveable vise jaw. To save cutting time, position the workpiece so that the longest stroke possible may be taken.
5. Tighten the vise just enough to grip the workpiece. Seat the workpiece on the parallels by striking it with a lead hammer. Tighten the vise until the workpiece is held securely. Strike the workpiece again with the lead hammer to seat it against the parallels. Mount rough castings in the vise as shown in Fig. 94-1.

(Cincinnati Incorporated)

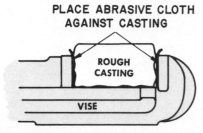

PLACE ABRASIVE CLOTH AGAINST CASTING

ROUGH CASTING

VISE

Fig. 94-1.   A method of holding rough castings tightly in the vise.

CAUTION
To avoid springing the workpiece, do not tighten the vise too much when clamping thin stock.

6. Install a sharp left-hand cutting tool in the toolholder.
7. Install the toolholder in the tool post so that it will have a short overhang, Fig. 94-2. Set the clapper box so that its top slants away from the cutting edge of the tool. The tool holder should be in a vertical position, Fig. 94-3.
8. Adjust the table until the part of the work to be machined is level with the point of the tool.
9. Determine the number of strokes per minute and set the machine accordingly (See Unit 93).
10. Adjust the shaper for the length of stroke desired (See Unit 93).
11. Next, adjust the position of the ram so that the cutting tool is about 6.35 mm (1/4") ahead of the front edge of the workpiece. Then tighten the ram in place.
12. Move the ram back to the end of the return stroke.
13. With the longitudinal feed screw, Fig. 93-1, move the table by hand until the workpiece is within 1.5 mm (1/16") of the side of the cutting tool.
14. Lower the cutting tool until it is in position to make a cut of the depth desired. When cast iron is to be machined, the first cut should be about 3 mm (1/8") deep. This will put the point of the tool below the hard surface scale on the metal.

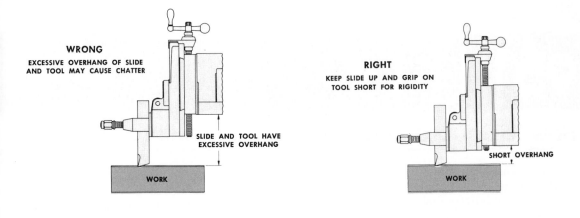

(Cincinnati Incorporated)

Fig. 94-2.   Use a short overhang for both the tool and the tool slide.

15. Turn on the machine and engage the clutch. At the end of the first stroke, feed the workpiece toward the cutter by turning the longitudinal feed screw. Repeat this movement until two or three cuts have been made. Then stop the machine, check the depth of cut, and make any adjustments necessary.

16. Engage the automatic longitudinal feed and restart the machine. The automatic feed should operate at the end of the return stroke, not at the end of the cutting stroke. If it does not operate at the correct time, reset the automatic feed so that it will.

17. When the surface has been machined, stop the automatic feed and turn off the machine.

18. Hand-crank the workpiece back to the starting side of the cut.

Procedure for Finishing Cut

1. Install a sharp left-hand finishing tool.
2. Adjust the tool to take a light cut of 0.25 to 0.76 mm (0.010" to 0.030").
3. Proceed as in steps 15 through 18 above.

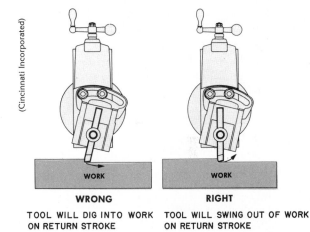

(Cincinnati Incorporated)

WRONG   **RIGHT**

TOOL WILL DIG INTO WORK   TOOL WILL SWING OUT OF WORK
ON RETURN STROKE   ON RETURN STROKE

Fig. 94-3.   The positions of the tool and the clapper box determine whether the tool will swing clear on the return stroke.

Unit

95

How to Square Rectangular Stock with a Shaper

In many cases, horizontal cuts may be used to machine rectangular stock on all its surfaces. In other cases, both horizontal and vertical cuts are required.

Fig. 95-1.   Testing the vertical plane of the workpiece with a square.

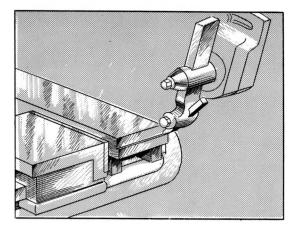

Fig. 95-2.   Machining the end of a long workpiece with vertical cuts.

Procedure

1. Position the vise so its jaws are parallel to the direction of ram travel. Place the workpiece in the vise with its long dimension parallel to the vise jaws. Plane the **first broad surface** true. See Unit 94.
2. Place the planed surface of the stock against the solid jaw of the vise. Then insert a round rod between the work and the movable jaw. This is done to assure that the machined surface will lie flat against the solid vise jaw.
3. Tighten the vise screw. Then, with a try square, test to see if the work is standing in a vertical plane, Fig. 95-1.
4. Machine the **first edge** as in step 1.
5. Release the vise, and turn the work over, placing the machined edge down and keeping the planed surface against the solid jaw of the vise.
6. Cut the **second edge** in the same manner as the first edge.
7. Next, place the first machined broad surface down in the vise. Use parallels if necessary. Then cut the **second broad surface** in the same manner as the first broad surface.
8. To machine the **first end,** stand the workpiece vertically in the vise if it is short. Check its alignment with a square. Then proceed as when shaping an edge.

    To machine a long workpiece, place it horizontally in the vise with the end projecting about 12.7 mm (1/2") past the vise jaws. (Swivel the vise 90° to place its jaws at a right angle to the direction of ram travel, Fig. 95-2.)
9. Cut the **second end** in the same manner as the first end.

Unit 96

How to Make Vertical and Angular Cuts with a Shaper

A vertical cut usually is made when cutting a groove, a shoulder or a keyway, and when planing the end of wide stock. An angular cut is made when cutting dovetails or bevels on the edges or ends of work. When such cuts are being made, the vertical feed is used.

Procedure for Vertical Cuts

1. Lay out on the workpiece the outline of the material to be removed.
2. Mount the workpiece securely in the vise.
3. Install the cutting tool in the toolholder, and the toolholder in the tool post.

(Cincinnati Incorporated)

Fig. 96-1. Using vertical cuts to make the narrow grooves necessary for a spline.

4. Position the tool slide so that enough down-feed is available to complete the cut without excessive slide overhang.
5. If a narrow groove is to be cut, do not swivel the clapper box. (Figure 96-1 shows the correct positioning.) For wide grooves or shoulders, however, swivel the clapper box 15° to 20° **away from the surface to be cut.**
6. Adjust the table until the part of the workpiece to be machined is just below the level of the tool.
7. Determine the length of stroke required and set the machine accordingly.
8. Position the ram so that the cutting stroke will end about 6.35 mm (1/4") beyond the workpiece. Lock the ram in this position.
9. Move the ram back to the starting end of the cutting stroke.
10. Move the work table until the point on the workpiece where the cut will be made is aligned with the cutting tool.
11. Determine the correct number of strokes per minute and set the machine accordingly.
12. Start the machine and engage the clutch.
13. At the end of each return stroke, move the tool down slightly until it just marks the workpiece. If the tool is located properly, finish the cut. If not, stop the machine and relocate the workpiece before finishing the cut.

Procedure for Angular Cuts

1. Proceed as in steps 1-3, "Procedure for Vertical Cuts."

2. Loosen bolts holding the tool head swivel plate. Then set the tool head at the required angle and tighten it, Fig. 96-2.

CAUTION
Do not run the ram back into the column with the tool slide set at an angle. The slide will strike the column and damage the machine.

3. Swivel the clapper box 15° to 20° away from the surface to be cut, Fig. 96-2.
4. Position the toolholder to present the cutting tool to the workpiece as desired.
5. Proceed as in steps 6-13, "Procedure for Vertical Cuts."

(Cincinnati Incorporated)

Fig. 96-2.   Using the tool slide to cut an angle.

Unit
97

Broaching Machines and Broaching Operations

Most broaching machines, like shapers and planers, move the cutting tool or the workpiece through a cutting stroke and a return stroke. Instead of using a single-point cutting tool, however, broaching machines use a multiple cutting edge tool called a **broach**, Fig. 97-1. With this tool the broaching operation is completed in one or two strokes. As a result, broaching is a very rapid, efficient process.

Broaching machines are of two main types according to the direction of tool orientation: **horizontal** or **vertical**, Figs. 97-2 and 97-3. On some machines the broach is fixed and the workpiece moves. Vertical machines may either push, pull down, or pull up the broach. Most horizontal machines pull the broach. To shorten the cycle time, the return (noncutting) stroke is often two or more times faster than the cutting stroke.

Pot-broaching machines, Fig. 97-4, are a type of vertical broaching machine made for use with **pot broaches**. The pot broach is a tubular broach with internal cutting teeth, Fig. 97-5. Pot broaches simultaneously cut grooves, gear teeth, and other shapes around the outside of the workpiece.

Continuous broaching machines use short broaches attached to an endless chain, Fig. 97-6. These machines rapidly and repeatedly cut slots, gear teeth, or similar surface shapes without moving the broach back and forth.

Rotary broaching machines, Fig. 97-7, are production machines that have broaches mounted around a column. The workpieces are mounted in fixtures on a rotary table that encircles the column. After each cutting stroke, the rotary table is automatically indexed to carry the workpieces to the next broach. The parts are thus gradually shaped by each broach as they move from station to station. When there is a workpiece at each station, each cutting stroke results in one finished workpiece.

Fig. 97-1. Three pull-type broaches of widely different size.

(Apex Broach and Machine Co.)

Fig. 97-2. A horizontal broaching machine.

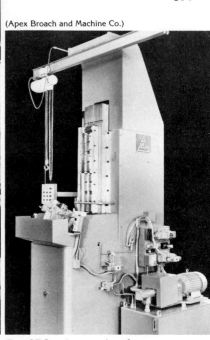

(Apex Broach and Machine Co.)

Fig. 97-3. A vertical surface-broaching machine.

(Detroit Broach & Machine Co.)

Fig. 97-4. A vertical pot-broaching machine.

(National Broach & Machine Co.)

Fig. 97-5. A pot broach for broaching a helical gear. Also shown are the gear blank and the finished gear.

Broaching Operations

Broaching operations are classified into two categories. **Internal broaching** is done inside a hole in the workpiece. It is used especially to form holes of complex shapes, but simple shapes

Fig. 97-6. A continuous broaching machine using chain-mounted broaches. The close-up shows one of the broaches.

Fig. 97-7. A rotary broaching machine has broaches mounted around a column.

Fig. 97-8. Common hole shapes made by internal broaching.

Fig. 97-9. Examples of parts shaped partly by surface broaching.

are also done this way, Fig. 97-8. Keyways, splines, gear teeth, and other holes with complex or irregular shapes are easily done by internal broaching. **Surface or external broaching** is done on the outside of the workpiece. Any shape from a flat surface to helical teeth can be made by surface broaching, Fig. 97-9.

Broach Design

Broaches are designed so that each tooth cuts a small amount of the total metal to be removed, as shown in Fig. 97-10. If a fine finish is required, the broach will be designed with roughing, semi-finishing, and finishing teeth, Fig. 97-11. Note

that the broach illustrated is provided with a **front pilot** and a **rear pilot**. The front pilot insures that the broach will be accurately aligned in the starting hole provided in the workpiece. The rear pilot supports the broach in the finished hole until the broach can be removed from the workpiece.

Each tooth on a broach is shaped much like any single-point cutting tool, Fig. 97-12. A shallow clearance angle of 1/2° to 4° provides

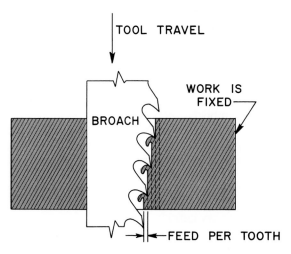

Fig. 97-10. Each tooth on a broaching tool removes a small amount of material.

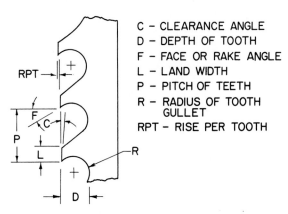

C – CLEARANCE ANGLE
D – DEPTH OF TOOTH
F – FACE OR RAKE ANGLE
L – LAND WIDTH
P – PITCH OF TEETH
R – RADIUS OF TOOTH GULLET
RPT – RISE PER TOOTH

Fig. 97-12. Broach tooth design and terminology.

(The duMont Corporation)

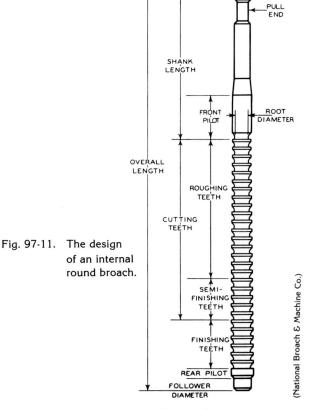

Fig. 97-11. The design of an internal round broach.

(National Broach & Machine Co.)

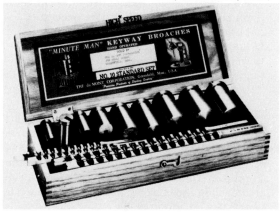

Fig. 97-13. A set of keyway broaches with guide bushings.

Using a Keyway Broach

Keyway broaches are used to cut slots called **keyways** in pulleys, gears, and other parts. Keyways have common widths such as 3.2 mm (1/8″), 4.8 mm (3/16″) and 6.4 mm (1/4″). Each keyway broach cuts a slot of correct width and depth for one standard size key. Two cutting strokes of the broach are normally required to cut the keyway to its full depth. For the second stroke, a shim is inserted behind the broach. The shim advances the broach the proper amount for the second cut.

When a keyway broach is used, a supporting bushing must first be placed in the workpiece hole, Fig. 97-13. The bushing has a groove of proper size to fit the broach, and also serves to align the broach accurately parallel to the hole.

maximum support for the cutting edge. A rake angle between 12° and 15° is used for cutting workpieces made of most steels. This angle increases somewhat for workpieces of softer metals. The radius at the bottom of the tooth provides great tooth strength and forms the chip into a coil that is easily removed.

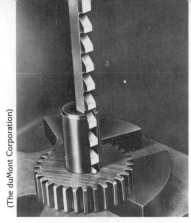

Fig. 97-14. Using an arbor press to force a keyway broach through a workpiece.

The following procedure is recommended for broaching a keyway:

1. Obtain the correct size broach, with its shim and the correct diameter bushing. Lubricating oil and cutting oil will also be needed.
2. Insert the bushing in the workpiece hole.
3. Lubricate the broach lightly all over so that it will slide easily in the bushing groove.
4. Insert the starting end of the broach in the bushing and place the assembly on the pad of an arbor press.
5. If the workpiece is steel, apply cutting oil to the broach teeth, then force the broach through the workpiece with the arbor press, Fig. 97-14.
6. Insert the shim in the bottom of the bushing groove. Then reinsert the starting end of the broach in the bushing and force it through for the second and final cut.
7. Wipe the broach, shim, and bushing clean and return them to proper storage.

NOTE: Broaches are expensive tools and must be handled carefully so they will not be damaged. They should be stored in a plastic or wood container while not in use.

Test Your Knowledge of Section 11

Units 92 and 93: Shaping and Planing Machines, and Horizontal Shapers.

1. Name three kinds of shapers and describe the kind of work done by each.
2. Name two kinds of planers and describe the kind of work done by each.
3. Name the two factors involved in designating shaper size.
4. List three ways in which shaper driving mechanisms may be powered.
5. Name two kinds of speed change mechanisms used on shapers to change the number of strokes per minute.
6. Name the two directions in which a shaper worktable can be moved.
7. List the main parts of a shaper tool head assembly.
8. Name two methods of controlling automatic longitudinal table feed on shapers.
9. Name the two basic kinds of shaper and planer cutting tools.
10. Why should the toolholder and the tool be used in the reverse position, **B** in Fig. 93-5, for heavy cuts?
11. For what purposes are extension toolholders used on a shaper?
12. Calculate the number of strokes per minute at which the machine should be set for shaping a piece of steel 127 mm (5″) long at 24.4 meters per minute (80 feet per minute).
13. Calculate the number of strokes per minute for shaping a workpiece of aluminum 254 mm (10″) long at 61 m/min. (200 ft./min.).
14. How much longer than the workpiece should the length of stroke be set?
15. Why should the ram never be run back into the column when the tool slide is set at an angle?

Unit 97: Broaching Machines and Broaching Operations

1. What is the name of the cutting tool used in the broaching process?
2. How many strokes of the cutting tool are usually used to complete a broaching operation?
3. Name the two main types of broaching machines.
4. What is the name given to tubular broaches that have their cutting edges on the inside of the tube?
5. Name the kind of broaching machine that uses short broaches attached to an endless chain.
6. Name the kind of broaching machine that uses several broaches mounted around a column.
7. Name the two categories into which broaching operations are classified.
8. What device is provided to guide a keyway broach during cutting?
9. What device is used to enable a keyway broach to cut the keyway to full depth?

Section 12

Abrasives and Grinding Wheels

Unit

98

Types of Abrasives and Their Uses

An abrasive is a hard and tough substance. It has many sharp edges. An abrasive cuts or wears away materials that are softer than itself. Abrasives are used as cutting tools or cutting materials. They are used in several forms which include **grinding wheels,** Fig. 98-1, **sharpening stones** or **sticks,** and **coated abrasives,** Fig. 100-1. Coated abrasive is cloth or paper with a **coating of abrasive grains cemented** to its surface. Coated abrasives used for machine grinding usually are in belt or disc form. They are used in sheet or strip form for hand polishing. Loose abrasive grains also are used for certain polishing and lapping operations.

Properties of Abrasives

Penetration hardness, fracture resistance, and wear resistance are the necessary properties of

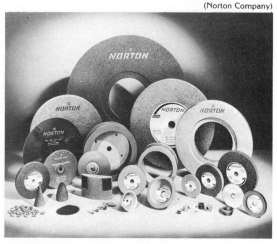

(Norton Company)

Fig. 98-1. Common types of tool grinding wheels.

Table 98-1
**Relative Hardness Values of Abrasives
and Other Materials**

	Moh Scale	Knoop Scale
Common glass (depending on composition)		300-500
Hard steel, Rockwell C. 60.5		740
Quartz	7	820
Synthetic blue spinel		1270
Topaz	8	1350
Garnet		1350
Cemented carbides		1400-1800
Tungsten carbide (not cemented)		1880
Aluminum oxide (Alundum) and corundum	9	2000
Silicon carbide (Crystolon)		2500
Boron carbide (Norbide)		2800
Diamond (mined or manufactured)	10	greater than 7000

(Norton Co.)

abrasives. **Penetration hardness** refers to the ability of the abrasive to scratch or cut a softer material.

Fracture resistance refers to the ability of an abrasive material to resist breaking or cracking under load. When an abrasive grain is fractured, sharp edges should appear without loss of the entire grain. Fracture of the grain should occur after the original point has started to dull, but before it becomes too dull. Excessive resistance to fracture causes excessive pressure and heat while grinding.

Wear resistance refers to the ability of the abrasive grain to maintain sharpness. Wear resistance is largely related to penetration hardness and tensile strength of the abrasive.

Mohs Hardness Scale

The **hardness** of abrasives often is rated according to the **Mohs hardness scale.** The scale ranges from a rating of No. 1 for talc to No. 10 for diamond, the hardest substance known. Except

for diamond, the natural abrasives rank below 9, and most artificial abrasives above 9. The following are the approximate Mohs hardness ratings for natural abrasives: Crocus, 6.0; flint, 6.9; garnet, 7.5 to 8.5; emery, 8.5 to 9.0. In the measurement of abrasives, the range from 9 to 10 actually is as broad and significant as the whole range from 1 to 9.

Knoop Hardness Value

A more recent method, developed by the National Bureau of Standards, is the **Knoop hardness value.** The material is tested with a microhardness tester, which in principle is similar to the Rockwell hardness tester for measuring the hardness of metals. Under a certain load, the tester presses a diamond point into the material being tested. The depth of the impression is indicated by a number value. As with the Mohs system, the higher the number, the harder the material. However, the Knoop number more clearly indicates the relative difference in the hardness of various abrasive materials. The hardness values (100-gram load) of several abrasives and other materials ground by abrasives are indicated in Table 98-1.

Natural Abrasives

Abrasives are grouped in two broad classifications — natural abrasives and artificial or manufactured abrasives. **Natural abrasives** are obtained from nature. They are being replaced rapidly by artificial abrasives. Those natural abrasives still used in industry include flint, garnet, emery, crocus, and diamond. Except for diamond, the natural abrasives are relatively soft in comparison with artificial abrasives. Flint and garnet are used in the form of coated abrasives in the woodworking industry. Emery, crocus, and diamond are used to work metals. For this reason, their properties and uses should be understood.

Crocus

Crocus is a reddish-brown oxide of iron and may be natural or synthetic. It is used in very fine powder form as a rouge, or as a coating on cloth known as crocus cloth. Crocus cloth or rouge is used for polishing corroded metals or rare metals where a minimum of base metal is to be removed.

Emery

Emery is one of the oldest natural abrasives used in the metalworking industry. It is black and is composed largely of a combination of corundum and iron oxide. Corundum is aluminum oxide, Al_2O_3. Emery used for abrasives usually is composed of about 60% corundum and 40% iron oxide and other impurities. Emery is used in making the coated abrasive, emery cloth. Although emery cloth still is manufactured in four grades of fineness, artificial abrasives rapidly are replacing it for use in grinding and polishing metals. Emery grains are not as sharp or as hard as artificial abrasive grains; hence they are slower cutting.

Diamond

Diamond, the hardest material known, is used in the form of grains bonded together to form an abrasive stick or grinding wheel. A diamond cluster abrasive stick or nib is used to cut or true other softer grinding wheels. Diamond grinding wheels are used to grind very hard materials such as cemented carbide cutting tools, ceramic cutting tools, glass, stone, and other types of ceramic materials. Industrial diamonds are relatively inexpensive when compared with the clear diamonds used for jewelry.

Artificial Abrasives

Artificial abrasives are also known as manufactured or synthetic abrasives. The commonly used artificial abrasives include **silicon carbide, aluminum oxide, boron carbide,** and **synthetic diamond.** The manufactured abrasives are harder and have greater impact toughness than any of the natural abrasives except diamond.

Silicon Carbide

The first artificial abrasive developed in the United States was crystalline silicon carbide. It was discovered by Dr. Edward G. Acheson about 1891. The principal ingredients in silicon carbide are silica sand, which contains the silicon, and coke, which provides the carbon. A small amount of sawdust is added to make the mixture porous.

Properties: The properties of silicon carbide depend upon its purity in manufacture. Its hardness and sharpness are ideal. Its Knoop hardness value is approximately 2500, in comparison with diamond rated at approximately 7000. However, silicon carbide is brittle, as compared with aluminum oxide, and its grain fracturing properties limit its use to grinding specific materials. It is hard enough to cut aluminum-oxide abrasive materials.

Uses: Silicon carbide abrasives are used for grinding wheels, abrasive stones or sticks, and coated abrasives. The grinding wheels are used for materials of low-tensile strength, including: cast iron, bronze, aluminum, copper, tungsten carbide, rubber, glass, marble, ceramics, pottery, plastics, magnesium, and fiber.

Aluminum Oxide

About 1897, several years after silicon carbide was developed, aluminum oxide was discovered by Charles P. Jacobs, an engineer in the laboratories of the Ampere Electro-Chemical Company at Ampere, New Jersey. The principal ingredient used in manufacturing aluminum oxide is bauxite ore. This is the same material from which metallic aluminum is derived. The bauxite ore is purified to crystalline form by heating to extremely high temperatures in large electric furnaces. Greater toughness is imparted to the aluminum oxide by adding titanium.

Properties: The properties of aluminum oxide are dependent on its purity in manufacture. Aluminum oxide of 99% purity is available with modern manufacturing methods. Since the addition of titanium imparts varying degrees of toughness, several types of aluminum oxide are available with slightly varying characteristics.

Aluminum oxide is not as hard as silicon carbide. The Knoop hardness value of a typical aluminum oxide material is 2000, in comparison with silicon carbide at 2500 or diamond at approximately 7000. However, aluminum oxide is tougher and more shock resistant than silicon carbide.

Uses: Aluminum oxide abrasives are used for grinding wheels, abrasive sticks and sharpening stones, and coated abrasives. The grinding wheels are tough and shock resistant. They are used for grinding materials of high-tensile strength, including: carbon steels, alloy steels, soft or hard steels, wrought iron, malleable iron, and tough bronze. Approximately 75 percent of all grinding wheels used today are manufactured from aluminum oxide.

Boron Carbide

A third kind of artificial abrasive material is boron carbide. A typical Knoop hardness value of boron carbide is 2800, which is harder than silicon carbide, but not as hard as diamond. It is produced from coke and boric acid at tremendously high temperatures in an electric furnace.

Boron carbide is used in stick form to dress grinding wheels 254 mm (10″) in diameter or smaller, Fig. 98-2. It is also used in powder form in place of diamond dust for lapping operations on very hard materials, such as hardened steels and cemented-tungsten-carbide materials. Norbide is very resistant to hard wear. It is also used in solid form for such items as linings for nozzles used in high-pressure sandblasting.

Synthetic Diamond

Synthetic diamonds were first produced in 1955 by the General Electric Company. They were made by compressing graphite to pressures of nearly 140,909 kg per square centimeter (2,000,000 lbs. per square inch) while heated to temperatures in excess of 2760 C (5000 F).

Synthetic diamonds are not of gem quality but are suitable for industrial applications. Most are so small that they are used as manufactured. They are used mostly for diamond grinding wheels with grit sizes from 80 to 500.

(Norton Company)

Fig. 98-2.   Using Norbide® (boron carbide) dressing stick to dress grinding wheel.

Unit
99

Grinding Wheels and Their Selection

Grinding wheels are made of thousands of crushed abrasive grains held together by a substance called a **bond.** Between the abrasive and the bonding material are pores or air spaces. These spaces provide clearance for chips removed in the grinding process, and they minimize wheel loading.

(Norton Company)

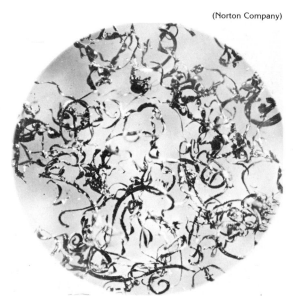

Fig. 99-1.   Magnified view of metal chips produced by abrasive wheel.

Cutting Action of a Grinding Wheel

Each abrasive grain in a grinding wheel is a cutting tool. Each has sharp cutting edges which cut off tiny particles from the metal being ground. Under a magnifying glass, the small particles appear as metal chips similar to chips removed by a shaper or lathe tool, Fig. 99-1. Because of the heat created by the speed of metal removal, the chips appear as sparks which are readily visible.

On **finish-grinding** operations, grinding wheels remove metal relatively slowly in comparison with other cutting tools. Finish grinding usually follows other rough-machining operations, and it generally involves machining to very close tolerances.

On **abrasive-machining** operations, metal is removed more rapidly than on finish-grinding operations. The term **abrasive machining** is used for grinding applications which involve the removal of a relatively large amount of metal, usually 1.5 mm (1/16") or more in depth. Both grinding wheels and coated abrasives are used. Abrasive machining involves machining castings, forgings, weldments, and bar stock to commercial tolerance and finish without previous machining operations. See Figs. 99-2 and 99-3. It is used in applications which are machined more profitably by grinding than by other machining methods.

The quality of the work achieved by grinding is controlled to a large extent by selecting a proper grinding wheel. As a grinding wheel is used, the cutting edges of the abrasive grains become

Fig. 99-2. Abrasive machining of worm screw from a solid piece.

dulled. When this happens, the grinding pressure should cause these dull edges to break off, exposing new sharp edges without breaking off the entire grain. Once the grain has been broken down sufficiently, the grinding pressure should cause the bonding material to release the remaining portion of the grain, thereby exposing a new, sharp grain. This process should continue repeatedly when the right grinding wheel is selected for the job.

Classification of Grinding Wheels

Grinding wheels are classified according to their size and shape, type of abrasive used, grain size, type of bond, grade or hardness, and structure.

Size and Shape

The size of a grinding wheel is given in terms of its outside diameter, the diameter of the spindle hole, and the width of the face. Grinding wheels are made in many standard shapes and sizes.

Manufacturers have adopted standard **type-number** designations for most of the basic shapes for grinding wheels. Cross-sectional views of shapes most commonly used in toolroom and cylindrical grinding are shown in Fig. 99-4.

Grinding wheel faces may be shaped for grinding contoured surfaces on cutting tools such as milling cutters, taps, and special tools. Grinding wheels which are manufactured with standard wheel faces are designated by letters, Fig. 99-5.

Kinds of Abrasives

The most common types of abrasive used in grinding wheels are aluminum oxide, silicon car-

Fig. 99-3. Abrasive machining on a rotary surface grinder.

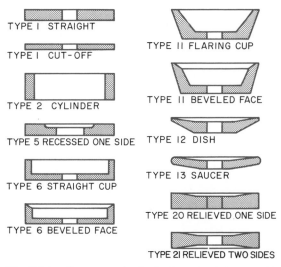

Fig. 99-4. Some standard grinding wheel shapes with type-number designations.

bide, and diamond. Diamond grinding wheels are treated separately later in this unit.

Aluminum Oxide

Aluminum oxide grains or crystals, although not the hardest artificial abrasive, are tough and are best for grinding materials of high-tensile strength. They are used to grind carbon steels,

(Norton Company)

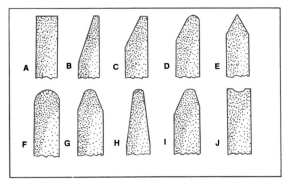

Fig. 99-5.   Typical standard wheel faces.

A

B

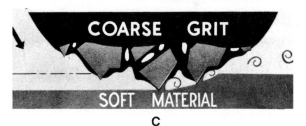

C

Fig. 99-6.   Comparison of grain size and rate of metal removal with grinding wheels.

alloy steels, soft or hard steels, cast-alloy cutting tools, wrought iron, and tough bronze.

Silicon Carbide

Silicon carbide abrasive grains are harder and more brittle than aluminum oxide. Its grain fractures more readily than aluminum oxide. Therefore, it is used to grind materials that are easily penetrated, such as copper, aluminum, rubber, plastics, magnesium, and fiber. It is also used to grind hard materials of low-tensile strength, such as cast iron, cast bronze, glass, marble, ceramics, and pottery. Cemented-tungsten-carbide cutting tools must be ground on either silicon carbide or diamond grinding wheels.

Grain Sizes

Grain refers to the size of the abrasive particles used in the manufacture of the grinding wheel. The grain size is determined by the mesh number of the finest screen through which the grain will pass. For example, a 36-grain wheel is one made of particles of abrasive which just pass through a 36-mesh screen, but which will be retained on a 46-mesh screen, the next finer screen. (A 36-mesh screen has 36 openings each lineal 25.4 mm (inch), or 200 openings per square centimeter (1296 openings per square inch). Grain numbers are sometimes called **grit** numbers.

Grains Available

Grain sizes vary from 10 to 600 and are classified as follows:

Coarse: 10, 12, 14, 16, 20, 24
Medium: 30, 36, 46, 54, 60
Fine: 70, 80, 90, 100, 120, 150, 180
Very Fine: 220, 240, 280, 320, 400, 500, 600

Uses of Various Grits

Fine-grain wheels are used on small-diameter work to produce small fillets or for fine finishes. Fine wheels also are preferred for grinding hard materials, since they have more cutting edges and, therefore cut faster than coarse-grain wheels. See view **A**, Fig. 99-6. Because coarse-grain wheels have fewer grains (view **B**), and the grains cannot penetrate hard material deeply without burning, they cannot cut as rapidly on hard

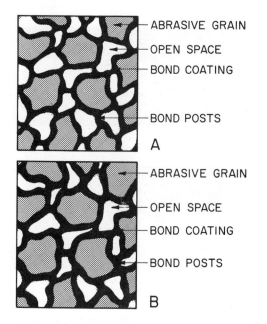

A

ABRASIVE GRAIN
OPEN SPACE
BOND COATING
BOND POSTS

B

ABRASIVE GRAIN
OPEN SPACE
BOND COATING
BOND POSTS

Fig. 99-7. Kind and amount of bonding material affects hardness or grade of grinding wheel.
A. Abrasive grains with light bond coatings and weak bond posts, as in a relatively soft-grade wheel.
B. Similar grains, but with heavier bond coating and thicker, stronger bond posts, as in a hard-grade wheel.

(Norton Company)

Fig. 99-8. Typical abrasive cutoff wheels. The first has a shellac bond; the others, a resinoid or rubber bond.

materials. **Coarse-grain** wheels are used for rapid metal removal on softer materials (view **C**). Coarse wheels also are used for grinding large workpieces. The grain size selected should be determined by the type of material to be ground, the finish desired, and the amount of metal to be removed.

Bonding Materials

The bond is the material which holds the abrasive grains together to form the grinding wheel. As the grains get dull, pressure on the wheel causes the bond to break down and release the dull grains, thus exposing new sharp grains. The bond holds the individual grain in much the same manner as a toolholder holds a tool bit. See Fig. 99-7. There are five basic types of bonds used in grinding wheels: vitrified, silicate, rubber, shellac, and resinoid. Additional modifications of these five materials are also produced by some manufacturers.

Vitrified

Approximately 75% of all wheels are made with vitrified or a modified vitrified bond. Vitrified-bond wheels are strong, porous, and are not affected by rapid changes in temperature, oils, acid, or water. These wheels are uniform in structure, free from hard spots, and hold their form well. The bond is formed when special clays are mixed with abrasive grains and heated to high temperatures. The mixture forms a molten glass which cements the grains together.

Silicate

Wheels bonded with silicate (silicate of soda) are known as silicate- or semi-vitrified-bond wheels. Silicate-bonded wheels release the grains more readily than vitrified bond. Hence, the wheel is softer and it breaks down more readily, thereby exposing new sharp grains. Silicate-bonded wheels are used for grinding edge tools, drills, reamers, milling cutters, and similar tools.

Rubber

Wheels which are rubber bonded are elastic, very strong, and shock resistant. This bond is used for very thin wheels, such as cutoff wheels for abrasive cutoff machines, Fig. 99-8. Cutoff wheels are used for cutting pipe, angle iron, or

Fig. 99-9.   Cutting pipe with an abrasive cutoff wheel.

Fig. 99-10. Shellac-bonded wheel used for grinding a mill roll.

bar stock, Fig. 99-9. For safety, rubber bond is used for high-speed grinding. It also produces a very good finish.

Shellac

Wheels with a shellac bond are also elastic in nature, resilient, and cool cutting. They produce a very fine finish. Hence, they are used to grind such items as mill rolls, camshafts, and fine cutlery. See Fig. 99-10.

Resinoid

Wheels with a resinoid bond have high strength and mechanical shock resistance. Resinoid bond is used for large, heavy-duty, high-speed wheels. They are used for rough grinding operations involving rapid stock removal, or for cutoff wheels. Resinoid wheels frequently are used in foundries and steel mills for snagging castings and for cleaning steel billets. (**Snagging** means to grind off the rough spots or surplus metal.)

Grade

Wheels from which the grit or abrasive is readily torn are termed **soft grade.** Conversely, wheels that do not readily release the grain are called **hard grade.** Hard-grade wheels generally are used for grinding soft metals such as mild steel. Soft-grade wheels generally are used for grinding hard metals such as high-carbon steel.

It should be remembered that the term **hard** as used with respect to grinding wheels has no relationship to the hardness of the abrasive, but rather to the ease or difficulty with which the worn particles of the abrasive are torn from the face of the wheel. With a given bond material, **it is the amount of bond which determines the hardness or softness of the wheel** — the more bond material, the harder the wheel. See Fig. 99-7.

The grade of grinding wheels is designated by letters of the alphabet, A being the softest and Z the hardest, Table 99-1.

Structure

The structure of a grinding wheel refers to the spacing between the grains, or the density of the wheel. Grains which are very closely spaced are more dense or **close,** while grains which are wider apart are less dense or **open.** See Fig. 99-7.

The structure of a wheel is rated with numbers from 1 (dense) to 15 (open). The rate of metal removal usually is greater for wheels with an open structure. However, those with dense structure usually produce a finer finish.

Table 99-1
Standard Marking System for Grinding Wheels
(Example Is a Typical Marking: 32A46 — H8VBE)

(Prefix)	(1) Kind of Abrasive	(2) Grain Size		(3) Grade or Hardness		(4) Structure		(5) Bond Type	(6) Manufacturer's Record
32	A	46		H		8		V	BE
Manufacturer's Symbol Indicating Exact Kind of Abrasive	A Aluminum Oxide	(Coarse) 10 12 14 15 20 24	(Fine) 70 70 90 100 120 150	Soft to Hard A Soft N B O C P D Q E R		Dense to Open 1 Dense 9 2 10 3 11 4 12 5 13		V Vitrified S Silicate R Rubber B Resinoid E Shellac	Manufacturer's private marking to identify wheel. May be a letter or number
	C Silicon Carbide	(Medium) 30 36 46 54 60	180 (Very Fine) 220 240 280 320 400	F S G T H U I V J W K X L Y M Z Hard		6 14 7 15 Open 8		O Oxychloride	or both to designate modification of bond or wheel characteristics.
(Use Is Optional)			500 600			(Use Is Optional)			(Use Is Optional)

Grinding Wheel Markings

The system of grinding wheel markings adopted by the American Standards Association is shown in Table 99-1. A separate system is used for diamond wheels, Fig. 99-12. Most manufacturers use this system for identification markings on grinding wheels.

The standard system for marking wheels includes six parts in sequence, as listed across the top of Table 99-1. Note that the prefix to item one in the sequence is optional for each manufacturer. For example, where several types of a given abrasive are available, such as several variations of aluminum oxide, the prefix number indicates the exact type of aluminum oxide. Also note that items four and six in the sequence are optional with the manufacturer.

The marking on the grinding wheel indicated in Table 99-1 is 32A46-H8VBE. This marking indicates that the abrasive is type 32 Alundum; with a 46 medium grain size; with H grade (which is rated between soft and medium); structure 8 (middle density); bond type V (which is vitrified); and BE represents the manufacturer's mark for

the specific type of vitrified bond. A grinding wheel of this type will do a good job in surface-grinding hardened carbon tool steel.

Several manufacturers may use the same number to identify a given type of grinding wheel, but this does not mean that all of the wheels so identified will produce equal results or equivalent grinding action. The physical properties of the materials used may vary with different manufacturers.

Grinding Wheel Speeds

Grinding wheels should be operated at cutting speeds as near as possible to those recommended by the wheel manufacturer. **Cutting speed** refers to the speed at which the circumference (the cutting face) of the wheel is traveling in **mpm** (meters per minute) or **fpm** (feet per minute). The cutting speed of a wheel is increased or decreased by changing the **rpm** of the grinding wheel spindle. For methods of calculating the cutting speed or rpm for grinding, see Unit 113.

The rpm of the wheel spindle may or may not be adjustable. Bench grinders and floor-model grinders used for offhand grinding generally cannot be adjusted. On these grinders, the wheel usually is mounted directly on the motor spindle. On some cylindrical grinders, tool grinders, and surface grinders, the rpm of the wheel can be varied through the use of step pulleys or a variable-speed mechanism.

The maximum speed at which a wheel should be operated is indicated on the wheel. This speed should never be exceeded. For example, a wheel designed for a maximum of 1800 rpm should not be used on a grinding machine which has a spindle speed of 3600 rpm. When used at speeds above those recommended, a grinding wheel may fly apart, and may cause serious injury to the operator.

The following are general recommended cutting speeds:

General offhand grinding with vitrified-bonded wheels, 1524 to 1829 mpm (5000 to 6000 fpm).

Surface grinding, 1219 to 1981 mpm (4000 to 6500 fpm).

Tool and cutter grinding, 1372 to 1829 mpm (4500 to 6000 fpm).

Cylindrical grinding, 1676 to 1981 mpm (5500 to 6500 fpm).

Hence, an average speed of about 1524 mpm (5000 fpm) is recommended.

Specific cutting speeds to be used with different types of grinding wheels on various grinding applications are available in standard handbooks for machinists.

Diamond Grinding Wheels

Diamond grinding wheels are in a class by themselves. They are used to grind cemented-tungsten-carbide cutting tools, ceramic-oxide cutting tools, wear-resistant die steel, ceramics, glass, granite, marble, and jewels. See Fig. 99-11.

Diamond grinding wheels are made of fine particles or grains of natural or manufactured diamond, which are held together with a bonding material. The diamond particles are graded in grain sizes ranging from 36 to 500. Wheels with 80-120 grain size often are used for rough grinding, 180-320 for finish grinding, and 120-150 for combination rough and finish grinding. For fine lapping operations, diamond abrasive in bulk form as fine as 2000-grain size is available.

(Norton Company)

Fig. 99-11. Commonly used shapes of diamond wheels for carbide grinding.

(Norton Company)

Fig. 99-12. Marking chart for diamond wheels and hones. Hand hones omit the grade designation.

Diamond wheels usually are made of a special composition material to which a layer of abrasive mixture is applied on the cutting surface. The mixture is made of diamond grains and bonding materials (either metal, resinoid, or vitrified). The abrasive layer is available in thicknesses from 0.8 to 6.35 mm (1/32" to 1/4"). It also is available in several different concentrations or proportions of diamond to bonding material — low, medium, and high. In addition, numerous grades of hardness are manufactured and are indicated by letters of the alphabet.

At the present time, there is no standard marking system used by all manufacturers of diamond grinding wheels. However, the system developed by the Norton Company is being adopted by several other manufacturers of diamond wheels. See Fig. 99-12.

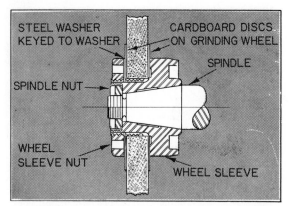

STEEL WASHER
KEYED TO WASHER

CARDBOARD DISCS
ON GRINDING WHEEL

SPINDLE

SPINDLE NUT

WHEEL
SLEEVE NUT

WHEEL SLEEVE

Fig. 99-13. Cross section of wheel mounted on arbor.

Since diamond grinding wheels have a high initial cost, care must be taken in selecting and using them. Cutting speeds from 1372 to 1829 mpm (4500 to 6000 fpm) usually should be used. A liberal supply of cutting fluid also may be used. The manufacturer's recommendations should be carefully followed. Several manufacturers of diamond wheels supply literature without cost concerning the selection, care, and use of their wheels. Recommendations concerning the selection of diamond grinding wheels for particular job applications are also available in standard handbooks for machinists.

Selecting and Using Wheels

The following factors must be taken into consideration in recommending and selecting a wheel for a particular job:

1. Type of grinding operation: offhand grinding, surface grinding, tool grinding, cylindrical grinding, internal grinding, etc.
2. Material to be ground.
3. Type of abrasive and bond to be used.
4. Amount of stock to be removed.
5. Finish required.
6. Area of wheel in contact with work: a wide wheel face may require a soft-grade wheel.
7. Wheel speed.
8. Work speed: for surface grinding and cylindrical grinding.
9. Whether grinding is wet or dry.
10. Machine condition: capacity and rigidity.
11. Abrasive grain size, grade, and structure.

Recommended grinding wheels for use on a number of basic grinding applications are indicated in Table 99-2. Recommendations for other grinding applications are available in standard handbooks for machinists and from manufacturers.

Ordering a Grinding Wheel

In ordering a grinding wheel, one may use the standard identification number to specify the type of abrasive, grain size, grade, structure, and specific bond type. In addition, the following specifications must be indicated: wheel type, wheel size (including diameter and width of face), type of face, diameter of hole or opening at the center, and the maximum rpm of the wheel.

Precautions in Using Wheels

1. The grinding machine should be rigid in order to prevent vibration or chatter.
2. Spindle bearings on the grinding machine should be adjusted properly in order to prevent vibration and chatter.
3. Wheels always should be mounted with proper cardboard or blotting paper discs between the wheel and properly relieved steel flanges or washers, Fig. 99-13.
4. Wheels should be mounted with the spindle nut fastened snuggly. If the nut is too tight, the wheel may crack.
5. Sound the wheel before installing it in order to test for cracks. The wheel will ring when struck very lightly with a nonmetallic object. A dull thud will be heard if the wheel is cracked.
6. The wheel should not be forced on the machine spindle. If the wheel is too tight, scrape the inside of the wheel bushing lightly and evenly all around until it will just slide on.
7. A new grinding wheel should be allowed to run at full speed, with the operator standing to one side, before it is used.
8. **Wheel glazing** is indicated by a smooth, glass-like appearance. It is caused when the abrasive grains wear too much before being released. This condition may be corrected by using a wheel of softer grade.
9. **Wheel loading** is caused by grinding a soft metal with the wheel or by using too heavy a grinding action. The wheel must be cleaned with a wheel dressing tool, as explained in Unit 28.
10. The wheel must be kept true and in balance as shown in Unit 112.

Table 99-2
Grinding Wheel Selection and Application

Suitable For	Wheel Material	Grain	Grade
External Cylindrical Grinding			
	Aluminox	2946	L
Good all-around wheels;	Alundum	3836	L
best adapted to soft steel	Aloxite	401	N
Hardened steel	Aluminox or Alundum	46	K
Soft steel of small diam.	Aluminox or Alundum	36	M[1]
Reamers, drills and general tool work	Aluminox or Alundum	80	K
Hard steel, dry grinding	Aluminox or Alundum	100	I
Cast iron and bronze	Crystolon	45	L
Facing Shoulders			
Ordinary work	Aluminox or Alundum	60	H or I
Fine finish	Aluminox or Alundum	80	I[1]
Surface Grinding			
	Alundum or Aluminox	46	H
Hardened steel	Alundum or Aluminox	46	G[2]
	Alundum or Aluminox	60	F[2]
Hardened high-speed steel or very thin	Aloxite	367	U
pieces of hardened carbon steel	Alundum or Aluminox	46	G
Cast iron	Carborundum or	36	M
	Crystolon	36	J
Disc Grinding			
Thick pieces, wet grinding	Aluminox or Alundum	30	K
Thin pieces, wet grinding	Aluminox or Alundum	30	J
High-speed steel, dry grinding	Aluminox or Alundum	60 or 80	H or I
Washers and similar pieces	Aluminox or Alundum	60	I
Internal Cylindrical Grinding			
Good all-around wheel	Aluminox or Alundum	46	2[1] I[1]
Roughing hardened steel	Aluminox or Alundum	46	J or K
Finishing hardened steel	Aluminox or Alundum	120	J or K
Ordinary finish without roughing	Aluminox or Alundum	80 & 90	J or K
Roughing brass	Crystolon	36	H or I
Finishing brass	Crystolon	80	H
Automobile cylinders	Crystolon	46	K
Automobile cylinders	Carborundum	36	M to P
Automobile cylinders, roughing or fair finish	Carbolite	36	H or I
Automobile cylinders, fine finish	Carbolite	60	H
Sharpening Carbon-Steel Cutters, Dry Grinding			
Milling cutters	Aluminox or Alundum	46 or 60	I
Formed and gear cutters	Aluminox or Alundum		
Sharpening High-Speed Steel Cutters, Dry Grinding			
Milling cutters	Aluminox or Alundum	46 or 60	I
Formed and gear cutters	Aluminox or Alundum		
Sharpening High-Speed Steel Cutters, Wet Grinding			
Milling and gear cutters	Aluminox or Alundum	46	I
Sharpening Carbon-Steel Cutters, Wet Grinding			
Formed cutters	Alundum	46	J
Tungsten-Carbide Grinding			
Tool grinding			
Roughing	Green Crystolon	60	I
Finishing	Green Grit Carborundum	80 or 100	S
Roughing	Diamond Grit	100	
Finishing	Diamond Grit	150	

[1]Elastic Wheel
[2]Silicate Wheel

(Brown & Sharpe Manufacturing Co.)

Unit 100

Coated Abrasives

A coated abrasive is composed of a flexible backing material to which abrasive grains are cemented. The coated abrasives used in the metalworking industry include emery, aluminum oxide, silicon carbide, and crocus. For information concerning the properties of these materials, see Unit 98. Emery has largely been replaced by aluminum oxide and silicon carbide.

Coated abrasives used for metalworking and other industrial jobs are available in the form of belts, rolls, sheets, discs, spiral points, and cones. See Fig. 100-1. Sheets usually are used for hand polishing, while the other forms are used for machine grinding and polishing.

(Norton Company)

Fig. 100-1.   Commonly used coated abrasives.

Selecting Coated Abrasives

Coated abrasives differ in the following characteristics: the abrasive material, the backing material, the bonding material, the method of coating, and the size of the grains. Each of these factors must be considered in selecting and purchasing the type to be used.

Abrasive Materials

The type of abrasive selected is determined by the type of material to be cut or polished, the amount of material to be removed, and the quality of finish desired. **Silicon carbide** abrasive is used on metals of low-tensile strength, such as cast iron, aluminum, brass, and copper. It also is used on plastics, glass, and marble.

Aluminum oxide abrasive is used on metals of high-tensile strength, such as carbon steel, alloy steel, stainless steel, and tough bronze. When considerable material is to be removed and a fine finish is desired, several grades of abrasive must be used — from coarse to medium to fine. **Crocus cloth** is made with a very fine crocus powder. It is used for polishing corroded metals or rare metals where a minimum of base metal is to be removed.

Backing Materials

The backing materials used for coated abrasives include paper, cloth, fiber, or various combinations of these materials. **Paper backing** is used largely for hand applications on woodwork. Cloth or fiber is used as backing on coated abrasives used in metalworking and for industrial machine grinding and polishing applications on other materials.

The **cloth backing** used for coated abrasives is of two types: a flexible lightweight cloth, called **jean,** and a heavier weight cloth, called **drill.** Drill is more stretch resistant and is used for machine grinding or polishing with belt or disc machines. **Fiber backing** is strong and durable and is used for tough disc applications.

Bonding Materials

Several types of adhesive materials are used to hold the abrasive grain to the backing material. Some are only suitable for use dry, while others may be used either wet or dry. Hide glue is used as a bond for dry, light, hand or mechanical polishing. Synthetic resins are used for water-proof cloth and for wet belt grinding and polishing. In addition, many modifications or combinations of glue, resin, and varnish are used for bonding materials. These materials add toughness, heat resistance, and moisture resistance to the abrasive cloth or belt.

Types of Coating

Coatings are of two basic types — open coating and closed coating. With an **open coating,** part of the backing surface is not covered with abrasive grains, thus leaving open space which resists filling or clogging when used with certain materials. With a **closed coating,** all of the backing surface is covered. Closed coating is used for cutting operations which involve a high rate of stock removal. Special heavy-duty coatings of adhesive and abrasive grain also are available for severe grinding operations.

Grain Sizes

The system used for designating abrasive grain sizes for coated abrasives is essentially the same as that used for grinding wheels. However, since coated abrasives have more of the grain exposed, the relative rating for coarseness is different. The following abrasive grain sizes, listed according to relative coarseness are used for coated abrasives:

Extra coarse: 12, 16, 20, 24, 30, 36
Coarse: 40, 50
Medium: 60, 80, 100
Fine: 120, 150, 180

Extra fine: 220, 240, 280, 320, 360, 400, 500, 600

Sizes Available

Rolls for hand use are manufactured in widths from 12.7 to 76.2 mm (1/2" to 3"). Sheets generally are 229 × 279 mm (9" × 11"). Belts, rolls, discs, spiral points, and cones are available in sizes to fit all standard machines.

Polishing with Abrasive Cloth

1. Tear a strip of abrasive from a sheet or roll.
2. For hand use, place the abrasive strip under a piece of wood. Apply a few drops of oil to the abrasive. Using straight strokes, polish the workpiece in the direction of the longest dimension of the surface.
3. When considerable stock is to be removed and a fine finish is to be produced, use several grades of abrasives. First, use medium, follow with fine, and finish with an extra-fine grade.
4. When polishing work on a lathe, use a strip of abrasive cloth under a stick of wood or a flat file. Apply several drops of oil to the abrasive, and polish with overlapping strokes. Another method of polishing is to loop the abrasive around the workpiece and, holding one end of the strip in each hand, pulling the strip back and forth with a slow sawing motion against the revolving workpiece. Be careful to avoid striking the lathe dog or the chuck with the fingers or hands. A high speed should be used.

Test Your Knowledge of Section 12

Unit 98: Abrasives and Their Uses

1. Explain what type of fracture characteristics an abrasive should have.
2. What does a comparison of Mohs and Knoop hardness values for aluminum oxide, silicon carbide, and diamond reveal?
3. List five natural abrasive materials. Which three are used in metalworking?
4. List four artificial abrasives.
5. What materials may be ground with silicon carbide?
6. Compare the properties of aluminum oxide with those of silicon carbide.

7. What materials may be ground on aluminum oxide grinding wheels?
8. What are the properties and uses of boron carbide?

Unit 99: Grinding Wheels and Their Selection

1. Explain the cutting action of a grinding wheel.
2. List several standard grinding wheel shapes.
3. What kinds of abrasives are used in making grinding wheels?
4. Explain how abrasive grains are sized and graded.
5. List five common types of bonding materials used in the manufacture of grinding wheels together with their properties.
6. Explain what is meant by the grade of a grinding wheel. How is it designated in wheel markings?
7. Explain what is meant by open- and close-grain structure. How is it designated?
8. Interpret the meaning of the following grinding wheel markings: 32A60-K5VBE, 39C60-E8VK.
9. What danger is involved if a grinding wheel is operated at a speed exceeding that recommended by the manufacturer?

10. What surface cutting speed is recommended for general offhand grinding with vitrified wheels?
11. On what materials are diamond grinding wheels used?
12. How are diamond grinding wheels constructed?
13. What factors must be specified in ordering a specific grinding wheel?
14. What precautions must be taken in mounting a grinding wheel properly?
15. What causes a grinding wheel to glaze?
16. What causes a grinding wheel to load?

Unit 100: Coated Abrasives

1. What principal types of coated abrasives are used in the metalworking industry?
2. In what forms are coated abrasive materials available?
3. On what materials are aluminum-oxide coated abrasives usually used?
4. For what applications is crocus cloth used?
5. What is meant by open coating and closed coating on coated abrasives?
6. Describe the procedure used for hand polishing with abrasive cloth.
7. Describe the procedure used for abrasive polishing of work mounted in a lathe.

Unit
101
Grinding and Grinding Machines

Grinding machines are precision machine tools. They machine metal parts to very close tolerances. They produce high-quality surface finishes. Grinding machines are available for grinding flat surfaces, external and cylindrical surfaces, tapered surfaces, and irregular surfaces.

It is common industrial practice to grind many mass-produced parts to tolerances of plus or minus 0.0025 mm (0.0001"). Special parts for precision instruments are ground to tolerances of plus or minus 0.000508 mm (0.000020", 20 microinches). Another distinct advantage of grinding is that it often is the only method by which parts hardened by heat treatment may be machined. Precision-ground parts range in size from the small parts in a wristwatch to the large rolls used for rolling sheet steel in steel mills.

Kinds of Grinding Operations

There are several basic kinds of precision grinding operations.

Surface Grinding

Surface grinding produces an accurate flat surface on a part, Fig. 101-1. Several types of surface grinders are made.

Cylindrical Grinding

Cylindrical grinding produces a cylindrical or conical shape on a workpiece. The workpiece is mounted between centers or in a chuck, and the face of the grinding wheel passes over the external surface of the revolving piece, Fig. 101-2.

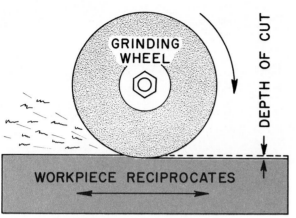

SURFACE GRINDING

Fig. 101-1.   Relationship of grinding wheel and workpiece with horizontal-spindle surface grinder.

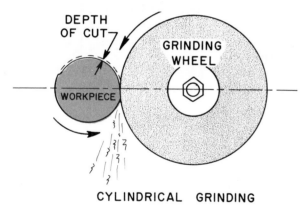

CYLINDRICAL GRINDING

Fig. 101-2.   Relationship of grinding wheel and workpiece in cylindrical grinding.

Cylindrical grinding is done on either plain or universal cylindrical grinding machines.

Internal Grinding

Internal grinding produces a smooth and accurate surface in a cylindrical hole. The surface may be straight, tapered, or irregular. Internal grinding is a form of cylindrical grinding, Fig. 108-1. This type of grinding may be done on universal grinding machines, internal grinding

machines, and with tool post grinders mounted on a lathe.

Form Grinding

Form grinding produces a smooth and accurate surface of a special shape. It is done with a grinding wheel which usually is shaped to conform to the contour of the surface it is designed to produce. An example of form grinding is the grinding of a thread from solid stock, Fig. 99-2. The grinding of fillets, rounds, or irregular shapes is another example. Form grinding may be performed with various types of grinding machines, including surface grinders, cylindrical grinders, internal grinders, and special grinding machines.

Plunge Grinding

Plunge grinding is another form of cylindrical grinding which may produce a straight, tapered, or formed surface on a workpiece. Instead of traversing the workpiece past the wheel as is usually the case, the wheel is fed directly into the revolving workpiece with little or no side movement. Automotive crankshafts and similar objects with deep shoulders often are ground by this method. A cylindrical grinding machine equipped with a plunge-type grinding wheel head generally is used for operations of this type.

Centerless Grinding

Centerless grinding is a form of cylindrical grinding. It produces accurately ground parts without requiring them to be mounted between centers. Parts are held in position on a workrest blade which is located between a grinding wheel and a regulating wheel, Fig. 101-3. The regulating wheel rotates the work which rests on the blade. At the same time, the grinding wheel grinds the surface. Straight or tapered objects may be ground in this manner.

Lathe centers, piston pins, roller bearings, and similar objects without center holes are examples of objects ground by the centerless grinding method.

Tool and Cutter Grinding

Tool and cutter grinding involves the grinding of milling cutters, counterbores, reamers, and many other kinds of cutting tools. This type of

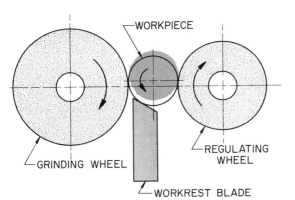

Fig. 101-3.   Relationship of grinding wheel and workpiece in centerless grinding.

(Norton Company)

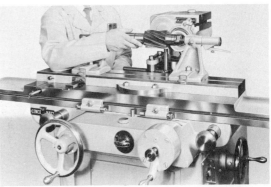

Fig. 101-4.   Grinding a plain milling cutter on a tool and cutter grinder.

grinding normally is done on a tool and cutter grinder, Fig. 101-4.

Offhand Grinding

Offhand grinding is the nonprecision type of grinding done on bench- or floor-model tool grinders, as shown in Fig. 28-1. Cold chisels, center punches, lathe tool bits, and shaper tool bits often are ground by offhand grinding. The tool or object being ground is held by hand, with or without a guiding device.

Kinds of Grinding Machines

Like other metalworking machines, grinding machines designed for many purposes are available. In this unit, the most common types of grinding machines, their applications, and their principal parts are described briefly. More detailed information concerning machine accessories, controls, and other operational factors is explained in succeeding units which include the procedures for specific types of grinding operations.

Grinding machines are often classified as to size by an arbitrary number assigned by the manufacturer. All, however, can be classified in terms of their maximum capacity to accommodate a workpiece.

Surface Grinders

Surface grinding machines are designed primarily for grinding flat surfaces. However, with special setups, angular and formed surfaces also

(Brown and Sharpe Manufacturing Company)

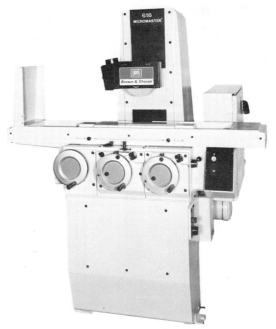

Fig. 101-5.   Surface grinder with horizontal spindle.

may be ground. Surface grinding machines are of two general types: the **horizontal-spindle** type (Figs. 101-5 and 101-6) and the **vertical-spindle** type (Fig. 101-7).

Machines with horizontal spindles are most common in tool- and die-shops, toolrooms, main-

(Norton Company)

Fig. 101-6.   Surface grinder with horizontal spindle and rotary table. Safety goggles are recommended.

(Norton Company)

Fig. 101-7.   Surface grinder with a vertical spindle.

tenance shops, and schools. With this type of machine, the work may be mounted on the table in a number of ways, often with a magnetic chuck. The work table reciprocates back and forth under the grinding wheel. With each succeeding table stroke, the work is fed crosswise under the wheel. Machines of this type are available with either manual or power longitudinal and transverse (crosswise) feeds. They may be operated dry or wet.

For dry grinding, an exhaust attachment generally is used to catch the dust. For wet grinding, a special attachment pumps fluid or a mist to the grinding area.

Further details concerning horizontal-spindle surface grinders are included in Unit 102. The parts, accessories, work setup procedures, and operational procedures are included in the unit.

A second type of surface grinder with a horizontal spindle is the **rotary** type, Fig. 101-6. This machine has a rotary table mounted on a supporting table which travels longitudinally. The workpiece revolves under the grinding wheel, and, at the same time, the table is fed longitudinally under the wheel. Workpieces of larger diameter may be ground on machines of this type.

Surface grinders of the type shown in Fig. 101-7 have a vertical spindle. As the grinding wheel revolves, the work is fed back and forth under the wheel. This type of grinder generally cuts much faster than those with horizontal spindles. Because of the amount of heat developed with this type of grinder, a cutting fluid always should be used.

Plain Grinders

Plain grinding machines (Fig. 101-8) are designed primarily for production grinding of external cylindrical surfaces. They are used for cylindrical grinding of straight surfaces, tapered surfaces, and shoulders. They also may be used for plunge grinding of formed surfaces which conform to the shape of the grinding wheel.

The principal parts on a plain grinding machine are: a heavy bed which gives the machine stability; a wheel head mounted on a slide base; a headstock mounted on a table which can be swiveled through 8°; a footstock or tailstock; a sliding table on which the swivel table is mounted;

Fig. 101-8.   Plain grinding machine.

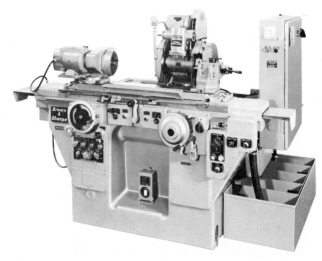

Fig. 101-9.   Universal grinding machine.

a longitudinal table feed mechanism; and a manual or automatic cross-feed mechanism.

The wheel head is set permanently at right angles with the table travel and cannot be swiveled. The headstock on many plain grinding machines cannot be swiveled. On some machines, such as the one in Fig. 101-8, the headstock may be swiveled up to 45° for grinding steep tapers. The headstock spindle has four step pulleys which provide work speeds ranging from about 200 to 800 rpm.

Longitudinal table travel may be operated manually or automatically. Six rates of power travel are provided, ranging from about 152 mm (6″) to 3378 mm (133″) per minute. The feeds are selected through the use of table speed-selector levers. Table reverse dogs may be set for automatic table reversal. Plain grinding machines often are equipped with features such as wheel slide rapid travel, independent automatic cross feed, and a wheel spindle reciprocating mechanism.

Universal Grinders

A universal grinding machine, Fig. 101-9, is far more versatile than the plain grinder. It is designed to perform both external and internal cylindrical grinding operations. It can grind straight surfaces, tapered surfaces, shoulders, steep tapers, and face grinding. Also, straight fluted reamers and milling cutters can be ground on a universal grinder. See Unit 114.

Universal grinding machines are made in sizes ranging in swing capacity from 254 mm to 355.6 mm (10″ to 14″) or more. Their capacity between centers ranges from 508 mm to 1524 mm (20″ to 60″) or more.

Principal Parts: The principal parts of a typical universal grinding machine include: a heavy base which gives the machine stability, a table which can be swiveled 8° for grinding tapers, a wheel spindle head which may be swiveled for grinding angles, a headstock which may be swiveled for grinding steep angles or faces, a footstock for holding work between centers, and an internal grinding unit which is mounted directly above the external grinding spindle.

A variety of work-holding devices and accessories also is available to extend further the versatility of this machine. Work may be mounted in a chuck on the headstock spindle, Fig. 107-4, for either face or angular grinding.

Further details concerning universal machines and their parts, controls, and operational features are included in Unit 105.

Internal Grinders

The internal grinding machine is used for finishing cylindrical or tapered holes. It is a highly specialized machine which rarely is found in schools or small commercial shops. In such shops, internal grinding usually is done with a universal grinding machine equipped with an in-

(Brown and Sharpe Manufacturing Company)

Fig. 101-10. Universal and tool grinding machine.

ternal grinding fixture, Fig. 108-2. Internal grinding also may be done with a tool post grinder mounted on a lathe.

Tool and Cutter Grinders

The tool and cutter grinding machine is designed for grinding milling cutters, reamers, taps, and other precision cutting tools used on milling and drilling machines, Fig. 101-4. When equipped with the appropriate accessories, tool and cutter grinding machines also may be used for accurately grinding single-point cutting tools. An introduction to the use of the tool and cutter grinding machine for sharpening milling cutters is given in Unit 114.

Universal and Tool Grinders

The universal and tool grinding machine, Fig. 101-10, is a general-purpose machine which may be used for an unusually large number of grinding applications. It is an extremely versatile machine which is particularly useful in toolrooms, small commercial shops, and schools. It will perform small and medium cylindrical grinding operations which normally are performed on a universal grinder, including both external and internal grinding operations.

The machine can also be used for grinding all of the common cutting tools (such as milling cutters, reamers, and taps) and many special cutting tools. Numerous accessories and holding devices are available for the machine, including a universal chuck, internal grinding attachment, wet grinding attachment, surface grinding attachment, collect chuck, magnetic chuck, index centers, end mill sharpening attachment, and many convenient accessories commonly used in conjunction with cutter sharpening.

Gaging Ground Parts

When many parts are to be ground to the same size on a cylindrical grinding machine, the size can be quickly checked with an **indicating grinding gage,** Fig. 101-11. This gage indicates the size of the work while the machine is running.

Several types of dial indicating snap gages may also be used to gauge the work while mounted in the machine, Figs. 10-11 and 10-12. These gages indicate size directly and are convenient for measuring work in the grinder. **Go and not-go** snap gages also may be used. However, gages of this type show only whether the workpiece is within specified tolerances.

(Federal Products Corporation)

Fig. 101-11. Dial-indicating grinding gage indicates size directly.

Table 101-1
Possible Grinding Troubles

Symptom	Probable Cause
Work shows chatter finish.	Grinding wheel out of balance or not clamped properly on the wheel sleeve. Grinding wheel dull, glazed, or loaded. Poor choice of wheel for material being ground. Work not well-supported; centers worn or need lubricant. Not a sufficient number of back rests used, or back rests not properly adjusted. Too high a work speed or rate of table travel. Cut too heavy, caused by excessive cross feed. Unbalanced workpiece (for example, a crankshaft) running at a speed which is too high or running away from the driving dog (too much momentum). A worn or defective driving belt; check headstock, spindle, and table belts. Machine located on an insufficiently rigid floor or a floor which transmits vibration to the machine.
Scratches on the work.	Using a dirty coolant. Grinding wheel not trued properly. Truing diamond dull, cracked, or broken; or not held rigidly in the holder; not clamped securely in the truing fixture; the fixture not rigidly clamped in position; or footstock spindle not clamped. Too rapid table feed, or a too deep cut when truing or dressing the wheel. Wheel too coarse for the work.
Spiral marks on the work.	Point of truing diamond too high. The wheel should be trued with the diamond point as near the heights of the work centers as possible.
Wheel burning the work.	Insufficient coolant used, or coolant not properly directed at the point of contact of the wheel and the work. Grinding wheel dull, glazed, or loaded; needs dressing. Wheel too hard, wheel speed too high, or work speed too low. Excessive cross feed.
Work not ground parallel.	Swivel table not set accurately at zero. Swivel table pivot shoe may need adjusting. Headstock or footstock not seated properly on the table. Centers not seated properly in the spindle, or center points worn out of round. Center holes in workpiece dirty, out of round, or do not fit the centers properly. Radial play in the footstock spindle. Spindle clamp not properly adjusted. Back rests needed, or if used, not properly adjusted.
Work not sizing uniformly.	Wheel slide rapid travel arrangement motor brake needs adjusting. Cross-feed screw thrust bearing needs adjusting.
Wheel spindle runs too hot or stalls.	Insufficient oil in spindle reservoir, or wrong kind of oil. Cross feed too heavy, beyond capacity of the machine. Spindle driving belts too tight.

General Preliminary Procedure

Some of the possible grinding troubles which might occur while performing grinding operations on various types of grinders are listed in Table 101-1, along with their probable causes. The following are general procedures which should be followed in preparation for grinding operations:

1. Make all settings and adjustments as explained in the units which follow.
2. Before starting the machine, see that all driving belts have proper tension. A V-belt is properly tensioned when it can be depressed about 25 mm (1") using only slight pressure. Excessive tension will cause belts to wear rapidly and may damage the spindle.
3. Make certain that all guards are in working condition and in place. See that the machine is properly lubricated.
4. Before starting a grinding operation, run the machine for about fifteen minutes. This will bring the machine to running temperature and thus eliminate variations due to expansion of machine parts.

Unit
102

How to Operate Surface Grinding Machines

As indicated in Unit 101, several different types of surface grinding machines are used for grinding flat surfaces. Since horizontal-spindle surface grinding machines are the most widely used, this unit is concerned with their operation. Other types of surface grinders, however, are operated according to the same basic principles.

Machine Features

Although the design and the control mechanisms vary with different manufacturers, the principles involved are similar.

Machine Size

The size of horizontal-spindle machines is designated by the maximum length and width of the workpiece which can be ground without shifting its position on the worktable. Thus a 150 × 450 (6 × 18) machine has a working area at least 150 mm × 450 mm (6" × 18"). That is, it has transverse (cross feed) table travel of at least 150 mm (6") and longitudinal table travel of 450 mm (18"). The maximum height of parts which may be surface ground on a given machine should also be known before a commitment to purchase is made.

Feed

Machines are furnished with either hand or hand and power feed. The power feed, available for either or both longitudinal and cross feeds, may be a mechanical-type feed mechanism or a hydraulically operated feed mechanism. The machine in Fig. 102-2 is the hand-feed type.

The vertical-feed handwheel and cross-feed handwheel, as in Fig. 102-1, have 0.005 mm (0.0002") graduations for accurate adjustments. An auxiliary adjusting knob for fine feed (Fig. 102-10) provides for vertical adjustments of the wheel elevation in 0.0025 mm (0.0001") graduations.

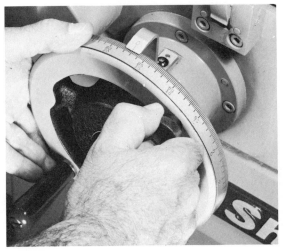

Fig. 102-1.   This cross-feed handwheel has 0.005 mm (0.0002″) graduations.

Fig. 102-2.   Hand-feed surface grinder with wet-grinding attachment.

Wet or Dry Attachments

For dry grinding, a vacuum exhaust attachment should be provided to remove the highly abrasive grit and dust produced. Otherwise, operators should be provided with efficient respirators. A wet-grinding attachment, Fig. 102-2, includes a pump, liquid container, splash guards, and suitable piping and hoses to deliver a stream of fluid to the grinding area. The cutting fluid reduces heat and, therefore, increases the rate at which metal may be removed. It also improves surface finish.

Depth of Cuts

When a considerable amount of metal is to be removed, roughing cuts should be made first, followed by finishing cuts. Roughing cuts may vary from 0.025 mm (0.001″) to 0.076 mm (0.003″) in depth, depending on the rigidity of the machine, whether a coolant is used, and the rate of feed. Finishing cuts generally should be 0.025 mm (0.001″) or less in depth.

Accessories

Numerous work holding and wheel dressing attachments are available for surface grinding.

They are described later in conjunction with the methods employed for making setups.

General Procedure

1. Select a grinding wheel suited to the work to be performed, and mount it on the wheel spindle. (See Unit 112.) Be sure the wheel is sound.

   CAUTION

   If the wheel is changed, it will be necessary to true and balance the new wheel before attempting to grind a workpiece. See Units 103 and 112. If the workpiece is to be form-ground, it will be necessary to shape the wheel accordingly. See Unit 104.

2. Mount the work on the table of the machine:
   a. **By Clamping:** Position, align, and clamp the workpiece, for example, as in Fig. 102-3.
   b. **By Using a Permanent Magnetic Chuck:** Mount the chuck on the table and clamp it

Fig. 102-3. Workpiece held in position with clamps.

Fig. 102-5. Workpiece held in adjustable swivel vise.

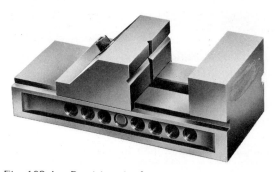

Fig. 102-4. Precision vise for use on magnetic chuck.

Fig. 102-6. Grinding a slot with workpiece mounted on magnetic chuck.

in position. Place the workpiece or pieces on the chuck, Fig. 102-6. With the work in position, move the control lever 180° to the right.

CAUTION

For safety, use back or end stops if the workpiece has only small contact with the chuck.

Flat pieces of metal that are thinner than the workpiece make good stops. See Fig. 102-7.

c. **By Using a Vise:** Plain, swivel, and tilting vises may be attached to the worktable or held on a magnetic chuck, Figs. 102-4 and 102-5. In this manner, workpieces may be held at any angle required.

d. **By Using Index Centers:** Mount the index head and footstock on the table of the machine, aligning them by means of the tongues on their bases. Clamp in position. Place a driving dog on the work. Then place it between centers and position it with respect to the grinding wheel, as in Fig. 102-8.

Fig. 102-7.   Method of blocking workpieces with small contact area on magnetic chuck to prevent their moving under pressure of grinding.

Fig. 102-8.   Workpiece held between index centers.

e. **By Using a Sine Plate or a Perma Sine:** A sine plate, Fig. 12-5, or a perma sine, Fig. 12-7 may be used when workpieces are to be ground to very precise angles. The work is attached to the sine plate by bolting or clamping; the perma sine holds the workpiece magnetically.

f. **By Using V-Blocks:** Round workpieces may be mounted in V-blocks on a magnetic chuck, Fig. 102-9, or be clamped directly to the machine table.

Fig. 102-9.   Work mounted in V-blocks on magnetic chuck.

3. Properly lubricate the machine before starting it. If the machine has been idle for some time, alternately press the start and stop buttons in rapid succession three or four times before running the machine at operating speed.

4. Protect eyes by wearing properly fitted goggles.

Procedure for Manual Feed

1. With the table handwheel, move the work under the grinding wheel. Then lower the wheel by means of the vertical-adjustment (elevating) handwheel, Fig. 102-10, until it almost touches the workpiece.

2. With the cross-feed handwheel, move the table outward until the far side of the wheel projects beyond the farthest edge of the workpiece a distance equal to three-fourths the width of the wheel.

3. Move the table to the left until the wheel clears the work by about 50 mm (2″).

On the machine illustrated, the power feed mechanism should be disengaged whenever continuous manual operation is desired. This is accomplished by turning the table throttle lever, Fig. 102-10, to the off position. The table handwheel always is engaged for manual operation. It is disengaged automatically when power table travel is engaged.

(Brown and Sharpe Manufacturing Company)

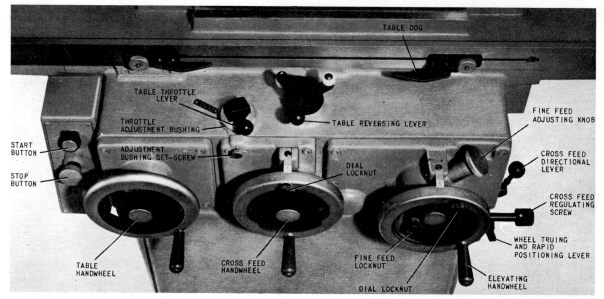

Fig. 102-10. Front operating controls on No. 618 surface grinding machine.

4. Lower the wheel to produce a cut of 0.025 to 0.076 mm (.001" to .003"). Lock the grinding head in position by tightening the clamping screw, if the machine is so equipped.

CAUTION
When dry grinding a flat surface, considerable heat will be generated, particularly if a heavy feed is used. Overheating should be guarded against by using a light cut and a light feed.

5. See that the **wheel guard is fastened securely in position.** Then, start the machine. With the longitudinal table handwheel, feed the workpiece steadily back and forth under the grinding wheel. Move the end of the work a sufficient distance past the wheel to allow adjustment of the cross feed before reversing the table. After each longitudinal stroke, turn the cross-feed handwheel clockwise one complete turn. This will give a 2.54 mm (0.100") feed. With experience, a heavier feed may be taken.

6. Continue as in steps 4 and 5 until the entire surface has been rough ground.

7. When rough grinding has been completed, move the grinding wheel back to the starting

point. Dress the grinding wheel if necessary. See Unit 103. Lower the grinding head sufficiently to make the finishing cut, usually 0.025 mm (.001") or less.

Procedure for Power Longitudinal Feed

1. With the table handwheel (Fig. 102-10), move the work under the grinding wheel. Start the machine. Then lower the wheel by means of the vertical-adjustment handwheel until it contacts the surface of the workpiece.

2. Move the table to the left until the work clears the wheel by about 25 mm (1").

3. Bring the right-hand reversing dog against the right side of the reversing lever. Fasten it in position by tightening the clamping bolt.

4. Move the table to the right until the grinding wheel clears the end of the work by about 25 mm (1")

5. Bring the left-hand reversing dog against the left side of the reversing lever, and clamp it in position.

CAUTION
When setting the table dogs for automatic reversing, be sure to allow sufficient

over-travel of the work in both directions for completion of the cross-feed action before the work comes back under the wheel.

6. Engaging the table throttle lever will now cause the table to feed automatically back and forth longitudinally.

Procedure for Power Cross Feed

The procedure for adjusting the power cross-feed mechanism varies for different makes of machines. Therefore, before using the machine, it is good practice to read the operator's handbook which usually is supplied with the machine. The procedure which follows is for the Brown and Sharpe No. 618 machine, Figs. 102-10 and 102-11.

By setting the cross-feed regulating screw, any cross feed from 0.254 to 6.35 mm (0.010″ to 0.250″) may be obtained at each reversal of the power longitudinal table travel. A continuous cross feed of 254 mm (10″) per minute also is

provided for wheel truing. The wheel also may be positioned rapidly over long distances, at a rate of 3.66 mpm (12 feet per minute), by using the selector lever provided.

1. To use power cross feed, first turn the selector lever, Fig. 102-11, to the **grind** position. Next, turn the cross-feed direction lever for the desired direction of cross feed, for example, **inward** or toward the rear of the machine.

2. Start the table travel with the table throttle lever, Fig. 102-10. Then adjust the cross-feed knob, Fig. 102-11, to the desired amount of cross feed for each stroke of the table. Cross feed should be set in accordance with the wheel width, the depth of cut, and the finish desired. For average cuts, a cross feed from 1.25 to 2.5 mm (0.50″ to 0.100″) is satisfactory. With experience, the rate may be increased, but it should not exceed one-half the width of the wheel face.

Some machines are equipped with trip dogs which can be positioned so that a cut automatically may be started or stopped at a particular point on the surface of the workpiece. After the cross feed of the table has been stopped by a trip dog, the cross-feed selector lever must be moved to change the direction of cross feed before again starting the table. Furthermore, the starting lever must be held in the engaged position until the trip dog has moved off the plunger sufficiently to permit the knob to remain engaged.

CAUTION
When using cross-feed stop dogs, under no circumstances attempt to force the machine to grind beyond the maximum permitted in either direction by the dogs.

3. With the grinding wheel at the right end of the workpiece and clear of the work, stop the table travel with the table throttle lever.

4. With the cross-feed wheel, position the workpiece where desired for starting the cut.

Procedure for Cutting

1. Turn the vertical-adjustment handwheel through sufficient graduations to produce a cut of 0.025 mm to 0.076 mm (0.001″ to

Fig. 102-11. Cross-feed controls on No. 618 surface grinding machine.

0.003"). Lock the grinding head in position by the clamping screw, if the machine is so equipped.

2. With the workpiece in position, the longitudinal table travel and cross feeds properly adjusted, and the machine and coolant turned on, engage the table throttle lever and make the cut.

3. When the roughing cut is completed, adjust the machine to take a finishing cut, usually 0.025 mm (0.001") or less.

High-Speed Attachment

By means of a high-speed surface-grinding attachment that can be mounted in the machine spindle of some machines, small diameter wheels may be used for grinding surfaces in confined areas. See Fig. 102-12.

(Brown and Sharpe Manufacturing Company)

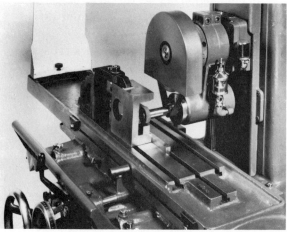

Fig. 102-12. High-speed surface-grinding attachment.

Unit

103

How to True or Dress Wheels on Surface Grinders

A grinding wheel should be trued each time it is put on the spindle. It should be dressed whenever it becomes dull, loaded, or glazed with use.

Truing refers to correcting an out-of-round condition of the wheel. A dressing tool is used to remove particles of the abrasive from the high part of the wheel. Truing also refers to forming the wheel to a particular shape, such as a concave or a convex face. To be in good condition, the wheel must be sharp and run true on both the periphery and the sides.

Dressing produces a sharp grinding surface. A diamond tool is used to remove the dull or loaded surface of the wheel. Dressing is necessary whenever the wheel cuts very poorly, usually resulting

in burning the work (the dark surface being caused by heat oxidation).

Procedure for Truing or Dressing

1. Protect the eyes by wearing properly fitted goggles.

2. Mount the wheel on the spindle as instructed in Unit 112.

3. Secure a wheel dressing and truing fixture such as that shown in Fig. 103-1.

4. Position the fixture on the table of the machine so that the diamond tool may be applied to the face of the wheel, as shown in Fig. 103-1. The contact point of the diamond tool should be slightly left of the vertical

(Brown and Sharpe Manufacturing Company)

Fig. 103-1.   Truing a wheel on a surface grinder.

center of the wheel (for clockwise rotation). The toolholder should also be inclined slightly in the direction of wheel travel. This is necessary to prevent gouging and a tendency to chatter.

5. Clamp the fixture in position. Check that the wheel guard is securely fastened and then start the machine. With the vertical-adjustment handwheel, lower the grinding head until the wheel is lightly in contact with the diamond tool.

6. With the cross feed, move the table so that the diamond tool clears the wheel. Then lower the grinding head about 0.0125 mm (0.0005″).

 To prevent chipping the edges of the wheel, it is good practice to round the corners of the wheel slightly before truing the face. To do this, bring the diamond tool into contact with the corner of the wheel. Then, by manipulating the cross and vertical feeds, remove a small amount of material from each corner of the wheel.

7. With the cross feed, pass the diamond tool across the face of the wheel two or more times. If a coolant is available, use it freely.

8. At the end of the second or third stroke, lower the head another 0.0125 mm (0.0005″), and take another stroke. Continue this procedure until the wheel is running true.

9. On rare occasions, it is necessary to true the sides of the wheel. To perform this operation, position the fixture so that a cut may be taken on respective sides of the wheel. Then taking a light cut, feed the grinding head downward. Continue with repeated cuts until the sides of the wheel are true.

10. When the operation is completed, remove the fixture and clean both it and the machine.

Unit

104

How to Shape Grinding Wheels on Surface Grinders

Often it is necessary to surface grind rounded fillets, grooves, or irregular surfaces. In these cases, the grinding wheel must be shaped according to the surface to be ground. In Fig. 104-1, a V-shaped wheel is being used to grind a V-shaped recess.

Procedure

1. Obtain and mount a wheel on the spindle, and true it in the usual manner, Unit 103.
2. Mount a radius and wheel truing attachment on the table of the machine, and clamp it in position, Fig. 104-2.
3. Mount a diamond tool in the toolholder at right angles to the slide, as in Fig. 104-2.
4. Loosen the clamping screw in the front of the base of the fixture. Then swivel the slide to

the angle required (for example, 60°), as indicated by the graduations on the base. Clamp the slide in position by tightening the screw.
5. With the vertical feed, lower the wheel until the tool contacts the wheel slightly below the axis.
6. With longitudinal and cross-feed hand-wheels, position the tool to make the cut.

(Brown and Sharpe Manufacturing Company)

(Brown and Sharpe Manufacturing Company)

Fig. 104-1.   Grinding a V-shaped recess.

Fig. 104-2.   Tool mounted at right angles to slide.

7. Start the machine. Then move the diamond tool across the edge of the wheel, taking a cut of about 0.0125 mm (0.0005″). Take two or three passes before advancing the tool.

8. With the longitudinal table feed, advance the tool for a second cut; again take at least two passes.

9. Continue as in steps 6 and 7 until the wheel has the desired shape.

10. When the operation has been completed, remove the attachment and return to the place where kept when not in use.

Figure 104-3 shows a radius and wheel truing attachment positioned for shaping a convex surface on the face of the wheel.

(Brown and Sharpe Manufacturing Company)

Fig. 104-3.   Shaping convex surface on a wheel.

Unit
105

The Universal Grinding Machine

Principal Parts

The principal parts and the operational controls for the No. 1 Brown and Sharpe universal grinding machine, Fig. 105-1, are illustrated in this unit. Although the controls and their locations may vary with different makes, the principles involved in their operation are similar. Nevertheless, it is advisable to study the operator's handbook supplied with the machine, before attempting to operate the machine.

(Brown and Sharpe Manufacturing Company)

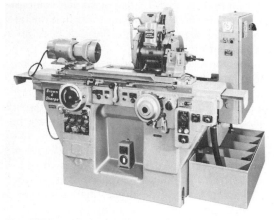

Fig. 105-1.   Universal grinding machine.

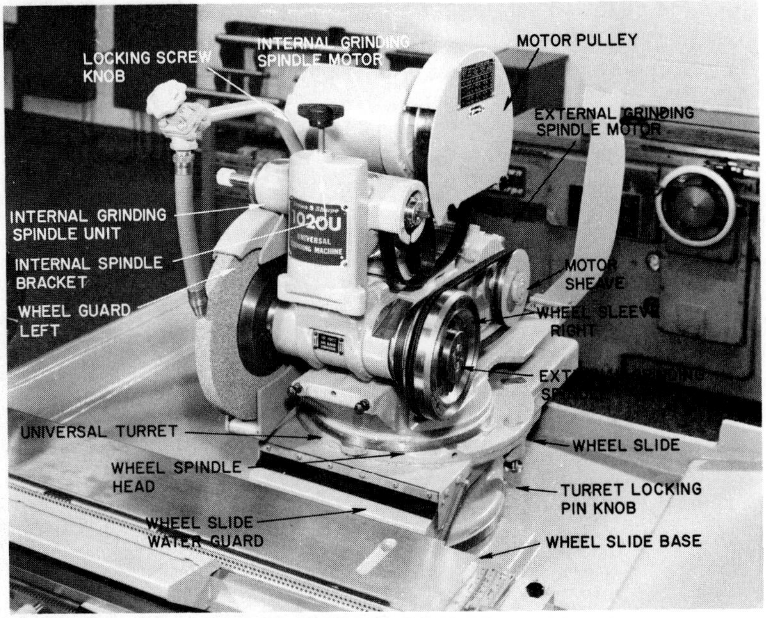

Fig. 105-2. Wheel stand unit with belt guards removed. Note that the external grinding spindle is in grinding position.

Wheel Stand

The wheel stand unit includes the **external grinding spindle**, which is shown in position for external grinding, Fig. 105-2. The **internal grinding spindle** is located immediately above the external spindle, and it is in a retracted position. For internal grinding operations, this spindle is brought forward and fastened in position as shown in Fig. 108-1. The external spindle is driven by a V-belt from its motor, while the internal spindle is driven with a flat belt from an independent motor.

The **wheel spindle head,** Fig. 105-2, supports both grinding spindles. It may be swiveled 90° in either direction from zero, as indicated by the scale graduated in degrees. The wheel spindle head, universal turret, and wheel slide are held secure by a single bolt. The head of the bolt is located directly below the internal grinding motor.

The **universal turret** may be swiveled and locked in four different angular positions, each 90° apart. In these positions, the wheel may be located closer or farther away from the sliding table, as desired. Each 90° position provides for a 50.8 mm (2″) difference in location of the wheel spindle head. This adjustment, together with the 127 mm (5″) maximum traverse movement along the **wheel slide,** makes it possible to properly locate the wheel for large- or small-diameter workpieces. Cross feed of the grinding wheel may be operated manually or automatically.

The **wheel slide base** has ways (bearing surfaces) on which the wheel slide traverses. This base may be swiveled up to 90° to either side of zero for face grinding or for grinding steep angles, Fig. 107-2. The base must be clamped securely in position for all grinding operations.

Headstock Unit

The headstock unit, Fig. 105-3, is a self-contained unit. It includes the headstock spindle, driving motor, work-driving plate with driving arm, and headstock base. The headstock may be swiveled up to 180° on the base which is graduated in degrees. It is swiveled for face grinding or for steep angular grinding of workpieces which are mounted on the headstock spindle or in a chuck, Fig. 107-4.

The headstock unit is aligned on the swivel table by a lip at the rear of the base. Two clamp bolts at the front of the base hold the unit securely in position. The table surfaces should be clean and free from nicks or dents when the unit is clamped in place. A variable-speed motor provides speeds ranging from about 60 to 600 rpm. Headstock speeds are controlled by the headstock speed-control knob, Fig. 105-4.

Headstock spindle rotation may be controlled with a **lever control** or a **handwheel control.** When the **lever-handwheel selector-switch knob** (Fig. 105-4) is set at lever control, headstock spindle rotation is started and stopped with the **headstock and table control lever,** Fig. 105-5. With this lever at **headstock running,** the spindle rotates and coolant flows. Then, when the lever is moved to **table running,** the table travels. When the lever-handwheel selector-switch knob is set for **handwheel control,** the headstock spindle is started and stopped by a switch which is controlled by the cross-feed handwheel (to be explained shortly).

Longitudinal Table Travel

The sliding table may be traversed manually with the table handwheel or automatically by a power-actuated hydraulic system. For power table travel, the **lever-handwheel selector switch,** Fig. 105-4, first is set at **lever control.** The table then is started or stopped with the headstock and table control lever. With the headstock spindle running, the table starts to travel when the lever is turned to the **table running** position. Table travel rates from 76.2 mm (3″) to 3810 mm (150″) per minute, as desired, may be selected with the **table speed-selector knob,** Fig. 105-5.

Table Dwell: The table may be set to dwell for a period of 0 to 2-1/2 seconds at one or at both ends of the work for each table reversal. Two knobs located near the table speed selector, Fig. 105-5, are used for this purpose.

Table reversal may be controlled manually or automatically with the **table reverse lever,** Fig. 105-5. For automatic operation, the **adjustable table dogs** may be set to actuate the table reverse lever for any desired length of table travel. For convenience in wheel truing, the knob on the table reverse lever may be pulled forward to permit the table to be traversed beyond the limit of the dogs.

(Brown and Sharpe Manufacturing Company)

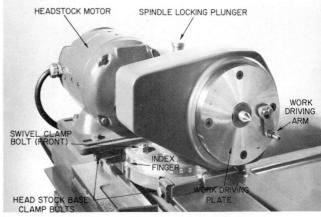

Fig. 105-3.   Headstock unit with work-driving plate in position with splash guard removed.

Cross-Feed Mechanism

The cross feed of the grinding wheel may be operated manually or by power, as determined by the position of the **cross-feed control lever,** Fig. 105-6. For power operation, the cross-feed handwheel is rotated slightly to the left, in order to release the **switch operating slide.** The feed is then engaged by throwing the cross-feed control lever to **start.** When grinding is complete, a fixed pin in the back of the cross-feed wheel contacts the switch operating slide, thus stopping the cross feed. The cross-feed control lever is then thrown to **stop,** and the wheel is withdrawn by rotating the handwheel to its starting position. The cross-feed control lever must be in the **start** postion in order to start the cross-feed mechanism in operation, in either handwheel or lever control. This movement may be stopped by turning the cross-feed control lever to the **stop** position.

For automatic operation, the **switch operating slide throw-out lever** must always be in its lower position. For hand cross feed over the complete range of wheel slide travel, this lever should be in its upper position.

Selecting Amount of Cross Feed

The cross-feed handwheel has 200 graduations, each representing 0.0125 mm (0.0005″). Hence, one complete revolution of the handwheel represents 2.54 mm (0.100″) on the diameter of

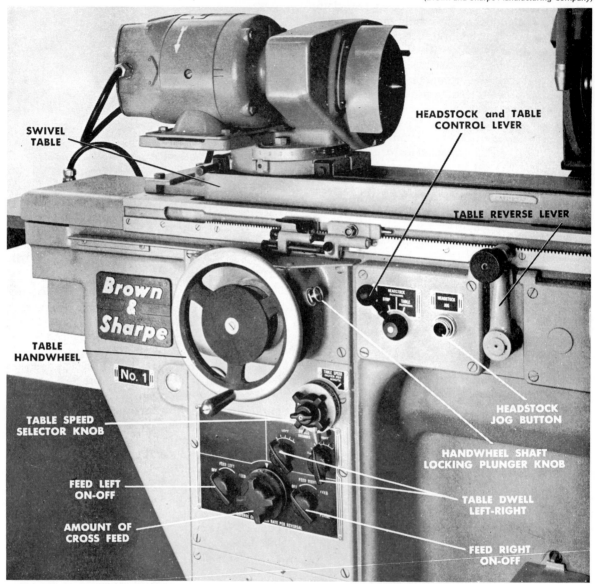

SWIVEL TABLE

HEADSTOCK and TABLE CONTROL LEVER

TABLE REVERSE LEVER

TABLE HANDWHEEL

Brown & Sharpe

No. 1

TABLE SPEED SELECTOR KNOB

HEADSTOCK JOG BUTTON

HANDWHEEL SHAFT LOCKING PLUNGER KNOB

FEED LEFT ON-OFF

TABLE DWELL LEFT-RIGHT

AMOUNT OF CROSS FEED

FEED RIGHT ON-OFF

Fig. 105-5. Handwheel for longitudinal table travel and controls at left front of machine.

the workpiece. Fine adjustments may be made with the index dial which has 0.0025 mm (0.0001″) graduations. Thus, one complete turn of this dial represents 0.1016 mm (0.004″) on the diameter of the work.

The amount of automatic cross feed is set with the cross-feed knob, Fig. 105-5. Cross-feed rates may be set for each or alternate table reversal in amounts ranging from 0.0025 mm (0.0001″) to 0.076 mm (0.003″) on work diameter.

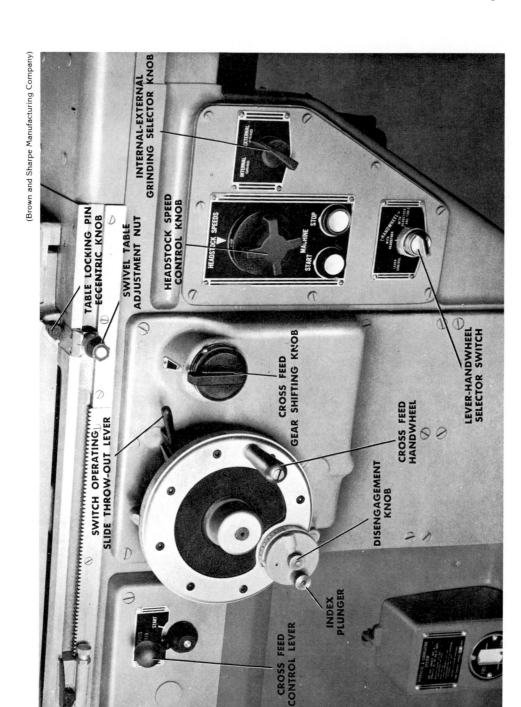

(Brown and Sharpe Manufacturing Company)

Fig. 105-4.   Cross-feed handwheel and controls at right front of machine.

(Brown and Sharpe Manufacturing Company)

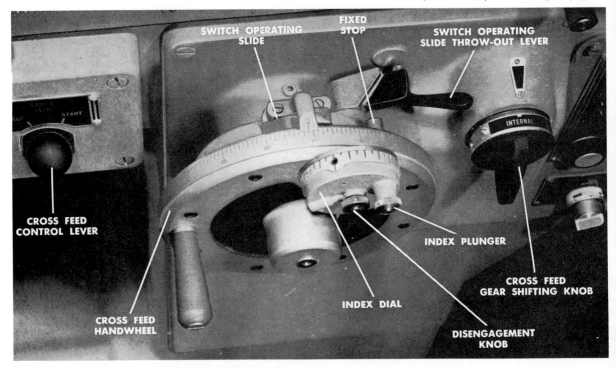

Fig. 105-6.   Cross-feed handwheel and positive stop mechanism.

The **internal-external grinding selector knob,** Fig. 105-4, should be set for the type of grinding operation being performed. By setting the knob at the internal position for internal grinding, the handwheel may be turned counterclockwise for advancing the grinding wheel into the work, the same as for external grinding. The grinding wheel then is positioned at the rear of the hole so that it may be easily observed by the operator, Fig. 108-1. The handwheel graduations and positive stop then are used in the same manner as they are for external grinding.

Setting Cross-Feed Positive Stop

Universal grinding machines generally are provided wih some type of cross-feed **positive-stop** mechanism. This mechanism may be set to cause further cross feeding to cease when the work is ground to the desired diameter. The particular method for setting the positive-stop mechanism may vary with different machines, but the principles involved are similar.

The procedure for setting the positive-stop mechanism, Fig. 105-6, for a Brown and Sharpe universal grinding machine will be described. The pin, located behind the index dial on the cross-feed handwheel, comes in contact with the right-hand side of the switch operating slide when the grinding wheel advances to the desired work diameter. This provides a positive stop for the cross feed when the handwheel is at zero. The stopping point for the positive stop is set for the desired work diameter in the following manner:

1. With the workpiece mounted and with all other setup adjustments made, use the cross-feed handwheel to bring the grinding wheel forward until it just touches the work.
2. Pull out the disengagement knob, thus freeing the handwheel. Turn the handwheel until the fixed stop pin comes in contact with the right side of the switch operating slide.
3. Again engage the engagement knob. Engage the cross feed, and allow the work to be ground to the size permitted by the stop. When the

grinding wheel has sparked out at the foot-stock end, disengage the cross feed, and turn the handwheel back to its starting point.

4. Carefully measure the work diameter, and set the index dial for the amount of stock still to be removed. When the index dial is turned clockwise, the diameter of the work is reduced. To avoid grinding the work undersize, set the stop to grind 0.025 mm to 0.05 mm (0.001" to 0.002") oversize for this stage of the initial setup.

5. Engage the cross feed, and again grind the work until cross feed is stopped by the positive stop. When the wheel has sparked out at the footstock end, disengage the cross feed.

6. Measure the diameter of the work, and make a final setting of the index dial for the amount of stock to be removed. As a final check, run through the grinding cycle again and measure the work. If further adjustments are necessary, make them as described under this step.

Accessories

Several accessory items are available for use on universal grinding machines. These include universal back rests for supporting long workpieces, Fig. 107-1; angle wheel-truing attachment, Fig. 110-3; and radius wheel-truing attachment, Fig. 110-2. Chucks and collets are used for internal grinding operations as shown in Fig. 108-1. They also are used for face-grinding operations, Fig. 107-4.

Unit 106

How to Operate
Universal Grinders — Cylindrical Work

The procedure in this unit describes how to set up a standard universal grinding machine to grind the exterior of straight cylindrical work. In general, the procedure is applicable to standard makes of universal and plain grinding machines.

CAUTION
To produce accurate work, care and precision must be exercised in setting up a grinding machine. For the same reason, the machine should be properly lubricated and warmed up before starting to grind.

When starting a universal grinding machine for the first time after it has been idle for several hours, press the start button; then almost immediately, push the stop button. Repeat this three or four times so that the bearings of the spindle will be well lubricated before running the spindle at operating speed.

General Procedure

1. Select a wheel suited to the work to be performed, and mount it on the wheel spindle. (See Unit 112.) Be sure the wheel is sound.

   If the wheel is changed, it will be necessary to true, and possibly balance, the new wheel before attempting to grind a workpiece.

2. Determine a desirable work surface speed. (See Unit 113.) Set the headstock speed-selector dial accordingly.

3. Determine the correct wheel rpm. If necessary, change the sheave on the spindle motor shaft.

4. Check all driving belts for proper tension. Adjust if necessary.

5. Select and set the rate of table travel, if any.

6. Check to make sure that the swivel table is set at zero. If adjustment is necessary, loosen the clamping bolts at each end of the table. Then make the adjustment by means of the adjusting nut at the front of the machine. See Fig. 106-1. Verify the setting by a trial cut at each end of the workpiece.

7. Thoroughly clean the table and ways. Then position the headstock and footstock to receive the workpiece. To avoid excessive wear at either end of the ways, the headstock and footstock should be an equal distance from the respective ends of the table. Align the headstock by means of the flange on the base, Fig. 105-3.

8. Examine the centers of the grinding machine to determine that they are clean, smooth, and ground at the correct angle. Test with a center gage. If the centers are not ground with an included angle of 60°, the angle should be corrected by regrinding.

9. Before starting to grind, attach table water guards, and position the coolant piping and nozzle.

Procedure for Mounting Stock

1. Secure the workpiece. Make certain that the center holes are clean, of the correct shape and depth, and well lubricated.

2. Secure a suitable driving dog, and place it on one end of the workpiece.

3. With the cross-feed handwheel, move the wheel slide forward. The universal turret should be so positioned that the wheel will come far enough forward to grind the work, yet will give sufficient clearance for inserting the workpiece between centers. If necessary to adjust the universal turret, loosen the clamping

(Brown and Sharpe Manufacturing Company)

(Brown and Sharpe Manufacturing Company)

Fig. 106-1. Table swiveled for grinding a taper.

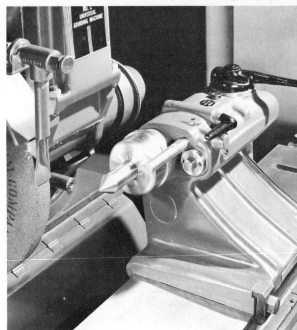

Fig. 106-2. Footstock of the universal grinder.

bolt; then make the adjustment and re-clamp, Fig. 105-2.

4. For straight cylindrical grinding, the wheel slide should be at right angles to the work-table. The grinding wheel spindle should be parallel with the worktable.

5. Place the drive end of the workpiece on the headstock center; then slide the footstock forward until it supports the other end of the workpiece. Draw the operating lever (Fig. 106-2) forward about a third of the way to withdraw the footstock center from the work-piece. Hold the operating lever in the above position by tightening the spindle clamp (Fig. 106-2). Now move the footstock forward until the center is seated in the center hole of the workpiece. Align the lip on the footstock with the edge of the swivel table, and fasten the footstock in position by tightening the clamp-ing bolts. Release the spindle clamp to permit the spring at the rear of the spindle to push the center firmly against the workpiece.

CAUTION:

Be sure the pressure of the spring is great enough to hold the center firmly against the work. Avoid excessive pressure.

6. With the cross-feed handwheel, Fig. 105-4, ad-vance the grinding wheel until it is within 6.35 mm (1/4″) of the workpiece.

Procedure for Setting Table Travel

Table travel may be accomplished by hand or power feed. When power feed is employed, stop dogs (Fig. 106-3) should be used. Stop dogs may be used with hand feed at the option of the operator.

1. To set the table reverse dogs, first move the table to the left until the right-hand edge of the grinding wheel is one-third of its width past the right end of the workpiece.

2. Push the table reverse lever to the left as far as it will go. Then move the right-hand dog along the rack until it comes against the table reverse lever. Clamp the dog in position.

3. With the longitudinal handwheel, move the table to the right until the wheel is located at the position where the cut is to stop; for exam-ple, near the driving dog or against a shoul-der. Set the left stop dog.

4. Back the wheel at least 25 mm (1″) away from the work. This will allow the wheel to clear the work and the work driving dog in case the table reversing dogs require further adjust-ment after the grinder is started.

Procedure for Manual Cross Feed

1. To feed the wheel manually over the com-plete range of wheel slide travel, turn the switch operating slide throw-out lever, Fig. 105-6, to its upper position.

2. Start the machine and engage the longitu-dinal table travel at the desired rate. Set the table dwell knobs for the desired table dwell at each reversal.

   Note whether the table stop dogs need fur-ther adjustment. If necessary, stop the ma-chine and adjust. The wheel should clear any work shoulders, the driving dog, and the tail-stock spindle.

3. With the cross-feed handwheel, advance the grinding wheel to the workpiece, taking a moderate feed. (See Unit 113.) Then engage the table travel.

CAUTION

Before starting to grind, the operator should wear properly fitted safety goggles for eye protection.

(Brown and Sharpe Manufacturing Company)

Fig. 106-3.   Table reversing lever and dog.

4. When the wheel has sparked out at the footstock end of the work, stop the table travel and rotation of the work. Then, without withdrawing the wheel, accurately measure the diameter of the work. Calculate the amount of material to be removed. Set the positive-stop mechanism so that further feeding will cease when the diameter is about 0.025 mm (0.001″) oversize. Unit 105 includes the procedure for engaging the positive-stop mechanism.

5. Engage the headstock spindle rotation, and advance the grinding wheel for the desired depth of cut with the cross-feed handwheel. Then engage the table travel for completion of the cut. When the wheel sparks out at the end of the workpiece, disengage the table travel. Continue this procedure, using manual cross feed, until the work has been reduced several thousandths of an inch, but not down to the positive-stop setting. Stop the table travel and work rotation when the wheel sparks out at the footstock end of the work.

6. Accurately measure the workpiece to determine the amount of stock still to be removed. Check to see that the positive-stop mechanism is still set to cease feeding when the work is at least 0.025 mm (0.001″) oversize. Readjust the positive stop if necessary.

7. Start the machine, advance the wheel, and continue grinding until further feeding is ceased by the positive-stop mechanism. When the wheel sparks out at the footstock end of the work, free of the work, stop the table travel and work rotation.

8. Measure the work diameter to see that it is down to about 0.025 mm (0.001″) oversize. If it is significantly more than this, reset the positive-stop mechanism accordingly and grind to 0.025 mm (0.001″) oversize as in steps 6 and 7.

9. Accurately measure the work to determine the amount of material still to be removed.

10. Start the work rotating, and with the fine hand-feed dial, set the feed to remove most of the remaining material. It is better to take several very light cuts than to grind the work undersize. Engage the table travel, and complete the cut.

(Brown and Sharpe Manufacturing Company)

Fig. 106-4. Universal back rests in position.

11. When the wheel sparks out (with the wheel at the footstock end of the work, and clear of the work), stop the table travel and work rotation. Accurately measure the size. If more material must be removed, proceed as in step 10. When the work is ground to the specified dimension, back the wheel away from the work. Then remove the workpiece. Long, slender pieces should be supported by one or more back rests, Fig. 106-4.

Procedure for Automatic Cross Feed

The procedure for grinding with automatic cross feed is identical to grinding with hand cross feed, the only exception being the need to engage the automatic controls. These are explained in sufficient detail in Unit 105.

Note: When grinding subsequent workpieces to the same size, mount the work in the usual manner. Grind with the use of the automatic cross feed until further cross feeding is stopped by the positive-stop mechanism. Any error in finish size will be due to wheel wear and may be compensated for by hand feeding to finish dimension with the fine hand-feed dial.

When a grinding gage is used, Fig. 101-11, the workpiece may be ground directly to finished dimension with the use of automatic cross feed.

Unit

107

How to Set Up Universal Grinders — External Tapers

The universal grinding machine may be set up for grinding either slight or steep external tapers. Whenever practicable, grind a taper with the wheel pressure toward the headstock.

(Brown and Sharpe Manufacturing Company)

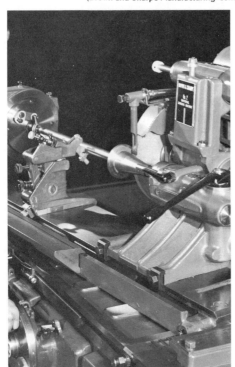

Fig. 107-1.   Grinding a taper with table swiveled. Long pieces are supported with one or more universal back rests.

Procedure for Slight Tapers

Tapers up to 8° usually are ground by swiveling the worktable, as in Fig. 107-1.
1. Unclamp the swivel table by releasing the bolts at each end, Fig. 106-1.
2. Swivel the table to the position required.
3. Fasten it by tightening the clamping bolts at each end.
4. Grinding may now proceed in the same manner as for cylindrical grinding, see Unit 106.

Procedure for Steep Tapers

Steep tapers may be ground by using several different kinds of setups. One common method

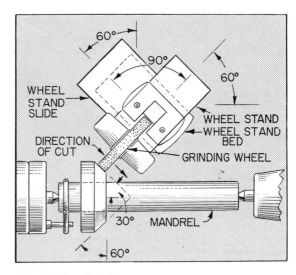

Fig. 107-2.   Grinding a steep taper.

Fig. 107-3.   Grinding a steep taper with headstock swiveled.

Fig. 107-4.   Headstock swiveled 90° for face grinding.

involves swiveling the wheel spindle head on its base to the required angle. This method works well if the surface to be ground is not wider than the wheel face. A second common method is to swivel the wheel spindle head 90° and, in addition, swivel the wheel stand slide base to the angle of the taper, Fig. 107-2. A third method in-

volves mounting the work on the headstock spindle with a chuck or other suitable device. The headstock then is swiveled to the desired angle of taper, as in Fig. 107-3. The latter method also may be employed for face grinding a workpiece; however, the headstock must then be swiveled 90°, as shown in Fig. 107-4.

The following procedure for grinding steep external tapers is for the second method, which involves swiveling the wheel head and the wheel stand slide base, Fig. 107-2.

1. Determine the angle of the taper, for example, 30°.
2. Loosen the clamping bolts, and set the wheel stand slide base at 60° (90° — 30°). This will put the wheel in position to grind a 30° angle on the work surface. (See Fig. 107-2.) Next turn the wheel head so that the spindle is parallel with the slide base. Then clamp the slide and spindle head in position by tightening the bolts.
3. Secure a suitable wheel. Change and dress it, if necessary.
4. Mount the workpiece in a chuck or other suitable holding device.
5. Adjust the headstock for revolving spindle grinding.
6. Determine the work and wheel speeds required. Make necessary adjustments.
7. Protect the eyes with properly fitted goggles.
8. Be sure the wheel guard is fastened securely in position.
9. Attach the table water guards, adjust coolant piping and nozzle, and then turn on coolant.
10. Start the machine. Then with hand table feed, bring the work into contact with the revolving wheel. Pass the wheel across the face of the work **with the cross-feed handwheel.**
    If the face of the wheel is the same width as, or wider than, the surface to be ground, no movement of the cross feed is necessary.
11. At the end of the first cut, advance the work slightly by moving the sliding table forward sufficiently for a cut of 0.025 mm (0.001″) or more, depending upon the amount of stock to be removed.
12. Continue with repeated cuts until the work is reduced to size.

Unit
108

How to Set Up
Universal Grinders — Internal Grinding

The procedures described in this unit may vary somewhat with different machines, but the principles are basically the same.

Some universal grinding machines are provided with internal grinding fixtures as standard equipment, Fig 105-2. This type of fixture is mounted on a hinged bracket which rests at the top of the wheel when not in use. It can be moved instantly into operating position, Fig. 108-1, and is locked in position with a screw knob. This type of spindle is driven with a flat belt from an independent motor without disturbing the external spindle.

When the internal grinding fixture is used, the operator sets the cross-feed selector knob to the internal grinding position. This permits the cross-feed handwheel to be turned counter-clockwise to advance the grinding wheel into the work, toward the rear of the hole where the grinding action may be easily observed by the operator. The internal setting on the selector knob also permits using power feeds, handwheel graduations, and the positive-stop mechanism in the same manner as for external grinding.

Some types of universal grinders are equipped with an internal grinding fixture which is attached to the wheel head of the machine as an independent accessory. The operator's handbook supplied with the machine explains how the fixture is mounted, as well as the procedure for changing and adjusting any necessary belts.

Procedure for a Straight Cylinder

1. Obtain the workpiece and a suitable chuck.
2. Mount the chuck on the headstock spindle.
3. Determine the desirable workpiece rpm and make the necessary adjustments. (See Unit 113 for steps 3-5.)
4. Determine the desirable wheel rpm and make the necessary adjustments.
5. Determine desirable rate of table travel, if any. Make adjustments accordingly.
6. Position the internal grinding spindle, and clamp or bolt securely. (Refer to the operator's handbook, if necessary.)
7. Mount the flat belt from the motor to the internal grinding spindle drive pulley.
8. On machines equipped with a motor selector knob, turn the knob to the **internal** position, so that the internal grinding motor will be operative.
9. On machines equipped with an internal-external grinding selector knob, turn the knob to the **internal** grinding position.
10. Disengage the power cross-feed control lever. For the beginning operator, hand cross feeding is recommended.
11. Be sure the swivel table and the headstock are set at zero degrees.
12. Mount the workpiece in the chuck. Be sure it is centered accurately; if necessary, test with a dial indicator.

13. Adjust the machine for table travel, if desired. When used, the table dogs should be so positioned that the wheel will pass only partly off the work at the beginning and end of the stroke.

CAUTION

If the workpiece has an internal shoulder or a closed end against which the wheel will come, the table stop dog at that point must be precisely located so that table travel will stop just as the wheel reaches the shoulder.

14. Protect the eyes by wearing properly fitted goggles.
15. Start the machine and dress the wheel if necessary, Unit 109. Then run the wheel into the opening by hand, and, at the same time, advance the wheel against the rear of the hole with the hand cross feed, as in Fig. 108-1, taking a light cut. At the end of the cut, withdraw the wheel, stop the machine, and measure the hole diameter.
16. Make repeated cuts in a similar manner until the hole is the correct diameter. When power table travel is being used and grinding has been completed, the wheel may be brought clear of the work by releasing the right-hand stop dog.
17. When changing for subsequent workpieces, loosen only two adjacent chuck jaws; then remove the finished piece and insert a new piece.

CAUTION

When grinding long pieces, support the outer end with a center rest or other suitable device.

18. When through, clean and restore the machine to its normal condition.

Procedure for Slight Tapers

1. Determine the angle of taper required.
2. Make certain that the wheel stand is set at zero.
3. Make certain that the headstock is set at zero degrees.
4. Swivel the table through the required number of degrees, for example, 5°. Clamp it in position.
5. Proceed as for internal grinding a straight cylinder, steps 1-10 and 12-18 above.

(Brown and Sharpe Manufacturing Company)

Fig. 108-1. Internal grinding fixture in position for grinding an internal surface.

(Brown and Sharpe Manufacturing Company)

Fig. 108-2. Grinding a steep internal taper with wheel slide base swiveled.

Procedure for Steep Tapers

Steep internal tapers generally are ground by swiveling the headstock on its base to the required angle.

1. Determine the angle of taper required.
2. Make certain the wheel stand is set at zero.
3. Make certain that the swivel table is set at zero.
4. Loosen the bolts which fasten the headstock to the base. Swivel the headstock to the desired angle, for example, 20°. Tighten the bolts.
5. Proceed as for internal grinding a straight cylinder, steps 1-10 and 12-18 above.

A second method for grinding steep internal tapers can be used. In this, the wheel slide base, Fig. 105-2, is swiveled to the desired angle of taper as in Fig. 108-2. The wheel spindle head then is swiveled so that the internal grinding spindle is parallel with the sliding table, Fig. 108-2. Finally, the grinding wheel is fed along the tapered surface with the cross-feed handwheel, or by power cross feed, as desired.

Unit
109

How to True or Dress Wheels on Cylindrical Grinders

Procedure for Truing or Dressing a Wheel with a Footstock Fixture

1. Obtain and mount a footstock type of wheel-truing fixture, as in Fig. 109-1.

   If preferred, a table fixture, Fig. 109-2, may be clamped to the table of a plain grinding machine.

2. Position the grinding wheel so that it will clear the footstock as the table is traversed, carrying the diamond tool across the face of the wheel.

3. Obtain a diamond tool, and clamp it in the fixture as shown in Fig. 109-1. The point of the diamond should touch the center of the face of the wheel, which usually is the high part. The diamond tool should contact the wheel very slightly below the axis of the wheel. This will prevent gouging and possible chattering. With the tool in position, tighten the footstock clamp.

   Note: Because a diamond wears through use, it is important that the tool always be positioned in a manner that will present a sharp edge to the face of the wheel. When the diamond has become cone-shaped, turn it in the holder so as to present it at a new angle to the wheel.

4. Position and adjust reversing dogs so that power table travel can be used. Be sure to provide enough travel so that the truing tool will pass a short distance beyond the edge of the wheel on each stroke.

5. Adjust the table travel for a moderately rapid rate for the roughing cut.

Fig. 109-1.   Wheel-truing fixture mounted on footstock.

Fig. 109-2.   Table wheel-truing fixture.

6. Set the grind-true switch at true. This will assure a constant flow of coolant.

7. Start the machine, and position the sliding table so that the diamond tool is in contact with one edge of the wheel. Round the corner of the wheel slightly by manipulating the cross and longitudinal hand feeds.

CAUTION

Be sure the wheel guard is in position and fastened securely.

Slight rounding of the corners prevents chipping the edges of the wheel when the tool is fed straight across its face.

8. Round the other corner of the wheel in like manner.

    If preferred, the edges of the wheel may be slightly rounded with a hand-held diamond tool or a radius wheel-truing attachment such as shown in Fig. 109-3.

9. Start the table travel by pulling the start-stop lever.

10. Advance the wheel to take a cut of 0.0125 to 0.025 mm (.0005″ to .001″).

11. Continue passing the tool across the face of the wheel until the wheel is running true. Then reduce the speed of table travel, and take finishing cuts of 0.0125 mm (0.0005″), followed by one or two of 0.0025 or 0.005 mm (0.0001″ or 0.0002″).

12. When through truing (or dressing) a wheel, remove the truing attachment, and thoroughly clean the machine to remove grit and other foreign matter. Return the attachment and other tools to their proper storage place.

Procedure for Truing a Radius

Radius truing is the process of rounding the corners of the wheel for the purpose of grinding up to shoulders where a fillet is required. It is accomplished by means of a **radius wheel-truing attachment,** Fig. 109-3. This device provides a convenient means of shaping either the right-hand or left-hand corner of the grinding wheel with a radius of up to 12.7 mm (1/2″).

1. Obtain, mount, and position a radius wheel-truing attachment on the swivel table, Fig. 109-3. Clamp the attachment in position with the knob at the front of the attachment.

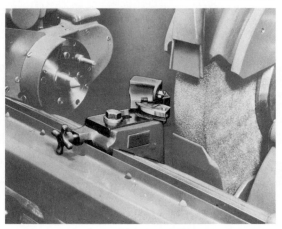

Fig. 109-3.   Radius wheel-truing attachment.

2. With the cross feed, bring the wheel to within a short distance of the radius attachment.

3. Determine the radial setting of the diamond tool by subtracting the desired radius from the figures stamped on the back face of the tool-holder. These figures show the exact radial distance from that surface to the center about which the diamond tool is rotated.

4. Using a micrometer or gage, set the tool at the determined radial distance.

5. Swing the diamond tool holder to its forward position (at right angles to the wheel face). Then loosen the clamping bolt at the front of the traverse adjustment slide, and move the attachment forward until the diamond touches the face of the wheel. Clamp the slide in position by tightening the nut on the clamping bolt.

6. Move the table sufficiently for the diamond tool to clear the wheel; then swivel the holder until the tool is at right angles to the side of the wheel. With the table handwheel, move the table forward until the diamond touches the wheel. The attachment is now correctly positioned to form the desired radius.

7. Withdraw the table slightly so that when the diamond-tool holder is swiveled, the tool will cut a small amount of material from the corner of the wheel. Swivel the holder back and

forth, removing a small amount of material from the wheel with each pass. Keep the diamond advancing toward the wheel by moving the table of the machine a small amount after each stroke.

8. When the operation has been completed, remove the attachment, and thoroughly clean both it and the machine. Return the attachment and other tools to their proper storage place.

Unit
110

How to Shape Wheels on Cylindrical Grinders

Form grinding requires the grinding wheel face to be shaped to a contour which will cut the desired workpiece profile.

This is accomplished by using wheel truing attachments equipped with diamond tools or by **crush truing.** Crush truing is done by forcing hardened steel or cast iron rollers of the desired shape into the revolving grinding wheel. This forces the grinding wheel to conform to the shape of the crush-

(Brown and Sharpe Manufacturing Company)

ing rolls. Crush truing can only be done on **very rigid grinding machines which will not deflect under the high pressure involved.** Crush truing is usually much faster than diamond truing, but only vitrified bond wheels can be successfully crush-trued. This unit is concerned only with procedures for shaping wheels using conventional wheel truing attachments with diamond tools.

(Brown and Sharpe Manufacturing Company)

Fig. 110-1.   Radius and angle wheel-truing attachment.

Fig. 110-2.   Radius wheel-truing attachment for concave radius.

Fig. 110-3.   Angle wheel-truing attachment.

General Procedure

1. True the grinding wheel, Unit 109. Be sure to protect the eyes with properly fitted goggles. Also be sure that the wheel guard is fastened securely in position.

2. Obtain a radius and angle wheel-truing attachment, such as that shown in Fig. 110-1. When appropriately adjusted, this attachment will form accurate convex or concave faces on wheels, as desired, up to a radius of about 25 mm (1").

   A second type of radius wheel-truing attachment is shown in Fig. 110-2. A second type of angle-truing attachment is shown in Fig. 110-3.

3. Position the fixture on the table of the machine; then tighten the clamping bolts.

Procedure for a Convex Face

1. Select a suitable wheel and mount it on the spindle.

2. Proceed as in steps 1 through 3 in "Procedure for Truing a Radius," Unit 109.

3. Place a diamond tool in the toolholder parallel with the slide, as in Fig. 110-1. The

diamond tool should contact the wheel slightly below the axis of the wheel. This will prevent gouging and possible chattering. Position the diamond point by means of the setting gage at the front of the toolholder. Clamp the tool in position.

4. Adjust the slide by means of the handwheel, withdrawing it to the **right of center,** by the scale on the slide, until the radius desired registers. Then tighten the clamping screw at the back of the slide.

5. With the longitudinal table feed, position the fixture so that the axis of the diamond tool is centered with respect to the center of the face of the wheel.

6. Start the machine. With the cross feed, advance the wheel toward the tool. At the same time, swivel the attachment on its base. Be careful to avoid a heavy first cut at the edges of the wheel.

7. Pass the tool across the face of the wheel two or more times by swiveling the fixture on its base. Then, with the cross feed, advance the wheel about 0.0125 mm (0.0005") and again take two or more strokes.

8. Continue with repeated strokes and wheel adjustments until the face of the wheel has the desired radius, leaving about 0.0125 mm or 0.025 mm (0.0005" or 0.001") for finishing.

CAUTION
Do not advance the wheel more than about 0.0125 mm (0.0005") at any single advancement.

9. Finish grinding with a series of light cuts, using a feed of about 0.006 mm (0.00025").

10. When the operation has been completed, remove the attachment and thoroughly clean both it and the machine. Return the attachment and tools to their proper storage place.

Procedure for a Concave Face

1. Proceed as in steps 1 through 3, "General Procedure."

2. To form a concave face on the wheel, clamp the diamond tool in the toolholder **parallel to the slide,** as in Fig. 110-1. Position the diamond point by means of the setting gage at the front of the toolholder. Clamp the tool in position.

3. Adjust the slide by means of the handwheel, advancing it to the **left of center** by the scale on the slide until the radius desired registers. Then tighten the clamping screw at the back of the slide.
4. With the longitudinal table feed, position the fixture so that the axis of the diamond tool is centered with respect to the center of the face of the wheel.
5. Start the machine. Then with the cross feed, bring the wheel into contact with the tool.
6. Proceed as in steps 7 through 10, "Procedure for a Convex Face."

Procedure for an Angular Face

1. Select a suitable wheel and mount it on the spindle.
2. Obtain an angle wheel-truing attachment, Fig. 110-3. Position and clamp the attachment on the table of the machine.
3. Swivel the attachment to the desired angle on its swivel base.
4. With the longitudinal table feed and the cross feed, position the wheel so that it just clears the diamond point.
5. With the table feed, traverse the diamond toward the wheel for a light cut, about 0.0125 mm (0.0005"). **The table remains stationary while the cut is being made.**

(Brown and Sharpe Manufacturing Company)

Fig. 110-4.   Grinding a diameter and a slight shoulder.

6. Start the machine. With the crank provided on the angle wheel-truing attachment, traverse the diamond past the wheel, Fig. 110-3.
7. Continue as in steps 5 and 6 until the wheel is formed to the proper angle and depth. Figure 110-4 shows a wheel shaped with an angular face for grinding a diameter and a slight shoulder.
8. When the operation is completed, remove the attachment and thoroughly clean the machine. Return the attachment and other tools to their proper storage place when finished.

Unit

111

How to Use a Spring Back Rest for Cylindrical Grinding

Whenever work is slender and has a tendency to spring away from the grinding wheel, it should be supported by means of a back rest or steady-rest. The rest shown in Fig. 111-1 will accommodate work up to 25.4 mm (1″) in diameter. Bronze shoes for the attachment are available for work from 3 mm (1/8″) to 25.4 mm (1″) in diameter, in increments of 1.5 mm (1/16″). The shoe remains in contact with the workpiece automatically, giving constant support and releasing its pressure when the work reaches the required size.

Procedure

1. Obtain a back rest, mount it on the sliding table, and clamp it in position with the clamping bolts.

    When grinding long, slender pieces, support should be given about every 100 mm (or several inches). For example, a piece 12.7 mm (1/2″) in diameter should be supported every 100 or 125 mm (4″ or 5″) of its length.
2. Place a workpiece between centers.
3. Back off the shoe adjusting screws until the shoe is clear of the workpiece.
4. Fit the shoe to the workpiece by hand, and hold the shoe in that position while turning the adjusting screws until they just bear against the shoe. The upper screw acts upon the shoe through the lever shown in Fig. 111-2.
5. Start the machine, advance the wheel, and take a light cut. As the wheel advances, turn the shoe adjusting screws just enough to keep the shoe in contact with the workpiece.

If more than one back rest is used, correction of the adjusting screws must be made on each rest.
6. With a micrometer, measure the diameter of the work at two or more points. If variation of size is found, correct it by slightly advancing the lower shoe adjusting screw at the large end of the work: Then grind as in step 5.
7. Again measure the work. Make adjustments as in step 6 until the workpiece is of the same diameter throughout its length.
8. Take a finishing cut of the required depth. As the wheel progresses, advance the shoe adjusting screws sufficiently to keep the shoe in contact with the workpiece.

(Brown and Sharpe Manufacturing Company)

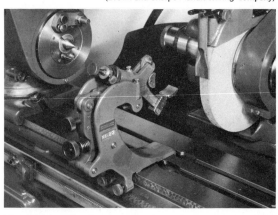

Fig. 111-1.   Back rest or work support.

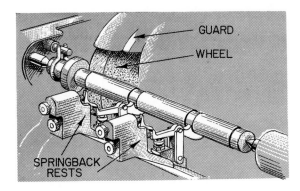

GUARD

WHEEL

SPRINGBACK
RESTS

Fig. 111-2.   Spring-type back rests in position.

9. With the workpiece reduced to size and the shoe in contact with the finished piece, each sliding nut (Fig. 111-2) will be seated against its positive stop.
10. With the shoe in contact with the work and the sliding nuts in position, tighten the clamping screws on both clamp collars.
11. Adjust the pressure on the shoe as necessary by means of the spring adjusting screws. The combined pressure of the two springs should be no greater than necessary to prevent chatter or springing of the work while it is being ground.
12. When a new workpiece is placed between centers, the back-rest springs will be compressed and the shoe (or shoes) held against the work to support it.

The type of spring back rest illustrated may be used as a solid type merely by tightening the spring adjusting screws until the springs are fully compressed.

When used as a solid back rest, the shoe must be lowered to permit placing the work between centers. This is accomplished by withdrawing the upper shoe adjusting screw. (See Fig. 111-2.) During the grinding operation, the shoe is kept in contact with the work by advancing the upper adjusting screw as the work is reduced in diameter.

Some operators spot grind the work when using either a solid or compensating type of back rest. When this procedure is used, short sections of the workpiece are ground straight in — to within about 0.05 mm (0.002″) of the finished diameter.

Unit
112

How to Change and Balance Grinding Wheels

A grinding wheel should fit easily, yet snugly, on the wheel sleeve. A loose wheel cannot be centered accurately and, consequently, will be out of balance. Such a wheel should not be used unless the core is recast. Neither should a wheel be used that has to be forced on the sleeve. Applying force may crack the wheel and thus make it unsafe for use. If the hole is only slightly too small, it may be enlarged by means of a half-round file of appropriate size. Simply remove a little metal uniformly around the hole. If the wheel is lead-bushed, sufficient metal may be removed by scraping with a pocket knife.

On some machines, the wheel is mounted permanently on a sleeve which, when change of wheel is made, is removed as a unit and a new

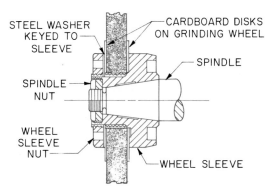

STEEL WASHER
KEYED TO
SLEEVE

CARDBOARD DISKS
ON GRINDING WHEEL

SPINDLE

SPINDLE
NUT

WHEEL
SLEEVE
NUT

WHEEL SLEEVE

Fig. 112-1.   Cross section of wheel mounted on arbor.

(Brown and Sharpe Manufacturing Company)

Fig. 112-2.   Puller for removing wheel sleeve unit
from spindle.

unit of sleeve and wheel is installed. See Fig. 112-1. This practice saves time, as the wheel need not be removed from the sleeve until a change is necessary due to wear or development of a flaw.

Procedure for Removing a Wheel and Sleeve Unit

1. Obtain an appropriate wheel sleeve puller and a pin wrench.
2. With the wrench supplied with the machine, loosen the spindle nut (Fig. 112-1) by turning it clockwise. The nut has a left-hand thread.
3. Thread the outer member of the wheel sleeve puller into the sleeve. Then tighten the cap screw, Fig. 112-2, against the end of the spindle, thus loosening the wheel sleeve from the spindle.

   On the No. 5 Brown and Sharpe plain grinding machine, a special T-handled wrench supplied with the machine has a threaded end intended for use as a wheel puller. To loosen the sleeve, tap the handle of the wrench with a hammer.

Procedure for Removing the Wheel from the Sleeve

1. With a pin wrench, loosen the wheel sleeve nut, Fig. 112-1.
2. Support the wheel on parallels or blocks of wood; then with the fingers, press the sleeve

off the wheel. If force is required, use a rawhide or wooden mallet, tapping the sleeve very gently.

Procedure for Mounting the Wheel on the Sleeve

1. Obtain an appropriate wheel.
2. Insert a finger in the spindle hole and suspend the wheel in the air. Very lightly tap the edge of the wheel with a mallet. If the wheel is sound, a clear ring will be heard. If no ring is heard, the wheel probably is cracked and should not be used.
3. Insert a rubber, leather, or blotting-paper washer between the wheel and each of the clamping flanges. Many makes of grinding wheels come with a heavy ring of blotting paper attached to each side of the wheel.
4. Screw the outer flange or sleeve nut onto the wheel sleeve, and tighten it with the wrench supplied for that purpose. Avoid exerting too much pressure.

Procedure for Mounting a Wheel and Sleeve Unit

1. Obtain an appropriate wheel and sleeve unit.
2. See that the hole in the wheel and the end of the spindle are clean. If necessary, wipe each with a clean cloth.

3. Slide the sleeve unit onto the spindle, and seat it by hand, aligning the keyway with the key in the spindle.
4. Draw the sleeve tight on the spindle by means of the spindle nut and wrench. The clamping nut should be tightened just enough to hold the wheel securely in place.

Procedure for Balancing

To avoid excessive vibration, it is essential that a grinding wheel be balanced. Most manufacturers balance their wheels before issuing them. Consequently, wheels 254 mm (10") or less in diameter rarely need further balancing. Larger wheels may need to be balanced because of wear and changes which may have developed within the wheel.

Balance may be achieved by: (1) Adding weight in the form of lead to the light side. This may be accomplished by removing small amounts of the wheel beneath the flanges and then filling the holes thus made with lead. (2) On some wheel units, balance is achieved by adjustment of segments attached to the inner sleeve flange, as in Fig. 112-3.

By Adding Lead

1. Remove some of the abrasive material beneath the flanges on the light side of the wheel. Avoid removing too much material in one spot. It is better to make two or three small cavities rather than one large one. Cavities should be placed a short distance apart so as to distribute the added weight.
2. Fill the holes thus made with melted lead.
3. Remove excess lead so that the flanges will fit properly against the sides of the wheel.
4. Mount the wheel on a tapered mandrel. Then place the mandrel on a support such as shown in Fig. 112-3. Locate the mandrel midway of the support.
5. Give the wheel a slight push, and allow it to roll back and forth until it comes to rest, which it will do with the heavy portion of the wheel at the bottom.
6. Continue adding or removing weight until the wheel is balanced. This will be evident when the wheel rolls to a gentle stop with no apparent tendency to roll backward.

(Brown and Sharpe Manufacturing Company)

Fig. 112-3.   Wheel balanced with segments.

Using Balancing Segments

1. Mount the wheel on a tapered mandrel; then place the mandrel on a support such as illustrated in Fig. 112-3. Locate the mandrel about midway of the support.
2. Give the wheel a very slight push, and allow it to roll back and forth until it comes to rest, which it will do with the heavy portion of the wheel at the bottom.
3. Move the balancing segments in the direction of the light side of the wheel; then test as before.
4. Continue adjusting the balancing segments until the wheel is balanced. This will be evident when the wheel rolls to a gentle stop with no apparent tendency to roll backward.

Unit

113

Cutting Speeds and Feeds for Cylindrical Grinding

In order to obtain good results when doing cylindrical grinding, it is necessary to use the correct wheel rpm, workpiece rpm, and table travel.

Grinding Wheel RPM

The recommended cutting speeds for most vitrified bond grinding wheels range from 1 219 to 1 981 meters per minute (mpm) or 4000 to 6500 feet per minute (fpm). For cylindrical grinding, the speeds generally range from 1 676 to 1 981 mpm (5500 to 6500 fpm).

CAUTION

Speeds above 1 981 mpm (6500 fpm) may be dangerous and should not be used.

Speeds below 1 676 mm (5500 fpm) are used when a soft wheel action is desired.

Cutting speed should not be confused with rpm. To determine the rpm necessary to produce a cutting speed which will not exceed 1 981 mpm (6500 fpm), find the circumference of the wheel and then divide the cutting speed desired by the circumference.

Example: The circumference of a 279.4 mm (11″) wheel =

Metric:
$$\frac{279.4 \times 3.141\,6}{1\,000} = \frac{877.76}{1\,000} = 0.878 \text{ meters}$$

Inch:
$$\frac{11'' \times 3.1416}{12''} = \frac{34.56}{12} = 2.88 \text{ feet}$$

Then:

$$\text{rpm} = \frac{\text{cutting speed}}{\text{wheel circumference}} =$$

Metric:
$$\frac{1\,981 \text{ mpm}}{0.878 \text{ meters}} = 2257 \text{ rpm}$$

Inch:
$$\frac{6500 \text{ fpm}}{2.88 \text{ feet}} = 2257 \text{ rpm}$$

The large pulley on a typical cylindrical grinder will produce a spindle speed of about 2077 rpm. With a 279.4 mm (11″) wheel, this will produce a cutting speed of 1 776.4 mpm (5828 fpm):

cutting speed = rpm × wheel circumference,

thus:

Metric:

cutting speed = 2077 × 0.878 m = 1 776.4 mpm

Inch:

cutting speed = 2077 × 2.88′ = 5828 fpm

This cutting speed is within the 1 676 to 1 981 mpm (5500 to 6500 fpm) limit recommended, and, assuming the appropriate type of wheel has been selected, it should produce work of good quality.

A higher cutting speed will result by using a larger diameter wheel — for example, a 304.8 mm (12″) wheel.

Metric:

$$\text{cutting speed} = 2077 \times \frac{304.8 \times 3.141\,6}{1\,000}$$

$$= 1\,989.8 \text{ mpm}$$

Inch:

$$\text{cutting speed} = 2077 \times \frac{12 \times 3.1416}{12}$$

$$= 6522 \text{ fpm}$$

This produces a cutting speed in excess of that recommended as a maximum, therefore, a lower rpm should be used, for example, 1772 rpm.

Then:

Metric:

$$\text{cutting speed} = 1772 \times \frac{304.8 \times 3.141\,6}{1\,000}$$

$$= 1\,697.6 \text{ mpm}$$

Inch:

$$\text{cutting speed} = 1772 \times \frac{12 \times 3.1416}{12}$$

$$= 5564 \text{ fpm}$$

This is only a little above the minimum cutting speed recommended, but it is a safer operating speed.

Workpiece RPM

Workpiece rpm must be calculated and set to produce the recommended work surface speed — that is, the distance a given point on the work-piece surface will revolve in one minute. Recommended work surface speed for cylindrical grinding is from 15.24 to 30.48 mpm (50 to 100 fpm) for most materials. Aluminum, brass, and other soft materials may be ground at a rate of 60.96 mpm (200 fpm). A work speed between 15.24 and 21.37 mpm (50 to 70 fpm) will produce good results. the slower the workpiece revolves, the harder will be the wheel reaction. On the other hand, too much speed will not accomplish more work and is likely to cause excessive wear and possible damage to the machine. Consequently, the work speed should not be lower than the minimum nor higher than the maximum recommended. For finishing cuts, work speed usually is increased about one third.

Example: Assume the diameter of the work-piece is 19.05 mm (3/4″) and a work speed of no more than 21.37 mpm (70 fpm) is required. Find the rpm of the workpiece by applying the following formula:

$$\text{rpm} = \frac{\text{work surface speed in mpm (or fpm)}}{\text{circumference of work in meters (or feet)}}$$

Calculate the workpiece circumference:

Metric:

$$\frac{19.05 \times 3.141\,6}{1\,000} = \frac{59.85}{1\,000} = 0.059\,85 \text{ m}$$

Inch:

$$\frac{.75 \times 3.141\,6}{12} = \frac{2.356\,2}{12} = 0.196 \text{ ft.}$$

Then figure rpm:

Metric:

$$\text{rpm} = \frac{21.37}{0.059\,85} = 357$$

Inch:

$$\text{rpm} = \frac{70}{0.196} = 357$$

For best results, set the workpiece rpm as near as possible to the calculated workpiece rpm.

Table Travel

The maximum recommended rate of table travel for rough grinding is about two-thirds of the width of the grinding wheel per revolution of the workpiece. For finish grinding, a rate of travel equal to one-third or less of the width of the wheel will produce good results. When a very smooth finish is desired, the rate of travel may be as low as one-eighth of the wheel width.

To determine the rate of travel, multiply the workpiece rpm by the distance the workpiece should travel per revolution.

Example: Given a wheel of 25.4 mm (1″) width, a workpiece rpm of 319 and a table travel per revolution of two-thirds the wheel width, the rate of table travel equals:

Table travel = rpm × 2/3 wheel width

Metric:

$$319 \times \frac{.67 \times 25.4}{1\,000} = \frac{5\,428.74}{1\,000} = 5.43 \text{ mpm}$$

Inch:

$$319 \times .67 \times 1 = 213.73 \text{ ipm}$$
$$\text{(inches per minute)}$$

Changes in longitudinal table travel are made by means of the table speed selector knob. On some grinding machines, the rate of travel may be changed while the table is traveling.

Depth of Cut

Assuming that a suitable wheel, correct wheel and work speeds, and proper table travel have been selected, then the depth of the cut is a matter of adapting the cutting load to the nature of the material and the power of the machine.

When making roughing cuts, beginners tend to take a series of light cuts instead of a heavy cut which will remove a major part of the material at a single pass. Experienced operators usually take a heavy roughing cut. Ordinarily a roughing cut should be from 0.025 to 0.100 mm (0.001" to 0.004"), and a finishing cut from 0.005 to 0.025 mm (0.0002" to 0.001"). The amount of stock left for grinding depends upon the character of

work and the nature of the material. On some materials or workpieces, as little as 0.127 mm (0.005") might suffice, while on others, as much as 1 mm (0.04") should be left.

Through experience, an operator can determine whether a cut is heavy or light by the volume of sparks that flow from the cut. A heavy volume indicates a heavy cut, and vice versa.

Unit

114

How to Sharpen Milling Cutters

Tool and cutter grinders are normally used for sharpening milling cutters, Fig. 101-4. However, plain milling cutters with straight teeth can be ground on plain or universal cylindrical grinders, as in Fig. 114-1.

Milling cutters are sharpened by using either the outer edge of a disc wheel or the face of a cup

wheel, Figs. 114-2 and 114-3. Form-relieved cutters, such as gear and radius cutters, are sharpened by grinding the face of each tooth with a saucer-shaped wheel, Fig. 114-4. Grinding wheel specifications for tool and cutter sharpening are included in Table 99-2.

(Brown and Sharpe Manufacturing Company)

(Cincinnati Milacron)

Fig. 114-1. Sharpening a plain milling cutter with straight teeth on a cylindrical grinder.

Fig. 114-2. Sharpening a plain, helical-toothed milling cutter with a disc wheel.

(Cincinnati Milacron)

(Cincinnati Milacron)

Fig. 114-3. Setup for grinding periphery of straight teeth on a slitting saw using a cup wheel. This type of setup also may be used on plain cutters and side milling cutters.

Fig. 114-4. Setup for grinding face of teeth on form-relieved cutters.

Table 114-1
Clearance Table for Sharpening Milling Cutters with Disc Wheels

Wheel Diameter mm (inches)	3° Clearance	4° Clearance	5° Clearance	6° Clearance	7° Clearance
76.2 (3)	2.01 (0.079)*	2.67 (0.105)	3.33 (0.131)	4.01 (0.158)	4.67 (0.184)
88.9 (3½)	2.34 (0.092)	3.10 (0.122)	3.89 (0.153)	4.67 (0.184)	5.46 (0.215)
101.6 (4)	2.67 (0.105)	3.56 (0.140)	4.45 (0.175)	5.33 (0.210)	6.22 (0.245)
114.3 (4½)	3.00 (0.118)	3.99 (0.157)	5.00 (0.197)	5.99 (0.236)	7.01 (0.276)
127 (5)	3.33 (0.131)	4.45 (0.175)	5.56 (0.219)	6.68 (0.263)	7.80 (0.307)
139.7 (5½)	3.66 (0.144)	4.88 (0.192)	6.12 (0.241)	7.34 (0.289)	8.59 (0.338)
152.4 (6)	3.99 (0.157)	5.33 (0.210)	6.65 (0.262)	8.00 (0.315)	9.35 (0.368)
165.1 (6½)	4.32 (0.170)	5.77 (0.227)	7.21 (0.284)	8.69 (0.342)	10.13 (0.399)
177.8 (7)	4.65 (0.183)	6.22 (0.245)	7.77 (0.306)	9.35 (0.368)	10.92 (0.430)

*mm (inches)

Table 114-2
Clearance Table for Sharpening Milling Cutters with Cup Wheels*

Cutter Diameter mm (inches)	3° Clearance	4° Clearance	5° Clearance	6° Clearance	7° Clearance
12.7 (½)	0.33 (0.013)**	0.43 (0.017)	0.56 (0.022)	0.66 (0.026)	0.79 (0.031)
19.05 (¾)	0.48 (0.109)	0.66 (0.026)	0.84 (0.033)	1.02 (0.040)	1.17 (0.046)
25.4 (1)	0.66 (0.026)	0.89 (0.035)	1.12 (0.044)	1.35 (0.053)	1.55 (0.061)
38.1 (1½)	0.99 (0.039)	1.35 (0.053)	1.68 (0.066)	2.01 (0.079)	2.34 (0.092)
50.8 (2)	1.32 (0.052)	1.78 (0.070)	2.21 (0.087)	2.67 (0.105)	3.12 (0.123)
63.5 (2½)	1.65 (0.065)	2.21 (0.087)	2.77 (0.109)	3.40 (0.134)	3.89 (0.153)
76.2 (3)	2.01 (0.079)	2.67 (0.105)	3.33 (0.131)	4.01 (0.158)	4.67 (0.184)

*For cutters larger than 76.2 mm (3″), use the figures for equivalent disc wheel sizes from Table 114-1.
**mm (inches)

Cutting Speeds and Depth of Cut for Sharpening Milling Cutters

Speeds of between 1 372 mpm (4500 fpm) and 1 981 mpm (6500 fpm) are recommended for grinding cutters of high-speed or cast nonferrous alloy. Sintered carbide cutters should be ground at 1 524 to 1 676 mpm (5000 to 5500 fpm). Dry grinding is recommended except when using diamond wheels.

A maximum depth of cut of 0.01 mm (0.0004″) should be used when grinding carbide cutters, Fig. 114-5. The cut should not exceed 0.076 mm (0.003″) per pass when grinding high-speed steel or cast alloy cutters.

Clearance Angles for Milling Cutters

Clearance angles should be as small as possible so as to provide as much metal as possible to support the cutting edge, carry away heat, and minimize the possibility of chatter. Recommended clearance angles vary from 12° or more to as little as 3°, depending on the cutter diameter and type of material to be cut. For general purpose use, cutters up to 76 mm (3″) should have 6° to 7° of clearance. Cutters over 76 mm (3″) should have 4° to 5° clearance. For best results, the primary clearance angle should be adjusted for the material being cut as follows:

Low carbon steels . . . . .5°-7°
High carbon and alloy steels . . . . .3°-5°
Cast iron, medium and hard bronze . . 4°-7°
Brass, soft bronze, aluminum,
   magnesium, plastics . . . . 10°-12°

When resharpening milling cutters, a secondary clearance angle of 3°-5° is normally provided so as to maintain the original land width. This prevents the heel of the land from interfering with the surface being milled. Original land widths are usually 0.4 mm (1/64″), 0.8 mm (1/32″), and 1.6 mm (1/16″) for small, medium, and large cutters respectively. End and side teeth are ground with less clearance, usually 2°.

(Cincinnati Milacron)

Fig. 114-5. Grinding periphery of tungsten-carbide face mill with diamond abrasive wheel.

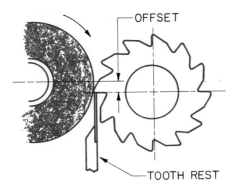

Fig. 114-6. Arrangement of tooth rest, cutter, and grinding wheel for disc wheel cutter sharpening.

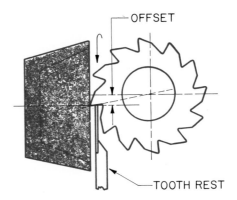

Fig. 114-7. Arrangement of tooth rest, cutter, and grinding wheel for cup wheel cutter sharpening.

Grinding Setup, Using a Disc Wheel

Figure 114-6 shows the arrangement of cutter, tooth rest, and grinding wheel for sharpening cutters with a disc wheel. Note that the tooth rest is set **to the same height as the cutter centerline.** The amount of offset between the cutter and grinding wheel centerlines determines the clearance angle ground on the cutter, see Table 114-1.

Grinding Setup, Using a Cup Wheel

Figure 114-7 shows the arrangement of cutter, tooth rest, and grinding wheel for sharpening milling cutters with a cup wheel. Note that the tooth rest is set **below the cutter centerline** the necessary distance for obtaining the desired clearance angle, see Table 114-2.

Procedure for Sharpening a Plain Milling Cutter on a Tool and Cutter Grinder

1. Check the table to see that it is set for zero degrees.
2. Protect the eyes with properly fitted goggles.
3. True the face of the wheel.
4. Mount the cutter on a suitable mandrel and place it between centers on the machine.
5. Fasten the tooth rest bracket to the machine table as in Fig. 114-3 for grinding a straight-fluted cutter. For a helical-fluted cutter, fasten the tooth rest bracket to the wheel head as in Fig. 114-8.

(Cincinnati Milacron)

Fig. 114-8. Setup for grinding an end mill with helical teeth.

6. Look up the amount of offset required to obtain the desired clearance angle. If a cup wheel is being used, lower the tooth rest below the centerline of the cutter the required amount. If a disc wheel is being used, set the tooth rest on the cutter centerline, then raise the centerline of the grinding wheel by elevating the wheel head the required amount.

7. Hold the first tooth to be sharpened down on the tooth rest. Then bring the wheel into contact with the cutter for a light cut. With the longitudinal hand wheel, move the cutter back and forth until the wheel sparks out.

8. Move the cutter past the wheel far enough to clear the wheel while the cutter is rotated over the flexible tooth rest to the next tooth to be ground.

9. Continue as in steps 7 and 8 until all teeth have been ground.

10. If the teeth require more grinding to become sharp, advance the wheel for a second cut, and proceed as in steps 7 and 8 until all teeth have been sharpened.

11. If necessary, grind the secondary clearance angle to reduce the width of the land.

12. When finished, return all tools and equipment to their proper storage place, and clean the machine thoroughly of all grinding grit.

Test Your Knowledge of Section 13

Unit 101: Grinding and Grinding Machines

1. To what dimensional tolerance are grinding operations frequently performed?

2. In addition to the close tolerances possible, list a second distinct advantage of grinding as compared with other methods of machining.

3. List two types of grinding machines which perform cylindrical grinding operations.

4. List three kinds of machines or methods which may be used to perform internal grinding operations.

5. Define the meaning of form grinding, and list three kinds of machines on which form grinding may be done.

6. Give an example of an object which is ground by the plunge grinding method.

7. Explain how centerless grinding is performed, and list several kinds of items which are ground by this method.

8. List several kinds of tools which are commonly ground on a tool and cutter grinder.

9. List two general types of surface grinding machines, and indicate the principal differences between them. Which type is commonly used in tool and die shops, toolrooms, maintenance shops, and school shops?

10. What kinds of operations are generally performed on a plain grinding machine?

11. Explain how an **indicating grinding gage** is used on cylindrical grinding machines.

12. List several possible causes for a chatter finish while grinding.

13. List several possible causes for burning the work while grinding.

14. List several possible causes for the wheel spindle running too hot on a grinding machine.

Unit 102: How to Operate Surface Grinding Machines

1. How is the size of a horizontal spindle surface grinder designated?

2. List six methods of holding workpieces for surface grinding.

3. Why must a newly installed wheel always be trued?

4. How much material may be removed in roughing cuts? Finishing cuts?

Unit 103: How to True or Dress Wheels on Surface Grinders

1. How does truing a wheel differ from dressing a wheel?

2. At what point on the grinding wheel should the diamond tool make contact?

3. How much should the grinding head be lowered for each pass of the diamond tool?

Unit 105: The Universal Grinding Machine

1. For what operations would it be necessary to swivel the wheel spindle head on a universal grinding machine?

2. For what purpose would it be necessary to swivel the headstock on a universal grinding machine?

3. What range of table travel speed generally is available on a universal grinding machine?

4. What devices are provided on a universal grinder for automatic control of the length of table travel?

5. For what purpose is a cross-feed positive-stop mechanism used on a universal grinding machine?

Unit 110: How to Shape Wheels on Cylindrical Grinders

1. Name the two principal methods of shaping grinding wheels and describe briefly how each is done.
2. What are the advantages and disadvantages of crush truing?

Unit 112: How to Change and Balance Grinding Wheels

1. Describe how a wheel and sleeve unit is removed from the grinding machine spindle.
2. Describe the two methods commonly used for balancing large grinding wheels.

Unit 113: Cutting Speeds and Feeds for Cylindrical Grinding

1. What is the range of cutting speeds recommended for grinding with most vitrified grinding wheels?
2. What is the recommended work-speed range for cylindrical grinding of most materials?
3. How do work speeds for finishing cuts compare with those recommended for roughing cuts on a cylindrical grinder?
4. Calculate the work rpm necessary for cylindrical grinding of a workpiece 31.75 mm (1-1/4") in diameter at a work speed of 18.29 mpm (60 fpm).
5. What is the maximum rate of table travel, per revolution of the work, for rough cylindrical grinding? What is a recommended rate of table travel, per revolution of the work, for finish cylindrical grinding?
6. What is the range in depth of cut recommended for rough grinding cuts? What is the range in depth of cut recommended for finish grinding cuts?

Unit 114: How to Sharpen Milling Cutters

1. What two grinding wheel shapes are used for sharpening plain milling cutters?
2. Describe the method of grinding and the shape of the wheel used for sharpening form-relieved cutters.
3. What cutting speeds are recommended for sharpening high-speed steel and cast alloy cutters? Carbide cutters?
4. Should milling cutters be sharpened wet or dry? Explain.
5. What is the recommended maximum depth of cut per pass for grinding carbide cutters? High-speed steel? Cast alloy?
6. What clearance angle is recommended for general purpose cutters under 76 mm (3") diameter? Over 76 mm (3") diameter?
7. How is the original land width on milling cutters maintained? Why is this desirable?
8. Find the amount of offset for: (a) 5° clearance on a milling cutter when using a disc wheel of 127 mm (5") diameter; (b) 5° clearance on a 12.7 mm (1/2") diameter end mill when using a cup wheel.

Section

14

Cutting Action and Cutting Fluids

Unit

115

The Action of a Cutting Tool on Metal

Cutting tools produce metal chips of various kinds and shapes. Some chips are broken into small pieces, some are continuous coils, and some are short, bent parts of a coil. The type of chip formation is a factor in determining how rapidly metal may be removed from the workpiece, and it influences the texture of the machined surface.

The type of chip formed is determined largely by the type of material being cut and the shape of the cutting tool. Since the type of material in a workpiece remains constant, the form of the chip may be changed by changing the shape of the cutting tool. The machinist can determine whether a cutting tool has the proper shape for cutting a given material by knowing the properties of the material and by observing the type of chip which is formed.

How the Tool Cuts

All metals in the solid state have a characteristic **crystalline** structure, frequently referred to as the **grain** structure. The crystals or grains may vary in size from very fine to very coarse, depending on the type of metal and whether or not it has been **heat-treated.** Each crystal is composed of groups of **atoms** or **molecules** clustered together. The crystals of a pure metal such as pure copper are composed of large clusters of atoms. The crystals of an alloy such as steel, which may be composed of several metallic or nonmetallic elements, have large clusters of either atoms or molecules.

Figure 115-1 is a diagram of a cutting tool, such as a shaper tool bit, forming a chip on a metal workpiece. As the tool advances against the workpiece, great forces are exerted on the

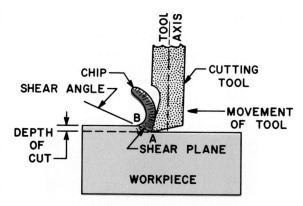

Fig. 115-1.   Action of cutting tool forming a continuous chip.

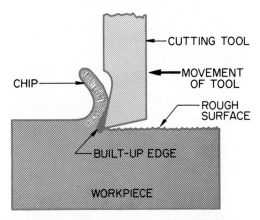

Fig. 115-2.   Continuous chip with builtup edge.

crystalline metallic grains in front of the tool face. These grains, in turn, exert similar pressures on the grains ahead of them, in the direction of the cut. As the tool continues to advance, the material at point **A** is sheared by the cutting edge of the tool, or it may be torn loose by the action of the bending chip which is being formed. As the tool advances, maximum stress is exerted along line **AB,** which is called the **shear plane.** This plane is approximately perpendicular to the cutting face of the tool. When the force of the tool exceeds the strength of the material at the shear plane, rupture or slippage of the crystalline grain structure occurs, thus forming the metal chip (or chips). The cutting edge of the tool tends to scrape or smooth the machined surface. This cy-

cle is rapidly repeated as the tool advances along the workpiece.

Types of Chips

Cutting tools form three basic types of chips, depending on the type of material being cut. The type of chip influences the amount of tool wear and determines the quality of surface finish.

Continuous Chip

The continuous chip is formed when ductile metals such as aluminum or free-machining steel are machined, Fig. 115-1. The chip has the form of a continuous coiled ribbon.

Ductile metals usually have a larger grain size and good plasticity, thus enabling the grains to withstand considerable distortion without fracture. As the tool advances, the metal ahead of the cutting edge is compressed. This compressed metal becomes **work-hardened** as it starts to form a chip. The hardened chip resists further compression and escapes along the tool face in the form of a continuous chip. This cycle rapidly repeats itself as the tool proceeds with the cut. The tool edge tends to smooth the surface which remains relatively soft.

The continuous chip is the ideal way of machining metal. It produces a good finish, causes less friction, and requires less power to remove a given amount of metal. Factors which tend to cause continuous chips on ductile materials are large rake angles on the tool, high cutting speeds, a sharp cutting edge with a highly polished tool face, and the use of a good cutting oil to reduce friction.

CAUTION

Continuous chips can be hazardous to machine operators. They also tend to clog automatic turning machines. For these reasons, it has become common practice to employ chip breakers which effectively break up the continuous chip into small curls in the shape of figure 9's. (See Unit 49.)

Continuous Chip with Built-Up Edge

The continuous chip with a built-up edge frequently is formed on tough metals such as medium-carbon steels, tool steels, and alloy steels. This type of chip, Fig. 115-2, takes the form of a somewhat-continuous ribbon, or short pieces of ribbon which break off frequently.

The crystalline grains of tough, ductile metals usually are moderate in size and are held together with a strong bond. As the tool compresses the metal ahead of the cutting edge, the metallic grains become work-hardened, resist further compression and start to escape along the tool face in the form of a chip. However, some of the tough, highly-compressed metal thoroughly cleans the tool face and forms a weld-like bond along the cutting edge and on the tool face. The metal bonded to the tool is called a **built-up edge.** As the chips slide over the tool face, the built-up edge slides off, and the cycle repeats itself.

Some of the metal immediately ahead of the tool, and ahead of the built-up edge on the tool, tends to fracture or tear apart at the shear plane. It then slides under the cutting edge of the tool, thus causing a rough surface which often is considerably work-hardened. This type of chip causes increased friction and increased heat, and it requires much more power for removal of a given amount of metal.

Factors which tend to minimize the formation of chips with a built-up edge are: proper rake angles on the tool, a sharp cutting edge, a polished tool face, correct cutting speed and feed, and the use of a good cutting fluid.

Discontinuous Chip

When a brittle metal such as cast iron or bronze is machined, the chips are broken up in the form of flakes along the shear plane and ahead of the cutting edge of the tool, as in Figs. 115-3 and 115-4. This type of chip is called a discontinuous

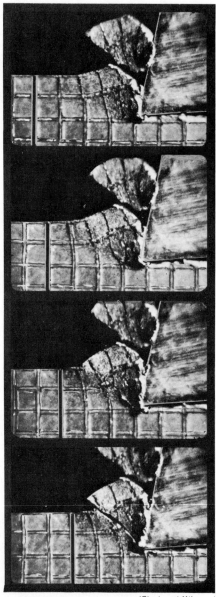

(Cincinnati Milacron)

Fig. 115-4.   Formation of typical discontinuous chip magnified about five times.

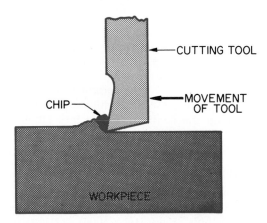

Fig. 115-3.   Discontinuous chip.

chip or segmental chip. Brittle metals have small, irregularly-shaped grains which are held together with a brittle glass-like bond. Since the grains are so hard, pressure from the tool transfers easily from one grain to another ahead of the tool. This causes the metal to fracture into short chips, which escape freely along the tool face and clear the cutting zone easily. This type chip, however, causes considerable tool wear and rounding of the cutting edge.

Factors which contribute to the formation of discontinuous type chips are: a small rake angle

on the cutting tool, low cutting speeds, thick chips, and vibration of the cutting tool.

The principles involved in the action of the cutting tool in forming the three types of chips described applies to the cutting action of many basic metal-cutting tools. They include (1) single-point tools such as lathe, shaper, or planer tool bits; and (2) tools with multiple cutting edges, such as drills, milling cutters, reamers, taps, threading dies, broaches, and other specialized cutting tools.

Unit 116

Cutting Fluids

In addition to the proper feed, cutting speed, and shape of the cutting tool, careful attention should be given to the selection and use of the proper cutting fluid. Cutting fluids prolong tool life, increase the rate of metal removal, aid in producing a finer finish, and enable machining to closer tolerances.

The term **cutting fluid** includes **straight cutting oils, water soluble cutting oils,** and **chemical** or **synthetic cutting fluids** which are not oils. Each of the basic cutting fluids is developed for certain machining applications. Therefore, in order to select the proper cutting fluid for a particular machining application, the machine operator must understand the purposes of cutting fluids, the properties and classification of cutting fluids, and the factors which should be considered in selecting cutting fluids.

Action of Cutting Fluids

The principal functions of cutting fluids include **cooling** the cutting tool and the work,

lubricating the face of the tool and the chip, and preventing the chip from welding to the cutting edge. In addition to these functions, cutting fluids should also provide a flushing action for chip removal, leave no stain or discoloration on the work, leave minimum sediment deposit on the machine, prevent or inhibit rusting of the machine and the work, resist smoking and fogging over the work, resist bacterial growth, and resist the development of an unpleasant odor as the fluid ages or becomes contaminated.

Works As a Coolant

Approximately 90% of the energy used in metal cutting is converted into heat. Heat results from the compression and friction generated as the metal chip is formed and escapes over the face of the cutting tool, Fig. 116-1. When the cutting speed and depth of cut are increased, the amount of heat developed also increases. When heavy cuts are made at high speeds, a cutting

(Texaco, Inc.)

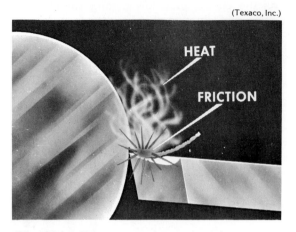

Fig. 116-1.   Heat generated at cutting edge.

(Texaco, Inc.)

Fig. 116-3.   Chip welded to tool face.

(Texaco, Inc.)

Fig. 116-2.   Built-up edge on tool face.

(Texaco, Inc.)

Fig. 116-4.   Complete breakdown of tool.

and the cycle repeats itself again and again. The work-hardened chip, while forming the built-up edge of the tool, leaves a rough surface on the workpiece and rapidly wears the flank of the tool. A heavy duty cutting fluid with **extreme pressure additives** retards this development.

Reduces Tool Wear

High cutting speeds with increased friction and heat result in rapid wear at the cutting edge of the tool. In some cases, complete tool breakdown results, Fig. 116-4. The work-hardened chip moves over the face of the tool too rapidly to weld to the tool face. High heat causes the tool to soften and wear.

The following tool materials begin to soften in the indicated heat ranges: carbon tool steel at 204.4 to 260° C (400° to 500° F); high-speed steel at 537.8 to 593.3° C (1000° to 1100° F);

fluid with good coolant properties is essential for cooling the workpiece and cutting tool.

Works As a Lubricant

A cutting fluid with good lubricating properties is an aid in the prevention of a built-up edge on the cutting tool. Such an edge frequently is developed when tough, ductile materials are machined at a cutting speed that is too low. High pressures and resultant friction at the edge of the tool cause the freshly cut metal to adhere or pressure-weld to the tool face, Figs. 116-2 and 116-3. As the metal is piled up, it slides off along the tool face, cast alloys at 648.9 to 815.6° C (1200° to 1500° F); and cemented carbides at 815.6 to 926.7° C (1500° to 1700° F). Therefore, the principal purposes of a cutting fluid at high speeds are to cool the tool and the work, to lubricate the tool face and the chip, and to resist the formation of a built-up edge on the tool.

Function in Grinding

Cutting fluids also are used for many grinding operations. A fluid with good cooling properties is necessary to prevent distortion of the work due to heat developed with heavy grinding. On light grinding, proper lubrication prevents wheel clogging and improves the smoothness of the finish. Cutting fluids also increase the life of the grinding wheel by reducing the frequency of wheel dressing.

Classification of Cutting Fluids

Numerous cutting fluids are available under different trade names. However, most of the commercial cutting fluids can be classified within three groups — the straight cutting oils, the emulsifiable oils, and the chemical or synthetic cutting fluids.

Straight Cutting Oils

Lard oil, derived from animal fats, is a good cutting oil for many machining applications at lower cutting speeds. However, it is relatively expensive when used undiluted in large quantities. Mineral oil may be added, and, when so diluted, it commonly is classified as mineral-lard oil — one of the basic straight mineral cutting oils.

There are several basic mineral cutting oils. Each type acquires its special characteristics or properties through additives, including fatty oils, fatty acids, sulfur, chlorine, phosphorous, and certain other chemicals. The basic ingredient, petroleum mineral oil, provides for the cooling property of the oil. It also provides for some lubrication of the cutting tool and the chip.

Lubrication properties are improved when fatty oils or fatty acids are added to mineral oil. The fatty oils include such oils as lard oil, sperm oil, or fish oil. These improve the wet-ability or **oiliness** of the cutting fluid, by enabling the oil to disperse evenly and cling to the cutting tool.

Under the heat and pressure developed while cutting, the fatty oils combine to form a metallic soap between the cutting tool and the chip, thus reducing friction, Fig. 116-5. Reduced friction results in improved tool wear, reduction in power, and improved surface finish.

Antiweld properties of mineral cutting oils are provided through the addition of sulfur, chlorine, or both sulfur and chlorine. When a clean metal chip, under heat of friction, is coated with a cutting fluid containing sulfur (or chlorine), a chemical reaction takes place between the surface of the tool and the chip, forming a sulfide (or chloride) film. This film has a lower shear strength than the metal being cut, thus reducing friction at the cutting edge of the tool. The film also aids in preventing the formation of a built-up edge on the tool. The chemical additives provide for antiweld properties at a higher temperature than can be provided for with the addition of fatty oils alone. The chemical reaction of the chlorine takes place at temperatures lower than that required for the sulfur reaction. Therefore, sulfur additives in cutting oils usually are recommended for heavy-duty machining operations where high heats are developed. Through various combinations of additives, a wide variety of cutting oils are produced for use in machining different metals under different conditions.

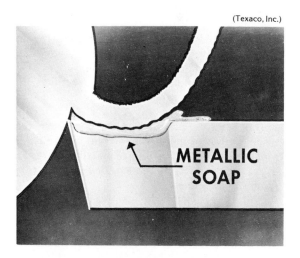

(Texaco, Inc.)

Fig. 116-5.   Fatty oil additives produce metallic soap film.

Fig. 116-6.   Transparent cutting oil.

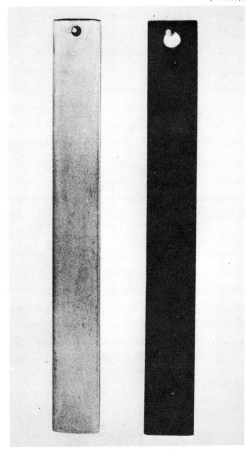

Fig. 116-7.   Copper strips from corrosion test at 100°C (212°F). Oil represented by strip on left contains no active sulfur; that on right contains active sulfur.

When heavily compounded cutting oils were first developed, they were usually very dark. These dark oils still are available. Through modern oil chemistry, **transparent** compounded cutting oils also are available and widely used today. The cutting tool and the work can be observed clearly through the transparent oils (Fig. 116-6). Shop cleanliness and working conditions also are improved through their use.

Active and Inactive Cutting Oils. The two basic types of mineral cutting oils are active and inactive. The **active** oils usually are recommended for use with ferrous metals on heavy-duty operations requiring extreme pressures. They cause discoloration of many copper alloys, including bronze bearings, due to the chemical reaction of additives, mainly sulfur. See Fig. 116-7.

The **inactive** cutting oils do not cause any discoloration of metal, and so can be used with machines which have bronze bearings or bushings. They are recommended for machining the nonferrous metals.

Both the active and the inactive types are available as transparent oils. Certain inactive, transparent cutting oils have been developed which are "tri-purpose" oils. That is, they can be used as a cutting fluid, as a lubricant for the machine, and in the hydraulic system of the machine. Hence, there is no problem of one oil contaminating the other where leaks or other forms of mixing occur.

Emulsifiable Oils

The emulsifiable oils also are known widely as **water soluble oils.** However, they are not really soluble, since oil does not dissolve in water. Because of an emulsifying base, usually in the form of a soap, the oil is dispersed in fine droplets throughout the water. These oils are mixed in various proportions according to the machinability of the metal and the severity of the operation. The mixtures vary from 10:1 (10 parts water to 1 part oil) for severe operations on metals of low machinability, to 50:1 or more for grinding operations. When mixed, the fluids usually are cloudy or milky white, Fig. 116-8.

The emulsified oil solutions have better coolant properties than the cutting oils. Actually, water is one of the best coolants known. However, water has little lubrication value, and it reacts readily with the work surfaces and machine surfaces causing rust. Therefore, rust inhibitors are compounded with the oil for prevention of such corrosion.

Other additives are compounded with emulsifiable oils to develop desired properties. Lubrication properties are developed through the addition of fatty oils or fatty acids. Antiweld properties are provided for through the addition of sulfur, chlorine, or both, depending on the characteristics desired.

Emulsified oil solutions have limited working life. They should be replaced periodically, and the machine should be thoroughly cleaned before the new solution is used. Emulsifiable oils are sterile as manufactured, and they frequently have germicidal substances compounded in the oil to combat bacterial growth. However, with age and heavy use, they become rancid and develop an unpleasant odor. The solution should then be replaced.

(Texaco, Inc.)

Fig. 116-8.   Use of soluble oil solutions. Note the cloudy white color.

Mixing Emulsifiable Oil Solutions. When emulsified oil solutions are mixed, **the oil always should be added to the water** (rather than water to oil) in order to form a proper solution. Soft water should be used when possible. If unavailable, an emulsifiable oil specially prepared for use in hard water should be used.

The proportion of oil mixed with water in preparing an emulsified oil solution varies according to the machinability of the metal and the severity of the operation. Various machining operations are listed according to severity in Table A-11, Appendix.

When oil manufacturer's recommendations are available, they should be followed. In their absence, however, the following general recommendations may be used as a guide:

Metal Machinability	Type of Operation	Dilution Water to Oil
1. Low machinability	medium severity	10:1
2. Low machinability	low severity	20:1
3. Medium machinability	medium severity	15:1
4. Medium machinability	low severity	30:1
5. High machinability	medium severity	30:1
6. High machinability	low severity	30:1
7. Average cylindrical and surface grinding operations		30:1
8. Very-fine-finish grinding operations (thread grinding)		40:1 or 50:1

The emulsifiable oil solutions are the most widely used of all the cutting fluids. They have wide application and may be used on most metals for all except the most severe operations; see Table A-11, Appendix. These oils usually are the most economical cutting fluids for use where they are recommended.

Avoiding Dermatitis

Dermatitis is an inflammation of the skin. Some people acquire dermatitis when in contact with emulsifiable cutting oils, particularly when the skin is dirty or contaminated. Therefore, hands should be kept clean by thorough washing with soap and warm water before and after working with cutting oils. Clothing also should be kept clean and free of cutting oil.

Chemical Cutting Fluids

A third group of cutting fluids are the chemical (or synthetic) cutting fluids. These fluids are diluted solutions of water and water-soluble chemical compounds. There are two principal types: (1) true solutions, and (2) surface-active types. The true solutions consist of inorganic and/or organic materials dissolved in water, mainly to inhibit rust. They have little lubricating value and thus serve mainly as coolants. These fluids are transparent and are usually dyed pleasing colors for identification purposes.

The surface-active type is a water solution which contains additives for lowering the surface tension of the water, and also for imparting lubricating qualities. To provide improved lubricating and extreme pressure qualities, sulfur, chlorine, or phosphorous compounds are added.

Chemical cutting fluids are replacing straight and emulsifiable cutting oils for many applications. For best results, the chemical concentrates must be mixed in correct proportions with deionized water. In all cases, the manufacturer's recommendations should be closely followed. When properly mixed and maintained, chemical cutting fluids can often provide longer life at less cost than oil base cutting fluids.

Other Cutting Fluids and Coolants

Several other types of cutting fluids may be used in machining metals:

Kerosene may be used on aluminum, aluminum alloys, and brass for machining operations of low severity. It has satisfactory cooling properties, and some lubricant properties for chip removal. It also may be mixed with lard oil as a cutting fluid for more severe applications on these materials. Nevertheless, the other cutting fluids listed in Table A-11, Appendix, usually are considered superior to kerosene and also present less of a fire hazard.

Compressed air sometimes is used as a coolant and for the purpose of removing chips when machining cast iron. Cast iron contains graphite, which tends to serve as a lubricant at the edge and face of the cutting tool. For this reason, cast iron may be machined dry, but cutting rates and tool life may be improved with the use of soluble oil or chemical cutting fluids.

Cutting wax fluids have been developed for use as cutting compounds. However, certain types also may be used as additives to be compounded with other cutting fluids, such as petroleum-based mineral cutting oils or emulsified oil solutions. The manufacturer's recommendations should be followed in mixing or using cutting waxes.

Selection of Cutting Fluids

There is no set rule which can be followed in the selection of a cutting fluid for a particular machining operation. The recommendations of suppliers for the use of cutting fluids are general, and their applications vary. It would be ideal if some physical property of the metal, such as its hardness, could be used to assign the metal a fixed **machinability rating** which would apply at all times in selecting a cutting fluid, but this is not possible. However, three principal factors can be used as a guide in selecting a cutting fluid for a particular application: (1) the machinability rating of the metal, (2) the severity of the operation being performed, and (3) the operating conditions.

Machinability

Machinability is a difficult word to define because its meaning is dependent on many factors, and not all authorities agree on its exact definition. In general, however, the term **machinability** means the ease with which a metal may be machined. A metal with a high machinability rating frequently machines with comparatively low power consumption, has a high rate of metal removal, produces a good surface finish, and causes minimum tool wear.

For the purpose of selecting proper cutting fluids, most of the commonly used metals have been classified in six groups according to their approximate machinability ratings as shown in Table A-11, Appendix. The machinability rating is expressed as a percentage, in comparison with A1S1 1112 cold-drawn steel, which has a rating of 100 percent. Metals which are more difficult to machine have a machinability rating of less than 100%. Metals which machine more easily have a rating of more than 100%.

In most machining applications, as the machinability rating of the metal increases, cutting speeds may be increased. Machinability ratings are based on exact comparisons of various metals which are machined under scientifically controlled conditions, Fig. 116-9. With metals having low machinability ratings, the use of an active cutting oil with heavy-duty lubricant and anti-weld properties generally is recommended.

Severity of the Operation

The severity of the machining operation being performed is a significant factor in the selection of a cutting fluid. Various metal-cutting operations are rated in Table A-11, Appendix, according to numbers 1 through 10. Broaching (rated number 1) is the most severe, while sawing and grinding (rated number 10) are the least severe. The heaviest-duty cutting fluids generally are recommended for the most severe machining operations.

Operating Conditions

Several operating conditions should be considered in the selection of cutting fluids. These include cutting speeds, feeds, depth of cut, tool geometry, tool material, heat treatment or work-hardness properties of the material, and rigidity of the work and the machine. Heavy cuts and slow speeds generally require excellent lubricating qualities. High speeds generally require greater coolant properties. High speeds, together with the high heat developed in machining steels of

low machinability, generally require good lubricity, cooling, and anti-weld properties.

Selection Guide

In the selection of cutting fluids, Table A-11, should be used as a general guide. Careful study of the table reveals that a specific cutting fluid often may give satisfactory results on several different metals of varying machinability ratings. It may also produce satisfactory results on several different operations.

Table A-11 indicates the basic types of cutting fluids. Each of the several major oil companies produces cutting fluids which are similar or equivalent to the basic cutting fluids recommended in the table. Each company generally has a trade name for the equivalent oil. A review of the company literature and recommendations will aid further in accurately selecting the proper cutting fluid. In the selection of a chemical cutting fluid, the manufacturer's recommendations should be reviewed carefully.

In summary, the following factors should be considered in the selection of a cutting fluid:

1. The machinability rating of the metal.
2. The severity of the operations being performed.
3. The operating conditions.
4. The properties of the cutting fluid.
5. The cost of the cutting fluid.
6. The recommendations of the manufacturer of the fluid.

(United States Steel Corp.)

Fig. 116-9.   Specially modified constant-pressure lathe is used to test the machinability of various steels.

Unit
117
Cutting Tool Materials

Metal cutting tools must possess a variety of different properties in order to cut the many different metals under varying conditions of severity. To meet these demands, tools have been produced from a variety of materials. In order to select the proper cutting tool for a given application, one must understand the basic properties required of cutting tools and the basic properties of each of the significant cutting-tool materials.

Properties of Cutting Tools

The most important properties of cutting tools are hardness at high temperatures, wear resistance, and impact strength.

Hardness at High Temperatures

As a tool cuts, high heat is developed as a result of compression and friction at the cutting edge of the tool. All metal cutting tools begin to lose hardness when heated to sufficiently high temperatures. As the tool softens due to heat, it wears and breaks down at the cutting edge or face. Various cutting materials have different degrees of initial hardness, and they begin to lose their hardness at different temperatures. Hence, the hardness of the tool and the degree to which it retains its hardness at high temperatures are important in the selection of a cutting-tool material.

Wear Resistance

A cutting tool is wear-resistant if it resists abrasion at the cutting edge and along the tool face. Wear resistance improves as cutting tool hardness increases. Increased wear resistance is also obtained by using correct tool geometry, rigid tool mountings, correct speeds and feeds, and cutting fluids. As indicated previously, wear resistance is also related to heat. When the temperature level is attained at which the tool starts to lose its hardness significantly, the metal rapidly looses its wear resistance. See Fig. 116-4.

Strength

Cutting tools must also have high strength in order to be vibration- and impact-resistant. Strength in cutting-tool materials is not always proportional to hardness. Some of the hardest tool materials lack strength because they are too brittle.

Materials Used in Cutting Tools

The various materials from which most metal cutting tools are made can be classified under the following principal headings:

1. Carbon tool steel
2. High-speed steel
3. Cast alloys
4. Cemented carbides
5. Ceramics
6. Diamonds

Carbon-Tool-Steel Cutting Tools

Many cutting tools are made from high-carbon tool steel. Some common examples include drills, reamers, center drills, forged boring bars, hand taps, and threading dies.

The chief advantage of carbon tool steel is its low cost. A principal disadvantage is its loss of hardness at relatively low temperatures. It begins to soften at 204.4° C to 260° C (400° F to 500° F). This temperature range is indicated by the heat-color range from brown to purple. When this range is exceeded, a blue or dark gray color appears, and the tool softens and wears rapidly.

Keen cutting edges can be produced on carbon steel tools, and they possess good shock resistance. Carbon-steel tools should be operated at one-third to one-half the cutting speeds recommended for high-speed steels. With lower cutting speeds, less heat is generated, and tool life is prolonged.

High-Speed Steel Cutting Tools

The principal advantage of high-speed steel (HSS) in comparison with high-carbon steel is its retention of hardness up through a dull red heat, ranging from approximately 537.8° C to 593.3° C (1000° to 1100° F). HSS tools begin to soften due to tempering in this range, but they do not soften significantly until heated above this range. The property of tool steels to retain hardness at high heat or red heat is called **hot-hardness** or **red-hardness.**

Tools which very frequently are made of high-speed steel are: drills, reamers, end mills, center drills, counter bores, milling cutters, taps, and lathe centers. High-speed steel also is widely used for tool bits used on lathes, shapers, and planers, Fig. 117-1. High-speed steel cutting tools generally will stay sharp for a much longer period of use than tools made of carbon tool steel.

The high-speed tool steels are the most heavily alloyed of all the steels. Their principal elements are tungsten and molybdenum. Other elements which also are alloyed with these steels in significant amounts are cobalt, vanadium, and chromium. Each of these elements imparts particular properties to the steel. Tungsten, molybdenum, chromium, and vanadium are carbide-forming elements; that is they combine with carbon to form carbides. The carbides cause the steel to resist softening at higher temperatures. They also impart high wear resistance to the steel. Cobalt is not a carbide former, but it combines with the iron in steel in a manner which increases the red-hardness of high-speed steels.

Fig. 117-1. A sharpened high-speed steel cutting tool.

Types of High-Speed Steels

There are several different types of high-speed tool steels. The most common types may be classified in the following manner:

1. **Tungsten base steels:** This type of steel commonly is accepted as the standard for use in comparing the properties of other types of HSS. One of the most commonly used steels of this type is the 18-4-1 steel (type T1). It is composed of 18% tungsten, 4% chromium, and 1% vanadium. Other tungsten steels with varying amounts of these elements are available.
2. **Molybdenum base steels:** One of the common steels of this type is 8% molybdenum, 4% chromium, and 2% vanadium (type M10). It has performance properties generally comparable to 18-4-1 tungsten steel.
3. **Tungsten-molybdenum base steels:** Steels of this class are similar to the tungsten steels, except that some of the tungsten is replaced with molybdenum. A common steel of this type is 5% molybdenum, 6% tungsten, 4% chromium, and 2% vanadium (type M2). It has performance properties similar to 18-4-1 tungsten steels.
4. **Cobalt steels:** Cobalt in amounts from 5% to 12% may be added to any of the above high-speed steels. The cobalt increases the red-hardness of the steel significantly. High-speed steel which includes a large percentage of cobalt is sometimes called **super HSS.** The super HSS frequently is used for tool bits or as tips on cutting tools. When the cobalt content is high, the hardness and wear resistance also are increased, but the impact toughness or resistance to shock generally is decreased.

Coated High-Speed Steel Tools

Cutting tools of high-speed steel are now available with coatings of several kinds. A thin metallic layer of **hard chromium** can double the life of cutting edges on taps. Coatings of **titanium carbide, titanium nitride,** and **aluminum oxide** provide even greater tool life. These coatings are formed by a chemical reaction between high-speed steel and the vapor of the coating material at 954-1054°C (1750-1930°F). The thickness of these coatings averages 0.0076 mm (0.0003").

Experience with these coatings shows that tool life can be increased from three to over six times, depending on the kind of tool.

Certain cutting tools are available in several types of high-speed steel, and with different coatings. When in doubt, follow the manufacturer's recommendations in making selections. Production efficiency is improved and tool costs are lowered through proper selection of cutting tools for each application.

Cast-Alloy Cutting Tools

A number of cast alloys have been developed for use as cutting-tool materials. Some common brand names are **Stellite, Rexalloy, Armaloy,** and **Tantung.** The cast alloys are used as brazed tips on tool shanks (Fig. 117-2), as removable tool bits, as inserts in toolholders, and as inserts in milling cutters. The cast alloys are nonferrous materials with a cobalt base. They do not contain iron, except that which is present in the form of an impurity in the raw materials used. Cast alloys used as cutting tools may contain various combinations of the following principal elements: cobalt 35% to 55%, chromium 25% to 35%, tungsten 10% to 20%, nickel 0% to 5%, and carbon 1.5% to 3%. Very small amounts of other elements sometimes are added. The cast alloys are cast slightly oversize and are ground to shape. They cannot be forged or machined successfully.

The principal advantage of the cast alloys, in comparison with high-speed steels, is their high red-hardness. Because of this property, higher

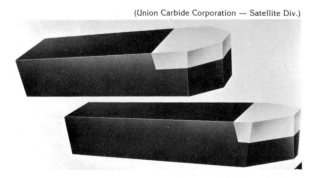

Fig. 117-2.   Tool bits with brazed cast-alloy tips.

cutting speeds may be used, and tool life is maintained at the resultant higher cutting temperature. Although cast alloys begin to soften slightly at temperatures from 648.9° C to 815.6° C (1200° to 1500° F), they are not seriously affected by temperatures below 815.6° C (1500° F). Any loss of hardness at these high temperatures is regained upon cooling. High-speed steels are slightly harder than the cast alloys at temperatures below 593.3° C (1100° F). Above this temperature, the cast alloys are harder and retain their hardness up to 815.6° C (1500° F). Thus, the cast alloys perform better for machining applications where temperatures ranging from 593.3° C to 815.6° C (1100° to 1500° F) are developed. These applications usually will occur at cutting speeds which exist between the highest cutting speeds for high-speed steels and the lowest practical speeds for carbide tools. The cast-alloy tools generally perform best at high speeds. They may be operated at cutting speeds approximately 50% to 75% faster than the maximum for high-speed steel tools.

Most cast alloys are more brittle and generally will not stand the heavy shock or impact pressures which carbon tool steels or high-speed steels will stand. They must, therefore, be well supported in a tool shank or toolholder.

Cast alloys are made by some manufacturers in several different grades which vary in impact-rupture strength. Certain grades of cast alloy have impact-rupture strength comparable to high-speed steel. It is best to follow the manufacturer's recommendation in selecting the grade of cast alloy used for a particular machining application.

Aluminum-oxide abrasive wheels are recommended for grinding or sharpening the cast alloys. They may be ground wet or dry, but they should not be quenched after grinding dry.

Cemented-Carbide Cutting Tools

Cemented-carbide cutting tools are used widely in production machining. Their principal advantages are high initial hardness, retention of hardness at red heats up through about 926.7° C (1700° F), and increased cutting speeds. Carbide cutting tools may be operated at speeds from two to four times the cutting speeds used for high-speed steel cutting tools. Figure 117-3 shows SAE 1137 steel being machined at about 140.2

(Cleveland Twist Drill; Armstrong Bros. Tool Co.)

A. Straight-shank drill

(Cincinnati Milacron)

Fig. 117-3. SAE 1137 steel being machined at 140.2 mpm (460 fpm) with a carbide-toothed milling cutter.

B. Straight-shank end mill

C. Shell end mill

D. Lathe center

E. Tool bit

Fig. 117-4. Carbide-tipped tools.

mpm (460 fpm) with a milling cutter which has carbide teeth.

The principal ingredients of the cemented carbides used for cutting tools are tungsten carbide and cobalt. Certain types of cemented carbides have titanium and tantalum carbides included to obtain specific properties. The carbide-tool materials are called **cemented carbides** because the carbide grains are cemented together during the manufacturing process with a binder, usually cobalt. Carbide tools are cast to shape and are very hard in the **as cast** condition. They do not require any further heat treatment. Carbides can only be shaped by grinding.

Carbide cutting tools frequently are used in the form of cutting tips brazed on the tool shank or on the body of a cutting tool as in Fig. 117-4. Carbide cutting tools are also used as tool inserts of the disposable type which may be rotated for use on each of the cutting edges or corners before they are thrown away, Fig. 117-5. Carbide inserts are made in many standard shapes, Fig. 49-3. Carbides are also used for disposable chip breakers on shank toolholders, as in Fig. 117-5. Smaller cutting tools such as drills, reamers, center drills, and end mills are available in solid carbide form, Fig. 117-6. Carbides also may be used as inserts on lathe or grinder centers, as in Fig. 117-4.

(Cleveland Twist Drill; Armstrong Bros. Tool Co.)

(Armstrong Bros. Tool Co.)

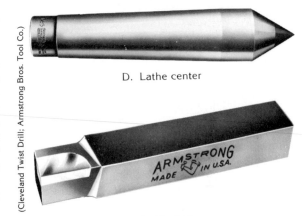

Fig. 117-5. A carbide throwaway-insert tool bit mounted in a lathe toolholder. Note the chip-breaker clamped on top of the insert.

A. Straight-shank drill

B. Four-flute double end mill

C. Two-flute double end mill
(Cleveland Twist Drill — an Acme-Cleveland Company)

Fig. 117-6.   Solid carbide cutting tools.

Cemented Carbide Groups

There are two basic groups of carbide materials used in cutting tools. **Group C** carbide is composed principally of tungsten carbide and cobalt, and it is used for machining cast iron and the nonferrous metals. **Group S** is composed principally of tungsten carbide, tantalum carbide, titanium carbide, and cobalt. It is used for machining the various steels. If Group C is used for machining steel, wear craters appear rapidly on the face of the tool. The titanium and tantalum carbide ingredients in Group S carbides improve the wear-resistance qualities.

The amount of cobalt in either of the two groups of carbide affects the hardness of the cutting tool. As the amount of cobalt is increased, the tool becomes harder. With increased hardness, the carbide has increased wear resistance and tool life, together with decreased shock resistance due to brittleness. With decreased hardness, there is a decrease in wear resistance, but there also is a corresponding increase in shock resistance or impact toughness.

A second factor which influences the hardness, wear resistance, and impact toughness of the carbides is the grain structure. A fine grain structure increases hardness, while a coarse grain structure decreases hardness. The manufacturer can, therefore, control the hardness and toughness of carbides through control of the ingredients and the grain structure.

Cemented-Carbide Classifications

There are many different machining applications on which carbide cutting tools are used. These vary from roughing cuts to light finishing cuts on materials which may be very hard or soft. Different **grades** of carbide materials have been developed by tool manufacturers to meet the conditions demanded. Some grades of carbide are recommended for use on very specific applications, while others may be used on a broad range of general applications. Attention should be given to the selection of the right grade for the particular job or machining application.

The **Carbide Industry Classification System** may be used as an aid both in classifying machining applications and in selecting the proper grade of carbide cutting tool. Many carbide tool manufacturers recognize this classification system. Each manufacturer recommends one or more specific grades of carbide for use in each machining application. In some cases, one carbide grade may be selected which will produce satisfactory results on several different machining applications. The following eight classifications are used in grouping machining applications for cemented-carbide cutting tools according to the Carbide Industry Classification System:

Cast iron and nonferrous materials —
  C-1: Finishing to medium roughing cuts
  C-2: Roughing cuts
  C-3: High-impact dies
Steel and steel alloys —
  C-4: Light high-speed finishing cuts
  C-5: Medium cuts at medium speeds
  C-6: Roughing cuts
  C-7: Light finishing cuts
  C-8: General purpose and heavy roughing cuts

Included in this system are six additional classifications concerned with **wear applications** and **impact applications.** However, these are not within the scope of this unit.

In selecting a specific grade of carbide tool for one or more of the above machining classifications, the manufacturer's recommendations should be consulted. Such recommendations are available in tool supply catalogs and manufacturers' bulletins.

Each manufacturing company has its own numbering or identification system for each carbide grade. There also may be a variation in the properties and performance of the carbide grades recommended by several different carbide tool manufacturers for a particular machining application.

Coated Carbides

Coated carbides are conventional or slightly modified carbide grades that are coated with a thin layer — 0.005 mm to 0.010 mm (0.0002" to 0.0004") — of a very hard heat-resistant material. The most commonly used coatings are titanium carbide, titanium nitride, hafnium nitride, aluminum oxide, and combinations of these materials. Coated carbides offer substantially greater tool life or higher cutting speeds but are somewhat less thermal- and shock-resistant than uncoated carbides of similar grade. Over 50% of the carbide cutting tools now being used are coated carbides. When carefully matched to the machining task, coated carbides offer substantial savings in machining time over uncoated carbides. Recommendations for coated carbide applications may be obtained from any carbide tool manufacturer.

Precautions in Use

Because of the special properties of carbide materials, particularly their hardness and brittleness, certain precautions should be observed with their use. Carbide tools must be rigidly supported in the toolholder or holding device. The machine and the work setup should be rigid and free from vibration. Interrupted cuts should be avoided when possible. The machine should not be stopped during a cut. Because of the pressure-welding characteristics of carbide tools at low speeds, they should be operated at the recommended cutting speeds, usually two to four times higher than those for high-speed steel tools. Proper cutting fluids also should be used.

Cermet Cutting Tools

Cermet cutting tools are blends of ceramic and metal powders which are formed into shape by pressing and sintering. One type which is commonly available is composed of 70 – 80% titanium carbide blended with a small percentage of molybdenum carbide, and with nickel as a binder.

Cermets are more brittle than carbides, which generally limits their use to light finishing cuts. Their hardness, however, is superior to carbides, which provides significantly greater tool life at the same or better cutting speeds. In turn, cermets are significantly outperformed by ceramic cutting tools.

Cermet cutting tools are available as throwaway inserts.

Ceramic Cutting Tools

Ceramic cutting tools are made of metal oxide powders which are formed into shape either by cold pressing and sintering, or by hot pressing. Hot pressed blanks are slightly stronger, but the strength of both types is lower than that of carbides. Most "straight" ceramic tools are made of aluminum oxide, although silicon and magnesium oxide are also used, either separately or in combination with aluminum oxide. Small percentages of metallic binders are added to improve impact strength.

Ceramic cutting tools, like cemented-carbide cutting tools, are used in the form of inserts or tool bits held in a toolholder mechanically. However, unlike carbide tools, they cannot be brazed. A ceramic cutting tool is shown in Fig. 117-7. Ceramic tool materials rank between sapphire and diamond in hardness; they are harder than cemented-carbide tools. They have a crystalline structure and are hard and brittle, and high in wear resistance, but they shatter quite easily because of low impact resistance or low rupture strength.

The hardness of ceramic tool materials is affected little by heat. Hot metal chips do not weld readily to the cutting tool when operated at the proper cutting speeds; therefore, cutting fluids generally are not needed. However, when coolants are needed to prevent distortion, the fluids should flow liberally over the cutting tool. A liberal flow will prevent intermittent cooling which may cause the tool to crack or shatter.

The principal advantages of ceramic cutting tools are increased cutting speeds and increased tool life per cutting edge. These tools also may be used for certain machining applications where it is necessary to cut heat-treated or very hard steel. Ceramic cutting tools may be operated at cutting speeds two to four times higher than those used

(General Electric — Metallurgical Prod. Dept.)

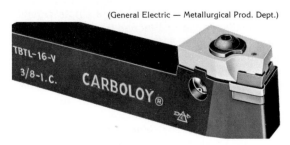

Fig. 117-7. A lathe toolholder with a throwaway ceramic insert.

for cemented-carbide tools. However, because of their low impact resistance, they generally should be used for fairly light finishing cuts at high speed. The high cutting speeds usually result in improved surface finish, which often eliminates the need for a ground finish.

A new type of ceramic tool material is made by blending aluminum oxide with about 30% titanium carbide. This new tool material is only slightly harder than the straight oxide tools, but it has higher thermal and shock resistance. This improved strength allows their use for milling steels as well as cast iron. They perform well at cutting speeds of 609.6 mpm (2000 fpm) for face milling gray iron, and are capable of many turning operations which cause straight oxide tools to fail.

Cubic boron nitride (CBN) is a new ceramic material that is nearly as hard as diamond. It can maintain its high hardness at nearly 1370°C (2500°F), and it oxidizes less rapidly than diamond. Cutting tools of CBN have reduced cutting times on tough nickel alloys as much as 500%. However, CBN tools are too brittle to use on interrupted cuts, and vibration will shatter the cutting edge.

Diamond Cutting Tools

Diamond is the hardest material known. This property, together with extreme heat resistance, permits cutting speeds up to 3048 mpm (10,000 fpm). Low strength and shock resistance, however, limit their use to machining soft, low strength, or highly abrasive materials such as graphite, plastics, ceramics, and certain aluminum and copper alloys.

Diamonds used for cutting tools are either natural diamonds of industrial quality or manufactured diamonds. Until recently, diamond cutting tools were available only as a single crystal or a cluster of crystals attached to a tool shank by brazing, Fig. 117-8. A disadvantage of these tools is that their strength, hardness, and wear resistance varies with the orientation of the crystal, making their performance somewhat unpredictable.

Sintered **polycrystalline diamond** tools are now available which eliminate the orientation problem. The tools are made up of fine crystals of diamond bonded together to form solid tool shapes. Random orientation of the diamond crystals provides uniform cutting behavior in all

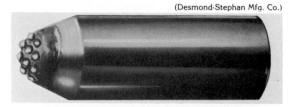

(Desmond-Stephan Mfg. Co.)

Fig. 117-8. A diamond-impregnated cluster nib for dressing and truing grinding wheels.

(Valeron Corporation)

Fig. 117-9. Sintered polycrystalline diamond cutting tools require rigid mounting in special toolholders.

directions, and with a marked improvement in toughness over single-diamond crystal tools. Polycrystalline diamond tools are available as brazed tips on various tool shanks, as tips bonded to carbide tool inserts, or as solid inserts as in Fig. 117-9. Wear resistance of diamond tools is excellent, outlasting carbides anywhere from 10:1 to as much as 450:1.

Comparative Cutting Speeds

There is no definite rule which can be used in recommending cutting speeds for cutting tools made from different materials. Recommended speeds will vary according to the following factors: the kind and hardness of material being cut, the rate of feed, the depth of cut, the finish desired, the rigidness of the machine, the rigidness of the work setup, the type of cutting tool, and the type of cutting fluid used.

In the absence of specific recommendations for each machining application, the following general rules may be used as a guide in selecting cutting speeds for cutting tools made of different materials.

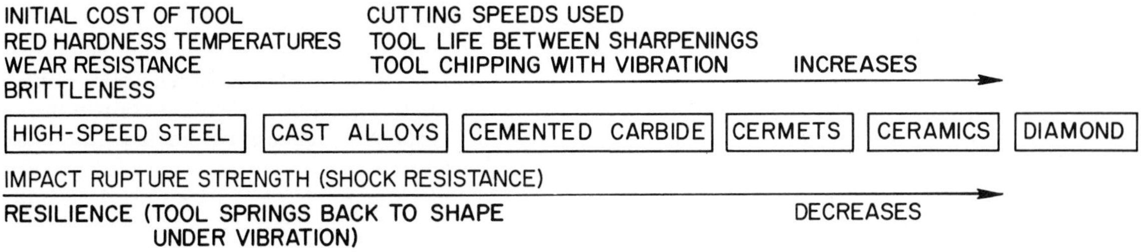

INITIAL COST OF TOOL   CUTTING SPEEDS USED
RED HARDNESS TEMPERATURES   TOOL LIFE BETWEEN SHARPENINGS
WEAR RESISTANCE   TOOL CHIPPING WITH VIBRATION   INCREASES
BRITTLENESS   ⟶

| HIGH-SPEED STEEL | CAST ALLOYS | CEMENTED CARBIDE | CERMETS | CERAMICS | DIAMOND |

IMPACT RUPTURE STRENGTH (SHOCK RESISTANCE)
⟶
RESILIENCE (TOOL SPRINGS BACK TO SHAPE   DECREASES
          UNDER VIBRATION)

Fig. 117-10.   General characteristics of cutting tool materials.

1. High-speed steel cutting tools may be operated at speeds about twice those recommended for carbon-steel tools.
2. Cast-alloy cutting tools generally may be operated at speeds approximately 50% to 75% greater than the maximum speed recommended for high-speed steel tools; these speeds are approximately three times greater than those for carbon-steel tools.
3. Cemented-carbide cutting tools may be operated at cutting speeds from two to four times faster than those recommended for high-speed steel tools. Cutting fluids generally are required at the high end of this speed range.
4. Cermets are capable of operating at cutting speeds between 10% and 100% higher than cemented-carbide cutting tools, depending on cutting conditions.
5. Ceramic cutting tools may be operated on certain light machining operations at speeds from two to four times greater than those recommended for the same application with carbide cutting tools.

6. Diamond tools may be operated in the same range of cutting speeds as carbide and ceramic tools, or at vastly higher speeds depending on cutting conditions.

General recommendations concerning cutting speeds for drilling, lathe work, shaper work, and milling are included in the sections of this book concerned with these machines and their operation. More specific recommendations for cutting speeds and feeds for various applications may be found in standard handbooks for machinists.

Summary

Cutting tools used for the machining of metals are made of several types of tool materials. Each material used has certain characteristics which are necessary in particular machining applications. The best tool material for a specific operation, such as light-duty milling of steel at low speeds, may not be the recommended material for high-speed milling of the same material. The relative properties of six commonly used cutting-tool materials are summarized in Fig. 117-10.

Test Your Knowledge of Section 14

Unit 115: The Action of a Cutting Tool on Metal

1. What basic factors determine the type of chip formed by a metal cutting tool?
2. Explain how a metal chip is formed at the cutting edge of a tool.
3. Why should the machinist be concerned with the type of chip formed by a cutting tool?
4. What type of metals usually form continuous chips?

5. What type of chip is considered the ideal way of machining ductile metals? Why is this type of chip a problem, and how is the problem remedied?
6. On what types of materials does the continuous chip with a built-up edge form?
7. What factors tend to minimize the formation of a built-up edge on a cutting tool?
8. Describe the type of chip formed when brittle metals (such as cast iron or bronze) are machined.

Unit 116: Cutting Fluids

1. What are the principal functions of cutting fluids?
2. How do good lubricant properties in a cutting fluid affect the formation of a built-up edge on the cutting tool?
3. What single factor is most likely to cause tool failure at high cutting speeds?
4. List the temperatures at which the different tool materials begin to soften.
5. Under what three classifications are most commercial cutting fluids grouped?
6. Of what value are fatty oil additives in mineral cutting oils?
7. What is the purpose of sulfur or chlorine additives in cutting oil?
8. What is the advantage of tri-purpose cutting oils?
9. Of what basic ingredients are emulsifiable oils composed?
10. What determines the proportions of oil and water used in an emulsifiable oil solution?
11. What are the advantages of emulsifiable oil solutions?
12. What procedure is used in mixing emulsifiable oil solutions?
13. What is dermatitis, and how can it be avoided?
14. Describe the two principal types of chemical cutting fluids.
15. On what materials may kerosene be used as a cutting fluid?
16. Are cutting fluids generally recommended for machining cast iron?
17. What three principal factors should be used as a guide in selecting a cutting fluid for a particular job?
18. Explain how machinability ratings were developed for various metals.
19. Explain the rating system which is used to rate the severity of a machining operation.
20. What factors involving "operating conditions" should be considered in selecting a cutting fluid?

Unit 117: Cutting Tool Materials

1. What are the most important properties that a cutting tool must possess?
2. What are the principal advantages of high-speed steel cutting tools?
3. What are the principal elements which are alloyed to produce high-speed steel? What properties does each element contribute?
4. List four basic types of HS steels used for cutting tools.
5. Name four kinds of coatings used on high-speed steel cutting tools to make their cutting edges last longer.
6. What are the principal elements from which cast-alloy tool materials are made?
7. What are the principal advantages of cast-alloy cutting tools when compared with HS steel tools?
8. What are the principal advantages in using cemented-carbide cutting tools?
9. What are the principal ingredients in cemented-carbide cutting tools?
10. Cemented-carbide cutting tools are classified into two groups. Give the group designations and the kind of metals each is designed to cut.
11. List four kinds of coatings used on carbide cutting tools.
12. Of the carbide cutting tools now in use, what percentage are coated?
13. Compare the performance of coated-carbide cutting tools to that of plain carbides.
14. What precautions must be considered in machining with carbide cutting tools?
15. What is the composition of cermet cutting tools, and how does their performance compare to straight ceramic cutting tools?
16. What are the principal ingredients in ceramic cutting tool materials?
17. What are the principal properties of ceramic cutting tools?
18. In what forms are diamonds used as cutting tools?
19. What are the properties of diamond cutting tools, and what materials can be successfully machined by them?
20. What principal factors determine the cutting speeds used for a particular machining application, such as turning?
21. Summarize the properties of each of the basic cutting tool materials in comparison with HS steel.

Unit
118

Electrical Discharge Machining

Electrical discharge machining (EDM) removes metal by controlled electrical arcing (sparking) between the tool and workpiece. Each of thousands of tiny arcs vaporizes a tiny amount of the workpiece, leaving a miniature crater in the metal's surface. A **dielectric** (insulating) fluid confines the arcing to the immediate vicinity of the tool. The arcing shapes the workpiece to a mirror image of the shape of the tool. Figure 118-1 shows the basic parts of an EDM system.

The EDM process is valued for its ability to machine complex shapes in metals of any hardness. It is used widely in making injection and compression molds for rubber and plastic molding, molds for die-casting metals, and dies for forging and metal stamping. EDM can also be used to remove broken taps and studs and to drill holes as small as 0.05 mm (0.002"). Because no tool pressures are involved, the process is ideal

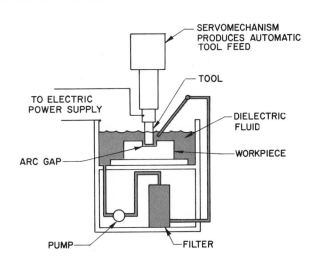

Fig. 118-1.   The basic parts of an EDM system.

(Photo Courtesy of Illinois State University)

Fig. 118-2.   An EDM machine typically found in tool and die shops. The cover of the dielectric fluid tank is removed to show the safety float, which stops the operation if the fluid is too low.

for machining delicate workpieces such as honeycomb structures.

EDM has two main disadvantages: (1) only materials that conduct electricity can be machined, and (2) machined surfaces are left with a thin [usually less than 0.025 mm (0.001″)], hard layer of metal. This makes it difficult to file and polish EDM-machined die cavities. However, it has been found that **electropolishing** (a form of electrochemical machining) can efficiently remove

this hard layer. Electrochemical machining is described in Unit 119.

Types of EDM Machines

There are two types of EDM machines: those that use solid electrodes and those that use a traveling-wire electrode.

Solid-electrode machines vary in size from small bench-top models to machines capable of handling huge workpieces. Figure 118-2 shows a typical EDM machine of medium size found in many tool and die shops. Basic machines of this type work by advancing a nonrotating tool directly into a workpiece. More versatile machines revolve the tool as it advances. This feature permits helical cutting operations such as tapping holes and cutting internal helical gears, Fig. 118-3. The advanced machines may also have worktables equipped with power feeds to permit sawing and slotting operations.

EDM machines with automatic tool changers and computer numerical control (CNC) are also available. CNC can provide numerous benefits, including:

1. Automatic tool centering over the workpiece.
2. Precise workpiece movements for accurate hole patterns, Fig. 118-4.

Fig. 118-3.   Helical gears and other helical parts can be cut on EDM machines equipped for tool rotation.

(Modern Machine Shop)

Fig. 118-4.   Precise hole patterns are easily programmed with a CNC-controlled EDM machine.

(Dieter-Hansen Corporation)

3. Programmed orbiting of the electrode (movement of the electrode in a tight circular path without rotating the electrode), which aids flushing of cutting debris, prolongs tool life, and improves the accuracy of the machined part.
4. Programmable tool withdrawal to promote flushing.
5. Automatic tool changing.

(Elox Division — Colt Industries)

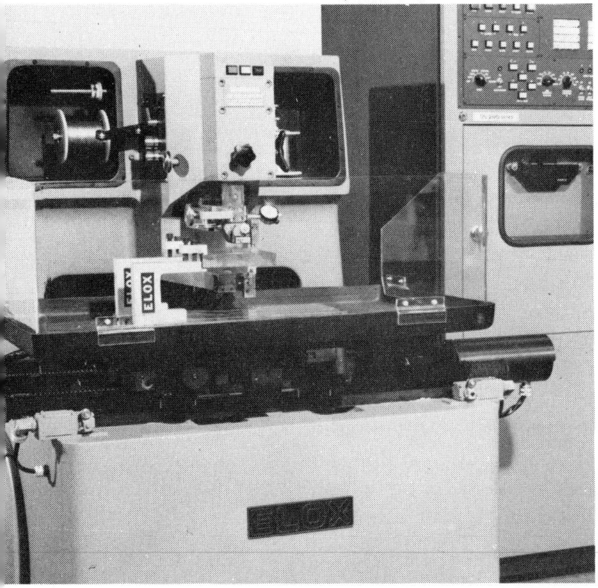

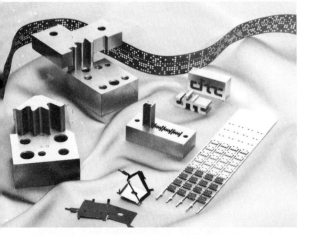

Fig. 118-5. A traveling-wire electrode EDM machine.

(Elox Division — Colt Industries)

Fig. 118-6. Shapes cut with a traveling-wire electrode EDM machine.

Traveling-wire electrode machines, Fig. 118-5, are also made in several sizes. The tool or electrode for this type of EDM machine is a round wire. As a part is machined, unused wire electrode is fed continuously from a supply reel through the workpiece to a take-up reel. Used wire is discarded.

Traveling-wire machines have revolutionized the making of blanking dies for sheet metal stamping. With a round wire electrode, cutting is possible in any direction, and any shape can be cut, Fig. 118-6. Die blocks may be hardened before being cut with a traveling-wire machine and the shapes can be cut exactly to size. Both punch and die can be made from the same block, the clearance being determined by the diameter of the wire used for cutting.

Production versions of traveling-wire EDM machines are also available. One such machine has five wire feeds, each with its own power supply. All of the wire feeds are under the control of one numerical control unit. With this machine, it is possible to stack and machine seven parts at each work station rather than machining only one part per station. This boosts production from five to 35 parts for each machining cycle, Fig. 118-7.

Fig. 118-7. This production EDM machine has five wire feeds and five power supplies. It is controlled with one numerical control unit.

(Modern Machine Shop)

EDM Electrodes (Tools)

EDM electrodes must be good electrical conductors. Commonly used materials include high-purity graphite, brass, copper, copper-graphite, copper-tungsten, and zinc alloys. The EDM process causes much greater tool wear than conventional machining processes. Therefore, tool materials which provide the best **wear ratio** (ratio of workpiece erosion to tool erosion) are used whenever possible. Graphite is widely used because it is inexpensive, is easily machined, and has an excellent wear ratio. Because of its brittleness, however, graphite cannot be used for electrodes that must have fine detail or thin sections. Its porosity also limits its use to parts not requiring extremely fine finishes.

At one time, the capacitor-type power supplies used in EDM caused high wear ratios. These high ratios in turn required that several identical EDM tools be made. One or more tools were used for roughing cuts, and one or more for finishing cuts. **Stepped cutting tools** may be used to accomplish the same result as several separate cutting tools, Fig. 118-8.

New transistorized power supplies now provide precise control of the cutting process. This makes possible lower wear ratios and better finishes. Some machines have a **no-wear** cutting cycle that allows the same electrode to be used for both the roughing and finishing cuts. However, the no-wear cutting rate is very slow. Using the no-wear cycle may require more time to finish a job than making and using a roughing and a finishing tool. Therefore, the no-wear cycle is often used only on the finishing cut, where it can provide improved accuracy and smoother finishes.

Since a gap must exist between the tool and the work, the tool produces a cavity slightly larger than itself. This size difference is known as **overcut** and generally amounts to only a few hundredths of a millimeter (thousandths of an inch). The breakdown voltage of the dielectric fluid, the voltage and amperage settings, and the cleanliness of the dielectric fluid all affect the amount of overcut.

When clean oil with a dielectric strength of 200 volts per 0.025 mm (0.001") is used with the voltage set at 200, arcing will begin when the tool is 0.025 mm (0.001") from the workpiece. If the voltage is reduced to 100, the gap must be reduced

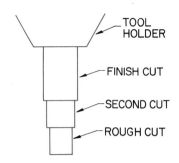

Fig. 118-8.   A stepped EDM electrode.

to 0.0125 mm (0.0005") before a spark will occur. However, when cutting begins, metal particles become suspended in the gap, forming a conductive path which effectively increases the overcut.

Other factors affecting overcut include the rigidity of the machine spindle to which the tool is attached, whether spindle travel occurs in a straight or curved line, and the resistance of the spindle to rotation when electrodes other than round electrodes are used.

Dielectric Fluids

The most commonly used dielectric fluids are low-viscosity petroleum oils. Other fluids sometimes used are kerosene, deionized water, silicone oils, and ethylene glycol/water solutions. To obtain uniform cutting action and good finishes, a flow of dielectric fluid must be directed through the arc gap so as to sweep away the chips. The fluid stream may be directed along the sides of the tool, through a hollow tool, or through a hole in the bottom of the workpiece. The fluid must be continuously filtered to remove chips and foreign matter.

Cutting Rates and Surface Finish

In EDM, the metal removal rate is directly related to amperage setting and arcing frequency. At a constant frequency, low amperage settings produce slow rates of metal removal and good finishes, while high amperage settings produce higher rates of metal removal and poorer finishes. See Fig. 118-9. At a constant amperage, low arcing frequencies produce rougher finishes than high arcing frequencies, Fig. 118-10. The best surface finishes are therefore obtained when the controls are set for the highest arc frequency and the lowest amperage. Finishes as fine as 0.25 micrometers (10 microinches) are possible with EDM.

Cutting rates vary widely when a given metal is cut with different electrode materials and when a given electrode material is used to cut different metals. Graphite electrode cutting rates exceed other electrode cutting rates from 2:1 to as much as 20:1. The average cutting rate for graphite electrodes cutting steel is generally less than 0.32 cubic centimeters (0.05 cubic inches) per ampere per hour, making EDM a very slow process. For this reason, conventional machining methods are used for the bulk of metal removal in order to minimize EDM machining time.

Safety

In EDM, cutting occurs by electrical arcing in what is essentially an open circuit, and electrical potential can reach 400 volts. Therefore, **under no circumstances should the operator touch the tool, toolholder, or workpiece while the machine is in operation.** When cutting has been completed and the machine has been turned off, the voltmeter should read **zero** before the operator handles the tool, toolholder, or workpiece.

While dielectric oils have low flammability, the gases produced in the breakdown of the oil by the electric arc contain a high percentage of acetylene, which is very flammable. **Since the acetylene can be ignited by the arcing between the tool and workpiece, cutting should never be started until the tool and workpiece have been flooded with at least 25.4 mm (1") of oil.** While industrial experience has shown that the fire hazard is very slight, it is strongly recommended that a class B and C fire extinguisher be placed near the machine.

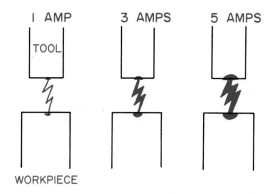

Fig. 118-9. The relationship of amperage setting to metal removal rate.

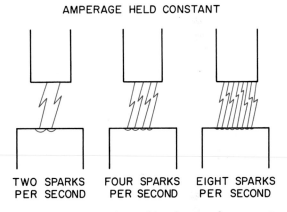

Fig. 118-10. The relationship of arcing frequency to surface finish.

Unit

119

Electrochemical Machining

Electrochemical machining (ECM) is based on the same principles as electroplating. However, instead of depositing metal on the workpiece, ECM reverses the process so that metal is **deplated** or removed from the workpiece. The basic parts of an ECM system are shown in Fig. 119-1. A general-purpose ECM machine is shown in Fig. 119-2.

In ECM, the tool is the cathode and the workpiece is the anode. A gap of 0.0254 mm (0.001″) to 0.76 mm (0.030″) is maintained between the tool and workpiece. This provides space for the flow of the electrolyte and keeps the electrical circuit from shorting out. A low-voltage, high-amperage direct current passes from the workpiece to the tool through the electrolyte. This current dissolves metal particles from the workpiece into the electrolyte by electrochemical reaction. The electrolyte is pumped through the gap between the tool and workpiece at pressures from 1 379 to 2 068 kPa (200 to 300 psi). The dissolved metal particles are swept away and filtered out. They are thus prevented from being deposited on the tool.

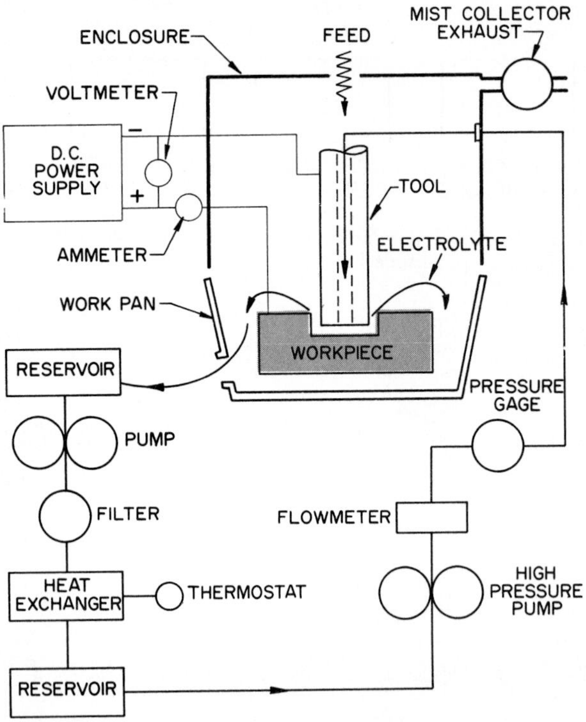

The accuracy of the ECM process and the life of ECM tools are very good for these reasons: (1) the tool never touches the workpiece, (2) the tool receives no build-up of metal from the workpiece,

Fig. 119-1. The basic parts of an ECM system.

Fig. 119-2.   This close-up of a general-purpose ECM machine shows a workpiece and the ECM cutting tool.

(Chemform)

and (3) the tool has almost no wear from the flow of the electrolyte.

ECM electrodes (tools) are most commonly made of copper and copper alloys, type-316 stainless steel, and titanium. Copper and brass are preferred except for thin tools requiring greater stiffness. Tools must be made undersize to adjust for overcut, which varies depending on the electrolyte flow and the required accuracy.

The electrolytes most commonly used are water solutions of sodium chloride, sodium chlorate, potassium chloride, sodium nitrate, and sodium hydroxide. Electrolytes must be continuously filtered to remove dissolved metal. Close control of electrolyte temperature is also required, since changes in temperature greatly affect the electrical conductivity, machining rate, and accuracy.

The ECM process can be used to machine any metal that conducts electricity, regardless of hardness. The absence of tool pressures on the workpiece makes the process ideal for machining thin metals and fragile workpieces.

Comparatively low metal-removal rates prevent ECM from competing with conventional machining methods when metals with good machinability are involved. ECM cutting rates range from 0.65 cubic centimeters (0.10 cubic inches) to 1.74 cubic centimeters (0.27 cubic inches) per minute per 1000 amperes, depending on the kind of metal being cut.

ECM excels in machining difficult-to-machine metals, especially when holes or cavities of complex shape must be made. No burrs are made by ECM. Surface finishes are bright and smooth, ordinarily $0.1 - 0.75\ \mu m$ ($4 - 30\ \mu in.$), and usually do not need polishing.

Electrochemical Deburring (ECD)

Electrochemical deburring is an adaptation of ECM. The tooling used in this process is designed for removal of burrs and sharp edges from parts machined by conventional methods. Figure 119-3 shows a typical ECD tooling arrangement.

Safety

Sodium nitrate and sodium chlorate compounds are powerful oxidizers, which accelerate combustion. They must never be exposed to combustible materials. These compounds are also dangerous to body tissue, thus requiring protection against inhalation of dusts, mists, and vapors. Protective clothing and face shields or masks are also required for handling acids and certain other chemicals used in ECM. Further, explosive hydrogen gas, released by the electrochemical reaction, must be vented from the top of the work enclosure to prevent explosion.

Electrochemical Grinding (ECG)

Electrochemical grinding applies the principles of ECM to the conventional grinding process. Figure 119-4 shows the basic parts of an ECG system. ECG grinding wheels are composed of conventional abrasives but must be made with an electrically conductive bond. ECG removes metal as much as 80% faster than conventional grinding without an appreciable loss in accuracy

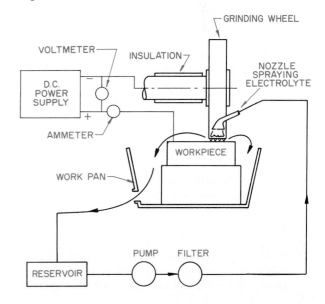

Fig. 119-4. The basic parts of an ECG system.

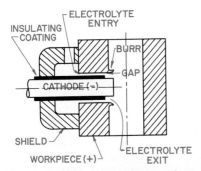

Fig. 119-3. A tooling arrangement for electrochemical deburring.

or surface finish. Grinding operations can often be completed in one pass of the grinding wheel. Only about 10% of conventional grinding wheel pressure is necessary. Since 90% of the metal is removed by electrochemical reaction (deplating), wheel dressing and wheel wear are sharply reduced.

The ECG process is preferred for grinding carbide tools because cutting edges can be ground with a single pass. With conventional grinding, several passes are normally necessary because heat build-up limits material removal to 0.025 mm (0.001″) per pass. Additional ECG benefits include smooth, burr-free cutting edges and wheel wear reduced to 15% of that for conventional grinding. Figure 119-5 shows an ECG tool grinder. ECG is also used to a limited extent for surface grinding and cylindrical grinding.

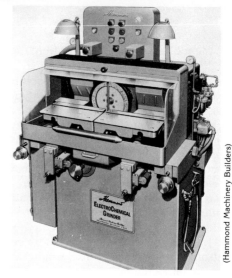

(Hammond Machinery Builders)

Fig. 119-5.   An electrochemical tool grinder.

Unit
120

Chemical Machining

Chemical machining (CHM) is a process of shaping metal by using strong acid or alkaline solutions to dissolve away unwanted metal. There are two types of chemical machining: **chemical blanking** and **chemical milling**. Chemical blanking is used for cutting out parts from thin sheet metals, Fig. 120-1. Chemical milling is used for selective or overall removal of metal from thick metal parts, Fig. 120-2.

Chemical Blanking

Parts to be made by chemical blanking are first accurately drawn up to 20 times their actual size, depending on the accuracy required. The enlarged drawing is then photographed and reduced to produce a negative exactly to required size. A multiple negative having many images of the part may then be made. An acid-resistant coating is applied to the metal to be cut. Then the negative is used to make a contact printing directly on the metal surface. An etching solution is used to dissolve the light-exposed portions of the metal, leaving the workpiece(s) blanked accurately to size and shape. Parts made by chemical blanking

Fig. 120-1.   A typical part cut by chemical blanking.

(Photo Chemical Machining Institute)

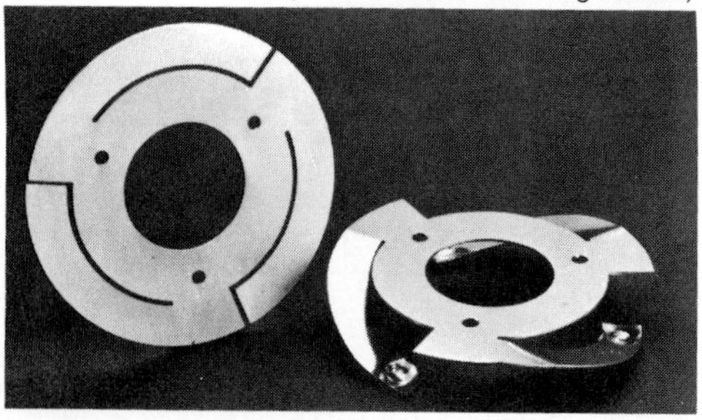

are usually etched from both sides simultaneously to speed processing.

Other techniques for applying the acid-resistant coatings, called **masking,** include silk screen and offset printing. These less expensive methods are used when the high degree of accuracy and detail obtainable with the photographic process is not required. Chemical blanking technology is used extensively in the production of printed circuit boards and integrated circuits.

Chemical Milling

Chemical milling is most widely used in the aircraft and aerospace industries. This process makes possible the efficient removal of unwanted weight from large, complex airframe parts without sacrificing strength, Fig. 120-2.

The procedure for chemical milling begins with the cleaning of the metal. Masking of neoprene rubber or vinyl plastic is then sprayed or flowed on and cured by baking. The areas to be chemically milled are then scribed with the aid of templates, and the masking is removed from these areas. Next, the part is submerged in the chemical milling solution until the unwanted metal is dissolved away. Rinsing with clean water and removing the rest of the masking complete the process.

The chemical milling process is basically very simple and requires neither highly skilled labor nor expensive equipment. Additional advantages include:

1. It does not cold-work the metal as conventional machining does.
2. Very large parts can be machined.
3. Any number of workpiece surfaces can be machined at the same time.
4. Removal of metal from complex surfaces is easily accomplished.

(General Dynamics)

Fig. 120-2.   An airframe section after chemical milling to reduce its weight without loss of strength.

5. Since there are no mechanical cutting pressures involved, thin, delicate workpieces can be safely machined.

Safety

Extreme care must be taken when mixing or working with metal etching solutions. Concentrated acid and alkaline solutions will instantly damage clothing and burn skin. Etching solutions are less dangerous but still very hazardous. A full face shield, rubber gloves, alkaline- and acid-resistant clothing, and rubber boots must be worn for complete protection against splashes and spills. An emergency shower should be readily available, and adequate ventilation must be provided to vent off toxic fumes.

Unit

121

Ultrasonic Machining

In ultrasonic machining (USM), also called **impact grinding,** fine abrasive particles suspended in a fluid (usually water) are pumped into a gap between the workpiece and the tool. The tool is made to vibrate a few hundredths of a millimeter (thousandths of an inch) at ultrasonic frequencies ranging from 19,000 Hz to 25,000 Hz. The rapid pumping action of the tool hurls the abrasive particles at the workpiece at high velocity, thus grinding the workpiece to the shape of the tool. The basic parts of a USM system are shown in Fig. 121-1.

The abrasives used in USM are aluminum oxide, silicon carbide, and boron carbide. Boron carbide cuts fastest and is the preferred abrasive despite its high cost. Grain sizes of 200 to 400 mesh are used for rough cutting, and sizes of 600 to 1000 mesh are used for finish cuts.

The accuracy obtained in USM is about 0.025 mm (0.001"), with 0.0063 mm (0.00025") possible when using the finest grain sizes. The surface finish varies directly with the abrasive grain size used, ranging from 0.75 μm (30 μin.) for 100 grit to 0.18 μm (7 μin.) for 800 grit.

Tool wear is high in USM, due to the abrasive cutting action. One or more roughing tools and a finishing tool are usually required for each job. Tools made of cold-rolled steel or stainless steel provide the best wear ratios.

USM tools are brazed or soldered to a stainless steel or Monel metal toolholder. The toolholder size and shape must be matched to the machining task to produce the desired resonance and resultant tool movement.

A major advantage of USM is that it can machine materials that cannot conduct electricity and which, therefore, cannot be machined by EDM or ECM. Glass, ceramics and precious and semi-precious stones are examples of nonconducting materials easily machined by USM. The USM process is also valued for its ability to machine tough alloys, hardened steels, and carbides. In addition, soft materials such as plastics and graphite may be cut as easily as hard materials.

The USM process is best suited to blanking various shapes from thin, hard, nonconducting

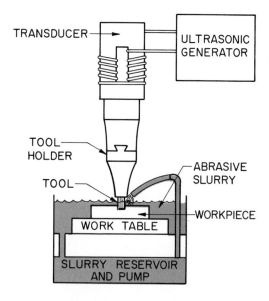

Fig. 121-1.   The basic parts of an ultrasonic machining system.

materials and machining shallow cavities of irregular shape. Figure 121-2 shows USM tools and the shapes produced by them. The principal disadvantages of USM are low rates of metal removal and high tooling and equipment costs.

Fig. 121-2. Examples of parts cut by ultrasonic machining. The tools used are shown above each workpiece.

Unit

122

Electron Beam Machining

Electron beam machining (EBM) is accomplished by focusing a high-speed beam of electrons on the workpiece, Fig. 122-1. The electron beam moves at more than half the speed of light. When it strikes the workpiece surface, its energy is transformed into heat sufficient to vaporize any known material. The amount of material removal, however, is very small.

EBM is mainly used for drilling very small holes and cutting narrow slots in difficult-to-machine materials up to 6.35 mm (1/4") thick. Holes as small as 0.0125 mm (0.0005") and slots as narrow as 0.025 mm (0.001") can be made in any material.

EBM is most efficient when done in a "hard" (near-perfect) vacuum. This permits electron beams of full power as narrow as 0.025 mm (0.001") to be directed to precise locations and focused on spots equally small. At any given location, the beam may be deflected into various patterns within a 6.35 mm (1/4") square by magnetic deflection coils, Fig. 122-2. This feature allows (1) the drilling of closely spaced multiple-

hole patterns from a single location, (2) the drilling of holes of various sizes and shapes, and (3) the cutting of slots of various widths and shapes.

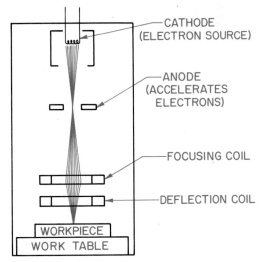

Fig. 122-1. Basic parts of an electron beam machining system.

Holes or slots only a few hundredths of a milli-meter (thousandths of an inch) in size are cut by focusing the electron beam. Larger holes may be cut by **trepanning** (moving the beam in a circular pattern) or by moving the beam off center and rotating the workpiece. Wider slots are made by moving the beam in a small circular path at the same time it is moved along the path of the desired slot.

The disadvantages of the EBM process include the need for skilled operators, high equipment costs, slow cycle times, and limits on workpiece size due to the size of the vacuum chamber available. The process also generates X-rays, thus requiring X-ray shielding of the work area. EBM uses the same equipment as electron beam welding (EBW), but the power settings for the two processes differ. An EBM machine is shown in Fig. 2-26.

BEAM DEFLECTION SYSTEM
FREQUENCY .005 TO 15,000 CPS

TYPE	WAVE FORM	BEAM PATTERN		
		X	Y	XY
SINE WAVE	OSCILLATION	—	|	/
SINE-COSINE WAVE	CIRCLE GENERATION	⬭	⬮	⬯
SQUARE WAVE	OSCILLATION	●—●	● ●	◆
DITHER	CIRCLE/ON CIRCLE	⌇⌇⌇	⌇	○
D.C.	OFFSET ANY ABOVE PATTERN	+|+	● ⊥	::

(Union Carbide Corporation)

Fig. 122-2. The electron beam cutting patterns that can be generated by magnetic deflection coils.

Unit

123

Laser Beam Machining

The energy source for laser beam machining (LBM) is a highly concentrated beam of light. (**Laser** is an acronym for **Light Amplification** by **Stimulated Emission** of **Radiation**). When the beam is focused to a small spot, its power density is raised to produce sufficient heat to vaporize any material. However, the rate of metal removal is very small, which limits the use of lasers to hole-drilling and cutting operations in relatively thin materials.

Two types of lasers are now in common use: (1) **solid lasers**, which are capable only of providing short bursts of power, and (2) **gas lasers**, which produce a continuous laser beam.

Figure 123-1 shows the basic parts of a solid laser system. The central part of the system is a

Fig. 123-1. The basic parts of a solid laser system.

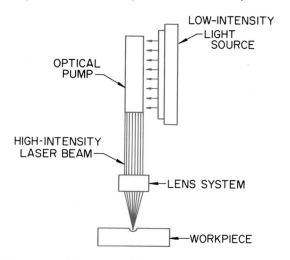

LOW-INTENSITY LIGHT SOURCE

OPTICAL PUMP

HIGH-INTENSITY LASER BEAM

LENS SYSTEM

WORKPIECE

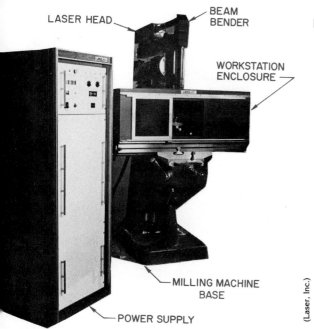

Fig. 123-2. A solid-laser drilling and welding machine.

special glass rod, which is known as the **lasing medium**. One end of this rod has a reflective coating. The other end is partially coated. A low-intensity lamp flashes light into the rod. The light reflects between the rod ends and builds up intensity. When it reaches the necessary intensity, the light escapes through the partially coated end.

Since solid lasers are limited to short bursts of power, they are best suited to hole drilling operations or to spot welding and spot heat-treating. Laser-drilled holes are slightly tapered, but have clean edges with little spatter. The surrounding heat-affected zone is slight. The power source can be precisely controlled, thus permitting precise control of hole sizes. Tolerances of ± 3 μm (118 μin.) are possible.

It is claimed that laser drillng is the least expensive way to drill holes up to 0.5 mm (0.020") diameter in thin, flat material.

For holes that enter on slanted or rounded surfaces, and for holes in difficult-to-reach spots, laser drilling is economical for diameters up to 1.25 mm (0.050"). Figure 123-2 shows a general-purpose laser drilling and welding machine.

At present, carbon dioxide (CO_2) gas lasers are most efficient for converting electricity into laser power. The basic parts of a gas laser are shown in

Fig. 123-3. Gas lasers operate in basically the same way as solid lasers, except that a gas serves as the lasing medium and can provide a continuous laser beam. This feature makes gas lasers best suited to continuous cutting, welding, or heat-treating operations.

Computer-controlled cutting systems that use oxygen to boost the heat of the laser beam are now in service. They can make fast, clean cuts in metals and ceramics up to 9.53 mm (3/8") thick and in nonmetals, such as plastics and wood, up to 25.4 mm (1") thick, Fig. 123-4. Such a system is shown in Fig. 2-28. These systems are also used for partial cutting operations, such as scribing

(Laser, Inc.)

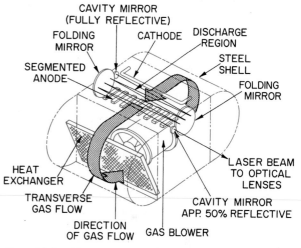

Fig. 123-3. The basic parts of a gas laser system.

Fig. 123-4. A gas laser cutting 0.5 mm (0.20") stainless steel. Cutting rates as high as 21.3 mpm (70 fpm) are possible.

and engraving, and for heat treating and welding. While the initial equipment cost is high, low operating costs and high productivity provide a quick return on investment.

Unit
124

Miscellaneous Metal Machining Processes

Plasma arc machining (PAM) is a high-temperature flame-cutting process. It is used mainly for profile-cutting of stainless steel, aluminum, and other metals that cannot be cut with oxyacetylene torches. Unit 2 has PAM equipment illustrations and gives further information on this process.

Abrasive jet machining (AJM) uses abrasive particles propelled by a high-velocity stream of air as the cutting tool. The stream of abrasive particles bombards the workpiece at nearly the speed of sound, but since the abrasive particles are very small, material removal is also very slow. See Fig. 124-1.

The abrasive particles used vary from less than 10 to more than 50 μm (394-1968.5 μin.) in diameter. Aluminum oxide and silicon carbide abrasives are most commonly used.

AJM is chiefly used for cleaning and deburring operations, etching of glass and ceramic materials, and trimming of electrical resistors. AJM is also used for scribing and cutting hard, brittle semiconductor materials. Accurate cuts as narrow as 0.127 mm (0.005″) are possible.

Abrasive flow machining (AFM) is also known as **abrasive flow deburring.** It is mainly used for deburring, rounding sharp edges, and surface polishing. The process involves pumping an abrasive slurry back and forth across the surfaces or edges to be machined. The work must be fixtured to restrict the abrasive flow to where

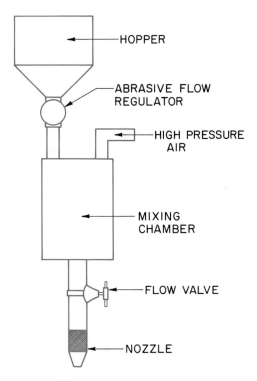

Fig. 124-1.   The main parts of an abrasive jet machining system.

HOPPER

ABRASIVE FLOW REGULATOR

HIGH PRESSURE AIR

MIXING CHAMBER

FLOW VALVE

NOZZLE

machining is required. See Fig. 124-2. The fixture surfaces exposed to the abrasive are provided with wear-resistant coatings to prolong their useful life.

Aluminum oxide and silicon carbide abrasives of 20 to 60 grit are commonly used in AFM. The viscosity of the abrasive slurry must be adjusted to obtain the desired velocity and, therefore, the cutting rate through a given size hole. Good control of pressure and the number of abrasive flow reversals provides consistent results. In many cases, AFM is the only way of deburring and finishing holes and surfaces in part interiors.

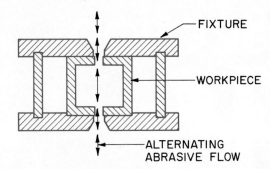

Fig. 124-2. An abrasive flow machining fixture for hole deburring.

Test Your Knowledge of Section 15

Units 118-124

1. Explain the EDM process and list its advantages and disadvantages.
2. Tell how the two types of EDM machines differ. Describe the kind of work done by each.
3. Explain the ECM process and list its advantages and disadvantages.
4. Describe the ECG process and list its advantages.
5. Explain how chemical blanking and chemical milling are alike and how they differ.
6. List the advantages and disadvantages of chemical milling.
7. Explain the USM process and list its advantages and disadvantages.
8. Describe the EBM process and list its advantages and disadvantages.
9. Name the two types of lasers used in metal machining and describe the kind of work for which each is best suited.
10. Describe the PAM process and the kind of work for which it is best suited.
11. Describe the AJM process and the kind of work for which it is best suited.
12. Describe the AFM process and the kind of work for which it is used.

Section 16

Numerical Control Programming

Unit

125

Introduction to Numerical Control

Numerical control (N/C) is a system of control that uses numerically coded instructions to operate motors and other devices that run a machine. The instructions are provided by either of two **binary coded decimal systems:** (1) the Electronics Industries Association (EIA) code, Fig. 125-1, or the American Society for Computer Information Interchange (ASCII) code. ASCII-coded machine control units will not accept EIA coded instructions and vice versa. Increasingly, however, control units are being made to accept instructions in either code.

Numerical control instructions may be provided through a keyboard on a **machine control unit** (MCU), in the form of a punched tape, Fig. 125-1, on magnetic tape, or directly from a computer. The current trend is to provide instructions directly from a computer. This method eliminates errors due to damaged tapes and faulty tape readers. It

also eliminates both the expense of preparing tapes and the need for tape punches, tape readers, and takeup reels.

Numerical Control Systems

Direct numerical control (DNC) systems use a large computer capable of controlling many machines simultaneously. In contrast, **computer numerical control** (CNC) systems use a microcomputer for the operation of each machine. In CNC, communication between the machine and the computer is provided through a computer terminal on the MCU. In DNC, communication is provided through a terminal at the large computer.

The use of computer terminals permits program changes to be quickly made at any time. This in turn greatly reduces the down-time formerly encountered while waiting for new tapes to be made. Programs are usually stored on magnetic tapes or discs. Some DNC and CNC systems, however, still use standard paper N/C tapes for permanent program storage. An N/C machine with a CNC control unit is shown in Fig 125-2.

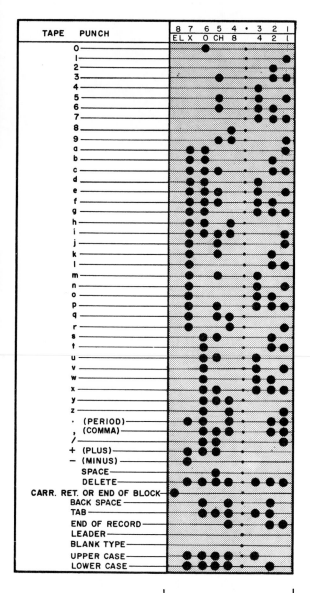

Fig. 125-1. EIA (Electronics Industries Association) standard code for 25.4 mm (1″) wide 8-track punched tape.

Importance of Numerical Control

Automatic operation by N/C is readily adaptable to the operation of all metalworking machines. Lathes, milling machines, drill presses, boring machines, grinding machines, turret punches, flame-cutting and welding machines, and even pipe benders are available with numerical controls.

Numerical control is also used with **coordinate measuring machines**, which take measurements of parts through the use of sensing probes. These machines display the actual part measurements on **digital readouts** and can even record the difference between actual and specified size if desired. See Fig. 125-3.

Numerical control is also used to operate drafting plotters, wire-wrapping machines, weaving machines, routers, and laser fabric-cutting machines. It is a proven system that has many applications in many industries.

Fig. 125-2. A vertical milling machine with a CNC control unit.

Fig. 125-3. An N/C coordinate measuring machine with digital readout equipment.

(Bendix Corporation)

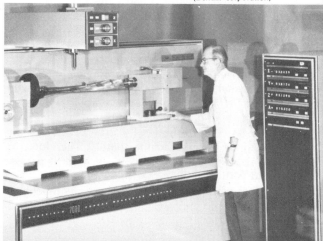

Advantages of Numerical Control

Of the many advantages of N/C machines, perhaps the following are the most important:

1. N/C machines can consistently produce close-tolerance work without the need for costly jigs to guide the tools. Usually only simple work-holding fixtures are required. This sharply reduces the recordkeeping and storage space formerly required for large quantities of jigs.
2. Often, one N/C machine can take the place of several conventional machines. The lathe in Fig. 125-4 is an example. This feature saves the floor space that would have been required by several conventional machines. It also sharply reduces the amount of materials-handling required.
3. N/C programs can be prepared much faster than jigs can be designed and built, thus saving **lead time.** Lead time is the time required for planning and tooling up to produce a product or a part. Design changes can be quickly made by modifying the N/C program, thus eliminating the need for costly jig redesigning and rebuilding.
4. Because N/C is accurate and reliable, fewer parts are spoiled because of human error. This feature also reduces the amount of inspection required to maintain the desired level of quality.
5. Since N/C provides a standard method of making a part, management is better able to plan production schedules and anticipate production costs.
6. Numerical control is ideal for small-lot repeat orders. Once written, N/C programs can be stored indefinitely. When needed, they can quickly be returned to the shop floor for production. This eliminates the need to replan the job and thus sharply reduces lead time.

Basis for N/C Measurement

A system of rectangular coordinates, called the **Cartesian coordinate system,** Fig. 125-5, is the basis for N/C measurements. It provides a system of identifying the axis of motion (X, Y, or Z) and the direction of motion (+ or −). The Z axis is assigned to the machine spindle. Therefore, for vertical-spindle machines, such as drill presses and vertical milling machines, Fig. 125-7, the Z axis is vertical. For horizontal-spindle machines,

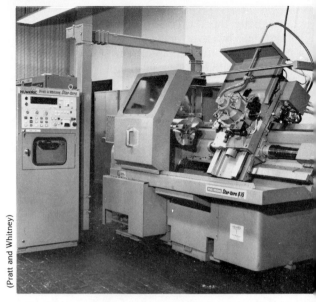

(Pratt and Whitney)

Fig. 125-4.   A lathe with a CNC control unit. One such machine can do the work of several conventional lathes.

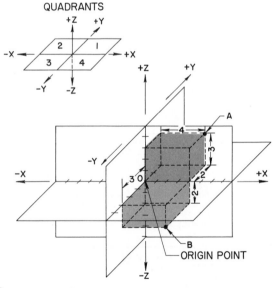

Fig. 125-5.   The Cartesian coordinate system used in N/C programming.

such as lathes and horizontal-spindle milling machines, the Z axis is horizontal, Fig. 125-8.

Once a reference point and a unit of measurement have been established, any point on a workpiece can be accurately located by specifying its coordinates. This method is illustrated in Fig. 125-5. With the intersection of the X, Y, and Z planes as the fixed reference point, Point A has

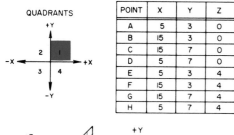

POINT	X	Y	Z
A	5	3	0
B	15	3	0
C	15	7	0
D	5	7	0
E	5	3	4
F	15	3	4
G	15	7	4
H	5	7	4

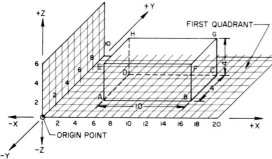

Fig. 125-6. A rectangular block positioned in the first quadrant. The chart shows the coordinate position of each corner.

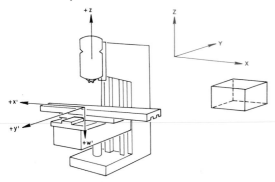

Fig. 125-7. The coordinate axes for vertical knee mills, drilling machines, and jig boring machines.

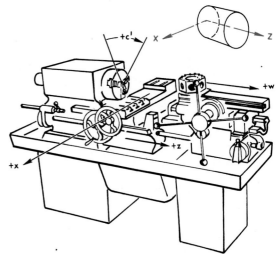

Fig. 125-8. The coordinate axes for a horizontal-spindle turret lathe.

coordinates of X + 4, Y + 2, Z + 3. Point B has coordinates of X + 4, Y − 3, Z − 2.

In Fig. 125-5, notice that the XY plane is divided into **quadrants.** Many N/C systems are designed so that all points on an object are assumed to be in the first quadrant. In these systems, all positions are designated positive (+). When the object is so located, the positive (+) sign may be omitted in the program manuscript. For example, the coordinates for Point A in Fig. 125-5 would be written as X4, Y2, Z3. Figure 125-6 outlines a rectangular block positioned in the first quadrant. The position of each of its corners, labeled A through H, is given on the accompany-ing chart. Study the diagram until you under-stand the method of describing the location of these points.

Kinds of Numerical Control

There are two principal kinds of N/C systems: (1) point-to-point and (2) contouring. Point-to-point control systems are simpler than contour-ing systems, both in design and in the amount of programming information required.

Point-to-Point N/C

A point-to-point N/C system is basically a posi-tioning system. Its primary purpose is to move a tool or workpiece from one programmed point to another. Usually the machine function, such as a drilling operation, is also activated at each point by command from the N/C program. Point-to-point N/C systems are suitable for hole-machin-ing operations such as drilling, countersinking, counterboring, reaming, boring, and tapping. Hole-punching machines, spot-welding machines, and assembly machines also use point-to-point N/C systems.

Most point-to-point N/C systems also permit straight-line cutting, such as milling of slots or grooves. The feed rate of the tool or workpiece is controlled by command from the N/C program. Straight-line cuts in either the X or Y axis may be programmed, and 45° angles are cut by moving both the X and Y axis drive motors simultaneous-ly. Programming of other angles and curves is possible, but this is not practical with point-to-point N/C systems.

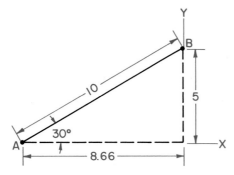

Fig. 125-9. Angular movement in contouring N/C systems is generated by programming the X and Y movements required. Programming X + 8.66 Y + 5 moves the tool 10 units along a 30° path from point A to point B.

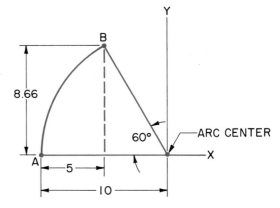

Fig. 125-10. Given an arc radius of 10 units in relation to point A, point B is located X + 5 Y + 8.66.

Contouring N/C

Contouring N/C systems can direct the tool or workpiece to move at any angle, and also along curved paths. Many contouring N/C machines, such as lathes and vertical milling machines, are of the **two-axis** type. This type allows continuous-path contours to be machined only in the XY plane. **Three-axis** machines are capable of simultaneous cutting movements in all three axes, such as is required for machining three-dimensional shapes in die and mold cavities.

Most contouring N/C systems can move the tool or workpiece only in straight lines parallel to the machine axes. Therefore, to produce angular or circular movement, the machine control unit must be capable of calculating and ordering movements in each axis which will most nearly approximate the angle or curve desired. Angular movement is generated by calculating the amount of movement required in the X and Y axes. See Fig. 125-9. A circular path is generated by determining the amount of X and Y movement required and locating the center of the arc, Fig. 125-10. Irregular curves are cut by breaking the curves into a series of straight lines that fit the curve to the degree of accuracy required, Fig. 125-11. The X and Y coordinates of each line must then be calculated and programmed. This task is greatly simplified by using computer-assisted N/C programming languages.

Production Steps in N/C Machining

The steps involved in preparation for machining parts on N/C machine tools are summarized

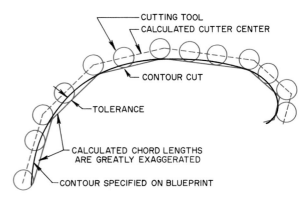

Fig. 125-11. Irregular curves must be machined as a series of straight lines.

in Table 125-1. The designer, usually an engineer, determines the part specifications and makes a sketch of the part. The engineer gives the sketch to a drafter, who uses it to prepare an engineering drawing. In some cases, an experienced drafter may make this drawing directly and have it checked by the engineer. In many companies, the engineer and the drafter use a computer-assisted design and drafting system to develop the sketch and create a finished drawing.

The N/C programmer then performs tasks that were formerly done by the machine operator. The programmer determines the sequence of machining operations and prepares a handwritten manuscript of the N/C program. He or she must specify tool or workpiece movement in each axis, tool changes, cutting speeds and feeds, coolant flow, and automatic machining cycle codes.

The programmer or a typist then types the manuscript information on a computer terminal, Fig. 125-12. The program is stored on a paper tape, a magnetic tape, or a disc. The programmer then checks the program for accuracy.

Table 125-1

Steps in the Production of a Part, Using N/C Machine Tools

Design Engineer
1. Determines part specifications.
2. Designs and makes sketch of part. May work directly with computer-aided design and drafting system (CAD) to create sketch.
3. Designs and makes sketch of work-holding fixture if needed.

Drafter
1. Makes engineering drawing of part. May also use CAD system to produce final drawing.
2. Makes engineering drawing of fixture if needed.

N/C Programmer
1. Studies engineering drawing to determine best sequence of machining operations; identifies position points needed for programming.
2. Determines location of part or fixture in relation to zero reference point on machine table.
3. Determines coordinate dimensions for position points.
4. Determines tool selection, cutting speeds, and feed rates.
5. Writes program manuscript.

Typist
1. Types program manuscript on computer terminal.
2. Stores program on disc or tape.
3. Verifies program.

N/C Machine Operator
1. Inserts N/C tape in MCU tape reader or enters N/C program into MCU memory.
2. Mounts fixture and workpiece on machine.
3. Installs tools in toolchanger or manually changes tools.
4. Adjusts feeds, speeds, tool-depth stops, and coolant on machines where these are not automatically controlled.
5. Removes machined workpiece from machine.

Usually, the first time a new program is used, the programmer and the machine operator make a **dry run**. A dry run is a trial run of the program on the machine to check whether the program is correct. The dry run may be done in two ways: (1) without a workpiece in position (cutting air) or (2) if critical dimensions must be checked, with an easily cut material in place of the workpiece. (A number of materials are made especially for this purpose.)

When a program has been approved for production, it is delivered to the machine operator along with a part drawing and a program manuscript. The program manuscript contains setup and operating instructions that must be followed before the machine operator can machine the part successfully. On simple N/C machines, the operator usually must make tool changes and may also need to adjust feeds, spindle speeds, tool depth stops, and turn the cutting fluid on and off. On more complex N/C machines, all of the work is done automatically except the initial fixture location and workpiece loading and unloading.

(Numeridex, Inc.)

Fig. 125-12.   A tape preparation center that includes a computer terminal, a plotter, a printer, and a tape punch.

Unit
126

Manual N/C Programming

Manual N/C programming is programming without using an N/C computer language. Many parts of simple shape can be programmed manually as fast as, or faster than, using an N/C computer language. Manual programming requires a programmer to prepare a written program manuscript containing the necessary tool movements, machine codes, and other instructions needed to operate the machine. The written program is then converted into machine-readable instructions in one of three ways: (1) the program is entered directly into a CNC memory from the machine's control unit, (2) a computer terminal is used to enter the program into the computer's memory for transfer to the machine tool when it is needed, or (3) a tape is prepared using a tape-punching typewriter or computer terminal.

Dimensioning Systems Used in N/C Programming

Tool movements are normally given in **absolute dimensions** for machines with absolute controls. **Incremental dimensions** are normally given for N/C machines with incremental controls. Some control systems, however, are capable of handling both absolute and incremental controls.

Absolute Dimensioning System

The part drawing in Fig. 126-1 is dimensioned by the absolute method. All points are dimensioned from a common reference point on the part called the **part reference point**. The part reference point should not be confused with the **machine table reference point. All tool movements on N/C machines with absolute control systems must be measured and programmed from the machine table reference point.**

When the part reference point is located directly over the machine table reference point, the dimensions programmed can be taken directly from the part drawing. However, when the part reference point is located elsewhere, as in Fig. 126-1, its distance from the machine table reference point must be added to all part dimensions in order to obtain the correct programming dimensions.

Since most absolute numerical control systems are designed so that the part is always located in the first quadrant of the coordinate system, all coordinate positions are positive numbers. **The plus sign and decimal points may be omitted in the programming of most N/C systems.**

Incremental Dimensioning System

The part drawing in Fig. 126-3 is dimensioned by the incremental method. With this method, the distance between points on a part is given without reference to a fixed point. Programmed dimensions for each tool movement are measured from the position of the tool at the starting point of each move. Incremental programming requires all moves in the negative direction to be preceded with the minus ($-$) sign, such as X $-$ 300.

N/C Language

The following paragraphs describe the terms used in manual numerical control programming.

Programming Format

Most N/C systems use either a **word-address format** or a **tab-sequential format.** "Words" consist of a letter followed by the numerical data, such as: N1G80X1750Y375F2000 (five words.) Each letter identifies a different machine function. The tab-sequential format uses numerical

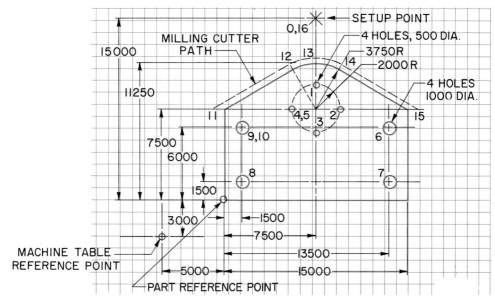

Fig. 126-1. A part drawing dimensioned by the absolute method. Note the difference between the part reference point and the machine table reference point.

N	G	X	Y	I	J	F	M	EOB	Instructions
N000								EOB	500 Dia. drill
N00	D							EOB	
N0							03	EOB	
N1	81		12500			2000		EOB	
N2		14500	10500					EOB	
N3		12500	8500					EOB	
N4		10500	10500					EOB	
N5							06	EOB	change to 1000 dia. drill
N6		18500	9000					EOB	
N7			4500					EOB	
N8		6500						EOB	
N9			9000					EOB	
N10	80						06	EOB	change to 1000 dia. end mill
N11		4000	10500					EOB	
N12		10375	14180			60	52	EOB	
N13		12500	14750	12500	10500			EOB	
N14		14625	14180		10500			EOB	
N15		21000	10500					EOB	
N16		12500	18000			2000	53	EOB	
N17							05	EOB	
N18							02	EOB	

Fig. 126-2. The absolute N/C program for the part shown in Fig. 126-1.

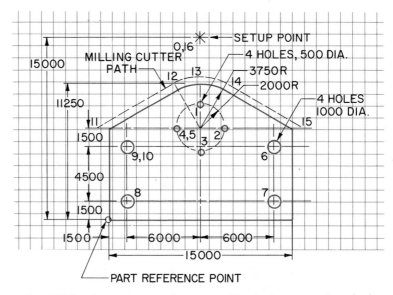

Fig. 126-3.   A part drawing dimensioned by the incremental method.

N	G	X	Y	I	J	F	M	EOB	Instructions
N000								EOB	500 Dia. drill
N00	D							EOB	
N0							03	EOB	
N1	81		−5500			2000		EOB	
N2		2000	−2000					EOB	
N3		−2000	−2000					EOB	
N4		−2000	2000					EOB	
N5							06	EOB	change to 1000 dia. drill
N6		8000	−1500					EOB	
N7			−4500					EOB	
N8		−12000						EOB	
N9			4500					EOB	
N10							06	EOB	change to 1000 dia. end mill
N11	80	−2500	1500					EOB	
N12		6375	3680			60	52	EOB	
N13		2125	570	2125	−3680			EOB	
N14		2125	−570		−4250			EOB	
N15		6375	−3680					EOB	
N16		−8500	7500			2000	53	EOB	
N17							05	EOB	
N18							02	EOB	

Fig. 126-4.   The incremental N/C program for the part shown in Fig. 126-3.

data separated by a space or tab code. The same information given in the word-address format above would appear as 1 80 1750 375 2000 in tab-sequential format.

Tab-sequential programming information must be entered in a fixed sequence determined by the way the machine control unit is wired. The number of tab codes preceding the numerical data determines which machine function will be activated. Word-address programming need not be entered in any specific sequence. The common practice, however, is to list the data in the sequence shown in the programming examples, Figs. 126-2 and 126-4.

Information Blocks

An N/C program is made up of **blocks** of coded information. The information on each line of a program manuscript constitutes a block. An **end-of-block (EOB) code,** Fig. 125-1, must be punched into the tape to separate the blocks of information.

The programming system may be either **full-block** or **variable-block.** Full-block programming requires that information be written in each category in each block of the program. This system cannot remember information from one block to another. Variable-block systems are capable of memorizing information that is the same for succeeding blocks in the program. This feature greatly reduces programming time.

Information in Blocks

A block of information for a two-axis contouring N/C program includes:

1. sequence number (N or H)
2. preparatory function (G)
3. X dimension
4. Y dimension
5. I dimension
6. J dimension
7. feed rate (F)
8. miscellaneous function (M)

More sophisticated machines also use an S number to specify spindle speed and a T number for automatic tool changing.

Sequence Number

The sequence number locates each block of information on the program manuscript and also on the tape. Many machine control units are equipped with a sequence number readout that displays the number of the operation being performed. A letter such as N or H is used as an address for the sequence number.

Preparatory Functions

G-code numbers are used to signal the MCU that a particular mode of operation is required. There are 99 G-codes available. Examples of G-codes are given in Table 126-1. G-codes sometimes differ for different N/C machines.

X, Y, Z, I, and J Coordinate Dimensions

Coordinate dimensions are written in terms of the smallest unit of movement for which the system is designed. Metric systems are commonly programmed in units of 0.02 or 0.01 mm. Inch systems are usually 0.001″ or 0.0005″. Thus, for a 0.02 mm system, 35.4 mm would need to be written as 3 540; 17.67 mm would have to be rounded to 1 766 or 1 768, since the system is not capable of less than 0.02 mm movement. For a 0.001″ system, 4-1/2″ would be written 4500; 7/16″ would have to be rounded to 437 or 438 (instead of 4375), since the system cannot move less than 0.001″. Decimal points in program dimensions are unnecessary.

X, Y, and **Z** are used in word address programming to designate moves in their respective axes. I and J are used in contour programming to describe the location of the center of an arc or circle to be machined. I identifies the distance to center in the X axis; J identifies the distance to center in the Y axis. For incremental programming, the center location is measured from the center line of the tool at the beginning of the contour cut.

Table 126-1
Example Codes of N/C Preparatory Functions

Cycle Code	Cycle Name
G70	Inch dimensions
G71	Millimeter dimensions
G78	Mill stop
G79	Mill
G80	Cancel
G81	Drill
G82	Dwell
G84	Tap
G85	Bore
G90	Absolute mode
G91	Incremental mode

See Fig. 126-4. For absolute programming, the coordinates of the center are given as in Fig. 126-2.

Feed Rate

The feed rate is programmable on most machines. Maximum feed rates are used for rapid workpiece or tool positioning between machining operations. Slower feed rates must be used during actual machining. Feed rates for milling operations may be in mm/min., m/min., or inches/min. For drilling and lathe turning, units of 0.02 mm/revolution or 0.001"/revolution are used. Some inch/min. systems allow programming to 1/10 inch, such as 6.5"/min. Since decimals are omitted, this would be written as 65.

Miscellaneous Functions

Miscellaneous functions, **M-codes,** are on-off N/C machine functions. There are 99 M-codes available. Examples of M-codes are given in Table 126-2. M-codes may differ for different N/C machines, but attempts are being made to standardize all code numbers.

Manual N/C Program Writing Procedure

The first step in manual program writing is to determine how the workpiece will be held for machining. Workholding fixtures sometimes provide obstacles to machine moves. The programmer must program around these obstacles to prevent collisions.

The location of the workpiece on the machine table should then be determined, so that a safe and convenient **setup point** can be selected. A setup point should be chosen far enough away from the workpiece so that sharp tools do not interfere with workpiece loading and unloading. **It is important that the workpiece be located so that the longest moves in each direction do not exceed the limits of machine travel.**

The most efficient sequence of tool or workpiece movements should then be determined, and the sequence numbers should be written on the part drawing. Zeros are used to number the program manuscript information blocks preceding the first tool or workpiece move. In this manner, the block numbers on the tape program will be the same as the sequence numbers on the part drawing.

Table 126-2
Examples of Miscellaneous Function Codes

Code #	Function
M02	End of program
M03	Spindle ON
M05	Spindle OFF
M06	Tool change
M52	Advance spindle
M53	Retract spindle
M56	Tool inhibit
M57	Index turret depth stop

The first line of every N/C program should contain only an EOB code. This information is necessary for carrying out the rewind stop command given on the next line of the program. The second line of every program should have only a **rewind stop code** and an EOB code. Rewind stop codes vary for different N/C systems. In the program examples shown in Figs. 126-2 and 126-4, the letter **D** is used.

Figure 126-2 gives the N/C program for the part shown in Fig. 126-1, which is dimensioned by the absolute method. The program is written for a variable-block word-address system with the part located in the first quadrant. Figure 126-4 gives the N/C program for the same part with incremental dimensioning, Fig. 126-3. This program is also written for a variable-block word-address system. Note that incremental programming requires the use of minus (−) signs to signal moves in the negative direction. **To determine whether a move is positive or negative, the programmer writes the program as though the tool were doing the moving, not the workpiece.**

After the program has been written, it should be checked for obvious errors or omissions. All N/C programs should start and stop at the setup point (sometimes called **home**). Therefore, incremental programs are partly self-checking, since the sum of all negative moves should cancel the sum of all positive moves in the X, Y, and Z axes. All programs are finished with an **end of program code,** followed by the final EOB code.

This will automatically rewind the tape or otherwise return the program to its starting point, ready to rerun for the next part.

Punching the N/C Tape

N/C tapes are prepared on typewriters or computer terminals equipped with tape punches, Fig. 125-13. After turning the tape punch on, the typist or keyboarder types the information on the program manuscript in the order that it was written. As each key is struck, its corresponding pattern of holes is punched into the tape. It is not necessary to leave spaces between information categories in word-address program tapes. This format saves typing and shortens the tape.

After the tape is punched, it may be run through the tape reader, which causes the typewriter or printing device to print out the information on the tape automatically. This step provides a way of checking the accuracy of the program tape.

The typewritten program obtained while punching the tape for the word-address program in Fig.

126-4 appears below (EOB codes do not print):

```
N000
N00D
N0M03
N1G81Y-5500F2000
N2X2000Y-2000
N3X-2000Y-2000
N4X-2000Y2000
N5M06
N6X8000Y-1500
N7Y-4500
N8X-12000
N9Y4500
N10M06
N11G80X-2500Y1500
N12X6375Y3680F60M52
N13X2125Y570I2125J-3680
N14X2125Y-570J-4250
N15X6375Y-3680
N16X-8500Y7500F2000M53
N17M05
N18M02
```

Unit

127

Computer-Assisted N/C Programming

Two types of computer-assisted N/C programming are now available. One type involves the use of a computer numerical control (CNC) machine control unit (MCU). The second type involves the use of a computer-based N/C programming language. Both types are designed to simplify and speed the N/C programming task.

CNC System Programming

CNC systems provide many of the programming capabilities formerly available only by using a computer-assisted N/C programming language. This is possible because CNC systems are designed around a microcomputer that is an integral part of the MCU. The microcomputer provides internal storage of the program, which eliminates the need for program tapes. However, the internal program storage capacity is usually stated as the equivalent of a given number of meters (feet) of punched tape.

Some CNCs are known as **manual data input** (MDI) systems because the program input must be made manually through the control panel. Other CNCs allow both MDI and program input from a remote tape reader or another computer. In these systems, MDI programs no longer needed can be transferred into another computer memory or converted into a tape for future use.

CNC programming is basically the same as manual programming, but the power of the microcomputer improves the programming in the following ways:

1. A more extensive library of "canned cycles" is provided. **Canned cycles** are pre-programmed machining sequences or different modes of machine operation. They are programmed with two-digit G-code numbers. See Fig. 127-1.
2. Custom canned cycles can be written and stored for easy recall and use.
3. Program segments may be repeated at any time without running through the entire program.
4. Program changes may be made at any time.
5. Parts of previous programs stored on tape or in another computer memory can be inserted into a program being run from the CNC memory.
6. Some CNC systems have graphics capability. The programmed tool paths are displayed as a line drawing on a television-like screen. This feature provides a convenient way of checking for certain program errors. It also assures that the correct program has been called into operation.
7. Both absolute and incremental dimensions may be used in the same program.
8. Switchable metric/inch programming is provided.
9. A cutter compensation feature allows different cutter diameters to be used for milling the same program.
10. If preset tools have different lengths, the program can be adjusted accordingly.

Because of their great versatility, CNC systems have become the new standard for operating N/C machines.

Computer-Assisted N/C Programming Languages

When using an N/C programming language, the programmer need only describe the shape of the part, specify the cutting tools, speeds, and

Fig. 127-1. Examples of "canned cycles", preprogrammed machining cycles that simplify CNC and computer-assisted N/C language programming.

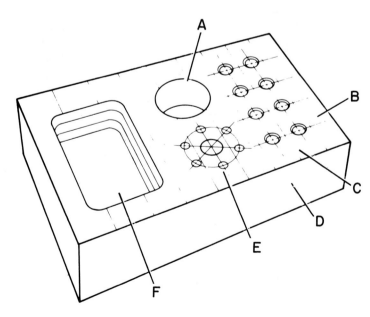

A. One command automatically mills a hole. The only inputs needed are feed rate and tool path radius.

B. Two "peck-feed" cycles can be performed. During a drilling operation one of the cycles pulls the tool from the hole to clear chips. The other cycle retracts the tool 1.27 mm (.05") to break the chip.

C. Patterns can be defined once and repeated a fixed number of times. They can also be automatically repeated for multiple operations on the same hole.

D. One command automatically mills an entire face. Inputs needed are X and Y coordinates, feed rate, and step data.

E. One command automatically produces bolt hole circles. Inputs needed are center location, hole spacing, and feed rate.

F. One command automatically mills an entire pocket. Inputs needed are X and Y coordinates, feed rate, and step data.

(Bridgeport Machines — A Textron Company)

feeds, and provide the desired machining sequence. The computer does all necessary calculations, checks for programming errors, and produces the program on a tape or other form of program storage.

The first N/C computer programming language was called Automatic Programming of Tools (APT). It was developed to assist the aircraft industry in machining large, complex forgings that required great precision. Evidence of the complexity of their N/C programs was provided in a report by the Boeing Aircraft Company that their N/C program tapes varied in length from 18.28 m (60') to 365.76 m (1200'). APT was of necessity a very powerful language, capable of handling up to five machine axes simultaneously. It required a large computer for its use, and it proved too complex and too costly for average N/C machine users.

Today there are dozens of N/C programming languages. In addition to powerful languages such as APT and Compact II® , many less powerful languages have been developed to operate particular types of machines, such as lathes, milling machines, and drill presses. In addition, a few very simple N/C languages have been written for making one kind of part on one particular type of machine.

Training periods required for developing basic proficiency in writing N/C programs vary from as little as one day to as much as six months, depending on the language complexity.

Computer-Assisted N/C Programming Example

The part that was used in Unit 126 to illustrate manual programmming methods, Figs. 126-1 and 126-3, will also be used to illustrate computer-assisted programming. Figure 127-2 shows the part labeled and dimensioned for Compact II programming.

The first three lines or **statements** in all Compact II programs must be the **MACHIN**e name, program **IDENT**ification, and **SETUP** location. See Fig. 127-3. Compact II programming then describes the part shape in terms of lines, points, and circles. A language of condensed English words and key letters, with the necessary dimensions, is used in writing the part description statements. **Major words** such as **DCIR**n (Define **CIR**cle number), **DLN**n (Define **Li**Ne number), and **DPT**n (Define **P**oin**T** number) are used to begin each program statement. **Minor words** such as **PARLN**n (**PAR**allel to **Li**Ne number), and **PERLN**n (**PER**pendicular to **Li**Ne number), and necessary dimensions complete each statement.

The rest of the program describes the movement of the tool(s) for machining the workpiece. Major words for straight-line moves are **MOVE** (for rapid moves between cutting operations), and **CUT** (for milling at a programmed feed rate). Major words for curved cuts are **CONT** (for **CON**Tour cutting with the center line of the cutter on the curve), **ICON** (for Inside **CON**tour cutting with

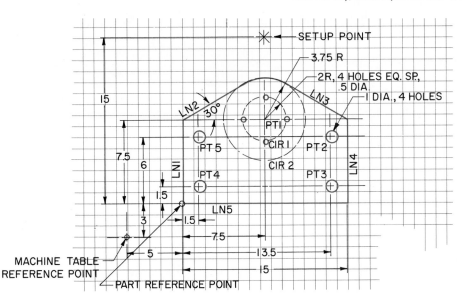

Fig. 127-2. A part dimensioned and labeled for Compact II computer-assisted N/C programming.

the cutter center line remaining half the cutter diameter from the inside of a circle or near side of a curve), and **OCON** (for Outside **CON**tour cutting with the cutter center line remaining half the cutter diameter outside a circle or on the far side of a curve). Minor words such as **TOLNn** (**TO** LiNe number), **ONLNn** (**ON** LiNe number), **PASTLNn** (**PAST** LiNe number), **TANCIRn** (**TAN**gent to **CIR**cle number), etc. are used to direct the movement of the tool.

The major word **DRL[n]** (**DRiL**) is used to begin a drilling statement. An **[n]** indicates the number of equally spaced holes on a line or around a circle. (The brackets indicate that the use of **n** is not required for a single hole). Minor words such as **nDP (DeeP)** indicate the full-diameter depth the hole is to be drilled; **nTHRU (THRoUgh)** indicates the thickness of the part to be drilled all the way through. **nTPA (Tool Point Angle)** states the included angle of the tool, which the system needs in order to calculate the length of the stroke needed to drill the hole.

Tool change command statements begin with the major words **ATCHG** (Automatic Tool **CHanGe)** or **MTCHG** (Manual Tool **CHanGe)**. This is followed by the minor words **TOOLn** (**TOOL** number), **nTD** (Tool Diameter), **nGL** (Gage Length of the tool — the distance the tool projects from the end of the spindle), **nRPM** (tool **R**evolutions **P**er **M**inute), **nMMPR** or **nIPR** (feed rate in **M**illi**M**eters or **I**nches **P**er **R**evolution), and **nTPA** (Tool Point Angle).

All programs are concluded with an **END** statement. The **END** statement causes the tool to return to the setup point. It also rewinds the tape in preparation for recycling the program.

Steps in Compact II N/C Programming

1. The programmer writes a program manuscript in Compact II language. Fig. 127-3 shows the program manuscript written for the part in Fig. 127-2.
2. The programmer or a typist types the program manuscript. This operation simultaneously punches a tape.
3. Using a computer terminal, the programmer uses the punched tape to load the program manuscript into the computer's working memory.
4. The programmer uses computer debugging routines to produce a program that is free of language errors.

```
MACHIN,MILL
IDENT,SAMPLE COMPACT II PART PROGRAM
SETUP,12.5LX,18LY,20LZ
BASE,5XA,3YA,4ZA
DLN1,XB
DLN2,XB,7.5YB,30CCW
DLN3,15XB,7.5YB,150CCW
DLN4,15XB
DLN5,YB
DPT1,7.5XB,7.5YB,ZB
DPT2,13.5XB,6YB,ZB
DPT3,13.5XB,1.5YB,ZB
DPT4,1.5XB,1.5YB,ZB
DPT5,1.5XB,6YB,ZB
DPAT1,2/5
DCIR1,PT1,2R
DCIR2,PT1,3.75R
ATCHG,TOOL1,12GL,.5TD,118TPA,1200RPM,.003IPR
DRL4,CIR1,S90,CCW,2THRU
ATCHG,TOOL2,12GL,1TD,118TPA,600RPM,.003IPR
DRL,PAT1,2THRU
ATCHG,TOOL3,12GL,1TD,3IPM
MOVE,OFFLN2/YL,OFFLN1/1XS,1ZB
OCON,CIR2,S(TANLN2),F(TANLN3),CW,-2.5ZB
CUT,PARLN3,OFFLN4/1XL
END
```

Fig. 127-3. The Compact II N/C program manuscript for the part shown in Fig. 127-2.

(Manufacturing Data Systems Incorporated)

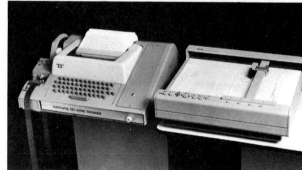

Fig. 127-4. A Compact II Graphics Systems Plotter used to check the accuracy of a Compact II N/C program prior to punching the machine control tape.

5. The programmer runs the debugged program to obtain a drawing on a plotter or a display screen. See Fig. 127-4. This reveals any X or Y axis tool-positioning errors. The programmer makes corrections if necessary.
6. The programmer records the corrected program manuscript, Fig. 127-3, for future use. The N/C program for operating the machine tool, Fig. 127-5, is also recorded.

```
N001G91
N002G00X0Y0Z0F2500000
N003T01M06
N004G00F2500000S1200M03
N005Y-55000
N006G01Z-23502F36000
N007G00Z23502F2500000
N008X-20000Y-20000
N009G01Z-23502F36000
N010G00Z23502F2500000
N011X20000Y-20000
N012G01Z-23502F36000
N013G00Z-23502F2500000
N014X20000Y20000
N015G01Z-23502F36000
N016G00Z23502F2500000
N017M05
N018T02M06
N019G00F2500000S0600M03
N020X40000Y-15000
N021Z-39000
N022G01Z-25004F18000
N023G00Z25004F2500000
N024Y-45000
N025G01Z-25004F18000
N026G00Z25004F2500000
N027X-120000
N028G01Z-25004F18000
N029G00Z25004F2500000
N030Y45000
N031G01Z-25004F18000
N032G00Z25004F2500000
N033Z39000M05
N034T03M06
N035G00X-30000Y12114F2500000
N036Z-30000
N037G01X68750Y39692Z-35000F30000
N038G02X21250Y56942I21250J36806
N039X21250Y-56940I0J42500
N040G01X68750Y-39693
N041G00Z65000F2500000
N042X-90000Y77887
N043M30
```

Fig. 127-5. The computer-generated N/C machine control program resulting from the Compact II program manuscript shown in Fig. 127-3.

7. The programmer checks the accuracy of the N/C program by using it to make a dry run on the machine tool.

Test Your Knowledge of Section 16

Unit 125: Introduction to Numerical Control

1. List four methods used to provide instructions to numerically controlled machines.

2. How do DNC and CNC systems differ? How are they alike?
3. Discuss the advantages of using N/C machines.
4. Explain the Cartesian coordinate system. Which axis is assigned to the machine spindle?
5. Of what advantage is an N/C system that assumes all points on the object are in the first quadrant?
6. Name the two principal kinds of N/C systems and explain how they differ.
7. Explain how irregularly curved lines must be programmed.
8. Outline the procedure involved in preparing and checking an N/C program.

Unit 126: Manual N/C Programming

1. Explain the difference between absolute and incremental dimensioning.
2. Explain the difference between the part reference point and the machine table reference point. From which point must all tool movements be measured?
3. In incremental programming, what symbol is used to indicate a move in the negative direction?
4. Explain the difference between word-address and tab-sequential N/C tape formats. Which is most efficient?
5. Explain what a block of information is on both a program manuscript and an N/C tape.
6. How do variable-block and full-block programming systems differ? Which is most efficient?
7. For what purposes are G-code numbers used in N/C programming?
8. Make a sketch showing the orientation of the X, Y, and Z axes for a vertical milling machine.
9. Explain the use of "I" and "J" in word address programming.
10. How are feed rates for milling operations specified?
11. For what kind of functions are M-code numbers used? Give several examples.
12. What factors should be considered when deciding where to locate the setup point for a part program?
13. Explain the partly self-checking feature of incremental programming.

In the study of metals, it is necessary first to understand basic metal classifications. Metals may exist in the form of a **pure metallic element,** such as aluminum, copper, lead, silver, and so on. Most pure metals, however, are useful for relatively few engineering applications. The great majority of metals used in engineering applications are mixtures of two or more metallic elements known as **alloys.** A modern automobile has more than one hundred different metals used in its manufacture, most of which are alloys. Thousands of alloys are used in modern industry.

Alloy

The term **alloy** can have several meanings, depending on whether it is used in a metallurgical sense or in a commercial sense.

In metallurgy, an alloy is a substance that possesses metallic properties and is composed of a combination of two or more elements. One of the elements must be a metal. The others may be either metallic or nonmetallic. In this sense, plain carbon steel is basically an alloy of iron and carbon. (It also includes certain other elements in the form of impurities.)

However, for **commercial** purposes, plain carbon steel is seldom classified as an alloy steel. Used in this sense, an alloy is a metallic substance composed of a combination of two or more metallic elements, one of which must be intentionally added.

Alloys are further classified as ferrous alloys and nonferrous alloys. The **ferrous** alloys contain iron as the **base** metal, and one or more other metallic elements. Nonmetallic elements also may be included. The base metal is generally the metal in largest proportion. For steels, the base metal is iron. Examples of ferrous alloys include

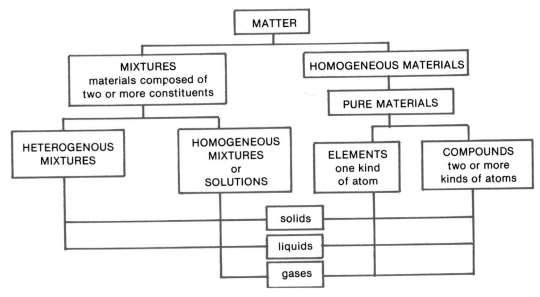

Fig. 128-1.   Classification of matter.

nickel steel, stainless steel, vanadium steel, and alloyed cast iron. All steels which have metallic elements intentionally added to change their properties are ferrous alloys.

Nonferrous alloys do not contain iron, except in very small amounts in the form of impurities. Examples of nonferrous alloys are brass, bronze, and monel metal. Brass is a copper-base alloy which is composed largely of copper and zinc. Bronze is also a copper-base alloy composed largely of copper and tin. Monel is a nickel-base alloy composed largely of nickel and copper.

Composition of Alloys

The classifications of alloys and the properties of alloys depend largely on their composition. Alloys generally are composed of a **mixture** of several ingredients. The mixture may be composed of **elements** or **compounds,** or a combination of both elements and compounds.

Some alloys are called **homogeneous** mixtures because they are uniform or alike throughout. Other alloys are called **heterogeneous** mixtures because they are not uniform throughout.

Since a clear understanding of these terms is important in understanding metals and alloys,

the terms will be reviewed briefly. They apply not only to metals and alloys, but also to all other materials. See Fig. 128-1.

Elements

All matter is made up of chemical **elements.** About 85% of these are natural elements — that is, they occur in nature. The remaining elements are laboratory-produced.

The chemical elements form the building blocks from which all substances are composed. The smallest particle of an element which can enter into a chemical combination to form a new substance is an **atom.** All elements are made up of atoms, and each element has atoms which are unlike those of another element. An atom is so small that it cannot be seen even with the most powerful microscope. The smallest crystalline grain of metal which can be seen with a powerful metallurgical microscope is made up of an extremely large number of atoms.

The chemical elements are classified as **metallic** or **nonmetallic** elements, depending upon whether or not they possess recognized metallic properties. About three-fourths of the elements are classified as metals.

Metallic elements and other metallic materials possess many properties, but several are sufficient to distinguish between metallic and nonmetallic materials. The recognizable metallic properties include: metallic luster, good electrical and heat conductivity, some degree of malleability, and a crystalline structure while in the solid state. All metallic elements are in the solid state at room temperature, except mercury.

Scientists determine whether an element is metallic or nonmetallic through the use of an **electrolysis test.** The element is dissolved in an acid, and an electric current is passed through the solution. The atoms of metallic elements show a positive charge and are attracted to the negative pole where the current enters the solution. Metals, therefore, may be scientifically defined as those elements which, in a pure state, in a solution, carry a positive charge and seek the negative pole in an electric cell. Hydrogen is a nonmetal which is an exception to this definition.

Compounds

Two or more elements may unite chemically to form a **compound,** which is a new substance. Water (H_2O) is a compound of hydrogen and oxygen. Iron and carbon unite to form **cementite,** which is iron carbide (Fe_3C), a compound. Cementite exists in some proportion in all types of steel and cast iron. In fact, through the control of the form and amount of cementite in steel and cast iron, the properties of these metals may be set at will.

A compound is a substance very different from the elements of which it is composed. An example is water. Independently, hydrogen will burn, and oxygen will support combustion. However, when chemically combined as water, the compound will extinguish most fires. Another example is cementite. Pure iron is too soft for most practical applications, but when it is combined with carbon in the form of cementite, it is very hard.

The smallest particle of a compound which can exist to show its properties is a **molecule.** A molecule of water (H_2O) is made up of two atoms of hydrogen and one atom of oxygen. A molecule of common table salt (NaCl) is composed of one atom of sodium and one atom of chlorine. The atoms are combined in specific proportions in the molecules of all compounds. Molecules, like atoms, are too small to be seen with the most powerful microscope. Their existence has been determined by x-ray studies. There are many thousands of molecules in one tiny grain of table salt.

Pure Substances

All subtances or materials which exist as a single element or a single compound are classified as **pure,** Fig. 128-1. Lead is a pure substance which exists as a single element. Pure water is a substance which exists as a pure compound.

Homogeneous Substances

Substances which are uniform in composition throughout are said to be **homogeneous.** All parts of a homogeneous substance are alike. Since all pure substances form in a consistent mathematical relationship, they are all homogeneous. Pure lead, pure iron, pure water, and pure oxygen are examples of solids, liquids, and gases which are homogeneous substances.

Many substances are not uniform throughout. Therefore they are **heterogenous** substances. Heterogenous has a meaning which is opposite from homogeneous.

Alloys Form as Mixtures

Most metals which are produced for commercial purposes are alloys which form as mixtures. Mixtures may be of two types — heterogenous mixtures and homogeneous mixtures. **Homogeneous mixtures** generally are called **solutions.** (These will be described later.) An understanding of mixtures will aid in understanding the properties of metals and alloys. It also will provide a good background for the study of heat-treatment processes.

Mixtures are composed of two or more ingredients which are mixed together without regard for specific proportions. **The ingredients in a mixture do not unite chemically as in a compound.** Rather, each retains its own identity. The individual ingredients in the mixture may be in the form of elements, compounds, or other mixtures.

Malleable cast iron is an example of a solid substance which is a mixture. A photomicrograph of the crystalline grain structure of low-

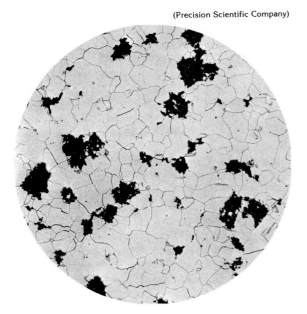

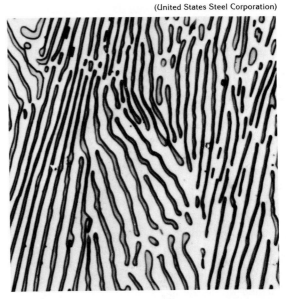

Fig. 128-2.   Photomicrograph of ferrite (light area) and temper carbon (dark area) in low-strength malleable cast iron magnified 100 times.

Fig. 128-3.   Pearlite, a microscopic structure, is shown magnified 2500 times. Dark streaks are cementite (iron carbide) and light areas are ferrite (nearly pure iron.)

strength malleable cast iron is shown in Fig. 128-2. The light area shows ferrite grains (nearly pure iron). The dark areas show aggregates of temper carbon in the form of graphite. Actually, there also may be small amounts of other constituents in the form of impurities which are mixed in with the ferrite. The impurities cannot be seen at the low magnification illustrated. **Pearlite** is a mixture which exists in the crystalline grain structure of many types of steel and cast iron. It is composed of tiny layers of ferrite (nearly pure iron) and cementite (iron carbide). A photomicrograph of pearlite is shown in Fig. 128-3. Pearlite is so named because it resembles mother-of-pearl. Steel with a high proportion of pearlite is harder and stronger than steel with a low proportion of pearlite.

The cast iron and the pearlite described above are **heterogenous mixtures** in the form of solids. They are heterogenous because the parts are not all alike throughout. Such mixtures also frequently are called **mechanical mixtures.**

Solutions

Many substances are in the form of a solution; that is, a homogeneous mixture of two or more substances whose proportions may vary within a wide range of limits. Normally, solutions are thought of as liquids, but there also are solid solutions and gaseous solutions. Since liquid solutions are most familiar, an example of this type will help describe the nature of solutions.

Liquid Solution. When sugar is added to water, the solid disappears. The sugar and water form a clear, homogeneous solution in which both ingredients are uniformly distributed. The sugar is called the **solute,** and the water is called the **solvent.** When only a small amount of sugar has been added and is dissolved, the solution is **dilute.** As more sugar is added, the solution becomes **concentrated.** When sugar is added to a point where it is no longer dissolved, but settles to the bottom, the solution is said to be **saturated.**

Solutions have several distinguishing characteristics. They are homogeneous mixtures. They are not chemical compounds, since they may form with the same ingredients in various proportions. They have properties of mixtures. However, the constituents differ from the aggregates which form heterogenous mixtures in that they are dissolved and are uniformly distributed. They remain uniformly distributed and do not settle out so long as the saturation point is not exceeded. Though solutions are not compounds, there is a loose chemical union between the solute and the solvent in most solutions.

Gaseous Solution. Air is an example of a gaseous solution. It is a homogeneous mixture of oxygen, nitrogen, water vapor, and small quantities of other gases.

Solid Solution. A solid solution is a solid substance which is a homogeneous mixture of two or more constituents. A number of alloys are of the solid solution type. An alloy may be formed where the alloying element or elements (solute) are dissolved in the base metal (solvent). The alloy thus formed is a solid solution. It is not chemically combined, yet it is alike throughout. The atomic arrangement of a solid solution is usually the same as that of the solvent.

Solid solutions may be formed with various proportions of solute and solvent. Some substances will dissolve in solids in very small proportions, and the saturation point will be reached. Other solid substances are mutually soluble in each other in all proportions.

A solid solution is formed in much the same manner in which sugar and water forms a liquid solution. Elements such as carbon, phosphorous, silicon, and manganese dissolve in iron in the unusually small proportions in which they are found in commercially pure iron. This form of iron is a solid solution.

Ferrite is nearly pure iron. Its composition is essentially the same as commercially pure iron. Ferrite exists in various proportions at temperatures below 721° C (1330° F) within the grain structure of the following: wrought iron, plain carbon steels which have not been hardened, gray cast iron, and in many alloy steels.

When sulfur is present in nearly pure iron, it combines with manganese to form manganese sulfide, a compound. The compound then exists as an inclusion in the iron. In this form, the iron alloy is a mechanical mixture of nearly pure iron and particles of manganese sulfide. The microscopic grain structure called **ferrite** is shown in Fig. 128-4. The tiny dark spots in the figure are inclusions in the form of impurities.

The following is a typical analysis for commercially pure iron: carbon 0.02%, manganese 0.01%, phosphorous 0.01%, silicon 0.01%, sulfur 0.02%, and iron 99.93%. Iron with a purity of 99.95% is available for use in chemistry laboratories.

Some alloying elements, such as nickel and copper, or gold and silver, are mutually soluble in each other. They dissolve in each other in all proportions in both the liquid and the solid states. The nickel-copper alloys and the gold-silver alloys are solid solutions.

Other alloying elements dissolve in the base metal in different proportions in the liquid and solid states. Furthermore, the proportion of the alloying element which remains in solid solutions may vary with the temperature and grain structure of the alloy formed. Carbon, the most significant alloying element in steel, behaves in this manner in steel.

Steel is basically an alloy of iron and carbon. Various heat-treatment processes are used in changing the properties of steel. These processes are possible largely because of the manner in which carbon is retained in solid solution with iron. The various heat-treatment processes are explained in Section 18.

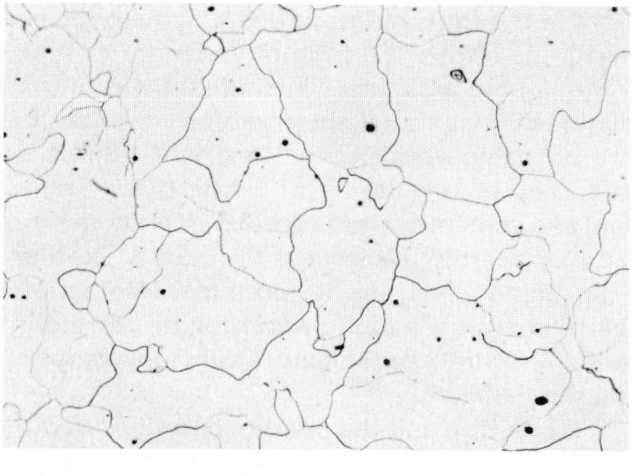

Fig. 128-4.   Ferrite (alpha iron) a microscopic structure is magnified 100 times. All grains have the same composition. The dark spots are impurities in the form of inclusions.

Unit
129

Properties of Metals

Each particular metal is developed or selected because of its characteristics or properties in relation to its use. Typical properties which are considered in the selection and production of metal products include: weight, hardness, toughness, brittleness, corrosion resistance, response to heat treatment, and machinability. A basic knowledge of various metals and their properties is important for workers engaged in the design, fabrication, and maintenance of metal products.

Metals can be hardened, softened, bent, twisted, pulled apart, or fractured, depending on the properties which they possess. Also, the ease with which metal may be formed, shaped, or machined depends on its properties.

The properties of metals and other materials may be classified in three main groups as follows: (1) **Chemical properties** include characteristics concerned with the chemical composition and chemical reactions of materials. (2) **Physical properties** include those characteristics used to describe a substance which is not being acted upon by external forces. It includes such items as color, density, heat conductivity, and electrical conductivity. (3) **Mechanical properties** include those characteristics which describe the behavior of metal which is being acted upon by external forces. It includes characteristics such as tensile strength, hardness, ductility, and machinability.

Mechanical Properties

In metal machining and in related areas of metalwork, an understanding of the mechanical properties of metals is most important. Some of the more significant mechanical properties which should be understood are described here.

Hardness

Hardness may be defined as resistance to penetration. Of necessity, metal-cutting tools are made of metals which rank high in hardness. Various types of instruments are available for measuring the relative hardness of metals. This property may be increased or decreased in many metals by various methods of heat treatment. Tools are machined to shape while soft. They are then hardened by heat treatment to provide strength and durability.

Hardenability

Hardenability is the property which enables a metal to harden completely through to its center when a heat-treatment method is used. Some metals rank low in hardenability. This means that they harden significantly on the surface layer only, while they remain relatively soft at the core or center.

Brittleness

Brittleness refers to the ease with which a metal or other material will crack or break apart without appreciable deformation. Gray cast iron is brittle in comparison with unhardened steel. Under a heavy blow, the cast iron will break apart like a cement block, while steel usually will bend instead of breaking. Brittleness is related to hardness in metals. Generally, as the hardness of a metal is increased, its brittleness also is increased.

Ductility

Ductility is the property which enables a metal to be bent, twisted, drawn out, or changed in shape without breaking. Metals which rank high in ductility, such as copper, aluminum, and soft steel, can be drawn out into wire without breaking. Ductile metals fail or break gradually with increased stress or load. In contrast, brittle metals, such as cast iron, break abruptly by cracking when placed under increasing loads. Ductility usually is expressed as a **percentage of elongation** and as a **percentage of reduction in area.** These factors are derived from specimens on which tests for tensile strength have been made.

Malleability

Malleability is the property which enables a metal to be permanently deformed by compres-

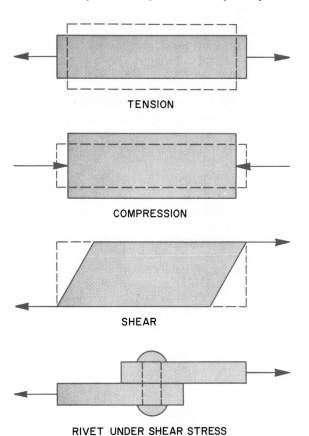

TENSION

COMPRESSION

SHEAR

RIVET UNDER SHEAR STRESS

Fig. 129-1.   Types of stress or load application.

sion, usually by rolling or hammering. Most ductile metals are malleable. However, there are some exceptions, such as lead. Lead lacks sufficient tensional strength to be stretched very far; therefore, it is not ductile.

Toughness

Toughness is the property which enables a metal to withstand sudden shock or impact forces without fracture. Impact tests measure the energy required to fracture a standard-size test specimen with one sharp blow. The testing devices generally utilize a falling weight or a swinging pendulum. Two common devices utilizing a swinging pendulum are the Charpy and the Izod tester. The energy required to fracture the test specimen is expressed in joules (J) (foot-pounds) and is a measure of impact toughness. Impact toughness and impact testing equipment are explained in greater detail in Unit 139.

Machinability

Machinability is a relative term which indicates the ease with which metals may be machined. Many factors are considered in arriving at machinability ratings. Some of the most important ones are rate of metal removal, quality of the finished surface, and tool life. Machinability ratings are expressed as a percentage, in comparison with AISI 1112 steel, which is rated at 100 percent. For a more detailed explanation of machinability, see Unit 116.

Strength

Strength of a metal is its resistance to deformation. There are several common types: **tensile** strength, **compressive** strength, **shear** strength, and **torsion** or twisting strength. Metals may be deformed when loads are applied which exert tensional forces, compression forces, forces producing a shearing action, and torsional forces producing a twisting action. See Fig. 129-1.

Tensile Strength

Tensile strength is that property of a material which resists tensional forces applied to pull the material apart. On a test specimen, tensile strength is the maximum load force per unit of original cross-sectional area before fracture of the speci-

(United States Steel Corporation)

men. Tensile strength of metals usually is expressed in terms of millions of pascals (MPa) [thousands of pounds per square inch (psi)]. The tensile strength of steel generally is related to its hardness. As the hardness is increased, the tensile strength is increased.

Tensile strength is determined through use of tensile-testing machines. Although metal test specimens of standard cross-sectional size generally are used, it is possible to test the tensile strength of specimens of many different shapes and sizes. A standard tensile specimen of 12.827 mm diameter (129.032 mm²) [0.505″ in diameter (0.2 square inches cross-sectional area)] commonly is used in many metallurgical and research laboratories.

An example will illustrate how the tensile strength for a test specimen is calculated. If a maximum load force of 88 964.432 N (20,000 psi) is required before rupture of a steel specimen 12.827 mm (0.505″) in diameter, what is the tensile strength of the steel?

Formula. The tensile strength is equal to the maximum load before rupture, divided by the original cross-sectional area of the specimen.

$$\text{Tensile strength} = \frac{\text{Load in newtons (pounds)}}{\text{Area in sq. mm (sq. inches)}}$$

$$\text{Tensile strength} = \frac{88\ 964.432\ \text{N (20,000 lbs.)}}{129.032\ \text{mm}^2\ (0.2\ \text{sq. in.})}$$

$$\text{Tensile strength} = 689.476\ \text{MPa (100,000 psi)}$$

It is evident that the tensile-testing machine used in this example must be able to withstand a tensile load of 137.895 MPa (20,000 psi). If a tensile-testing machine of less capacity were used, the cross-sectional area of the test specimen necessarily would be smaller. It would have to be within the capacity of the machine.

The general procedure involved in making a tensile strength test on steel is illustrated in Fig. 129-2. The original length of the test specimen is noted, and the specimen is inserted in the tensile-testing machine. A steadily increasing load then is applied. When the **yield point** is reached, the specimen exhibits a rather sharp increase in length and a decrease in diameter, as shown in Fig. 129-2B. When the tensile strength is exceeded under a continued increase in load, the specimen is fractured, as shown in Fig. 129-2C.

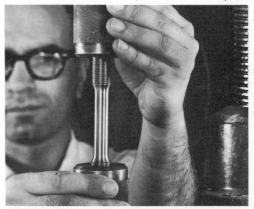

A. Inserting specimen in testing machine.

B. Effect on specimen when yield point is reached.

C. Specimen fractured when ultimate tensile strength is exceeded.

Fig. 129-2. Making a tensile strength test on a steel specimen.

Yield Point

The yield point is determined in the process of making a tensile strength test on certain ductile metals, such as low-carbon steel. It usually is expressed in terms of MPa [thousands of pounds per square inch (psi)]. When the yield point is reached, the specimen will continue to elongate or stretch a certain amount without a further increase of the tensional load.

Some ductile and brittle metals do not behave in this manner under tensile tests. Ductile metals such as copper and aluminum deform steadily with increased tension until they fail. The **yield strength** for these metals is the stress at which the test specimen exhibits a specified amount of permanent elongation or distortion. Some brittle metals, such as cast iron, also behave in a similar manner in tensile tests, except that they exhibit much less elongation before failure.

Elongation

Elongation is the amount of permanent extension in length in the vicinity of the fracture on a tensile strength specimen, as shown in Fig. 129-2. It is expressed as a percentage of the original gage length of the specimen. The original gage length is usually 50.8 mm (2"). Elongation also is a measure of ductility. The greater the percentage of elongation, the greater the ductility of the metal.

Stress

Stress in a material or structural member is its internal resistance to an external force or load. Actually, it is equal to the load or force applied, but it is specified in terms of force per unit of cross-sectional area.

The difference between load and load per unit of cross-sectional area can be understood more clearly with an example. A common-size, standard steel specimen for a tensile strength test has a diameter of 12.827 mm (0.505"), the cross-sectional area being 129.223 mm² (0.2 sq. in.). When a tensional load of 4 448.222 N (1000 lbs.) is applied, the specimen will be under a stress of 34.423 MPa (5000 psi). The stress is equal to the load force divided by the cross-sectional area, or in this case, 4 448.222 (1000) divided by 129.223 (0.2").

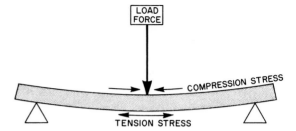

Fig. 129-3.   Combination of forces acting on a structural member.

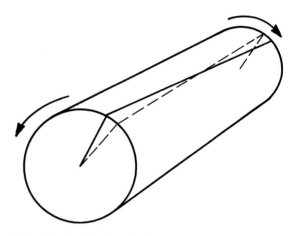

Fig. 129-4.   Torsional load.

There are three general types of stress — tension, compression, and shear. See Fig. 129-1. In **tension** stress, the forces tend to pull the material apart. In **compression** stress, the forces tend to squeeze or shorten the material under stress. In **shear** stress, there is a tendency to cause one part of the material to slide by the other part. When bending loads are applied, a combination of tension stress and compression stress is involved, Fig. 129-3. When torsional loads are applied, all three types of stress are involved to some degree. A torsional load is shown in Fig. 129-4.

Reduction of Area

Reduction of area is the difference between the original cross-sectional area of a tensile strength test specimen and the final cross-sectional area at the point of rupture. During a tensile strength test

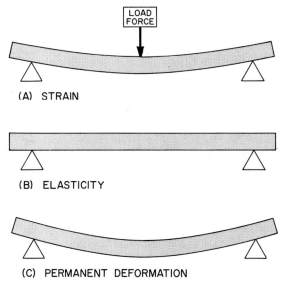

(A) STRAIN

(B) ELASTICITY

(C) PERMANENT DEFORMATION

Fig. 129-5.   Effects of load force on a steel beam.

on ductile metals, at stresses greater than the yield point, the test specimen becomes deformed. The deformation is in the form of elongation and reduced cross-sectional area, as shown in Fig. 129-2.

The reduction of area usually is expressed as the percentage of the original area. There is a much greater percentage reduction of area for ductile metals than for brittle metals. The reduction of area also is a measure of the amount of ductility.

Strain

Strain is the deformation resulting from an external force applied to a material or structural member. The structural member in Fig. 129-5A is said to be under strain, since it is deformed from an external load. In this case, the deformation is in the form of a bend. When the member returns to its original form with the removal of the load, it is no longer under a strain.

The types of deformation which take place when materials are under strain include elongation, compression, shear, torsion, and bending. The type of deformation depends on how the load is applied. Methods of applying loads and the direction in which the forces are exerted are shown in Figs. 129-1, 129-4, and 129-5.

Elasticity

Elasticity is the property which enables a material to return to its original size and shape after an external, distortional force has been removed. Rubber is elastic because it returns to its original shape after deformation, not because it stretches. In this sense, steel is more elastic than rubber. When a steel beam returns to its original form after removal of the load, as in Fig. 129-5B, it exhibits elasticity.

Elastic Limit

Elastic limit is the maximum load, per unit of area, which may be applied to a material without permanent deformation. The elastic limit occurs at a stress just short of that which causes permanent deformation. When the load force on a structural member exceeds the elastic limit, the member will be deformed permanently. The steel beam in Fig. 129-5C is an example of permanent deformation caused by a load which was greater than the elastic limit.

Plasticity

Plasticity is the property in a material which permits permanent deformation to occur without rupture. It is the opposite of elasticity. Modeling clay has excellent plasticity. Many metals exhibit plasticity when stressed beyond their elastic limits. Ductile metals generally rank high in plasticity, while brittle metals rank low. When ductile metals are machined, continous-type chips are formed due to plasticity. See Fig. 115-1.

Fatigue

Fatigue is the property that causes metal to fracture under a repeated load which is considerably below the tensile strength of the material. By repeated bending, a piece of metal may be broken due to fatigue. Springs, propeller shafts, gear teeth, and other parts of machines subjected to repeated vibration sometimes fail due to fatigue. Fractures due to fatigue often occur as a result of tool marks or sharp edges on shafts or other machine parts. Elimination of machine marks and provision for a fillet at intersecting surfaces will aid in preventing **fatigue failures.**

Testing machines have been devised to exert various types of cyclic stresses on different types of metal parts, as well as on standard test specimens of various metals and alloys. On the basis of the standard tests, various metals are assigned fatigue strength ratings, usually referred to as **endurance limit ratings.** This rating is defined as the greatest stress which will not produce fatigue failure in a specified number of cycles. A typical endurance limit for one type of steel is approximately 275.79 MPa (40,000 psi) for 10 million cycles. Similar tests usually will reveal a lower load rating for an increased number of cycles, such as the load for 100 million cycles.

Fusibility

Fusibility is that quality which enables a metal to join readily with another metal when heated to a liquid state. Fusibility is important in welding. Metals which rank high in fusibility generally can be welded with ease. **Weldability** is the specific term which indicates the relative ease with which metal may be welded by practical means.

Chemical Properties

The chemical properties of metals are concerned largely with chemical composition, melting temperatures, and corrosion resistance. A detailed discussion of the chemical properties of metals is not within the scope of this book. This type of information can be found in standard handbooks for engineers and machinists, chemistry handbooks, and manufacturer's and metal wholesaler's catalogs.

Unit
130

Metal Designation and Identification

Thousands of standard metal alloys are available, each of which has unique properties making it the best choice for certain applications. In order for different manufacturers to produce the same alloys, it was necessary to standardize the alloy compositions. For the convenience of manufacturers and consumers, it was also necessary to develop standard numbering systems so that each alloy could have its own unique designation.

Steel Numbering Systems

The two main steel numbering systems were developed by the **Society of Automotive**

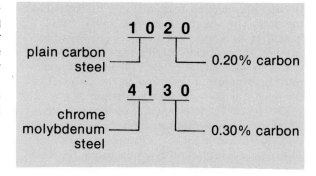

Fig. 130-1.   SAE and AISI code numbers.

Table 130-1
Series Designations in SAE-AISI Steel Code

Series	Types		
10xx ...Nonsulphurized carbon steels			
11xx ...Resulphurized carbon steels (free machining)			
12xx ...Rephosphorized and resulphurized carbon steels (free machining)			
13xx ...Mn 1.75%			
*23xx ...Ni 3.50%			
*25xx ...Ni 5.00%			
31xx ...Ni 1.25%	Cr 0.65%		
33xx ...Ni 3.50%	Cr 1.55%		
40xx ...Mo 0.20 or 0.25%			
41xx ...Cr 0.50 or 0.95%	Mo 0.12 or 0.20%		
43xx ...Ni 1.80%	Cr 0.50 or 0.80%	Mo 0.25%	
44xx ...Mo 0.40%			
45xx ...Mo 0.52%			
46xx ...Ni 1.80%	Mo 0.25%		
47xx ...Ni 1.05%	Cr 0.45%	Mo 0.20 or 0.35%	
48xx ...Ni 3.50%	Mo 0.25%		
50xx ...Cr 0.25, 0.40 or 0.50%			
50xxx ...C 1.00%	Cr 0.50%		
51xx ...Cr 0.80, 0.90, 0.95, or 1.00%			
51xxx ...C 1.00%	Cr 1.05%		
52xxx ...Cr 1.00%	Cr 1.45%		
61xx ...Cr 0.60, 0.80 or 0.95%			
V 0.12%, 0.10% min., or 0.15% min.			
81xx ...Ni 0.30%	Cr 0.40%	Mo 0.12%	
86xx ...Ni 0.55%	Cr 0.50%	Mo 0.20%	
87xx ...Ni 0.55%	Cr 0.05%	Mo 0.25%	
88xx ...Ni 0.55%	Cr 0.50%	Mo 0.35%	
92xx ...Mn 0.85%	Si 2.00%	Cr 0 or 0.35%	
93xx ...Ni 3.25%	Cr 1.20%	Mo 0.12%	
94xx ...Ni 0.45%	Cr 0.40%	Mo 0.12%	
98xx ...Ni 1.00%	Cr 0.80%	Mo 0.25%	

*Not included in the current list of standard steels.
Abbreviations:

C	Carbon	**Mo**	Molybdenum
Cr	Chromium	**Ni**	Nickel
Mn	Mangenese	**Si**	Silicon
		V	Vanadium

First Digit

In the SAE and AISI code classification systems, the first number frequently, but not always, indicates the basic type of steel as follows:

1 — Carbon
2 — Nickel
3 — Nickel-chrome
4 — Molybdenum
5 — Chromium
6 — Chromium-vanadium
7 — Tungsten
8 — Nickel-chromium-molybdenum
9 — Silicomanganese

All Digits

The first two digits together indicate the **series** within the basic alloy group. There may be several series within a basic alloy group, depending on the amount of principal alloying elements. Hence, the second digit very often, but not always, indicates the approximate percentage of the principal alloying element. The third, fourth, and fifth digits are intended to indicate the approximate middle of the carbon range. The carbon content is indicated in points — 1-point carbon is 0.01%, 45-point carbon is 0.45%, and 100-point carbon is 1.0%.

Two examples of the SAE/AISI numbering system are explained in Fig. 130-1. The series designations and the types of steel which they designate are summarized in Table 130-1.

Many alloy steels of constructional grade are produced to meet certain hardenability standards. The letter **H** is used after the SAE or AISI number to designate this type of steel. An example is 1340H steel. Additional information concerning H steels is available in standard handbooks for machinists.

A special numbering system is used to designate commonly used tool and die steels. There are seven major groups, some of which contain several subgroups, Table 130-2. Each group is designated with a letter, followed by numerals identifying each specific tool steel.

Aluminum Numbering Systems

Wrought aluminum and its alloys are also designated by a system of four-digit numbers adopted by the **Aluminum Association** (AA) in

Engineers (SAE), and the **American Iron and Steel Institute** (AISI). These systems designate standard constructional grades of carbon and alloy steels according to their basic chemical composition. Both the SAE and AISI systems use a four-digit series of code numbers. Occasionally, a five-digit series is used for certain alloys.

Table 130-2
Tool and Die Steel Designation System

W	Water-hardening tool steel
S	Shock-resisting tool steel
O	Cold-work tool steel, oil-hardening
A	Cold-work tool steel, medium alloy air-hardening
D	Cold-work tool steel, high-carbon, high-chromium
H1-19	Hot-work tool steel, chromium types
H20-39	Hot-work tool steel, tungsten types
H40-59	Hot-work tool steel, molybdenum types
M	High-speed tool steel, molybdenum types
T	High-speed tool steel, tungsten types
L	Special-purpose tool steels, low-alloy types
F	Special-purpose tool steels, carbon-tungsten types
P1-19	Mold steel, low-carbon types
P20-39	Mold steel, other types

1954. The first digit identifies the principal alloying element as follows:

1. Aluminum with 99% or greater purity
2. Copper
3. Manganese
4. Silicon
5. Magnesium
6. Magnesium and silicon
7. Zinc
8. An element other than mentioned above
9. Unassigned

The second digit indicates control over one or more impurities, with 0 indicating no control. In the 1000 series, the last two digits indicate the purity of aluminum in hundredths of a percent beyond 99%. For example, 1025 indicates an aluminum with a purity of 99.25%. However, in the remaining series, the last two digits are used only for identification of different alloys in each series.

Most aluminum alloys are available in several degrees of hardness. In order to indicate this hardness, a **temper designation** follows the aluminum alloy number, separated by a dash. The temper designations and their subdivisions are given in Table 130-3. Thus, 1100-0 indicates an alumi-

num with 99% or better purity in an annealed condition; 2011-T3 indicates an aluminum/copper alloy that has been solution heat treated and then cold-worked.

Other Metal Numbering Systems

Additional numbering systems in common use include a three-digit system for stainless steels, a three-digit system for copper alloys, and separate systems for identifying zinc alloys, cast irons, titanium alloys, magnesium alloys, super-strength space-age alloys, and precious metals. In most cases, several numbering systems exist for each type of metal, creating needless confusion among all concerned.

The Unified Numbering System for Metals and Alloys, a joint project of SAE and ASTM (American Society for Testing and Materials) begun in 1967, is intended to provide a single, comprehensive system for all metals. The unified

Table 130-3
Aluminum Temper Designations

F — as fabricated
O — annealed (wrought alloys only)

H — strain hardened (wrought alloys only)
   H1, + one or more digits, strain-hardened only
   H2, + one or more digits, strain-hardened, then partly annealed
   H3, + one or more digits, strain-hardened and stabilized

W — solution heat-treated only, unstable temper

T — thermally treated
   T2, annealed (casting alloys only)
   T3, solution heat-treated and cold-worked
   T4, solution heat-treated
   T5, artificially aged
   T6, solution heat-treated and artificially aged
   T7, solution heat-treated and stabilized
   T8, solution heat-treated, cold-worked, and artificially aged
   T9, solution heat-treated, artificially aged, then cold-worked
   T10, artificially aged and cold-worked

Table 130-4
Unified Numbering System
for Metals and Alloys

Axxxxx	Aluminum and aluminum alloys
Cxxxxx	Copper and copper alloys
Exxxxx	Rare earth and rare earthlike metals and alloys
Fxxxxx	Cast irons
Gxxxxx	AISI and SAE carbon and alloy steels
Hxxxxx	AISI and SAE H-steels
Jxxxxx	Cast steels (except tool steels)
Kxxxxx	Miscellaneous steels and ferrous alloys
Lxxxxx	Low-melting metals and alloys
Mxxxxx	Miscellaneous nonferrous metals and alloys
Nxxxxx	Nickel and nickel alloys
Pxxxxx	Precious metals and alloys
Rxxxxx	Reactive and refractory metals and alloys
Sxxxxx	Heat- and corrosion-resistant stainless steels
Txxxxx	Tool steels, wrought and cast
Zxxxxx	Zinc and zinc alloys

numbering system (UNS) consists of 16 series of numbers, each series identifying certain metals and alloys. The UNS numbers are made up of a single letter followed by a five-digit number. Table 130-4 lists the UNS categories and corresponding designations.

Whenever possible, the UNS uses numbers from previous systems. For example, the aluminum alloy designated 1100 by the Aluminium Association is assigned the UNS number A91100; carbon steel designated 1090 by AISI is assigned the UNS number G10900, and so on.

Color Code for Steel

The ends of steel bars frequently are painted at the steel mill for identification purposes. However, not all steel producers use the same color-code system. The color code used generally is listed in each steel producer's catalog. Therefore, two factors must be known in order to identify steel bars by color code — the name of the steel producer and the color code used by him.

The stock always should be removed from the unpainted end, so that the painted end remains, and the type of steel always can be identified for future use.

Spark Test

The identity of unknown types of steel often can be determined by a **spark test.** When steel is ground on a grinding wheel, sparks are produced. Different types of sparks are produced by different types of steel. Therefore, an unknown type of steel may be identified when its spark is the same as that produced by a known type of steel.

Some practice usually is required in the identification of steel or other metals by spark testing. The pressure applied to the piece against the grinding wheel should be about the same for all specimens. The type and color of sparks produced by each type of steel or other metal should be studied carefully. Table 130-5 may be used as an aid in the general identification of unknown types of metals.

The volume of sparks produced with carbon steel is quite large. The quantity of sparks depends on the carbon content. Thus, wrought iron, which has very low carbon content, produces only a few sparks. High-carbon tool steel produces many sparks. Careful study of figures 1 through 3 in Table 130-5 will aid in determining whether a carbon steel is of the low-, medium-, or high-carbon type.

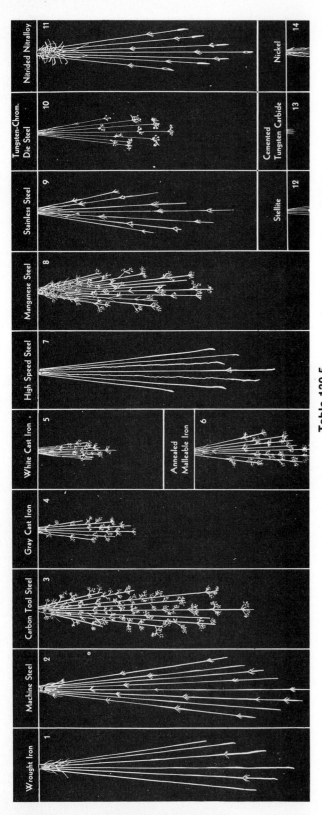

Table 130-5

Characteristics of Sparks Generated by the Grinding Metals

Metal	Volume of Stream	Relative Length of Stream, mm (in.)†	Color of Stream Close to Wheel	Color of Streaks Near End of Stream	Quantity of Spurts	Nature of Spurts
1. Wrought iron	Large	1651 (65)	Straw	White	Very few	Forked
2. Machine steel (AISI 1020)	Large	1778 (70)	White	White	Few	Forked
3. Carbon tool steel	Moderately large	1397 (55)	White	White	Very many	Fine, repeating
4. Gray cast iron	Small	635 (25)	Red	Straw	Many	Fine, repeating
5. White cast iron	Very small	508 (20)	Red	Straw	Few	Fine, repeating
6. Annealed malleable iron	Moderate	762 (30)	Red	Straw	Many	Fine, repeating
7. High-speed steel (18-4-1)	Small	1524 (60)	Red	Straw	Extremely few	Forked
8. Austenitic manganese steel	Moderately large	1143 (45)	White	White	Many	Fine, repeating
9. Stainless steel (Type 410)	Moderate	1270 (50)	Straw	White	Moderate	Forked
10. Tungsten-chromium die steel	Small	889 (35)	Red	Straw*	Many	Fine repeating*
11. Nitrided Nitralloy	Large (curved)	1397 (55)	White	White	Moderate	Forked
12. Stellite	Very small	254 (10)	Orange	Orange	None	
13. Cemented tungsten-carbide	Extremely small	51 (2)	Light Orange	Light Orange	None	
14. Nickel	Very small**	254 (10)	Orange	Orange	None	
15. Copper, brass, aluminum	None				None	

†Figures obtained with 305 mm (12") wheel on bench stand are relative only. Actual length in each instance will

(Norton)

Unit
131

Plain Carbon Steels

Steels may be classified into two basic groups — plain carbon steels and alloy steels. Those steels which are composed basically of iron and carbon are classified as **plain carbon steels.** Steels which have other elements intentionally added for the purpose of changing their properties usually are classified as **alloy steels.**

In studying the various types of steel, the properties of **commercially pure iron** should be understood first. It then is easier to understand the effects of alloying elements in producing the varieties of steel developed from this base metal.

Commercially Pure Iron

Commercially pure iron is iron which is made as nearly pure as practical applications in industry demand. Usually it does not contain more than two-hundredths of one percent (0.02%) of its weight in impurities. The carbon content generally does not exceed 0.03%. This type of iron actually represents a very small percentage of ferrous metal production.

The tensile strength of commercially pure iron averages about 275.79 MPa (40,000 psi.) It is very ductile, possessing an elongation rating of about 40%. It does not harden significantly through heat treatment, but it does harden a small amount through hammering, rolling, or other forms of cold-working. It can be welded or forged readily.

Commercially pure iron becomes highly magnetized when subjected to a magnetic field. Similarly, it loses its magnetism readily. Hence, it frequently is used as a magnetic material. Magnetic chucks on machine tools often are made of this type of iron.

The mechanical properties of commercially pure iron may be changed readily by the addition of small amounts of alloying elements. These may be either metallic or nonmetallic elements.

Wrought Iron

Wrought iron is a mechanical mixture of commercially pure iron and slag. Its internal structure is characterized by threadlike fibers formed by the mixture. The fibers run in the direction in which the iron was rolled.

Wrought iron is probably the first form in which iron was used. It is so named because of its distinguishing properties. It is malleable, ductile, tough, fatigue-resistant, and corrosive-resistant. It has good weldabilitiy, but it ranks relatively low in machinability. It can be bent, twisted, and hammered into various shapes while hot or cold. The following are average properties of wrought iron in common use today: tensile strength 344.74 MPa (50,000 psi); yield point 206.84 MPa (30,000 psi); elongation, 25%; and reduction in area, 40%.

Although it has been replaced largely by low-carbon steel, wrought iron still is used for a number of products. Frequently it is used in the production of water pipe, gas pipe, and pipe joints. It also is used with a galvanized coating for culverts, metal roofing, fencing, and similar items which require corrosion resistance.

Carbon Content of Steel

Carbon is the most important element which is alloyed with iron to form steel. It is largely through control of the form and amount of the carbon content that the properties of various steels are controlled.

The amount of carbon combined to form steel is rated as a percentage of the total weight. Except for very-high-carbon tool steels, the carbon content of steel is usually less than 1%. A typical low-carbon steel has 0.18% (eighteen-hundredths of one percent) carbon. The carbon content of steel usually is expressed in terms of **points.** Thus, steel with 0.18% carbon has 18-point carbon content.

Tensile Strength of Soft Steel

The approximate tensile strength of annealed (unhardened) steel, with carbon content up to 0.80%, can be calculated. It should be remembered that commercially pure iron has a tensile strength of approximately 275.79 MPa (40,000 psi). For each point (0.01%) of carbon content combined with iron, the tensile strength increases approximately 6.89 MPa (1000 psi). Thus, the following formula may be used to calculate the approximate tensile strength of annealed carbon steel with up to 0.80% (80-point) carbon content:

Approximate tensile strength =
275.79 MPa (40,000 psi) +
[6.89 MPa (1000 psi) × points of carbon]

Example: Determine the approximate tensile strength (TS) of steel with 45-point carbon content.

Metric Solution:
TS = 275.79 + (6.89 × 45)
TS = 275.79 + 310.05
TS = 585.84 MPa

Inch Solution:
TS = 40,000 + (1000 × 45)
TS = 40,000 + 45,000
TS = 85,000 psi

Tensile Strength of Hardened Steel

The tensile strength of steel can be increased further through heat-treatment processes. The amount of carbon largely determines the degree to which steel may be hardened by heat treatment. The tensile strength increases in proportion to the amount of increased hardness. However, as the hardness and tensile strength of steel increase, the ductility of the steel generally decreases.

The approximate tensile strength of steel may be calculated easily if its Brinell hardness is known. For an explanation of Brinell hardness, see Unit 143. The approximate tensile strength in MPa (psi) is equal to the Brinell hardness divided by 2, and the quotient multiplied by 6.89 (1000).

Example: Determine the approximate tensile strength (TS) of steel which has a Brinell hardness of 170.

Metric Solution:
TS = (Brinell hardness ÷ 2) × 6.89
TS = $\frac{170}{2}$ × 6.89
TS = 585.65 MPa

Inch Solution:
TS = (Brinell hardness ÷ 2) × 1000
TS = $\frac{170}{2}$ × 1000
TS = 85,000 psi

Plain Carbon Steels

Plain carbon steel is essentially an alloy of iron and carbon. In addition, it includes small amounts of silicon, sulfur, phosphorous, and manganese, usually in the form of impurities.

The carbon content in plain carbon steels may range from 0.05% to 1.7%. However, it seldom exceeds 1.5%. The following is a typical analysis of the range of elements in plain carbon steels:

Carbon . . . . 0.05 to 1.5%
Manganese . . . . 0.25 to 1.65%
Phosphorus . . . . 0.040% maximum
Sulfur . . . . 0.050% maximum
Iron . . . . 97% (the remainder, usually more)

The carbon content in the plain carbon steels is the most important factor in determining its properties. With significant **increases** in the carbon content, the following properties are influenced:

1. Tensile strength is increased.
2. Hardness is increased.
3. Response to various heat-treatment processes is increased. Greater hardness may be obtained.
4. Ductility decreases.
5. Malleability decreases.
6. Weldability decreases.

The plain carbon steels may be classified as low-, medium-, or high-carbon steel, according to the relative amount of carbon content.

Low-Carbon Steel

Low-carbon steel, sometimes called **machinery steel** or **mild steel,** has a range of carbon content from 0.05% to 0.30%. The low-carbon steels with less than 0.15% carbon sometimes are classified further as the **very-low-carbon** steels.

The properties of the low-carbon steels include good ductility, malleability, weldability, and good cold- and hot-working characteristics. The very-low-carbon steels rate lower in machinability than those with carbon content above 0.15%. The tensile strength for the low-carbon steels, in the as-rolled condition, varies from 310.26 to 517.11 MPa (45,000 to 75,000 psi), depending on the carbon content.

Direct hardening by heat treatment depends largely on the carbon content. Therefore, low-carbon steels may be hardened only a very slight amount by direct hardening. However, for special applications requiring a hard, wear-resistant surface and soft but tough inner core, the low-carbon steels may be **casehardened.** This is a special heat-treatment process by which the outer surface, or skin layer, may be carburized and hardened. Hence, these steels are also called **carburizing grades.**

The low-carbon steels have many uses. They are used for nails, screws, bolts, nuts, washers, wire fence, structural parts for machinery, and for forged parts. They also are used for many pressed or deep-drawn sheet metal parts, such as automobile fenders, bodies, and oil pans.

The low-carbon steels are available in the form of bars, strips, sheets, plate, shapes of various types, and wire. These steels are available in either the hot-rolled, cold-rolled, or cold-drawn condition.

Medium-Carbon Steel

Medium-carbon steels range in carbon content from 0.30% to 0.60%. These steels have less ductility than the low-carbon steels, but they are harder and possess greater tensile strength. The tensile strength in the as-rolled condition ranges from 482.63 to 689.48 MPa (70,000 to 100,000 psi), depending on the carbon content.

The strength of the medium-carbon steels also may be increased significantly by heat treatment. The tensile strength of steel with 0.45% carbon can be increased to 1 034.21 MPa (150,000 psi) with proper heat treatment. The medium-carbon steels may be hardened directly in water or brine, at appropriate temperatures. The maximum degree of hardness obtainable will vary with the carbon content. In comparison with high-carbon steels, it can be stated that a moderate degree of hardness may be obtained with the medium-carbon steels.

A typical medium-carbon steel with 0.45% carbon is SAE 1045 steel. A maximum hardness value of about Rockwell C-59 can be obtained by hardening this steel with a water quench at a temperature of 815.6° C (1500° F). Of course, the hardness is reduced when the steel is tempered. The effect of various tempering temperatures on the hardness of SAE 1045 steel is shown in Fig. 139-1. Rockwell hardness values are shown in Table A-6, appendix.

Medium-carbon steels are used for many parts on various types of machinery. Their use frequently involves heat treatment to control such properties as tensile strength, hardness, wear resistance, and toughness. They are used for parts which require medium strength or wear-resistance surfaces. The parts may be forgings, structural elements, or machined parts. Although they do not weld easily by the oxyacetylene process, they may be brazed or electric-welded. They are used for gears, truck axles, steering arms, crankshafts, connecting rods, shift levers, and similar items. They also are used for stampings which are not heat-treated.

The medium-carbon steels are available in rods, bars, strips, and forging stock. They may be hot-rolled, cold-rolled, cold-drawn, or annealed.

High-Carbon Steel

High-carbon steels may range in carbon content from 0.60% to 1.70%. However, the maximum carbon content seldom exceeds 1.50%. Most high-carbon steels have less than 1.30% carbon.

The plain high-carbon steels may be divided into two basic groups for ease in understanding their properties and uses. The two groups include the standard **high-carbon constructional steels** and the **plain carbon tool steels.** The high-carbon constructional steels have carbon content ranging from 0.60% to 0.95%. The plain carbon tool steels have carbon content which usually ranges from 0.60% to 1.40%.

The high-carbon steels are harder and possess greater tensile strength than the carbon steels with lower carbon content. They generally are used for products which require heat treatment. The heat treatment improves both the hardness and the tensile strength. In the as-rolled condition, the tensile strength may range up to 965.27 MPa (140,000 psi) or higher. With heat treatment, this strength may be increased beyond 1 378.95 MPa (200,000 psi).

Maximum hardness values of about Rockwell C-66 may be obtained with the high-carbon steels. However, these values usually are decreased several points by tempering to increase toughness. The effects of various tempering temperatures on the hardness of high-carbon steel, such as SAE 1095 steel, are shown in Fig. 139-1.

A Rockwell hardness of C-60 or higher is generally required for metal cutting tools such as drills, reamers, taps, dies, and similar tools.

The maximum hardness obtainable by heat treatment in plain carbon steels increases with increased carbon content up to about 0.80%. Thereafter, additional carbon content does not increase wear resistance appreciably.

The maximum hardness obtainable increases rapidly with increased carbon content up through 0.45%. Thereafter, the increase in hardness tapers off gradually to a maximum hardness obtained with 0.70% to 0.80% carbon. The plain high-carbon steels usually are quenched in water or brine for maximum hardness.

Constructional Grades: The constructional grades of high-carbon steel are used where carbon steels with improved strength and wear-resistant characteristics are required. These applications include parts used in the construction of automobiles, farm machinery, railroad equipment, and mining equipment. They also are used widely for many types of springs, including coil and flat springs. Practically all of these products are heat-treated. These products are made from the constructional grades of high-carbon steel: plow beams, plow shares, cultivator shovels, rake teeth, scraper blades, disks, hammers, cutlery,

chisels, gouges, strongest steel wire, axes, saws, wear plates on machines, and many other machinery parts.

Plain Carbon Tool Steel: This is one of several widely used types of tool steel. Actually, any steel which is used to make metal-cutting tools may be called a **tool steel.** However, as the term is commonly used in the tool and die making phase of industry, plain carbon tool steels are one group of tool and die steels in common use. The other types, although they may have high carbon content, are alloyed with other metals. Hence, they are **alloy tool steels.** The carbon tool steels usually are hardened in water and therefore may be called **water-hardening tool steels.** The alloy tool steels usually are oil-hardened or air-hardened.

There are several grades of plain carbon tool steel, based on both carbon content and quality. The grades include **special, extra, standard,** and **commercial.** The special and extra grades are the highest quality. These grades are used for tools and dies requiring higher quality material. The standard and commercial grades are produced according to less rigid production controls, and they are used for general-purpose tools or other products which require the properties of high-carbon tool steel.

Carbon tool steels are used for the following products: punches, dies, taps, threading dies, forming dies, drills, reamers, cold chisels, razors, files, and many other special cutting tools for cutting metals or other tough materials.

The high-carbon steels are available in the hot-rolled condition in the form of bars, plates, or other special forms specified by the purchaser. Some bars or plates are available with machined or ground surfaces. Drill rod is an example of this material with a ground surface. It is available in various diameters.

The cost of the plain carbon tool steels varies with the grade or quality. The cost per kilogram (pound) may vary from four to eight times the cost of cold-drawn low-carbon steel.

Free-Machining Carbon Steels

In carefully controlled amounts, sulfur improves the machinability of steel. Normally, the sulfur in carbon steels is held to a maximum of 0.05%. However, sulfur in amounts from approximately 0.08% to 0.33% sometimes is added to certain grades of carbon steel to improve machinability. These grades of steel are designated **resulfurized carbon steels.** They are used widely in making cylindrical and threaded parts in turret lathes, hand screw machines, automatic screw machines, and engine lathes.

The machinability of the resulfurized carbon steels is superior to that of the plain carbon steels. The resulfurized steel designated AISI 1112 is given arbitrarily a machinability rating of 100%. Other steels and many other metals are given approximate machinability ratings in comparison with AISI 1112 steel. The corresponding plain carbon steel, AISI 1012, has a machinability rating of approximately 53%. The resulfurized carbon steel designated AISI 1113 has a machinability rating of about 130% to 140%. The tensile strength and the hardness of the resulfurized steels are somewhat higher than the corresponding plain carbon steels.

Lead sometimes is added to the resulfurized carbon steels to further improve their machinability. The lead usually is included in amounts from 0.15% to 0.35%. Generally, this is less than 151 grams per 45 kg (one third of a pound per one hundred pounds) of steel. The lead is evenly distributed in the steel in extremely fine particles. It does not affect the mechanical properties of the steel, but it does improve the machinability. Modern resulfurized and leaded steels are available with machinability ratings as high as 300% or higher. Hence, these steels can be machined many times more easily and efficiently than the plain carbon steels.

The resulfurized steels do have some shortcomings which must be considered when selecting steel for a specific use. The sulfur, when included greatly in excess of that used in the plain carbon steels, affects the hot-working properties of the steel. It produces an effect called **red shortness.** The steel becomes brittle and unworkable at high temperatures. Also, the grades with high amounts of sulfur do not have good forging or welding properties. Hence, they should not be selected where these hot-working operations are involved. The resulfurized steels, however, may be heat-treated.

Unit
132

Alloy Steels

Alloy steels include steels such as manganese steels, nickel steels, nickel-chromium steels, and molybdenum steels. Steels which are classified as alloy steels have one or more alloying elements other than carbon. These elements are intentionally added, and usually one or more of the alloying elements is a metallic element.

Like plain carbon steels, alloy steels contain small amounts of impurities. These include silicon, sulfur, phosphorous, and manganese. The alloy steels are classified according to the predominating alloying element(s). An example of an alloy with three principal alloying elements is nickel-chromium-molybdenum steel.

Properties from Alloying Elements

The principal properties which may be developed through the addition of alloying elements in steel include the following:

1. Increased hardenability.
2. Decrease in rate of cooling during the hardening operation, thus reducing distortion and the development of cracks.
3. Resistance to becoming soft when being tempered.
4. Retention of hardness and strength at higher temperatures without softening.
5. Increased strength through heat treatment.
6. Increased wear or abrasion resistance at both normal and increased temperatures.
7. Increased toughness.
8. Increased strength as manufactured.
9. Increased corrosion resistance at both normal and high temperatures.
10. Increased machinability.

Of the above properties, the most important are increased hardenability, retention of basic properties at higher temperatures, and resistance to corrosion. The alloy steels are selected for applications which require properties not obtainable with the plain carbon steels. In most applications, the alloy steels must be heat-treated in order to develop their best potential properties.

The cost of alloy steel generally averages about three times that of equivalent types of plain carbon steel. Hence, it would be uneconomical to select alloy steels for applications where plain carbon steels would be adequate.

Not all properties acquired through the addition of alloying elements are desirable. Sometimes it is necessary to accept some undesirable characteristics. For instance, in selecting an alloy tool steel which ranks high in abrasion resistance, it may be necessary to accept a material more brittle than desired.

Classification of Alloy Steels

Alloy steels can be classified into three categories: (1) constructional alloy steels, (2) alloy tool steels, and (3) special alloy steels.

Constructional Alloy Steels

Constructional alloy steels include various types of steel used in the construction of parts for machines. These include items such as gears, levers, shafts, bolts, springs, pistons, and connecting rods. This group of alloys also includes steels used for large structural members for bridges, buildings, railroads, automobile frames, and ships. The constructional alloy steels usually contain relatively low alloy content, as compared with that of alloy tool steels. The total alloy content of the constructional steels generally ranges from 0.25% to 6%.

Alloy Tool Steels

The alloy tool steels are used in making cutting and forming tools. They are used to produce items such as drills, reamers, milling cutters, punches, dies, and wear plates. Although several types of alloy tool steel may be hardened in water, most must be hardened in oil or in air. For this reason, they often are called **oil-hardening** or **air-hardening** tool steels. Generally they rank higher in hardenability than the plain carbon tool steels; that is, they harden more deeply.

The alloy tool steels generally include higher alloy content than the constructional grades. The total alloy content usually ranges from 0.25% to more than 38%. There are hundreds of different alloy tool steels. These generally are classified in several categories according to their basic properties. Each category then contains numerous grades. See Table 130-2.

The basic types of tool and die steels usually are listed according to the AISI numbering system in handbooks for machinists. Such lists include data concerning the basic properties and heat treatment of the various steels. This information also frequently is available in tool steel manufacturer's and suppliers catalogs.

Special Alloy Steels

As the name implies, the special alloy steels are designed for special purposes. Included in this group are steels such as those which require very high heat resistance, or corrosion resistance. The group also may include steels which become tougher and harder with service. Steels possessing the latter characteristic often are used in making teeth on power shovels, lugs on crawler-type tractors, and jaws of rock crushers.

Effect of Alloying Elements

About 26 elements are used, either singly or in combination with other elements, in the production of alloy steels. It is not within the scope of this book to discuss all of these elements. However, the most important ones and their influence on the properties of alloy steels are included.

Carbon

Carbon is one of the most important elements in steel. When carbon is added in amounts up to about 0.80%, the steel becomes harder, increases in tensile strength, and increases in response to heat treatment. Carbon content in greater amounts does not cause the steel to harden significantly more, but it does increase wear resistance.

Manganese

Manganese is next in importance to carbon. Normally, it is present in amounts from 0.30% to 1.50% in the constructional grades of steel. It acts as a deoxidizer and increases responsiveness to heat treatment. It increases hardenability, strength, toughness, and shock resistance.

Constructional grades of **manganese steel** usually have from 0.90% to 1.50% manganese. These steels are used for machine parts which must withstand severe duty.

Steel with 1.5% to 5% manganese is so brittle that it is useless, but with a further increase in manganese, its strength improves. Special manganese steels have from 10% to 14% manganese. These steels possess extreme hardness and ductility. The ductility is acquired through quenching in water at a high temperature, a procedure opposite from that used with carbon steels. These steels become **strain-hardened** by cold-working. With continued use, they become extremely hard and tough, and are used for rock-crushing, grinding, and railroad equipment.

Nickel

Nickel increases strength, toughness, wear resistance, and corrosion resistance of steels. It is not very effective in increasing hardenability. The usual alloying quantity is 3% to 3.7%.

The **nickel steels** have high impact resistance, especially at low temperatures. They are used for machine parts which are subject to repeated shock and stress. These steels also are used for structural purposes such as bridge construction, buildings, rails, armor plate, and heavy machinery.

Nickel often is alloyed with chromium to produce **nickel-chrome** steels of constructional grade. These steels are used for items such as spline shafts, crankshafts, steering knuckles, rear axles, parts of farm machinery, and parts of earth-moving equipment subject to severe work conditions.

Nickel is a principal alloying element in the nickel-chromium **stainless steels.** In these steels, nickel is alloyed in amounts up to 22%, together with chromium.

Chromium

Chromium is alloyed with steel to produce **chromium steels** and certain other alloy steels. Chromium steels also are called **chrome steels.** Chromium increases corrosion resistance, toughness, wear resistance, hardenability, and response to heat treatment. Like other steels, the tensile strength depends on the carbon content.

The chromium content in constructional grades ranges from about 0.30% to 1.60%. The carbon content usually ranges from about 0.20% to 1.30%. With proper hardening and tempering, the tensile strength of these steels may be doubled and the yield strength tripled in comparison with corresponding annealed steel. The ductility is reduced by heat treatment. Without proper heat treatment, these steels are not superior to carbon steels.

The chrome steels are used for machine parts such as ball bearings, races for bearings, coil springs, flat springs, gears, and shafts.

Chromium may be alloyed alone with steel, or it may be alloyed in combination with other elements such as nickel, vanadium, tungsten, and molybdenum. It is used in larger amounts in the stainless steels. In these, the chromium content generally ranges from 11% to 26%. Chromium is also one of the principal alloying elements in tool and die steels and high-speed steels.

Molybdenum

Molybdenum improves the heat-treatment properties of steel. It increases hardenability, and it increases resistance to softening at high temperatures. It also increases toughness and shock resistance. The **molybdenum steels** are used for machine parts such as bolts, coil springs, differential gears, leaf springs, and propeller and transmission shafts.

Molybdenum may be alloyed alone or in combination with other alloying elements in steel. It is alloyed with chromium in the **chromium-molybdenum steels.** It is alloyed with nickel and chromium in the **nickel-chromium-molybdenum steels.** It is alloyed with nickel in the **nickel-molybdenum** steels. In all of these constructional steels, molybdenum is alloyed in comparatively small amounts, from 0.15% to 0.40%. On the other hand, it is alloyed in tool and die steels in amounts up to 9%.

Vanadium

Vanadium promotes the development of fine grain structure in steels. In the construction grades, it is alloyed in small amounts, usually from 0.03% to 0.20%. Vanadium in amounts less than 0.20% increases tensile strength, yield strength, wear resistance, and impact toughness. These properties are increased without a corresponding loss of ductility. In amounts greater than 0.20%, vanadium generally reduces strength of steel.

In steels of constructional grade, vanadium usually is alloyed in combination with chromium. These grades, therefore, are called **chromium-vanadium** steels. They are used for items such as heat-treated and forged parts, flat springs, coil springs, valve springs, piston rods, and spline shafts.

Vanadium is included in amounts up to 5% in high-speed steels to impart stability at high temperatures. With stability, steels tend to maintain their properties without change. The vanadium retards tempering or softening of high-speed steel cutting tools at high working temperatures. The high-speed steels are used widely for cutting tools such as drills, reamers, milling cutters, and lathe tool bits. In alloy tool steels, vanadium usually is alloyed in combination with other elements such as chromium, tungsten, molybdenum, and cobalt.

Tungsten

Tungsten is one of the principal elements in many alloy tool steels. It is alloyed in these steels in amounts from 2% to 20%. It produces a fine, dense grain structure, and it improves the heat-treatment qualities in the steel. The steels with high tungsten content possess high wear resistance. Keen cutting edges can be produced on these steels, and they retain their hardness at high temperatures after being hardened.

Tungsten is a principal ingredient in cemented-carbide cutting tools. In these tools, grains of tungsten carbide are cemented together with cobalt. The carbide grains are very hard and wear-resistant. In some grades of cemented carbide, tungsten is alloyed with certain other elements. Tungsten carbide is discussed in greater detail in Unit 117.

Cobalt

Cobalt is an important alloying element used in cutting tool materials. These materials include high-speed steels, cast alloys, and cemented carbides. The outstanding characteristic of cobalt is its ability to improve the **hot-hardness** or **red-hardness** of cutting tool materials. With high cobalt content, these materials retain their hardness at the lower red-heat temperatures. The cobalt also improves the wear resistance of these materials.

The cobalt content in high-speed steels varies from 5% to 12%. In the cast alloys, it is used in amounts from 35% to 55%. In both materials, it frequently is alloyed in combination with chromium, vanadium, and tungsten.

Cobalt also is used as an alloying element in making metal used for permanent magnets. **Alnico,** a nonferrous alloy of aluminum, nickel, and cobalt, often is used for high-quality permanent magnets.

The properties of cobalt are discussed further in Unit 117.

Other Elements

Copper in small amounts sometimes is alloyed with steel to improve corrosion resistance. It also tends to improve tensile and yield strengths of steel.

Lead is alloyed with steel to improve machinability. When finely divided and evenly distributed, it has no effect on the strength properties. It generally is used in amounts from 0.10% to 0.35%.

Sulfur is alloyed with steel in amounts from 0.05% to 0.33% to improve machinability.

However, in larger amounts, sulfur is detrimental to the hot-working properties of steel, and thus it is detrimental for welding and forging purposes.

Silicon is used as a deoxidizer in steel production. It exists in many steels in small amounts, ranging from 0.10% to 0.40%. When the silicon is held within specified limits from 0.60% to 2.20%, the steel is considered an alloy steel and is called **silicon steel.**

Silicon may be alloyed alone or in combination with manganese or chromium in steel. In alloying amounts, silicon improves tensile strength, hardenability, elastic limit, and the electrical properties of steel. It also improves resiliency. The silicon steels often are used for cores in electrical machinery and for various types of springs.

Forms of Alloy Steel

Constructional Grades

Alloy steels of constructional grade generally are available from steel suppliers in standard bar shapes: squares, rounds, flats, hexagons, and octagons. Some grades are available in the form of structural members such as angles and channels. Certain grades also are available in the form of sheets, strip, and plate. Steels for forging purposes are available in the form of bars, billets, or in special forms.

Tool Steels

Alloy tool steels are available in standard bar shapes, including squares, rounds, octagons, hexagons, and flats. They also are available in special shapes for machining or forging applications.

The alloy tool steels generally are available with the following types of finish: hot-rolled, cold-drawn, rough-machined, ground, and ground and polished. They may be ordered in the as-rolled condition or in the annealed condition. Some suppliers will furnish these steels with heat treatment as ordered.

Unit
133
Nonferrous Alloys

An endless number of alloys is possible. When an alloying element is mixed in different proportions with a given base metal, different alloys result. Each alloy possesses properties somewhat different from other alloys. Two, three, four, or more alloying elements may be mixed with the base metal to produce a large number of alloys of one base metal. This is how the many copper alloys, aluminum alloys, and alloys of other types of metal were developed.

Common Nonferrous Alloys

The general characteristics and some of the uses of several of the most important groups of nonferrous alloys are included in this unit. The most important groups of alloys are identified by the following headings, which indicate the base metal: copper, aluminum, zinc, magnesium, nickel, lead, tin, and special alloys.

A detailed explanation of specific properties of each alloy within each group is not within the scope of this book. Detailed information concerning the chemical composition, physical properties, and mechanical properties of these alloys is readily available in sources such as handbooks for machinists and handbooks of metals properties.

Copper-Base Alloys

Bronze is composed largely of copper and tin. Brass is composed basically of copper and zinc. Both of these alloys also may include a number of other alloying elements.

Bronze was one of the first alloys produced. It was used during the Bronze Age, about 3500 to 5000 years ago, to make hammers, hatchets, saws, chisels, files, and arrowheads. Tools such as these have been found and are on exhibit in museums. Some of the early bronze was made by alloying copper with zinc, lead, and tin. The bronze thus formed was harder than any of the metals from which it was made.

Today there are more than 300 standard copper-base alloys, including both wrought and cast alloys. The principal alloying elements in these alloys are tin, zinc, and lead. However, small amounts of the following alloying elements also frequently are included: iron, phosphorous, antimony, aluminum, manganese, and nickel. Silver is alloyed with copper in several special copper alloys. The melting temperatures for various copper alloys may range from about 704.4 to 1037.8° C (1300° to 1900° F), depending on the kind and amount of alloying elements. Table A-2, Appendix, lists some of the common wrought copper alloys.

Aluminum-Base Alloys

There are more than 350 types of wrought- and cast aluminum alloys used today. Within each of these types, there also may be several grades with different properties, depending on the kind of heat treatment employed. The tensile strength of aluminum, like that of most metals, is increased through the use of alloying elements and heat treatment. This property varies from about 89.63 MPa (13,000 psi) for soft, pure aluminum to about 558.48 MPa (81,000 psi) for special hardened alloys.

The characteristics of specific aluminum alloys vary considerably. Each alloy is developed to

possess special properties such as hardness, ductility, machinability, corrosion resistance, and weldability. The hardest aluminum alloys generally cannot be bent cold without breaking. However, the harder grades generally machine better. Some of the wrought-aluminum alloys have greater tensile strength than low-carbon steel, and they can be machined about three to four times faster.

The following elements are used as alloying elements in various aluminum alloys: copper, manganese, silicon, magnesium, zinc, iron, chromium, nickel, lead, bismuth, and titanium.

The melting temperatures for most aluminum alloys range from about 482.2 to 657.2° C (900° to 1215° F). Table A-3, Appendix, lists some of the commonly used wrought aluminum alloys.

Zinc-Base Alloys

The zinc-base alloys are used widely in making die castings for items such as engine blocks for small gas engines, housings for small engines, carburetors, automobile door handles, parts for typewriters, and many parts for electrical appliances. Zinc alloys also are available in the form of rolled products such as sheets. This material is used for items such as nameplates, weatherstrip, and photoengravings.

There are 16 standard zinc-base alloys in common use. The following elements may be alloyed with zinc in producing these alloys: aluminum, and very small amounts of copper, magnesium, iron, lead, cadmium, and tin. The tensile strength for zinc alloys varies from about 172.4 to 324 MPa (25,000 to 47,000 psi), depending on the type of alloy. The melting temperatures of most zinc alloys range from about 386.1 to 419.4° C (727° to 787° F).

Magnesium-Base Alloys

Magnesium is noted particularly for its light weight. It is about two-thirds the weight of aluminum. Magnesium-base alloys are available in the form of sheets, wire, extruded bars, and ingots for casting. The tensile strength of magnesium alloys may range from 82.7 MPa (12,000 psi) for pure magnesium sand castings to 344.7 MPa (50,000 psi) for certain alloys which are used in extruded bars. Magnesium alloys often are used for making die castings. The tensile strength of common magnesium die-casting alloys ranges from about 206.8 to 255.1 MPa (30,000 to 37,000 psi).

The following elements are used in making various magnesium alloys: aluminum, zinc, silicon, copper, nickel, manganese, tin, and iron. The melting temperatures for magnesium-base alloys range from about 460 to 676.7° C (860° to 1250° F).

CAUTION
Magnesium burns at 315.6° C (600° F), a temperature easily reached during machining operations. Therefore, only sharp cutting tools should be used, and they should be designed to minimize heat buildup due to friction.

Nickel-Base Alloys

There are about 60 nickel-base alloys. The nickel content in these may range from about 62% to 99%, depending on the particular alloy.

Nickel is a silver-colored metal, which is hard and tough. Its outstanding characteristic is its resistance to corrosion by weather, sea water, many chemicals, and a number of acids.

Pure nickel is used for plating other metals. Alloys high in nickel content are used for chemical-handling equipment, and for other items which must resist the corrosive action of chemicals. **Monel** is a common nickel alloy composed of about 65% nickel and 30% copper. The remaining 5% includes small quantities of several other elements. Monel is used for items such as nonmagnetic parts of aircraft, valve stems, pump rods, and valves and pipes used in corrosive and chemical environments. Nickel is also the base metal used in several high-strength, high-temperature-resistant space age alloys such as Inconel, Rene 41, Hastelloy, and Waspaloy.

The alloying elements used in the nickel-base alloys include copper, aluminum, iron, silicon, manganese, titanium, chromium, and tungsten. The melting temperatures of most nickel-base alloys range from 1287.8° to 1426.7° C (2350° to 2600° F.)

Titanium-Base Alloys

Titanium is a silver-gray metal with high strength and heat resistance. It weighs about 44% less than steel alloys, but its tensile strength

is equal to or greater than common structural alloys. Its melting point is 1 675° C (3047° F). Temperatures up to 426.7° C (800° F) do not weaken the metal, and it will tolerate up to 1 093.3° C (2000° F) for short periods of time. Because of these properties, it is used for many supersonic aircraft and space vehicle parts.

Titanium is relatively inert and is used to replace bone and cartilage in surgery. It is also used as a liner for pipes and tanks in food processing. There are about 30 titanium alloys. The principal alloying elements used are aluminum, tin, zirconium, manganese, molybdenum, vanadium, chromium, and columbium.

Special Alloys

A number of alloys have been developed for special purposes. These special alloys sometimes are called **super alloys,** because of the special properties which they possess. Their properties usually are concerned with wear resistance and with maintaining strength at high temperatures. The special alloys are used for purposes such as blades for gas turbines, ratchet components, conveyor mechanisms in large heat-treating furnaces, wear plates on machines, and cutting tools used in the machining of metals.

The special alloys include two major classifications — the **wrought** alloys and the **cast** alloys. The special wrought alloys generally are not as hard as the cast alloys. The cast alloys are so hard that they usually are not machinable except by grinding. However, the special wrought alloys generally are very tough and difficult to machine with standard cutting tools. Hence, they also are machined most frequently by grinding.

The special alloys generally are considered to be nonferrous alloys because they contain very little iron. Although some wrought alloys may include up to about 35% iron, the cast alloys usually contain much less. The cast alloys which are used for cutting tools contain no iron except that which exists in the raw materials used in their manufacture, usually less than 1%.

The principal alloying elements used to produce the various special alloys include nickel, cobalt, molybdenum, tungsten, chromium, and iron. Small amounts of carbon, manganese, silicon, titanium, columbium, aluminum, and tantalum also may be included in the special alloys.

For the machinist, a general understanding of the special cast alloys which are used for cutting tools is important. Some of the common trade names of cast alloy cutting tools include **Stellite, Rexalloy, Armaloy,** and **Tantung.** These materials are explained in greater detail in Unit 117.

Unit
134

Types of Cast Iron and Their Uses

Molten cast iron is poured into molds to form castings which are made into machinery parts. There are several basic types of cast iron, including gray cast iron, malleable cast iron, white cast iron, and ductile cast iron. Other special classes, which are subdivisions of the basic types, include alloy cast iron, chilled cast iron, and ferritic malleable and pearlitic malleable cast iron.

Cast iron is one of the most important products of the foundry industry. It is used to make many types of castings which range in weight from a few dekagrams (ounces) to many Tonnes (tons). The major output of many production machine shops is the machining of cast iron castings.

All types of cast iron are basically alloys of iron and carbon. Any iron-carbon alloy containing more than 1.7% carbon is cast iron. The carbon may range from 1.7% to 4.5%, but the more common range is from 2% to 4%, averaging 3.25%. Other elements totaling 2% to 5%, in varying amounts, include silicon, manganese, phosphorous, and sulfur. Minute quantities of other elements may exist in the form of impurities. Additional metallic elements sometimes are added to make an alloy which has special properties. The iron content of various types of cast iron may range from 91% to 97%.

Alloying Elements

The most important alloying elements in cast iron are carbon and silicon. How they are controlled determines the properties of cast iron.

Carbon

Carbon appears in cast iron in two forms. The **first** is free carbon in the form of **graphite.** Graphite occurs in gray cast iron, malleable cast iron, and ductile cast iron. Examples of flakes or aggregates of graphite as they appear in three types of cast iron when viewed through a metallurgical microscope which magnifies 100 times are shown in Fig. 134-1.

The **second** form in which carbon appears in cast iron is as a chemically-combined carbon. In this form, the carbon combines chemically with iron to form **cementite** or iron carbide (Fe_3C). Cementite is a very hard substance, and, in large amounts, it makes ferrous metals hard to machine. Cementite exists in steel as well as cast iron.

Carbon exists in white cast iron in the combined form as cementite. A photomicrograph of white cast iron, under very high magnification in Fig. 134-2, shows the light regions as **cementite** and the darker regions as **pearlite.** The pearlite is

(Precision Scientific Company)

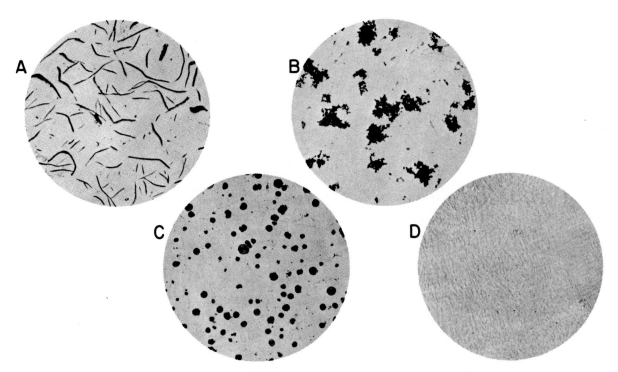

Fig. 134-1. Types of cast iron as distinguished by graphite structure magnified 100 times.
  A. Flake graphite in gray cast iron.
  B. Temper carbon aggregates in malleable cast iron.
  C. Spheroidal graphite in ductile cast iron.
  D. No free graphite in white cast iron.

a form of steel which is made up of alternating layers of cementite and **ferrite.** The ferrite is nearly pure iron.

Since white cast iron has an extremely large proportion of cementite, it is virtually non-machinable. It can be machined only by grinding. Consequently, control of the **form** of carbon in cast iron is important in controlling the properties of cast iron.

Silicon

A second alloying element which is important in controlling the properties of cast iron is silicon. It acts as a **graphitizer** in the production of cast iron. As such, silicon causes carbon to separate from iron more readily at elevated tem-

(Precision Scientific Company)

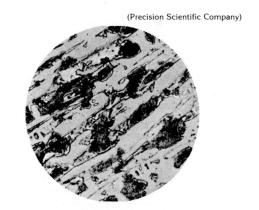

Fig. 134-2. Fine pearlite (dark area) and iron carbide known as cementite (light area) in white cast iron, magnified 500 times.

peratures, thus forming flakes or globules of free graphite. This reduces the proportion of cementite formed, which, in turn, improves the machinability of the cast iron.

A reduced amount of silicon, coupled with a reduced amount of carbon, causes an increased percentage of combined carbon or cementite. Most white cast iron is produced as a first step in the process of making malleable iron castings.

The properties and uses of various types of cast iron, therefore, depend largely on the form and relative amount of carbon content. The properties also may be modified by the addition of other alloying elements, by variations in casting procedures, and by heat-treatment processes.

The various types of cast iron generally are classified by their mechanical properties rather than their chemical compositions. This system is used because castings of varying chemical composition may be produced with similar mechanical properties (made possible by varying the casting procedures used). Also, castings of the same chemical composition may vary in mechanical properties because of the variation in casting, cooling, and heat-treatment procedures used. Hence, foundries can produce castings with specified properties by modification of either the **chemical composition** or the **casting and heat-treatment** procedures used.

Gray Cast Iron

Gray cast iron is so named because of the characteristic gray color at its fracture. This color appears because of the excess carbon content which is in the form of flakes. Gray cast iron contains from 1.7% to 4.5% carbon, usually averaging about 3.25%. The silicon content may vary from 1% to 3%. Other elements, totaling from 1% to 2%, include manganese, sulfur, and phosphorous.

Gray iron contains excess carbon in the form of graphite flakes, as in Fig. 134-1A. The graphite is supported in a matrix or envelope of iron or steel. Low-strength gray iron usually has a matrix composed largely of ferrite, which is nearly pure iron. High-strength gray iron usually has a matrix composed largely of pearlite, which is a form of steel.

Properties and Uses

As it is cast, gray iron is relatively brittle. When it is fractured, cracks generally follow along the line of the graphite flakes. The flakes form sharp cavities which enable the cast iron to fracture easily. Gray cast iron also lacks ductility and malleability. However, these properties may be improved by various methods of heat treatment.

Gray iron must be cast or machined to shape. It has good machinability due to the flakes of free graphite. The graphite acts as a lubricant for cutting tools, and it enables the machined chips to fracture and escape easily from the edge of cutting tools. Gray iron is a relatively inexpensive metal which is cast easily into many intricate shapes without the problems which arise in making steel castings.

There are many uses for gray cast iron. It is used chiefly for parts which do not require toughness or extremely high tensile strength, such as foundation bases for machinery, pistons, engine blocks, various structural parts on farm machinery, and similar machinery parts.

Grades

Gray iron is available in seven standard grades established by American Society for Testing Materials (ASTM). The grades are identified by numbers which represent the tensile strength. The tensile strength for the seven grades range from 137.9 to 448.2 MPa (20,000 to 65,000 psi). Seven standard grades for automotive gray iron castings also have been established. Specifications for the various grades of gray iron are included in standard handbooks for machinists. In the selection of gray iron for castings, it should be remembered that with the higher hardness and tensile strengths, brittleness is increased and machinability generally is decreased.

Alloy Cast Iron

The mechanical properties of cast iron may be modified by the addition of small amounts of certain alloying elements, such as nickel, silicon, chromium, vanadium, molybdenum, and copper. Alloyed castings are used widely in the automobile industry. Toughness, strength, and hardness are increased by the addition of the proper alloying elements.

Alloy castings have been developed which have tensile strengths up to 689.5 MPa (100,000 psi.) Some of their uses include brake drums, plow shares, forging dies, piston rings, and similar items requiring high strength.

White Cast Iron

White cast iron is so named because of the silvery-white appearance of its surface where fractured. It has practically no free carbon. Rather the carbon is in the form of cementite, as in Figs. 134-1D and 134-2. Since cementite is an extremely hard substance, white cast iron is very hard and virtually nonmachinable, except by grinding.

Types and Uses

The direct use of white cast iron is limited to castings requiring hard wear surfaces, such as rolls used in rolling steel. The major use of white cast iron applies to **malleable iron** castings. The first step in their production is the production of white cast iron castings.

White cast iron may be produced accidentally when gray iron castings are cooled too rapidly. When this occurs, the casting may harden in spots, to a surface depth only or throughout, depending on the rate of cooling. Hard spots of white cast iron so formed in gray iron castings are difficult to machine and can cause damage to cutting tools. Through careful annealing or other forms of heat treatment, the castings may be machined readily.

Chilled Cast Iron

Chilled cast iron is a term used for gray iron castings which have a very hard, wear-resistant surface layer of white cast iron. It is produced by rapid cooling of the outer surface of gray iron as it is poured into molds. The molds may be metal, or they may be equipped with metal chill plates. Such rapid cooling causes cementite to form, thus producing the hard, white cast iron surface.

Malleable Cast Iron

White cast iron can be converted to malleable cast iron by a heat-treatment process called **malleablizing.** The process involves **graphitization,** the reduction of carbon from its combined form as cementite (Figs. 134-1D and 134-2) to free carbon in the form of aggregates which are called **temper carbon** (Fig. 134-1B).

At high temperatures, the carbon combined to form cementite in white cast iron breaks down to form iron and free carbon. The carbon becomes rounded aggregates, while the soft iron forms a matrix around these aggregates. This new material is malleable iron, with properties very different from white cast iron or gray iron.

Malleable castings have many of the tough characteristics of steel. They possess good ductility, strength, and machinability. There are several standard types of malleable cast iron, and each type has several standard grades. The properties vary according to the type and grade. When greater ductility and malleability are desired, softer grades are selected. However, these grades have lower tensile strength. The various grades of malleable cast iron may be classified according to two major types — **ferritic** and **pearlitic.**

Ferritic Type

The physical grain structure of ferritic or standard malleable cast iron is largely ferrite surrounding the temper carbon. Ferrite is nearly pure iron. Hence, ferritic iron ranks high in ductility, impact resistance, and malleability. Ferritic malleable cast iron has a tensile strength of 344.7 to 379.2 MPa (50,000 to 55,000 psi).

Pearlitic Type

The malleablization method used in the production of pearlitic malleable cast iron is modified somewhat from that used for ferritic malleable iron. The graphitization process is stopped before all of the carbon has broken down into free temper-carbon aggregates. As a result, some cementite remains to form a mechanical mixture with the ferrite. The mixture thus formed is pearlite, which surrounds the aggregates of temper carbon. With an increased proportion of pearlite, the cast iron becomes harder. It also increases in tensile strength and becomes more difficult to machine.

Tensile strength of pearlitic malleable cast iron ranges from 448.2 to 689.5 MPa (65,000 to 100,000 psi), depending on the particular grade.

Uses

Malleable cast iron usually is limited to castings of small cross section. This is due to the need for rapid cooling of the initial white iron castings in the first step of making the malleable cast iron. Its properties enable it to be used for purposes which otherwise would require steel. It is useful for auto parts such as gear housings, camshafts, differential housings, and brake pedals. It also is useful for many types of machinery parts.

Ductile Cast Iron

Ductile cast iron also is known as **nodular** cast iron or **spheroidal graphite** iron. Its distinguishing feature is that it contains free carbon in ball-like form as shown in Fig. 134-1C. Each nodule of graphite is surrounded by a matrix of ferrite, or ferrite and pearlite, depending on the grade.

Ductile iron is produced by adding magnesium alloys, and sometimes certain other elements, to a ladle of molten gray iron before it is poured into molds. The additional elements, together with special processing, causes the carbon to form as nodular graphite during the solidification and cooling process.

Grades and Uses

There are several grades of ductile cast iron. This type of cast iron may be heat-treated for further modification of its properties. Hence, there is a wide range of variation in its properties and uses.

The properties of ductile cast iron lie in a range between gray cast iron and steel. Basically, ductile cast iron is tough, shock-resistant, high in tensile strength, and easily machined.

The tensile strength varies from 413.7 MPa to 551.6 MPa (60,000 to 80,000 psi). The tensile strength also can be increased to 689.5 MPa (100,000 psi) by heat treatment. Hence, properties similar to pearlitic malleable cast iron can be obtained.

Various forms of heat treatment may be applied to ductile iron, including annealing, hardening, induction hardening, and flame hardening. Of course, as the hardness and tensile strength are increased, there is a corresponding decrease in ductility and impact or shock resistance.

Ductile cast iron has many uses in modern industry. It is used in the automobile industry for cylinder heads, crankshafts, camshafts, pistons, and similar parts. It is also used for control levers on heavy machinery, clamps, wrenches, lathe chuck bodies, and for numerous parts on farm machinery.

Semisteel

Semisteel is a term which has been used for cast iron which includes a high percentage of scrap steel. The use of this term should be discouraged. Cast iron with a high percentage of scrap steel has properties closely related to the properties of steel castings.

Test Your Knowledge of Section 17

Unit 128: Classification of Metals

1. Why were alloy metals developed?
2. Explain the meaning of an alloy, as the term is used in the commercial classification of metals.
3. Explain the difference between ferrous and nonferrous metals. Give several examples of each.
4. Explain the difference between atoms and molecules.
5. Define a mixture, and list the forms in which it can exist.
6. Define a solution, and list the forms in which it can exist.
7. Explain the difference between a solid solution and a mechanical mixture.

Unit 129: Properties of Metals

1. List several properties which are considered in the selection of a metal for a particular use.
2. Explain the meaning of chemical properties, physical properties, and mechanical properties as applied to metals.
3. What is meant by ductility of metals?
4. Define malleability as applied to metals.
5. Define toughness as applied to metals. How is it measured?
6. What is meant by the machinability of metals?
7. List four types of strength possessed by metals.

8. What is the difference between the yield strength and the tensile strength of a metal?
9. Explain the meaning of elongation as applied to tensile strength tests.
10. Explain the meaning of reduction in area as applied to tensile strength tests.
11. Explain the meaning of stress as a property of metals.
12. Explain the meaning of strain as a property of metals.
13. List three general types of stress which may be applied to a structural member.
14. Explain the meaning of elastic limit, as applied to a structural member.

Unit 130: Metal Designation and Identification

1. List the basic types of steel which usually are represented by the first digit of the SAE number code for steel.
2. What determines the difference between two different **series** of steel within a basic type of alloy steels?
3. What does the letter **H** designate when used with the SAE or AISI steel code number?
4. Explain the special numbering system used to designate tool and die steels.
5. Explain the numbering system used to designate wrought aluminum and its alloys.
6. Explain the system used to designate different degrees of hardness for wrought aluminum alloys.
7. Explain the Unified Numbering System and tell why it was developed.
8. What two factors must be known when identifying steel by color code?
9. Explain how some metals can be identified with a spark test.

Unit 131: Plain Carbon Steels

1. What are the basic ingredients in plain carbon steels?
2. Describe the composition and properties of commercially pure iron.
3. Describe the composition and properties of wrought iron.
4. What material has largely replaced wrought iron?
5. What is the tensile strength of commercially pure iron?

6. What relationship exists between the carbon content and the tensile strength of steel in the unhardened condition?
7. Calculate the tensile strength of unhardened carbon steel with 70-point carbon content.
8. What element in steel largely determines the hardness which can be obtained through heat treatment?
9. What properties are changed in steel when it is hardened by heat treatment?
10. Determine the approximate tensile strength of steel that has a Brinell hardness of 240.
11. What is the range of carbon content for plain carbon steels?
12. What properties are influenced in steel when the carbon content is significantly increased?
13. List the three basic groups of carbon steel, and indicate the range of carbon content for each group.
14. List several uses for low-carbon steels.
15. List several uses for medium-carbon steels.
16. What hardness is obtainable when SAE 1045 steel, a medium-carbon steel, is hardened by quenching in water?
17. What hardness values are obtainable when high-carbon steels are hardened?
18. What hardness value generally is required for metal-cutting tools?
19. What purpose is served in increasing the carbon content beyond 0.80% in carbon steel?
20. What quenching mediums generally are used to harden carbon steels?
21. List several uses for constructional grades of high-carbon steel.
22. List several uses for high-carbon tool steel.
23. List several grades of carbon tool steel.
24. How does the cost of high-carbon tool steel compare with that of cold-drawn low-carbon steel?
25. How does the addition of sulfur and lead affect the machinability of steel?
26. For what purposes are free-machining screw steels widely used?
27. List several shortcomings or disadvantages of resulfurized steels.

Unit 132: Alloy Steels

1. Describe three of the most important properties which are developed through the addition of alloying elements in steel.

2. How does the cost of alloy steel compare with the cost of plain carbon steel?
3. List three basic categories of alloy steel.
4. What are some uses of alloy steel of constructional grade?
5. What is the range of alloy content in alloy steels of constructional grade?
6. List several uses for alloy tool steels.
7. What is the range of alloy content in alloy tool steels?
8. List seven classifications for tool and die steels.
9. In what sources can you find information concerning tool and die steels?
10. How does manganese content affect steel?
11. List several uses for manganese steel.
12. How does nickel content affect steel?
13. List several uses for nickel steels.
14. List several uses for nickel-chrome steels.
15. How does chromium content affect steel?
16. List several uses for chromium steel.
17. How does molybdenum content affect steel?
18. List several uses for molybdenum steel.
19. How does vanadium content affect steel?
20. List several uses for chromium-vanadium steels.
21. List several uses for high-speed steels.
22. How does tungsten content affect steel?
23. Why are carbide cutting tools called cemented carbide?
24. What is the outstanding characteristic of cobalt?
25. List three types of cutting-tool material in which cobalt is used.
26. How does silicon content affect steel?

Unit 133: Nonferrous Alloys

1. What are the principal alloying elements which are alloyed with copper to form copper-base alloys?
2. Approximately how many common types of aluminum alloys are available?
3. How does the tensile strength of the strongest aluminum alloys compare with that of low-carbon steel?
4. List several elements which are alloyed with aluminum in producing aluminum alloys.
5. List several uses for zinc-base alloys.
6. What is the tensile strength range and melting temperature range for zinc-base alloys?
7. What is the principal property of magnesium?
8. List the tensile strength range and the melting temperature range for magnesium alloys.
9. What potential hazard is associated with a machining magnesium?
10. What are the principal characteristics of nickel-base alloys?
11. List several uses for nickel-base alloys.
12. List the principal properties of titanium, and give its melting point.
13. List several uses for titanium and its alloys.
14. List the principal properties of the special or super alloys, together with some of their uses.

Unit 134: Types of Cast Iron and Their Uses

1. List four basic types of cast iron.
2. What is the appproximate range of carbon content which may exist in cast iron?
3. What elements, other than carbon, exist in cast iron?
4. What two alloying elements are most important in controlling the properties of cast iron?
5. In what types of cast iron does free carbon exist?
6. How does cementite affect the machinability of ferrous metals?
7. How is white cast iron machined?
8. Explain how increased silicon content helps in controlling the properties of cast iron.
9. List several additional ways in which the properties of cast iron may be modified or changed.
10. Why is cast iron generally classified according to its mechanical properties rather than by chemical properties?
11. Explain the characteristics and properties of gray cast iron.
12. How does the graphite in gray iron affect its machinability?
13. List several uses for gray cast iron.
14. What are the properties and uses of white cast iron?
15. Describe chilled cast iron and explain several of its uses.
16. Explain how malleable cast iron is produced.
17. What are the characteristics and properties of malleable cast iron?
18. Explain the characteristics and properties of ferritic malleable cast iron.
19. Explain the characteristics and properties of pearlitic malleable cast iron.
20. List some typical uses of malleable cast iron.
21. Explain the properties of ductile cast iron and list several of its uses.

Heat-treatment processes involve heating and cooling of metals in their solid state for the purpose of changing their mechanical properties. Depending on the heat-treatment procedure used, steel may be made hard and wear-resistant, or it may be made soft so that it can be machined easily.

Heat Treatment and Metallurgy

The theory underlying heat treatment of metals is actually one phase of the study of metallurgy. The study of heat treatment is most concerned with the branch of metallurgy called **physical metallurgy** — that is, the physical and mechanical properties of metals which are affected by composition, mechanical working, and heat treatment. It includes study of both the internal grain structure of metals and the properties which are affected by changes in the grain structure. Heat-treatment processes affect the grain structure

and the mechanical properties of metals. Metals may be made harder, stronger, tougher, or softer through various heat-treatment processes.

Tools, such as drills, milling cutters, punches, and dies, possess properties which enable them to cut other metals. The materials used to produce cutting tools and other machined parts are selected because of the properties which they possess or because of the properties which they are **capable** of possessing after heat treatment. The type of steel and its chemical analysis must be known before heat treatment takes place. Small amounts of certain elements, particularly the carbon content, greatly affect the properties after heat treatment.

Heat-Treatment Processes

Although the properties of nonferrous metals may be changed by heat-treatment processes, the content in this section of the book is limited to the heat treatment of ferrous metals. The common processes used with steel and cast iron are included, the most common being hardening, tempering, and annealing.

Hardening and Tempering

The hardness of certain types of steel may be increased greatly by the heat-treatment process called **hardening.** For example, before they can be used, the cold chisel, punch, file, and milling cutter must be hardened after being machined. However, after hardening, these tools are too brittle for immediate use. They must be further heat-treated by a process called **tempering** (or drawing) to relieve internal stresses produced in the steel during the hardening process. Tempering also increases toughness. Thus, a chisel which has been hardened and tempered withstands heavy blows when used to cut other metals.

Annealing

Hardened steel may be softened by a heat-treatment process called **annealing.** Before high-carbon tool steel may be machined efficiently, it usually must be annealed. Annealing, therefore, is the opposite of hardening.

Understanding the Processes

Hardening, tempering, and annealing include only three heat-treatment processes. Numerous others often are used on steel and cast iron products. All heat-treatment processes consist of heating metal according to a **time-temperature cycle** which includes the following three steps:

1. Heating the metal to a certain temperature.
2. Holding the metal at an elevated temperature (soaking) for a certain period of time.
3. Cooling the metal at a certain rate.

During the above three steps, the properties of iron or steel may be altered in various ways, depending on the chemical content. These operations generally alter the internal structure of the iron or steel in some way.

It is possible for the average metalworker to obtain satisfactory results in heat-treating steel, even though the **metallurgical** theory related to heat treatment is not understood. However, it is necessary to follow the correct procedures. By following the procedures outlined in Unit 140,

the metalworker may harden and temper steel with satisfactory results. The procedure there explains how to select the hardening temperature, how long to soak the steel at the hardening temperature, how to select the quenching media, and how to quench the steel for hardening.

Skilled workers engaged in the designing, machining, and heat-treating of metal products, however, often are required to possess an understanding of the basic metallurgical theory related to heat treatment. Knowledge of the following factors is essential for understanding the metallurgical theory related to heat treatment: grain structures of steel, effect of temperature changes on grain structures, effect of carbon content on obtainable hardness, effect of temperature changes on the form of carbon in steel, effect of the severity of the quench on grain structure and hardness, and effect of tempering on the toughness of steel. The metallurgical theory underlying these factors is included in the next four units of this section.

In addition to the processes and procedures which are included in succeeding units, the equipment used in the operations and the methods for determining the hardness of metals also are included in the various units of this section.

Furnaces and Temperature Control

Metal parts are heated in special furnaces during heat-treatment processes. The furnaces may be heated with gas, oil, or electricity. Modern heat-treatment furnaces usually are equipped with indicating and control devices which maintain temperature within a few degrees of that selected.

Temperature-indicating and control devices usually consist of an indicating pyrometer and certain accessory switches, valves, or solenoids. The pyrometer actuates the accessory parts as necessary to control the fuel or current which heats the furnace. The temperature sensing, controlling, and indicating apparatus is used to control temperatures in gas heat-treatment furnaces, Fig. 135-1.

Types of Furnaces

Small electric furnaces such as that shown in Fig. 135-2 are often used in toolrooms for hardening, tempering, and other heat-treatment

(Johnson Gas Appliance Company)

TEMPERATURE –
INDICATING
CONTROLS

Fig. 135-1.   Gas-fired heat-treatment furnace equipped
with indicating temperature controls.

(Thermolyne Corp.)

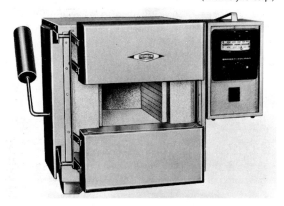

Fig. 135-2.   Electric heat-treatment furnace with
temperature-indicating control unit.

operations. They are usually equipped with temperature-indicating and control equipment which automatically regulates temperatures selected in the range from 148.9° to 1260° C (300° to 2300° F).

(BTU Engineering Corporation)

Fig. 135-3.   Continuous   conveyor-type   production
heat-treating furnace.

Production heat-treating furnaces may be either batch-type or continuous-conveyor-type. Batch-type furnaces are often equipped with **programmable temperature controllers.** This allows the furnace to be programmed to hold different temperatures for different periods of time to suit the requirements of the particular parts being heat-treated. Conveyor furnaces, Fig. 135-3, are designed for continuous pass-through operation. Temperature zones in the furnaces are varied to match the rate of heating and cooling required. In most cases, production heat-treating is done in **atmospherically controlled furnaces.** When steel is heated in the presence of oxygen, carbon dioxide, and water vapor, the result is undesirable surface rust, scale, or decarburization. This can be prevented by flooding the furnace with various combinations of gases which effectively protect the steel. Carbon dioxide, carbon monoxide, nitrogen, and hydrogen are used in various combinations, depending on the kind of metals being heat-treated.

Metals may be heated for certain hardening operations in hot, liquid baths. Parts heated in this manner are not in contact with air at elevated temperatures. Thus, oxidation or scaling is reduced or prevented. Commonly used liquid baths include molten salts of various types and molten lead.

Parts which are casehardened by a process called **cyaniding** are heated in a bath of molten

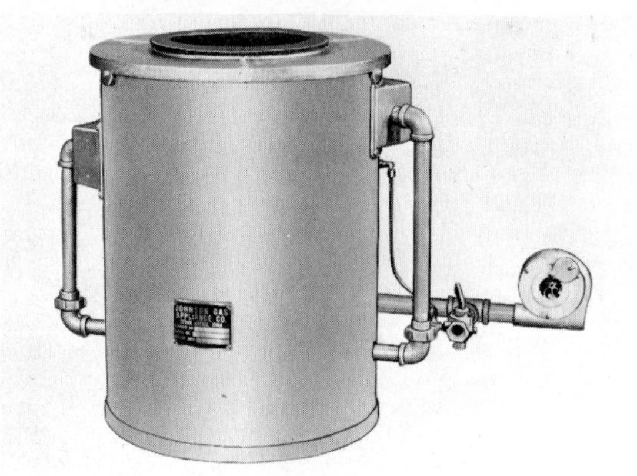

Fig. 135-4.   Pot-type liquid hardening furnace.

Fig. 135-5.   Temperature-indicating products: pellets, crayons, and liquid.

cyanide-carbonate-chloride salts. The parts then are quenched in water, brine, or mineral oil. A pot-type liquid hardening furnace which may be used for salt, lead, and cyanide baths is shown in Fig. 135-4.

Small bench-type gas furnaces, propane torches, or Bunsen burners may be used for hardening and tempering small tools or parts. However, since there is no means for controlling temperatures, the temperature of the tool or part must be estimated according to its color.

Temperature Colors

When clean, bright steel is heated to about 193.3° C (380° F), its color starts to change due to oxidation. First, a very pale yellow appears. With further increases in temperature, other col-

ors emerge. At red heat and beyond, the colors are due to energy being radiated from the steel. The approximate colors of steel at various temperatures are shown in Table 135-1 and Fig. 137-8.

Old-time blacksmiths and heat-treaters often determined the temperature of steel by its color. However, this is not a very accurate method, even for skilled heat-treaters. For example, estimates within the tempering range from 190.6° to 315.6° C (375° to 600° F) may easily be off 11.1° to 16.7° C (20° to 30° F). Steel appears dull red at about 537.8° C (1000° F). At higher red-heat temperatures, estimates based on color may be in error as much as 93.3° to 148.8° C (200° to 300° F).

An inexpensive method for estimating temperatures of metals is through the use of temperature-indicating pellets, crayons, or paints, as shown in Fig. 135-5. These materials are designed to melt at various temperatures, as specified, within the range from about 37.8° to 1371.1° C (100° to 2500° F). One simply selects the crayon or other material for the temperature desired and marks the workpiece. The material melts when the specified temperature is reached.

Steel loses its magnetism when heated above temperatures indicated by line A_2 in Fig. 137-6. Hence, a magnet may be used for estimating temperatures required for hardening or annealing medium- or high-carbon steels.

Hardness Designations

In most heat-treatment operations, the hardness of the metal must meet the specifications designated. The hardness of steel and other metals may be indicated by hardness numbers according to several different hardness number systems. The most commonly used systems include Rockwell C scale, Rockwell B scale, Brinell hardness numbers, and Shore Scleroscope hardness numbers. When the hardness is designated according to one of these number systems, it may be converted to an approximate reading in any of the other equivalent hardness number values by reading horizontally across Table A-6, Appendix. The method for determining the hardness of metals is explained in Unit 143.

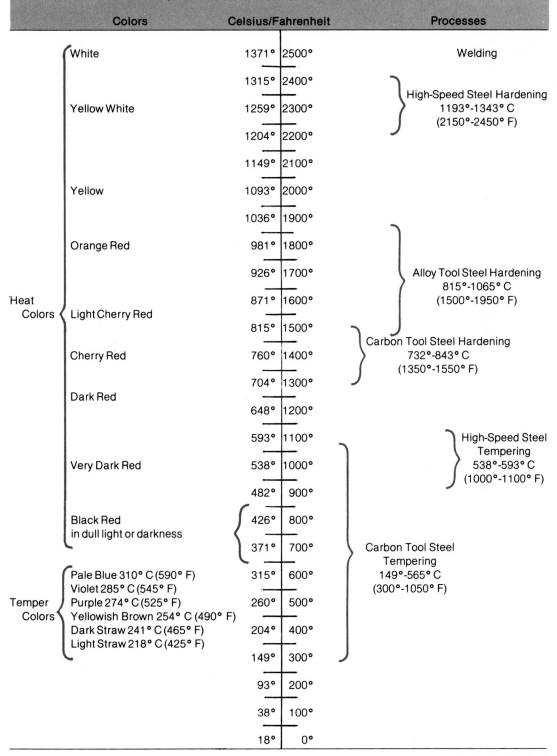

Table 135-1
Temperatures, Steel Colors and Related Processes

Colors	Celsius/Fahrenheit	Processes
Heat Colors		
White	1371° / 2500°	Welding
	1315° / 2400°	High-Speed Steel Hardening 1193°-1343° C (2150°-2450° F)
Yellow White	1259° / 2300°	
	1204° / 2200°	
	1149° / 2100°	
Yellow	1093° / 2000°	
	1036° / 1900°	
Orange Red	981° / 1800°	Alloy Tool Steel Hardening 815°-1065° C (1500°-1950° F)
	926° / 1700°	
Light Cherry Red	871° / 1600°	
	815° / 1500°	Carbon Tool Steel Hardening 732°-843° C (1350°-1550° F)
Cherry Red	760° / 1400°	
	704° / 1300°	
Dark Red	648° / 1200°	
	593° / 1100°	High-Speed Steel Tempering 538°-593° C (1000°-1100° F)
Very Dark Red	538° / 1000°	
	482° / 900°	
Black Red in dull light or darkness	426° / 800°	Carbon Tool Steel Tempering 149°-565° C (300°-1050° F)
	371° / 700°	
Temper Colors		
Pale Blue 310° C (590° F)	315° / 600°	
Violet 285° C (545° F)		
Purple 274° C (525° F)	260° / 500°	
Yellowish Brown 254° C (490° F)		
Dark Straw 241° C (465° F)	204° / 400°	
Light Straw 218° C (425° F)		
	149° / 300°	
	93° / 200°	
	38° / 100°	
	18° / 0°	

(Allegheny Ludlum Steel Corp.)

Unit 136

Grain Structure of Steel

Metals in the solid state have an internal crystalline structure. The crystals in commercial metals are called **grains.** They are formed as the metal cools and changes from a liquid to a solid state. This change in state is called **freezing.** Thus, metals freeze at high temperatures to form a solid, just as water freezes at a lower temperature to form a solid.

The rolled or cast surface of metal does not indicate its internal grain structure. However, when the metal is fractured, the surface shows a crystalline or granular structure. The grains in various metals may vary considerably in size. In some metals, the grains may be seen with the unaided eye. In other metals, they cannot be seen without magnification.

Using a Metallurgical Microscope

The grain structure of metals may be observed through a metallurgical microscope. Common types can magnify from 100 to 500 diameters the microstructure of the surface of a highly-polished metal specimen. This magnification is designated as $100\times$ to $500\times$. Some metallurgical microscopes magnify up to $2500\times$. With an electron microscope, magnification of $250,000\times$ is possible. For the study of most metals, however, the magnification of $100\times$ to $500\times$ is satisfactory.

Some metallurgical microscopes are equipped with devices for photographing the microstructure of metallic materials. A modern instrument called a **metallograph,** which is basically a metallurgical microscope equipped with a camera, is shown in Fig. 136-1. It is used for investigating and photographing the microstructure of metals. A photomicrograph of microstructures which are common in steel is shown in Fig. 137-2. The study of microscopic structures of metals is called **metallography.**

A metal specimen must be specially prepared for examination with a metallurgical microscope. The specimen is prepared by grinding a flat surface first. The surface then is polished with finer and finer abrasives until a mirror finish is produced. As a result of the polishing procedure, a fine film of impurities is produced on the surface and embedded between the grains. This film usually is dissolved by an etching solution so that

(Bausch and Lomb)

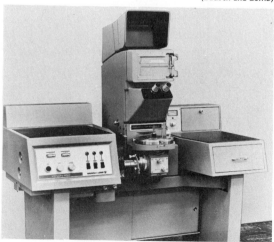

Fig. 136-1.   Balphot metallograph.

Fig. 136-2.   Using hot-stage microscope to study and record photographically transformations occurring in steel at elevated temperatures.

the grain boundaries may be observed. **Nital,** a solution of 5% nitric acid and alcohol, often is used as an etching solution for carbon-steel specimens.

After etching has been completed, the grain size, grain boundaries, and the form and distribution of the carbon can be investigated. All of these factors influence the properties of steel or cast iron. Also, all of these factors may be altered by the temperature changes which take place in certain heat-treatment processes.

Through the use of a hot-stage microscope, metallurgists can study the transformations which occur in steel or other metals at elevated temperatures. With this type of instrument, the changes also can be recorded photographically. Such an instrument is shown in Fig. 136-2.

It is not necessary to observe the grain structure of metal through a microscope in order to obtain good results with most heat-treatment operations. However, detailed study brings about a better understanding of heat-treatment processes. The grain structure of metals is affected by elevated temperatures and by the rate of cooling. It is largely because of these effects that the properties of steel and cast iron may be changed by heat-treatment processes.

Space Lattices

The crystalline grains in metals are composed of atoms arranged in orderly geometric patterns in three dimensions. The graphic representation of this systematic arrangement of atoms is called a **space lattice.** The atoms in a space lattice are represented by dots or spheres. The lines actually do not exist, but they are shown graphically to help visualize the arrangement of the atoms.

The individual atoms in a metallic grain are so small that they cannot be observed directly by the most powerful microscope. Through the use of X-ray studies, however, scientists have determined that the atoms in metals and other crystalline materials have definite space lattice arrangements. Each type of metal has its own characteristic arrangement of atoms in its space lattice.

The smallest repetitive group of atoms which makes up a space lattice system is called a **unit** or **cell.** The type of unit varies with the kind of metal. The space lattice units of most common metals are of the following four types of systems:

1. Body-centered cubic unit.
2. Face-centered cubic unit.
3. Body-centered tetragonal unit.
4. Hexagonal close-packed unit.

Body-Centered Cubic Unit

The body-centered cubic unit, Fig. 136-3A has nine atoms, one at each of the eight corners of the cubic arrangement and one at the center of the cube. With this arrangement, each atom has eight other atoms in its **field of influence.** This means that each atom is held in its relative position by the force of eight other atoms acting upon it.

The body-centered cubic arrangement is not as closely packed as the other unit arrangements which will be discussed. The atoms do not touch one another. In fact, in comparison with their actual size, they are spaced at great distances apart. However, in order to visualize them more easily, we can assume that they are touching one another in the form of spheres, as shown in Fig. 136-3.

Repetitious body-centered cubic units unite and grow in three directions, thus forming the space lattice arrangement of a single metallic grain of metal. The grains form when molten metal cools from a liquid to a solid.

Several metals which have the body-centered cubic arrangement include tungsten, molybdenum, vanadium, and columbium. Ferrite, which also is called **alpha iron,** has this arrangement at

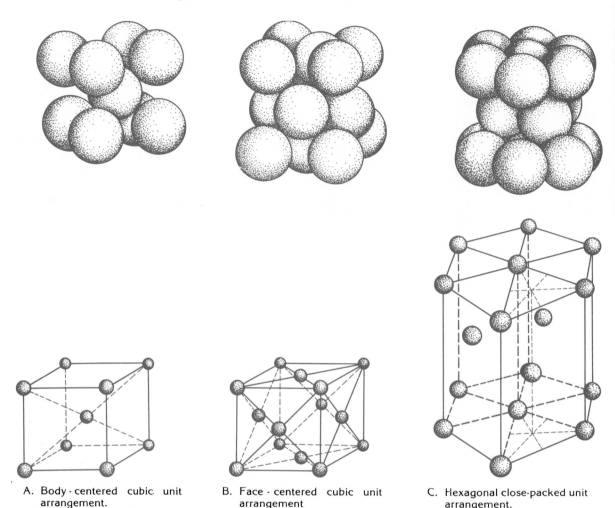

A. Body - centered cubic unit arrangement.

B. Face - centered cubic unit arrangement

C. Hexagonal close-packed unit arrangement.

Fig. 136-3.   Atom arrangement in a space lattice system.

temperatures below the hardening temperature range. (The **hardening temperature** is the temperature at which steel is quenched in the hardening process.) At the hardening temperature, the lattice arrangement transforms to another form — the face-centered cubic form. Metals with the body-centered cubic arrangement generally rank higher in strength and lower in cold-working characteristics than those with the face-centered cubic arrangement.

Face-Centered Cubic Unit

The face-centered cubic unit, as illustrated in Fig. 136-3B, has 14 atoms — one at each of the eight corners of the cubic unit and one at the center of the six faces of the cube. Hence, the atoms in this arrangement are more dense (closely packed) than in the body-centered arrangement. Metals which have a face-centered cubic lattice structure include nickel, aluminum, copper, lead, gold, and silver.

When steel is heated to the hardening temperature, the space lattice units in the grain structure transform from the body-centered cubic form to the face-centered cubic form. In this form, it is called **austenite** or **gamma iron,** as shown in Fig. 137-7. At the elevated temperature at which austenite forms, the carbon in the steel decom-

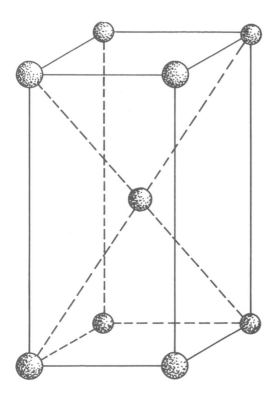

Fig. 136-4.   Body-centered tetragonal unit arrangement of atoms.

poses from its combined state as cementite (iron carbide) to free carbon. The free carbon then dissolves into the solid hot iron to form a solid solution of uniformly dispersed carbon in iron. This form of iron will dissolve up to a maximum of about 2.00% carbon. In contrast, the body-centered (alpha) form of iron (ferrite) will dissolve a maximum of about 0.05% carbon.

Hexagonal Close-Packed Unit

The hexagonal close-packed unit, Fig. 136-3C, has a total of 17 atoms — one at each of the six corners of the two end faces, one at the center of each end face, and three equally spaced between the two end faces. Some metals which have this type of space lattice arrangement include zinc, cadmium, cobalt, magnesium, and titanium. Metals with this type of lattice arrangement generally lack plasticity. Hence, they usually cannot be cold-worked.

Body-Centered Tetragonal Unit

The body-centered tetragonal unit is shown in Fig. 136-4. This type of space lattice has nine atoms arranged in a manner similar to the body-centered cubic unit in Fig. 136-3A, except that the unit is not square but oblong. It is essentially a body-centered cubic system with one axis elongated.

When steel is heated to the hardening temperature, the grain structure transforms to **austenite.** Austenite has the face-centered cubic lattice arrangement shown in Fig. 136-3B. Then when the steel is quenched at the hardening temperature and cooled rapidly to a temperature below 204.4° C (400° F), it transforms to another form of grain structure called **martensite,** as shown in Fig. 137-9. The space lattice arrangement of martensite is made up of body-centered tetragonal units. Martensite is a supersaturated solid solution of carbon in iron. Martensite is the hardest and most brittle form of steel. The formation of martensite and other transformations which take place in the grain structure of steel during heat treatment processes are discussed further in Unit 137.

Of the four common forms of atomic space lattice arrangements, the cubic forms are the most important. Knowledge of these will aid in understanding heat-treatment processes used with steel and cast iron. Heat treatment of steel is possible largely because of the form and amount of carbon content. In turn, the form and distribution of the carbon content is dependent upon the change which takes place in the lattice structure at the hardening temperature.

Allotropy

Some metals may exist in more than one space lattice form. Iron, as previously indicated, may exist with either the **body-centered** or with the **face-centered** cubic structure, depending on its temperature. These forms are called **allotropic forms.** When the lattice structure changes from one form to another, it is called an **alloptropic change.**

Pure iron, at room temperature, has the body-centered cubic lattice form and is called **alpha** iron. Upon heating above 910° C (1670° F), it changes to the face-centered cubic form, called **gamma** iron. Upon further heating to about

1398.9° C (2550° F), it again changes to a body-centered cubic form called **delta** iron. The delta form of iron is very similar to the alpha form. Upon still further heating to about 1537.8° C (2800° F), the iron becomes liquid. In the liquid state the atoms lose any specific lattice arrangement and thus gain fluidity. Iron has a stable cubic lattice arrangement in each of the above forms within a specific temperature range.

When nearly pure iron is allowed to cool slowly from a molten state, the same changes which took place on heating will again take place, but in reverse order. However, upon cooling, the changes take place at slightly lower temperatures in each case.

The temperatures at which allotropic changes, or changes in phase, take place are called **transformation temperatures.** The changes from one form of iron to another are not often instantaneous at a specific temperature. Rather, they generally take place within a range of temperatures called the **transformation temperature range.** The temperature at the lower end of the range is called the **lower transformation temperature.** It is represented by line A_1 on the iron-carbon phase diagram in Figs. 137-5 and 137-6. The temperature at the upper end of the range is called the **upper transformation temperature,** and it is represented by line A_3. The transformation temperature range and its relationship to various heat-treatment processes is shown in Figs. 137-6 and 137-8.

When iron or steel is heated, the change from one cubic lattice form to another takes place more rapidly at the upper transformation temperature. Since the change in form involves the rearrangement of atoms throughout, considerable time is necessary for the change, even at the proper temperatures. Therefore, when heating for heat-treatment purposes, steel generally should be allowed to **soak** at the proper temperature for a period of time. The minimum soaking period should be about one hour for each 25.4 mm (inch) of metal thickness. This will provide adequate time for complete transformation. For thicknesses of 6.35 mm (¼″) or less, the change takes place almost instantaneously. For metal thickness between 6.35 and 25.4 mm (¼″ and 1″), the soaking period may range from several minutes to one hour, depending on the thickness.

It was stated previously that heat-treatment operations generally have a time-temperature cycle which involves three factors: heating metal to a given temperature, holding (soaking) it at the prescribed temperature for a given period of time, and cooling it at a given rate. The second factor, **the soaking period,** provides time for the internal structural changes to take place as described above.

When steel is heated to the proper hardening temperature, the atomic lattice arrangement is transformed from the body-centered cubic type to the face-centered cubic type. At the proper temperature for hardening, the grain structure will be very fine. A fine grain structure usually is required for the best combination of hardness and toughness. The grain structure increases in coarseness at temperatures significantly below or above the proper hardening temperature which is about 27.8° to 55.6° C (50° to 100° F) above the upper transformation temperature, Fig. 137-6. The relative coarseness of the grain at this temperature is shown to the left in Fig. 137-8.

Grain Formation

The crystals or grains in metal form as the metal cools from a liquid to a solid state. The individual grains start to form in the coolest areas, usually along the edges of the molten metal. With a very slight drop in temperature, grains start to form throughout the metal in many locations. With more rapid cooling, the grains form rapidly, and the metal freezes (solidifies) quickly.

Crystallization starts with the formation of individual space lattice units in various locations throughout the molten metal. The individual units are called **seed** crystals or **unit cells.** The seed crystals grow through the addition of other lattice units which become attached and extend in three directions from the individual seed crystals. With a very slight drop in temperature, new seed crystals develop throughout the body of metal. The crystals grow by extending in three directions until their growth is obstructed by adjacent grains. In this manner, the enlarged crystals develop irregular grain boundaries as crystallization of the metal becomes complete.

The initial grain size is largely dependent upon the rate of cooling from a liquid to a solid state. When the metal cools slowly, the grains are fewer

(American Society for Testing Materials)

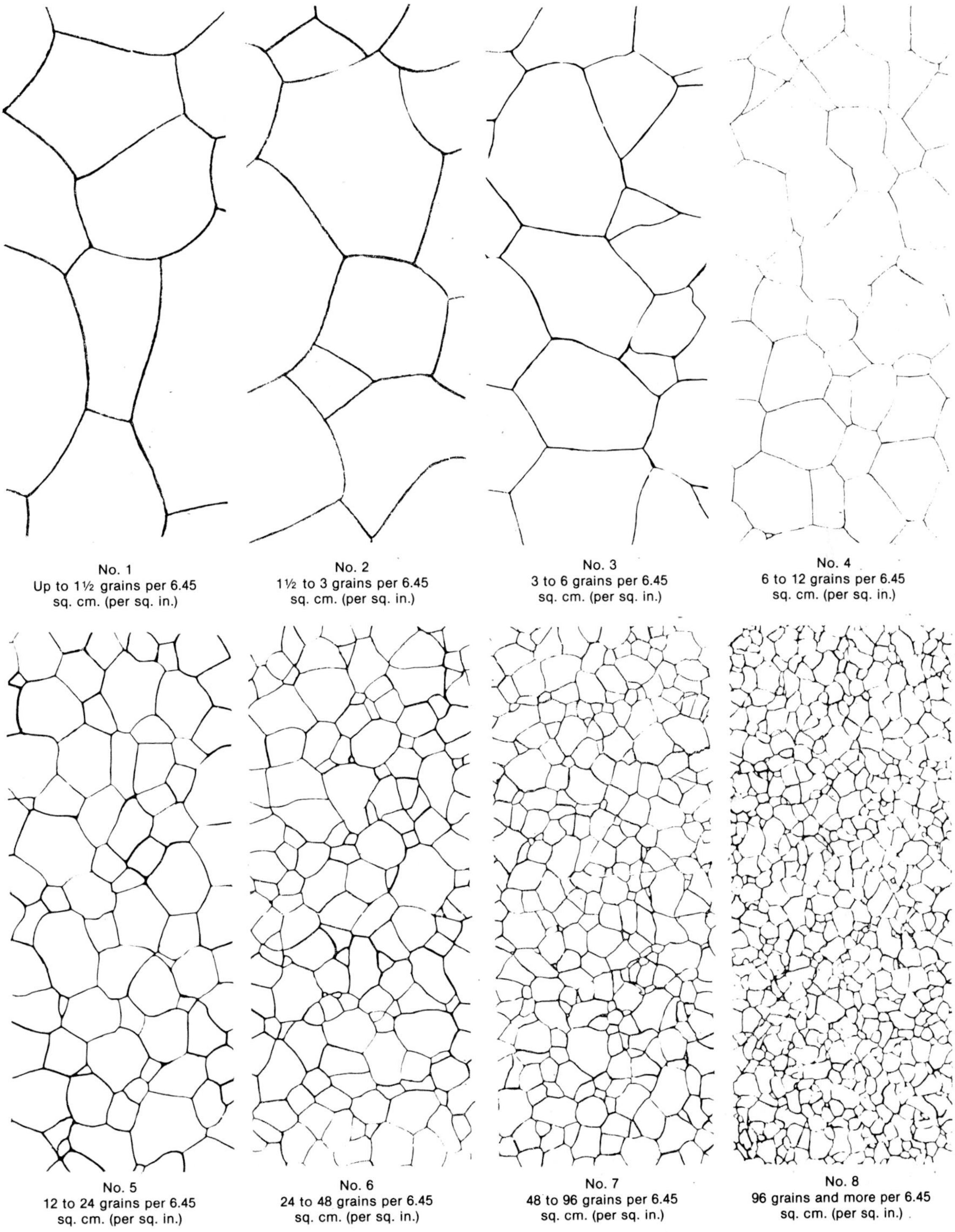

No. 1
Up to 1½ grains per 6.45
sq. cm. (per sq. in.)
No. 2
1½ to 3 grains per 6.45
sq. cm. (per sq. in.)
No. 3
3 to 6 grains per 6.45
sq. cm. (per sq. in.)
No. 4
6 to 12 grains per 6.45
sq. cm. (per sq. in.)
No. 5
12 to 24 grains per 6.45
sq. cm. (per sq. in.)
No. 6
24 to 48 grains per 6.45
sq. cm. (per sq. in.)
No. 7
48 to 96 grains per 6.45
sq. cm. (per sq. in.)
No. 8
96 grains and more per 6.45
sq. cm. (per sq. in.)

Fig. 136-5 ASTM Grain-Size Chart, Untwinned Grains (100 ×) (Amer. Society for Testing Matrials)
The average grain size of steel after standard heat-treatment processes is measured at 100 magnifications by comparision with this chart.

in number, but larger in size. When the metal cools more rapidly, there is less time for grain growth, and grain size is smaller. The properties of various metals often vary with the grain size. When molten steel is allowed to cool slowly to room temperature, the grain structure generally is quite coarse.

The grain size may change from coarse to fine, or vice versa, with certain temperature changes. The variation in grain size for carbon steel upon being heated to increasing temperatures is shown to the left in Fig. 137-8. Because of the change in grain size at different temperatures, the temperature of metals must be carefully controlled in heat-treatment operations.

At the proper temperature for hardening steel, the grain structure is very fine. If steel at this temperature is cooled rapidly by quenching in water or oil, the fine grain structure will be trapped throughout, and the steel will be hardened. Thus, the third important factor involved in the time-temperature cycle, **the rate of cooling,** is applied for the purpose of controlling grain size and condition.

Fig. 136-6.   Ferrous grain-size chart mounted on magnaviewer of metallograph.

Grain Size

The average grain size in steel generally is determined after steel has been heat-treated by a standard process. Several methods can be used for estimating the average grain size. One common method involves the use of a **grain-size comparator.** This may be in the form of a simple chart (as shown in Fig. 136-5) or in the form of printed, transparent, plastic sheets -- one sheet for each grain-size range.

Grain sizes are designated by a standard number system. The most common range from 1 through 8. In estimating grain size, the grain is projected on the magnaviewer or screen of a metallurgical microscope at a magnification 100 times the actual size. See Fig. 136-6. The magnified grain is then compared with the grain sizes on the comparator chart. The estimated average grain size corresponds with the nearest equivalent comparator grain size. Grain size also may be determined by comparing the grain size from a photomicrograph which is at the same magnification as the comparison specimens on the chart.

A No. 1 grain size means that the microstructure, as viewed at $100\times$, has up to $1\frac{1}{2}$ grains per 6.45 sq. cm (per square inch) or an average of one grain per 6.45 sq. cm (per square inch). Similarly, the No. 8 size has more than 96 grains per square 6.45 cm (per square inch), as shown in Fig. 136-5. Steels with grain sizes ranging from 1 to 5 generally are considered coarse-grain steels; those ranging from 5 to 8 generally are considered fine-grain steels. For the best combination of hardness and toughness in hardened tool steel, a very fine grain structure usually is desirable.

The general range of grain sizes which may exist in a given type of steel, under various conditions, depends largely on the chemical constituents used in its manufacture. Therefore, some steels may be classified in a general way as coarse-grained steels, while others are classed as fine-grained steels. However, actual grain size for a given kind of steel may vary with the temperature, with heat treatment, with hot working such as forging, and with cold working such as rolling, twisting, or machining.

Unit
137

How Carbon Content Affects Hardening

The steel hardening process involves the following three steps:

1. Heating the steel to the proper temperature for hardening.
2. Holding the steel at the hardening temperature for a period of time (soaking).
3. Cooling the steel at the proper rate by quenching in either water, oil, or air, depending upon the type of steel.

The factors involved in determining the hardening temperature, the soaking period, and selection of the quenching medium will be discussed shortly. First, however, it is important to understand how the amount and the form of carbon in steel affect its hardness. The form of the carbon and the grain structure of the steel change during each of the above three steps involved in hardening steel.

Amount of Carbon

The maximum degree of hardness obtainable in steel by direct hardening is determined largely by the amount of carbon content. Pure iron is relatively soft and cannot be hardened directly by heat treatment. Low-carbon steels can be hardened only very slightly. Medium-carbon steels may be hardened considerably by direct hardening, but generally they cannot be hardened sufficiently to cut other metals. High-carbon steels and tool steels may be hardened sufficiently to cut steels and many other metals.

The maximum obtainable hardness in most alloy steels also is determined largely by the amount of carbon content. The use of certain alloying elements in steel, however, does improve hardenability. Hardenability, it will be recalled, refers to the depth of hardness.

The carbon steels generally are considered to be shallow-hardening steels. The depth of high hardness for carbon tool steels varies from about 3.2 to 7.9 mm (4/32″ to 10/32″), depending on the hardenability rating of the particular tool steel, as indicated in Fig. 137-1. Certain alloy tool steels, on the other hand, can be highly hardened to depths of several inches. Some alloying elements also enable alloy steels to retain their hardness at elevated temperatures where carbon steel would become softened. Whether steel is carbon steel or alloy steel, generally it cannot be hardened significantly if the carbon content is not sufficiently high.

Form of Carbon

In addition to the amount of carbon in steel, the form of carbon is important in hardening and in other common heat-treatment processes. At temperatures below approximately 721.1° C (1330° F), the carbon in unhardened steel is combined chemically with iron, thus forming cementite. The chemical term for cementite is iron carbide (Fe_3C). In unhardened steel, the cementite normally exists as a mechanical mixture with ferrite (nearly pure iron). This mixture is called **pearlite.** The pearlite, it will be recalled, is composed of tiny platelike layers of ferrite and cementite in specific proportions. Under special heat-treatment conditions (which will be discussed later), the cementite may exist in the form of **spheroidite,** instead of pearlite, as shown in Fig. 137-2.

Fig. 137-1.   Penetration-fracture classifications of carbon tool steels.

Hardenability Rating	P	F	Temperature	P (Depth of Case)	F (Fracture Rating of Case)
SHALLOW			788° C (1450° F)	3.2 mm (4/32") or less	9
	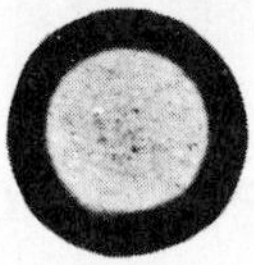	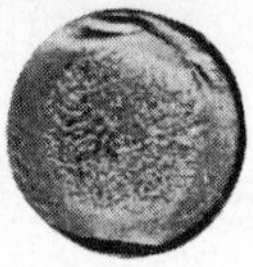	843° C (1550° F)	3.2 mm (4/32") or less	8
MEDIUM SHALLOW		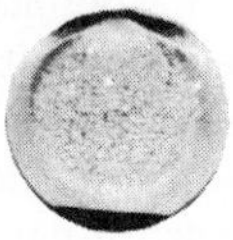	788° C (1450° F)	3.2 mm (4/32")	9
		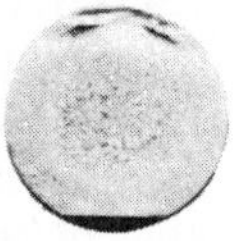	843° C (1550° F)	3.9-4.7 mm (5/32"-6/32")	8
MEDIUM DEEP	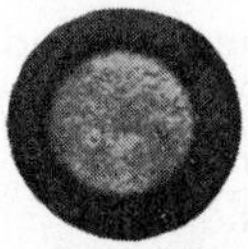	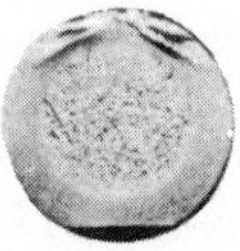	788° C (1450° F)	3.9 mm (5/32")	9
	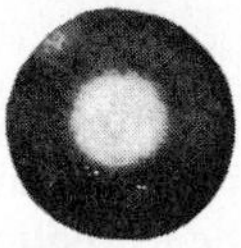	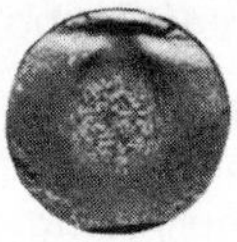	843° C (1550° F)	5.5-7.1 mm (7/32"-9/32")	8
DEEP	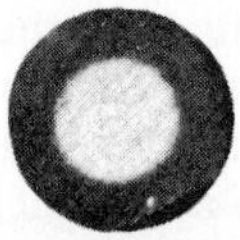	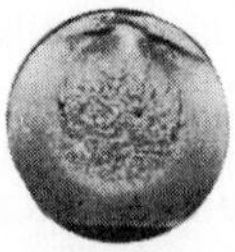	788° C (1450° F)	4.7 mm (6/32") or over	9
			843° C (1550° F)	7.9 mm (10/32") or over	8

Note: Figures above are based on test pieces 19 mm (3/4") round 76 mm (3") long, hardening time 45 minutes in electric muffle, quenched in water at 21° C (70° F). Unless otherwise specified, shallow hardening is applied on rounds up to 25 mm (1") or equal, medium shallow on pieces from 25 to 76 mm (1" to 3"), medium deep on pieces 76 to 102 mm (3" to 4"), and deep on pieces over 102 mm (4") round or equal.

(United States Steel Corporation)

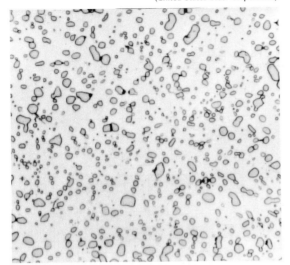

Fig. 137-2. Microscopic structure called spheroidite magnified 1000 times.

(Union Carbide Corporation — Linde Division)

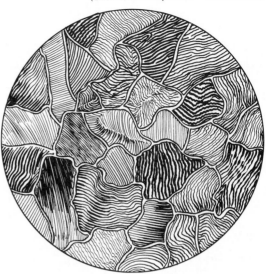

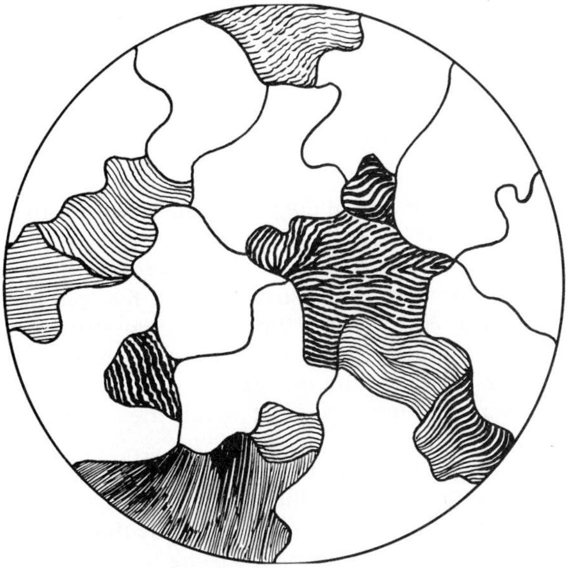

Fig. 137-3. Sketch of a photomicrograph of unhardened low-carbon steel. White area represents grains of ferrite; the shaded grains are pearlite.

Solid cementite is very hard and virtually nonmachinable, except by grinding. However, when cementite is mixed with the soft ferrite, as in pearlite or spheroidite, it can be machined readily. Steel usually must have more than 0.03%

(Union Carbide Corporation — Linde Division)

Fig. 137-4. Sketch of photomicrograph of unhardened steel containing about 1.1% carbon. Grains of pearlite are surrounded by cementite grain boundaries, shown as white lines.

(three-point) carbon before pearlite will form. Carbon in lesser amounts is dissolved in iron, thus forming the solid solution called ferrite, which is nearly pure iron, as shown in Fig. 128-4.

As the carbon content in unhardened steel is increased up to about 0.80% to 0.85%, the proportionate amount of the grain structure which is pearlite also increases. As the proportion of pearlite increases, the strength of the unhardened steel is increased. When hardened by heat treatment, steel which has a high percentage of pearlite will harden to a higher hardness value than steel with a lesser proportion of pearlite.

Steel with 0.80% to 0.85% carbon is called **eutectoid** steel. In the unhardened condition, it has a grain structure which is all pearlite. Steel with less than 0.80% to 0.85% carbon is called **hypo**eutectoid steel; in the unhardened condition, this steel has a grain structure composed of a mixture of pearlite grains and ferrite grains, Fig. 137-3. Steel with more than 0.80% to 0.85% carbon is called **hyper**eutectoid steel. In the unhardened condition, this steel has a grain structure composed of all pearlite grains and excess cementite; the latter is located in the grain boundaries, between the pearlite grains, as shown in Fig. 137-4. Since hypereutectoid steels

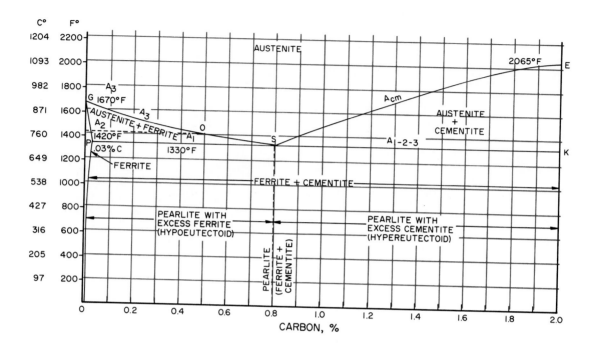

Fig. 137-5.   Phase diagram for carbon steels.

have excess iron carbide existing in their grain boundaries, they are harder and more wear-resistant and rank lower in machinability than the hypoeutectoid steels.

Perhaps a good way to remember the difference between the two classifications of steel is to remember that **hyper,** which has five letters, represents the steel with the higher carbon content. **Hypo,** with only four letters, represents steels with the lower carbon content. **Hyper** is a prefix meaning "above" or "excessive" — much like **super. Hypo** has an opposite meaning of "under" or "down" — much like **sub.** See the designations at the bottom of Fig. 137-5.

Iron-Carbon Phase Diagram

The relationship between the amount of carbon and the grain structure in steel can be understood more clearly by studying the **iron-carbon phase diagram** in Fig. 137-5. This particular diagram is only a portion of one which lists all ferrous metals, including cast iron.

However, the partial diagram is sufficient for the scope of this book.

When steel is heated to certain temperatures, the form and the distribution of the carbon changes. At the same time, the grain structure also undergoes certain transformations, called **phase changes.** The iron-carbon phase diagram shows the temperatures at which these various changes occur.

It will be noted in the diagram that eutectoid steel, which is all pearlite at temperatures below 721.1° C (1330° F), is represented by the dotted line extending downward from the letter **S.** The line A_1, designated **PSK,** represents the **lower transformation** temperature or **lower critical** temperature of 721.1° C (1330° F). The line A_3, designated **GOSK,** represents the **upper transformation** temperature or **upper critical** temperature. The upper transformation temperature varies with the carbon content in the steel; it is lowered as carbon content increases toward 0.80% to 0.85%. The temperature range between lines A_1 and A_3 is called the **transforma-**

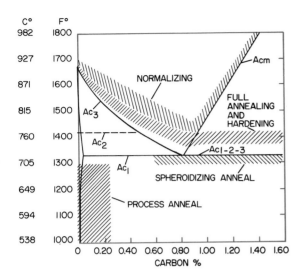

Fig. 137-6.   Recommended temperature ranges for heat-treating plain carbon steels.*

At temperatures below line A_1, the iron in steel has the body-centered cubic lattice structure called **alpha iron.** Upon heating steel to temperatures above line A_1, the grain structure begins to transform to the face-centered cubic lattice structure called **gamma iron.** This form of iron also is called **austenite,** Fig. 137-7. Transformation to austenite continues as the temperature of the steel is increased from line A_1 to line A_3. At line A_3, transformation to austenite is complete.

Austenite is a solid solution of carbon in face-centered cubic iron. When steel transforms to austenite, the carbon which was in a combined state as cementite breaks down into pure carbon and iron. The carbon then is absorbed in the iron in the form of a solid solution. In this form, the carbon is distributed uniformly, just as sugar is when dissolved in water. This condition usually is necessary for obtaining maximum hardness by quenching.

When steel is heated to the proper temperature for hardening, the internal structure not only should be transformed to austenite, but it also must possess a fine grain structure. In stress-relieved steel, the grain size is usually coarse at temperatures below line A_1. As the temperature is increased toward line A_3, the grain structure becomes finer. The finest grain structure exists at

tion temperature range. It is so named because of the structural change which takes place in steel within this range. Also, it is called the **critical temperature range.**

The dotted line, designated A_2, indicates the **magnetic point,** about 771.1° C (1420° F). Above this temperature, steel is no longer magnetic. The magnetic point also continues along line **OSK.** The hardening temperature of medium- and high-carbon steel, therefore, may be estimated with a magnet if more accurate equipment is not available. Note that for hypereutectoid steels, line **SK** represents line A_1, A_2, and A_3. The hardening temperatures of various steels are shown in Table A-4, Appendix. High-carbon steels usually are quenched at temperatures from 27.8° to 55.6° C (50° to 100° F) above the magnetic point, as in Fig. 137-6.

*Thomas G. Digges, Samuel J. Rosenberg, and Glenn W. Geil, **Heat Treatment and Properties of Iron and Steel,** National Bureau of Standards Monograph 88, Washington D.C.: U.S. Government Printing Office, 1966.

(United States Steel Corporation)

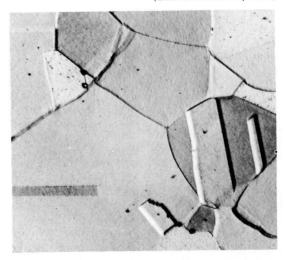

Fig. 137-7.   Microscopic structure called austenite magnified 500 times.

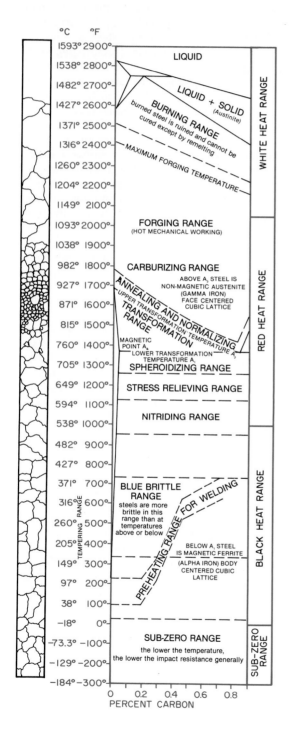

°C   °F

1593° 2900° — LIQUID

1538° 2800°

1482° 2700° — LIQUID + SOLID (Austinite)

1427° 2600° — BURNING RANGE
burned steel is ruined and cannot be cured except by remelting

1371° 2500°

1316° 2400° — MAXIMUM FORGING TEMPERATURE

1260° 2300°

1204° 2200°

1149° 2100°

1093° 2000° — FORGING RANGE
(HOT MECHANICAL WORKING)

1038° 1900°

982° 1800° — CARBURIZING RANGE

927° 1700° — ANNEALING AND NORMALIZING
ABOVE A₃ STEEL IS NON-MAGNETIC AUSTENITE (GAMMA IRON) FACE CENTERED CUBIC LATTICE

871° 1600° — UPPER TRANSFORMATION TEMPERATURE A₃
TRANSFORMATION RANGE

815° 1500°

760° 1400° — MAGNETIC POINT A₂
LOWER TRANSFORMATION TEMPERATURE A₁

705° 1300° — SPHEROIDIZING RANGE

649° 1200° — STRESS RELIEVING RANGE

594° 1100° — NITRIDING RANGE

538° 1000°

482° 900°

427° 800°

371° 700° — BLUE BRITTLE RANGE
steels are more brittle in this range than at temperatures above or below

316° 600° — PREHEATING RANGE FOR WELDING

260° 500°

205° 400° — BELOW A₁ STEEL IS MAGNETIC FERRITE

149° 300° — (ALPHA IRON) BODY CENTERED CUBIC LATTICE

97° 200°

38° 100°

−18° 0°

−73.3° −100° — SUB-ZERO RANGE
the lower the temperature, the lower the impact resistance generally

−129° −200°

−184° −300°

TEMPERING RANGE

WHITE HEAT RANGE
RED HEAT RANGE
BLACK HEAT RANGE
SUB-ZERO RANGE

0   0.2   0.4   0.6   0.8
PERCENT CARBON

Fig. 137-8.   Basic guide to ferrous metallurgy.

A_3, as shown at the left in Fig. 137-8. When the temperature of the austenite is increased above line A_3, the grain size again starts to increase gradually. At temperatures more than 37.8° C (100° F) above line A_3, the grain size becomes quite coarse.

Hardening Temperature

The recommended hardening temperature is usually 27.8° to 55.6° C (50° to 100° F) above the upper critical temperature, as shown in Fig. 137-6. Generally, the grain size is not the finest obtainable when quenched from 27.8° to 55.6° C (50° to 100° F) above A_3. However, at these hardening temperatures, the grain size and structure usually give the most desirable combination of hardness and toughness. The finest obtainable grain structure is not always the most desirable.

The hardening temperature varies for different steels, depending on the carbon content. It also varies according to special alloying elements which are included in alloy steels.

For hypereutectoid steels, the recommended hardening temperature usually is in the range from 746.1° to 787.8° C (1375° to 1450° F), Fig. 137-6. For hypoeutectoid steels, the hardening temperature generally is higher. It is advisable to follow the steel manufacturer's recommendations concerning hardening temperatures and temperatures for other heat-treating operations on specific types of steel. If these are unavailable, the hardening temperatures recommended for the various steels represented in Fig. 137-6, or in Table A-4, Appendix, may be used. The hardening temperatures for various steels also are listed in handbooks for machinists and in handbooks concerned with the properties of metals.

Martensite Formation

When high-carbon steel is quenched at the critical temperature and cooled rapidly to a temperature below about 204.4° C (400° F) the austenite transforms to a structure called **martensite,** as shown by the photomicrograph in Fig. 137-9. Martensite is the hardest and most brittle form of steel. It is a super-saturated solid solution of carbon in iron which has a body-centered tetragonal lattice arrangement (Fig. 136-4). Under magnification, this type of grain structure has an extremely fine, needlelike or acicular ap-

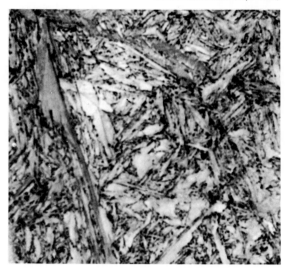

Fig. 137-9.   Microscopic structure called martensite magnified 2500 times.

pearance. Pure martensite does not contain carbon in the form of cementite. For maximum hardness after quenching, the grain structure of steel usually must be completely martensitic.

Carbon steel must be cooled rapidly from the hardening temperature for complete transformation from austenite to martensite — in fact, usually from the hardening temperature to below 93.3° C (200° F) during a period of several seconds. With slower cooling, some pearlite will form and will be mixed in with the martensite. When this happens, the steel will not be hardened completely.

When steels are cooled from the hardening temperature, austenite begins to transform to pearlite at temperatures below 721.1° C (1330° F). The rate of this transformation depends on the temperature. In carbon steels, austenite decomposes rapidly to pearlite after one second in the temperature range from 537.8° to 593.3° C (1000° to 1100° F). After three seconds in this temperature range, transformation to pearlite is nearly complete. Consequently, when carbon steels are quenched at the hardening temperature, they must be cooled below the 537.8° C (1000° F) temperature range during a period of one second or less. So long as cooling through

this range takes place in less than one second, several additional seconds may be allowed to cool the steel below 121.1° C (250° F). Therefore, it is evident that the hardening process is essentially the **avoidance of pearlite grain formation** in steel while it is being quenched.

Depth of Hardness

The depth to which steels may be uniformly hardened depends largely on the speed of cooling and the thickness of the metal. Carbon steels less than 6.35 mm (¼") in thickness will transform from austenite to martensite almost immediately when quenched. The result is complete and uniform hardening. Carbon steels with a larger cross-sectional area, such as 25.4 mm (1") thickness, do not transform to martensite uniformly from the surface to the center. Where the transformation is incomplete, a softer structure will exist as a mixture with the martensite, toward the center of the material.

An example will illustrate the lack of uniform hardness in larger cross sections of carbon steel. When 25.4 mm (1") diameter carbon steel, with 0.95% carbon content, is quenched in water at 787.8° C (1450° F), hardness indicated by the following Rockwell C numbers may result: surface hardness 63, hardness at one-half radius 47, hardness at center 43. These hardness numbers are more meaningful if it is understood that a metal-cutting tool, such as a tap or drill, generally should have a minimum hardness of Rockwell C-60. A Rockwell C hardness value of 63 to 66 would be more desirable.

The depth of hardness for carbon tool steels with various hardenability ratings is shown in Fig. 137-1. These steels (compared to alloy tool steels) harden to relatively shallow depths.

It is apparent from our discussion up to this point that rapid cooling is important for depth of hardness when quenching carbon steels. However, when the rate of cooling is too fast during quenching, warpage occurs, internal stresses develop, and cracks often appear. Of course, when the quenched part cracks, it is ruined. For these reasons, certain precautions are necessary in quenching both high-carbon steels and alloy steels for hardening. The proper quenching media and appropriate quenching procedures should be used.

Unit

138 ||||||||||

Quenching Media and Their Applications

Selection of the quenching medium and use of proper quenching procedures are important factors in hardening steel. Various quenching media absorb and remove heat from steel at different rates. Cooling too rapidly often produces internal stresses which may cause cracks, particularly in high-carbon steels. Cooling too slowly may result in incomplete hardening.

Common Quenching Media

The common quenching media (solutions) include water, brine, and oil. Air, either still or compressed, also is used to quench special types of tool and die steel. Some quenching solutions produce a more severe quench than others. (**Quenching severity** refers to the rate at which heat is removed from the article being quenched.) The severity of the quenching operation depends on the thickness of the metal, the kind of quenching medium, the temperature of the quenching solution, and the degree of agitation of the metal in this solution.

Severity of Quenching Media

Each basic quenching medium cools with a different degree of severity. The rate of cooling is most severe with brine, less rapid with water, slow with oil, and slowest in air. The rate of cooling for each quenching solution can be increased considerably by agitation of the article in the solution or by circulation of the solution. In some instances, the solution is flooded over the steel under pressure, thus producing a more severe quench than with normal agitation.

The relative severity of the common quenching solutions, in comparison with the rate of cooling by water, may be designated by numbers. For this purpose, water is assigned a rating of No. 1 when no agitation or circulation occurs. In comparison, under the same conditions, air is 0.02; oil 0.30; and brine 2.0. With good circulation or agitation, the above severity ratings may be increased by approximately 50%. With strong circulation or agitation, they may be increased by 100% or more.

The temperature of the quenching solution also affects the severity of the quench. Water or brine should be maintained at a temperature of about 15.6° C (60° F). Because of its increased viscosity, oil, unlike water, cools best when it is maintained at temperatures from about 37.8° to 60° C (100° to 140° F).

Various quenching oils are available. Some cool more slowly than others. Vegetable oils, animal oils, mineral oils, and various combinations of these have been used as quenching media. Straight mineral oil with a Saybolt viscosity rating of about 100 at 37.8° C (100° F) frequently is used. When metals are quenched at the lower temperatures, such as 148.9° to 260° C (300° to 500° F), oil will cool at only about 10% the rate of water.

Brine is a solution of 5% to 10% salt (sodium chloride) and water. Rock salt usually is used. Brine cools about twice as rapidly as water. It also has the ability to help **throw** the scale away from steel during quenching, thus enabling the steel to cool more uniformly.

Selecting Proper Quenching Medium

The proper quenching solution should be determined for each type of steel. The steel manufacturer's recommendations should be followed in selecting the quenching medium, particularly when quenching tool and die steels. If these are unavailable, the recommended solutions for various steels listed in Table A-4, Appendix, may be used.

The carbon steels, which are relatively shallow-hardening steels, generally require severe quenching solutions such as water or brine. Very thin sections of high-carbon steel often can be hardened satisfactorily by agitating them in oil. This procedure also is recommended in cases where the severity of water quenching causes cracks in materials of irregular cross-sectional thickness.

While some alloy steels are designed for water quenching, the majority are designed for oil quenching. Where an oil quench is recommended, it should be used, because there is less distortion and less danger of cracking. However, when an oil quench is used for water-hardening steel greater than 6.35 mm (¼") thickness, maximum hardness generally cannot be obtained.

Agitation

Proper quenching usually involves moderate agitation of the work in the quenching bath. When steel is quenched, the liquid in contact with the hot surface vaporizes and reduces the rate of cooling. Thus, the piece should be agitated to remove the vapor film. An **up-and-down** movement or a **figure-eight** movement often is used. Where a spray quench is used, agitation is provided. Since increased agitation increases the severity of the quench, care must be taken not to agitate the work too vigorously, particularly in water or brine. The increased severity of the quench may cause cracks; this is especially true when parts with thin or irregular cross-sectional areas are quenched in water or brine.

Special Solutions

Special solutions are used in quenching special steels or steels for special purposes. Examples of special solutions include hot oil, molten salt baths, and molten lead. High-speed steels often are hardened at temperatures of 1148.9° to 1343.3° C (2100° to 2450° F); they usually are quenched in molten lead or salt baths which are held at temperatures ranging from 426.7° to 648.9° C (800° to 1200° F). High-speed steels are relatively expensive, and their properties sometimes vary with different manufacturers. Consequently, it is advisable to follow the manufacturer's recommendations in heat-treating these steels.

A special molten salt solution is used for austempering in the temperature range from about 204.4° to 426.7° C (400° to 800° F). (Austempering is explained in Unit 142.) The salt melts at about 137.8° C (280° F) and may be used at working temperatures in the range from about 162.8° to 537.8° C (325° to 1000° F).

Several kinds or combinations of salt are available commercially for use as quenching media or as heating media for heat-treating operations. These salts are available with working temperatures ranging from about 162.8 to 1315.6° C (325° to 2400° F). By comparison, lead has a total working temperature range from about 343.3° to 926.7° C (650° to 1700° F). The salts are water-soluble, thus permitting quenched parts to be cleaned easily in hot water after quenching. The salt is very corrosive when left on steel parts.

Development of Stress

When steel is hardened by quenching, internal stresses are developed. The amount of stress often is related to the severity of the quenching solution. The stresses, therefore, usually are greater when water or brine solutions are used. They occur because martensite forms near the surface first, then toward the center. The temperature difference between the center and the surface of the steel also causes uneven rates of expansion and contraction. Both factors cause internal stress.

Because of the internal stresses in hardened high-carbon steel, the material usually is so hard and brittle that it is of little practical use until it has been tempered. In this condition, the steel is often referred to as being **glass hard.** A piece of high-carbon steel which has been water-quenched often will fracture when struck lightly with a hammer. Hence, in most applications, before hardened steel may be used, the internal stress must be removed by a heat-treatment process called **tempering.** The hardened steel should be tempered as soon after hardening as possible, preferably before the piece has cooled to room temperature. When completely hardened steel gets cold, the internal stresses are further increased. If left in this condition, the steel sometimes will crack by itself.

Tempering is discussed in Unit 139. The procedures for hardening and tempering of steel are outlined in Unit 140.

Unit
139

Tempering, Annealing, and Normalizing

The Tempering Process

Tempering is a heating and cooling process which relieves some of the brittleness and internal stress developed in steel during the hardening operation. This process causes some loss of hardness and tensile strength, but it increases toughness. Tempering also is called drawing.

The tempering process follows hardening. It involves heating hardened steel to a temperature somewhere below the lower critical temperature; this is below the temperature at which steel begins to harden when quenched. The heated part then is allowed to cool in air, or it may be quenched. The rate of cooling is of little significance. However, cooling in still air often is preferred when the nature of the workpiece will permit this procedure.

The primary purposes of tempering are to relieve internal stresses and to increase toughness in hardened steel, not to reduce hardness and strength. If it were possible to increase toughness without reducing hardness and tensile strength, the ideal combination of properties would be attained. As the hardness of hardened steel is decreased by tempering, the tensile strength also is decreased. Toughness, however, is increased.

When high-carbon steel is fully hardened, its grain structure is fully martensitic. Pure marten-

site is a supersaturated solution of carbon in ferrite. This is an abnormal arrangement for carbon and iron at low temperatures. Martensite is extremely hard and brittle, and it also is under great internal stress and strain. When parts in this condition are used, there is danger of fracturing. Hence, the parts must be tempered to increase toughness before use.

Tempering Temperatures

When hardened steel is tempered, the grain structure is called **tempered martensite.** There are varying degrees of tempered martensite, depending on the tempering temperature. At the lower tempering temperatures, some of the carbon and iron in martensite start to combine, thus forming tiny particles of cementite. This is a more normal atomic arrangement for carbon in steel at lower temperatures. Consequently, some internal stress and strain is relieved, and the steel becomes tougher. At the same time, there is some loss of hardness. With an increase in the tempering temperature, the cementite particles increase in number and size, causing a further increase in toughness and a further loss of hardness.

Steel which has toughness also has some degree of plasticity. This means that it will deform to some extent before fracturing. The amount of plasticity depends on the degree of toughness.

The temperature to which steel is heated for tempering depends on the following factors:

1. The type of steel (carbon steel or special alloy steel).
2. The carbon content.
3. The hardness required.
4. The toughness required.

The effect of various tempering temperatures on the hardness of plain carbon steels with 0.45% and 0.95% carbon content is shown in Fig. 139-1. The hardness values indicate the surface hardness only, since carbon steels are relatively shallow-hardening. The major factors affecting maximum obtainable hardness include carbon content, hardening temperature, quenching medium, and cross-sectional size. The data concerning these factors are indicated below the figure just described.

Figure 139-1 shows there is little loss of hardness at temperatures below 121.1° C (250° F). At higher temperatures, the rate at which hardness decreases is more rapid. Suggested tempering temperatures for various tools are shown in Table 139-1.

Temperatures in the range from about 148.9° to 593.3° C (300° to 1100° F) are used for various tempering applications with carbon steels (Table 135-1). Temperatures as low as 148.9° C (300° F) are used in tempering items which require a combination of very high hardness and some toughness, such as ball bearings and roller bearings. Lathe centers require very high hardness and may be tempered at temperatures as low as 193.3° C (380° F). Temperatures within the range from about 193.3° C to 315.6° C (380° to 600° F) are used for tempering many cutting tools. Articles which require very high toughness, such as steering knuckles for automobiles, are

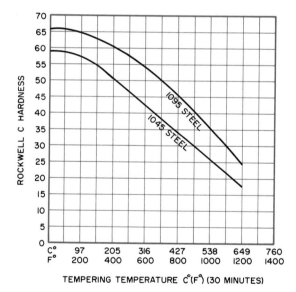

TEMPERING TEMPERATURE C°(F°) (30 MINUTES)

Fig. 139-1.   Effect of various tempering temperatures on the hardness of carbon steel. Surface hardness of carbon steel bars, SAE 1045 steel 19 mm (3/4″) square, and SAE 1095 steel 12.7 mm (1/2″) diameter, after tempering at various temperatures. Both steels were hardened in a water quench, the 1045 steel at 815° C (1500° F) and the 1095 steel at 788° C (1450° F).

Table 139-1
Typical Tempering Temperatures
for Various Tools

Degrees	Temper Color	Tools
193.3° C (380° F)	Very light yellow	Tools which require maximum hardness: lathe centers and cutting tools for lathes and shapers
218.3° C (425° F)	Light straw	Milling cutters, drills, and reamers
240.6° C (465° F)	Dark straw	Taps, threading dies, punches, dies, and hacksaw blades
254.4° C (490° F)	Yellowish brown	Hammer faces, shear blades, rivet sets, and wood chisels
273.9° C (525° F)	Purple	Center punches and scratch awls
285° C (545° F)	Violet	Cold chisels, knives, and axes
310° C (590° F)	Pale blue	Screwdrivers, wrenches, and hammers

tempered in the range from 426.7° to 593.3° C (800° to 1100° F).

High-speed tool steels generally are tempered at temperatures much higher than those used for carbon steels. Usually they are tempered in the range from 537.8° to 593.3° C (1000° to 1100° F). It always is advisable to follow the manufacturer's recommendations in selecting temperatures for hardening and tempering high-speed steels and special tool and die steels.

Determining Toughness

The toughness of tempered steel often is interpreted or estimated in terms of hardness. Hardness and toughness vary indirectly with each other. As the hardness is reduced by higher tempering temperatures, toughness is increased. The hardness may be measured with hardness-testing instruments as outlined in Unit 143.

A tool such as a cold chisel may be tempered at the suggested temperature indicated in Table 139-1. For high-carbon tool steel, this temperature should result in a combination of hardness and toughness required to produce satisfactory results for average applications. If the chisel is too hard, the cutting edge may crack or chip when used. If it is too soft, the edge will dull rapidly. Cold chisels, depending on their particular applications, generally are hardened to values ranging from Rockwell C-45 to C-57. The tempering temperatures may be increased or decreased to produce the combination of hardness and toughness desired.

A tool such as a cold chisel or center punch should have a hard cutting edge and a soft head. The cutting end should be hardened and tempered for proper hardness and toughness. The head should be softer and tougher to withstand hammer blows. When only a portion of a workpiece is heat-treated, the procedure is called **selective** heat treatment. It is possible to apply any of the common heat-treatment processes to only a portion of a tool or part.

Notched-Bar Toughness

The toughness of notched-bar test specimens of the type shown in Fig. 139-2, is determined by laboratory testing equipment. The most commonly used standard-size specimen is 10 mm (0.3937") square, the type shown to the left in Fig. 139-2. A pendulum is allowed to swing

Fig. 139-2.   Notched-bar specimens for impact-toughness tests.

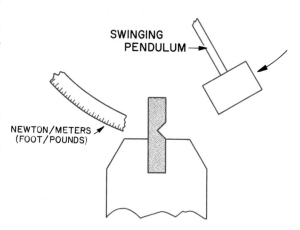

Fig. 139-3.   Specimen mounted for Izod impact-toughness test.

against the specimen, as shown in Figs. 139-3, 139-4, and 139-5. The difference in the height of the pendulum at the beginning and at the end of the swing is registered on a gage which indicates the energy absorbed in fracturing the specimen. The toughness of notched-bar specimens is indicated in **newton/meters** (foot-pounds) of energy required to fracture the notched specimen.

Two standard tests, the Charpy and the Izod tests, utilize test specimens of the same size. A V-type notch is standard for the Izod test specimen. Although other notch designs often are used for the Charpy test, the V-type may be used for the Charpy specimen also. It then is called the **Charpy V-notch** specimen. The V-type notched-bar specimen is shown in Fig. 139-2.

For the Izod test, the specimen is mounted in the machine by clamping it at one end, as shown in Fig. 139-3. For the Charpy test, the specimen

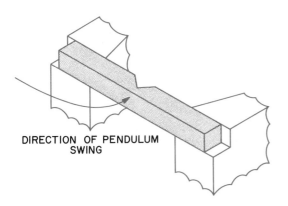

Fig. 139-4. Specimen mounted for Charpy impact-toughness test.

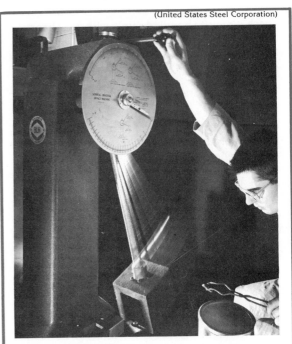

A. Making Charpy test.

is mounted as shown in Fig. 139-4. A knife-shaped edge on the pendulum strikes the specimen on the side opposite from the notch. The pendulum impact-testing machine in Fig. 139-5 is equipped for the Charpy impact test. Certain machines of this type also may be equipped with Izod test fixtures for use in making Izod tests.

Brittle Temperature Range

The temperature range between 204.4° and 371.1° C (400° and 700° F) often is called the **brittle temperature** range. It is so named because notched-bar specimens tempered in this range actually exhibit less toughness than those which are tempered at temperatures below or above this range. However, these toughness ratings sometimes are misleading. They usually do not apply to the toughness of specimens of similar cross-sectional size which do not possess sharp notches or fillets. This factor is not clearly understood by some heat-treaters. Consequently, the **brittle tempering** range from 204.4° and 371.1° C (400° to 700° F) too often is avoided in tempering applications.

Notched-bar tests of tempered specimens do illustrate clearly, however, that the presence of sharp notches, corners, or fillets affects the toughness of hardened and tempered parts. Therefore, parts which possess these features generally should not be tempered within the brittle temperature range, if greater toughness is required. Rather, they should be tempered at temperatures either below or above this range.

B. Specimen mounted for Charpy test.

Fig. 139-5. Pendulum impact testing machine.

Tempering should follow immediately after hardening, particularly with high-carbon steels. After quenching, hardened steel should be allowed to cool only to the highest temperatures at which it may be handled safely with bare hands. If tempering is delayed until the part becomes cold, internal stresses increase; and there is greater possibility of cracking.

Time at Temperature

The time at which a part is held (soaked) at the tempering temperature is important. The soaking period permits time for atomic rearrangement and for relief of internal stresses. The article to be tempered should be heated uniformly to the desired temperature. It should then be soaked for a period of about one hour for each 25.4 mm (inch) of thickness. Thus, a part which is 76.2 mm (3") thick should be soaked at the tempering temperature for a minimum of three hours. Items 6.35 mm (¼") or less in thickness need not be soaked beyond the period necessary for bringing them up to the uniform required temperature.

Annealing

Annealing is a process which involves heating and cooling of steel to produce softening and to improve machinability. Annealing also relieves internal stress and strain caused by previous heat treatment, by machining, or by other cold-working operations.

The temperature to which steel is heated for annealing operations depends on the type of steel and the purpose for which the annealing is being done. Three types of annealing are in common use: full annealing, process annealing, and spheroidizing.

Full Annealing

Full annealing is a process utilized primarily to produce maximum softness in steel. It also improves machinabilty and relieves internal stress. It may be used to soften hard steel or to soften work-hardened steel for remachining.

Procedure for Full Annealing:
1. Heat the part uniformly to the full annealing temperature. This is about 27.8° C (50° F) above the upper critical temperature. The annealing temperature also is shown in Table A-4, Appendix.
2. Soak the part at this temperature for about one hour per 25.4 mm (inch) of thickness.
3. Allow the part to cool slowly. It may be removed from the furnace and packed in lime or

ashes for slow cooling. **Or** the furnace may be shut off and the part permitted to cool slowly in the furnace.

Process Annealing

Process annealing, often termed **stress-relief annealing,** is utilized primarily for relieving internal stresses due to machining or other cold-working processes. Most frequently it is applied to low-carbon steels.

Procedure for Process Annealing:
1. Heat uniformly to about 16.7° C (30° F) below the lower critical temperature. Temperatures as low as 537.8° C (1000° F) may be used if no further machining is involved.
2. Allow the part to soak at the desired annealing temperature for a period of about one hour per 25.4 mm (inch) of thickness.
3. Remove the part from the furnace and allow it to cool in air.

Spheroidizing

Spheroidizing is a process that involves heating and cooling of steel to produce a special kind of grain structure that is relatively soft and machinable. This special microstructure has cementite in rounded globules called **spheroids,** Fig. 137-2. The spheroids are surrounded by ferrite. A microstructure of this type generally is tougher and more ductile than pearlite which is produced by full annealing.

Spheroidizing generally is applied to high-carbon steels to improve machinability. It also is used to prepare high-carbon steel for cold-drawing into wire.

Procedure for Spheroidizing: The procedure for this process varies considerably with different types of steel. It also varies with the different conditions of heat treatment existing in steel parts. The following is one of several procedures which may be utilized with plain high-carbon steel which is in a fully-hardened (martensitic) condition:

1. Heat the part uniformly to a temperature just below the lower critical temperature.
2. Allow the part to soak for several hours.
3. Allow the part to cool slowly in the furnace, about 27.8° to 55.6° C (50° to 100° F) per hour, to a temperature below 537.8° C (1000°F).
4. Remove the part from the furnace, and allow it to cool in air.

Normalizing

Normalizing is a process that involves heating steel to slightly above the hardening temperature, soaking it at this temperature, and cooling to room temperature in air. Normalizing relieves internal stresses due to machining, forging, and cold working. It removes all previous effects due to heat treatment. Normalizing also softens hardened steel and improves its machinability. However, it does not result in the uniformity and degree of softness which is produced by full annealing.

Parts of irregular cross section may lack uniformity of hardness and ductility after normalizing, particularly where thin sections are involved. The grain structure of normalized steel is generally somewhat harder, less ductile, and has a finer pearlitic grain structure than fully-annealed steel.

Procedure for Normalizing Steel:

1. Heat the part uniformly to the normalizing temperature indicated in Fig. 137-6, or in Table A-4, Appendix.
2. Soak the part at this temperature for about one hour per 25.4 mm (inch) of diameter.
3. Remove the part from the furnace, and allow it to cool to room temperature in air. **Note:** High-carbon steel 3.2 mm (1/8″) or less in thickness may cool rapidly enough in circulating cool air to become partially hardened. The rate of cooling may be slowed, and normalizing improved, by packing the part in lime or ashes.

Unit
140

How to Harden and Temper Steel

The metallurgical theory and the principles involved in hardening and tempering were discussed in the four preceding units. The more advanced student, or the student seeking a clear understanding of these two processes, should study these preceding units first. However, by carefully following the procedure outlined in this unit, steel tools or parts can be hardened and tempered with good results.

Preliminary Planning

Before attempting to harden and temper steel tools or parts, you should know the type of steel with which you are working. This information usually is included on the drawing or blueprint for the part, or it may be identified by your instructor.

When you have identified the kind of steel, you can determine the following information required for proper hardening and tempering of the part: the hardening temperature, the quenching medium, the combination of hardness and toughness desired, and the tempering temperature.

Hardening Temperature

The hardening temperature usually is from 27.8° to 55.6° C (50° to 100° F) above the upper critical temperature. Thus, the hardening temperatures for various plain carbon steels can be interpreted from Fig. 137-6. The hardening temperatures for various heat-treating grades of steel also are shown in Table A-4, Appendix. For high-speed steels and tool steels, the manufacturer's recommendations should be followed.

Quenching Medium

The proper quenching medium for various heat-treating grades of steels is listed in Table A-4, Appendix. Plain carbon steels generally are quenched in water or brine. Some alloy steels may be quenched in water, but most must be quenched in oil. Water should be at a temperature of about 15.6° C (60° F). Oil cools best when held at a temperature from 37.8° to 60° C (100° to 140° F). The average brine solution has about 8% salt, by weight.

Hardness and Toughness Desired

The combination of hardness and toughness desired will depend upon the application or desired use of the heat-treated part. Generally, maximum hardness associated with adequate toughness is most desirable. By studying Table 139-1 and Fig. 139-1, you can determine the desirable degree of hardness and toughness. The hardness of metals can be determined by methods outlined in Unit 143.

Tempering Temperature

After determining the degree of hardness and toughness, you can determine the tempering temperature from Table 139-1 and Fig. 139-1.

Soaking Period

When steel is heated to either the hardening or the tempering temperature, it should be allowed to soak at this temperature long enough for required internal changes to occur in the grain structure. Parts with a thickness of 6.35 mm (¼") or less do not require a soaking period. Parts of greater thickness usually should soak for about one hour per 25.4 mm (inch) of thickness. For parts from 6.35 mm to 25.4 mm (¼" to 1") in thickness, the period should be from 3 minutes to one hour, depending upon the thickness.

Heating Precautions

The following precautions should be taken when heating steel to the hardening temperature:

1. Heat parts slowly and uniformly to avoid excessive internal stresses and distortion. The rate of heating is most rapid in a molten bath (such as a salt or lead bath), less rapid in a gas furnace, and least rapid in an electric furnace.

2. When two furnaces are available, preheat large parts in one furnace to about 537.8° C (1000° F). Then transfer the parts to the second furnace which is at the proper hardening temperature. More uniform heating and less distortion will result.

3. Support long parts on fire bricks or other suitable refractory material to prevent sagging or bending.

4. Avoid excessively long soaking periods. They cause decarburization (loss of carbon) of the surface, scaling, and coarseness of grain structure.

5. Adjust the fuel and air mixture so that the furnace atmosphere is neutral to slightly oxidizing, rather than carburizing. An oxidizing flame has a slight excess of air.

6. Avoid or reduce surface oxidation (scaling) and decarburization by providing a protective atmosphere in the furnace. Several methods are used for this in large industrial heat-treatment furnaces. Oxidation and decarburization also may be reduced by heating in a molten salt or lead bath. Molten baths are described in Unit 138. The rate of heating is much more rapid in molten baths than in furnaces. Certain alloy steels will crack when placed directly into molten baths at hardening temperatures.

7. Reduce surface oxidation and decarburization by **pack heating.** With this method, steel parts first are placed in a container. They then are packed with cast iron chips and are heated to the hardening temperature. After the proper soaking period, they are removed from the container for quenching.

8. If steel inadvertently is heated significantly above the proper hardening temperature, allow the part to cool in the air to below 537.8° C (1000° F). Then reheat it to the proper hardening temperature. Otherwise, the grain will become too coarse for the best combination of hardness and toughness.

9. Quench parts immediately upon removal from the furnace or heat bath. A significant delay will permit a drop in temperature and increase the coarseness in the grain structure.

10. In general, heat alloy steels to the hardening temperature more slowly than plain carbon steels. Also, for the best surface appearance, heat them in a controlled atmosphere, either in a furnace or by the pack method.

Procedure for Hardening Steel

1. Determine the hardening temperature, and light the furnace as directed by your instructor. Allow the furnace to reach the hardening temperature.
2. With a pair of tongs, place the part in the furnace, and allow it to heat uniformly to the hardening temperature. If the furnace is equipped with a temperature control, the procedure is simple and the temperature will not be exceeded. Without temperature controls, however, care must be taken not to overheat the part. The temperature can be estimated through the use of the temperature color chart in Table 135-1 and Fig. 137-8. A magnet also should be used to check the temperature of the part.
3. Allow the part to soak at the hardening temperature for the proper period of time.
4. Select the proper quenching medium, and quench the part. Agitate the part in the solution, using an **up-and-down** movement. If the part is thin and flat, use a cutting motion, as with a knife, to prevent warping due to uneven cooling.
5. Determine whether the part is properly hardened. This can be done by testing with the **corner** of a file — not the flat face, since hardened steel will dull the teeth and ruin the file. If the part is properly hardened, the file either will not cut the metal or will cut with great difficulty. A properly-hardened piece is very hard and brittle, and it is under great internal stress. It should be tempered as soon as possible to prevent cracking.

Procedure for Tempering

With Temperature Controls

To temper the whole part in a furnace equipped with temperature-indicating controls:

1. Light the tempering furnace as directed by your instructor, and allow it to heat to the tempering temperature.
2. Place the part in the furnace, using a pair of tongs. Allow time for the work to heat uniformly to the tempering temperature. Then soak the work at the temperature for the desired period of time.

3. Remove the part from the furnace with tongs, and allow it to cool in still air or by quenching. Cooling in air usually is recommended.
4. Determine the hardness of the part as outlined in Unit 143.

Without Temperature Controls

To temper the whole part in a furnace or over an open flame without the use of temperature controls:

1. Using a piece of abrasive cloth, remove the scale from the part until it is clean and bright, so that temper colors can be seen.
2. Light the furnace or flame, as directed by your instructor. A low flame is best for tempering.
3. Using a pair of tongs, hold the work over the flame, and heat slowly. Rotate the piece so that it will heat uniformly to the proper tempering temperature. The temperature is identified by the temper color, as in Table 139-1.
4. Remove the part from the furnace or flame, and allow it to cool as desired, either in still air, or by quenching.
5. Determine the hardness of the part as outlined in Unit 143.

Small Cutting Tools

To temper small cutting tools, such as a small center punch or cold chisel:

1. Harden the whole tool as outlined above.
2. Using a piece of abrasive cloth, remove the scale until the tool is clean and bright.
3. Heat a scrap piece of steel red-hot.
4. Lay the tool on the hot piece of steel, being careful to extend the cutting end beyond the end of the block. Thus the opposite end will heat first.
5. Watch the temper colors. When the desired color reaches the cutting end, quench the tool in water. This will prevent the cutting end from being overheated due to the excess heat in the opposite end of the tool.

This procedure also may be used for tempering small tools over a small flame (such as that produced by a Bunsen burner, soldering furnace, or propane torch), instead of a scrap steel block. Care must be taken to heat only the area or end opposite from the cutting end. As the tool is heated, the heat gradually will conduct toward the cutting end of the tool until the proper temper color appears for quenching.

Unit

141

Casehardening Processes

Casehardening is a process which produces a hardened layer on the surface of ferrous alloys, as shown in Fig. 141-1. The inner core below the hardened surface remains relatively soft. The steels most frequently used for casehardening include plain low-carbon steels and certain low-carbon, low-alloy steels. The low-alloy steels usually are alloyed with nickel, chromium, and molybdenum. The casehardened surface provides a combination of increased wear resistance and increased strength over a softer and tougher inner core.

Many tools and parts require only **casehardening,** rather than **through hardening.** Consequently, such items may be made from the less-expensive casehardening grades of steel. Items which frequently are casehardened include small hand tools, large wrenches, toolholders for machine tools, bolts, setscrews, pins, bearing surfaces, gears, pinions, splines, and shafts.

A casehardened surface may vary in depth from several hundredths of a millimeter (thousandths of an inch) to 3.2 mm (1/8"). The depth depends upon the casehardening process and

procedures used. The surface hardness is usually in the range from Rockwell C-60 to C-66. The core hardness depends largely on the carbon content of the steel. It usually is from C-20 to C-30 for steels with low-carbon content.

Effects of Casehardening

Casehardening alters the chemical content and the grain structure of the surface layer of ferrous metals. Steels with low-carbon content cannot be hardened significantly by heating and quenching. Therefore, substances must be added to the surface layer to cause it to harden when properly heated and quenched. These substances include carbon, nitrogen, or various combinations of carbon and nitrogen. They can be introduced into the surface of a ferrous alloy by various casehardening processes.

When carbon is introduced in the surface layer of low-carbon steel, by a process called **carburizing,** the surface layer then becomes high-carbon steel. When heated to the proper hardening temperature and quenched, this layer will become hardened.

When nitrogen is absorbed in the surface of steel, nitrides form; these are very hard and aid in producing a hardened surface. Both nitrogen and carbon are absorbed in the surface of ferrous alloys in certain casehardening processes which will be discussed shortly.

General Procedures

Casehardening processes, like other heat-treatment processes, include several basic pro-

Fig. 141-1.   Cross section of casehardened, carburized, outboard motor crankshaft.

cedures according to a time-and-temperature cycle. There may be variations for specific case-hardening processes, depending on the materials and equipment used. However, the following basic procedures usually are involved in most casehardening processes:

1. The parts are heated to the proper casehardening temperature.
2. The parts are soaked at this temperature while in contact with the casehardening substance being absorbed into the surface.
3. The parts are quenched, at the proper hardening temperature, in water, brine, oil, or air. Plain carbon steels generally are quenched in brine or water. Certain alloys are quenched in oil. Cooling in air sometimes is utilized for nitriding special steels.
4. When parts are to be tempered, this is done soon after hardening.

Although many casehardened parts are used without being tempered, the majority are tempered before use. Tempering is performed at low temperatures, usually in the range from 121° to 204° C (250° to 400° F). At these temperatures, there is little loss of hardness, and internal stress is adequately relieved. Deeply hardened surfaces sometimes are tempered at higher temperatures to avoid surface cracking or chipping.

The Casehardening Processes

The following casehardening processes commonly are used in industrial applications:

1. Carburizing processes
   a. Pack carburizing
   b. Gas carburizing
   c. Liquid carburizing
2. Carbonitriding processes
   a. Cyaniding (liquid carbonitriding)
   b. Carbonitriding (gas atmosphere used)
3. Nitriding processes
   a. Gas nitriding
   b. Liquid nitriding

The principal difference between carburizing, carbonitriding, and nitriding is in the materials introduced into the surfaces. In carburizing processes, the materials introduced include only carbon. In nitriding, only nitrogen is introduced. In carbonitriding, both carbon and nitrogen are introduced.

Carburizing Processes

Carburizing is a process which introduces **carbon** into the surface of solid ferrous alloys. It involves heating the steel while in contact with a carbonaceous substance. Carburizing temperatures are above the upper critical temperature of the steel, usually from 899° to 927° C (1650° to 1700° F).

In carburizing operations, the steel must be soaked at the carburizing temperature long enough to permit the carbon to penetrate to proper depth for casehardening. The time required for penetration varies with the carburizing temperatures, the carburizing substance used, and the depth of penetration desired.

The carbonaceous substances used for introducing carbon into steel include solid materials, liquids, and gases. Three common carburizing processes, therefore, often are designated according to the kind of carburizing material used. These processes include **pack, gas,** and **liquid** carburizing.

Pack Carburizing

For pack carburizing, the parts to be carburized are placed in a steel box. A carbonaceous substance then is packed around them. There usually is about 12.7 to 25.4 mm (½" to 1") of material between the parts. The box generally is made of sheet or plate steel.

Several kinds of carbonaceous substances in the form of grains or fine pellets are used for pack carburizing. These materials are mixtures usually composed of various proportions of the following materials: carbonates, coke, hardwood charcoal, and binders. The binders may consist of materials such as oil, tar, or molasses. Mixtures of charred leather, bones, and charcoal sometimes are used. Carburizing materials may be reused a number of times. However, since the materials decompose with repeated use, fresh materials should be added to the used materials as necessary.

Carburizing materials of the type listed above are inflammable. Therefore, the box in which they are packed must be covered and sealed tightly enough to prevent air from entering and burning the material. However, the cover also must provide a means for exhausting gas pressure which may develop within the box. The cover may be sealed with fire clay or other suitable refractory cement.

The packed box of parts then is placed in a furnace where it is heated to the carburizing temperature. The box is allowed to soak at this temperature until the parts are carburized to the required depth. Carburizing temperatures in the range from 816° to 982° C (1500° to 1800° F) may be used. However, temperatures from 899° to 927° C (1650° to 1700° F) are used most often. At the higher temperatures, carbon penetrates the surface more rapidly. In fact, at 927° C (1700° F), the rate of carburization is about double that at 816° C (1500° F). At 927° C (1700° F), the approximate time required to carburize to various depths is 4 hours for depths from 0.75 to 1.14 mm (0.030" to 0.045"), 8 hours for 1.6 mm (1/16") depth, and 24 to 30 hours for a depth of 3.2 mm (1/8"). At the carburizing temperature, carbon monoxide is released from the carburizing compound. Carbon from the carbon monoxide is absorbed by the surface of the steel parts.

When the parts are carburized to the desired depth, the box is removed from the furnace and allowed to cool. The parts then are removed from the box and cleaned. They may be cleaned by several methods, including sandblasting, tumbling with abrasive materials in tumbling machines, or hand cleaning with a scraper and wire brush.

After Carburizing

Several different heat-treatment procedures may follow carburizing. However, each procedure involves quenching the carburized part to harden the surface. The simplest procedure is to quench the part directly from the carburizing temperature. This hardens both the surface and the core, so far as the core is capable of being hardened.

Another simple treatment is used widely. The parts are allowed to cool slowly from the carburizing temperature. They then are reheated to the hardening temperature of the high-carbon steel case, about 777° C (1430° F), and quenched. This procedure hardens the surface, but not the low-carbon steel core.

A third procedure is to **double quench.** The part is first heated to the hardening temperature of the low-carbon steel core, about 899° C (1650° F), and quenched. This hardens both the

surface and the core, but the surface grain structure is relatively coarse. Hence, the part is then reheated to the hardening temperature of the surface, about 777° C (1430° F), and quenched. This refines the grain structure of the core, and it hardens the surface at a temperature which will produce a more desirable combination of hardness and toughness. The casehardened parts may be tempered, if desired.

How to Carburize

For the school shop, one of the safest and most common methods of casehardening involves pack carburizing with a commercial material such as Kasenit, a nonpoisonous, noninflammable, carbonaceous substance. Pack carburizing with a material of this type is done in an open or well-vented container. The container may be made of sheet or plate steel construction. The following procedure generally is used.

Pack Method — For case depths up to 0.38 mm (0.015"):

1. Place the part in an open, shallow container, and cover it with Kasenit or other equivalent carburizing compound. A well-vented cover may be placed on the container if desired.
2. Place the part in a furnace, heat to 899° C (1650° F), and soak at this temperature for 15 to 60 minutes, depending on the depth of case desired. A case depth from about 0.13 to 0.5 mm (0.005" to 0.020") can be obtained in this manner.
3. Use dry tongs to remove the part from the molten compound, and quench in clean cool water.
4. The part may be tempered if desired. A tempering temperature of 149° C (300° F) is satisfactory for many applications.

Dip Method — For a shallow case, several hundredths of a millimeter (thousandths of an inch) in depth:

1. Heat the part uniformly to about 899° C (1650° F). At this temperature, the steel will have a bright red color.
2. Dip or roll the part in Kasenit, or in an equivalent casehardening compound, until a crust is fused to the surface.
3. Reheat to about 899° C (1650° F), and quench immediately in clean cool water.

4. To increase the depth of the case, repeat steps 2 and 3 before quenching.
5. Temper if desired. Tempering very often is omitted when parts are casehardened to very shallow depths.

Gas Carburizing

In gas carburizing, parts are heated to the carburizing temperature in a gas atmosphere. Common fuel gases (such as natural gas, commercial gas, or propane gas) are used. The gas provides the source of carbon which is absorbed into the surface at the carburizing temperature.

Special furnaces are required for gas carburizing. The parts are placed in a sealed gas container which serves as the carburizing chamber. An inlet for circulating gas and an outlet for exhaust gases are provided for the chamber. The gas chamber is heated from an external source, usually by fuel gas or oil. An industrial-type gas-carburizing furnace is shown in Fig. 141-2.

The parts are soaked in the carburizing chamber at the carburizing temperature for a period of time which depends upon the depth of case desired. With a carburizing temperature of about 927° C (1700° F), case depths of 0.5 to 0.75 mm (0.020" to 0.030") are obtained during a period of four hours.

The parts may be quenched directly from the carburizing temperature when the design of the furnace permits this arrangement, as in Fig. 141-2. Or, the parts may be allowed to cool first. Then they are reheated to the hardening temperature and quenched. Quenching directly from the carburizing chambers is preferred, as it prevents buildup of oxidation or scale on the surface. Thus, the need for surface-cleaning operations is eliminated.

Liquid Carburizing

Liquid carburizing is a process of adding carbon to the surface of steel parts by soaking them in molten salt baths. Two types of baths can be used for this process: salt baths that contain cyanide and cyanide-free salt baths.

Salt baths that contain cyanide are mixtures composed largely of cyanides, chlorides, and carbonates. They provide both carbon and nitrogen for absorption into the steel surface. For liquid carburizing, the composition of the bath is regulated so that carbon is the principle substance being absorbed. Carburizing temperatures for salt baths containing cyanide range from about 843° to 927° C (1550° to 1700° F). A case depth of about 0.5 mm (0.020") can be obtained at a carburizing temperature of 899° C (1650° F) in about two hours. Case depths of 1.5 mm (1/16") require more than 24 hours.

Cyanide-free salt baths provide a safer alternative to the use of salt baths that contain cyanide. They eliminate the hazards of working with a highly poisonous material, as well as the difficulty and expense of disposing of waste materials to conform with Environmental Protection Agency standards. Because of these problems, most carburizing is now done in cyanide-free salt baths or in gas carburizing furnaces.

Cyanide-free salt baths are mixtures of neutral salts and graphite. They must be stirred constantly to ensure that the carbon-containing graphite comes into direct contact with the entire surface

(American Gas Furnace Company)

Fig. 141-2.   Rotary gas-carburizing furnace being discharged into quenching tank.

Fig. 141-3.   Parts being liquid-carburized in salt bath.

of each workpiece being carburized. The carburizing temperature range for one cyanide-free salt bath is 927° to 954° C (1700° to 1750° F). A case depth of 0.3 to 0.45 mm (0.012 to 0.018″) can be obtained in about one hour. The parts shown in Fig. 141-3 are being liquid-carburized in a cyanide-free salt bath.

When the parts have been carburized to the desired depth, they may be quenched directly from the molten salt bath. Liquid-carburized parts generally are quenched in water or brine, although those steels requiring it may be quenched in oil. Since the molten salts are water-soluble, a water quench aids in cleaning parts. The parts may be tempered as desired. Low tempering temperatures are generally used.

Safety Precautions

CAUTION
Since all molten salt solutions are potential hazards, the following safety precautions should be followed when using carburizing, cyaniding, and nitriding salt baths:

1. Violent explosions or spattering will result when water or moisture comes in contact with molten salts.
2. Workpieces, tongs, and other metal to be submerged in a molten salt bath must be clean and dry. Preheating of all such metal to a temperature near that of the molten salt bath will remove all moisture and will also help keep the salt bath at its correct working temperature.

3. The salt bath and rinse tanks must be properly vented to the outdoors to remove harmful fumes.
4. Workers and others in the working area of molten salt baths should wear safety shields, protective gloves, and protective aprons.
5. Cyanides are fatally poisonous if taken internally. Both cyaniding and nitriding salts normally contain cyanide.
6. Cyanides are violently poisonous when brought into contact with an open wound.
7. Fatally poisonous fumes are developed when cyanide is brought into contact with acids.
8. If nitrate-nitrite salts come into contact with molten nitriding salts, a violent explosion will result.
9. The remelting of a solidified cyanide salt bath can be very dangerous because of the expansion of gases when the salt is reheated. This problem does not arise in furnaces equipped with immersed electrodes because the salts melt from the top down. However, when molten cyanide is allowed to solidify in other types of furnaces, a steel or cast iron wedge should be placed in the center of the pot with the tapered end down and touching the bottom. The top of the wedge should extend about 127 mm (5″) above the level of the salt. Before the salt is reheated, the wedge should be removed. The wedge may be loosened by tapping it on alternate sides with a hammer. As the solidified cyanide is reheated, the gases can escape through the cavity provided. **Never** should the wedge be removed before the salt is completely solidified; molten salt may be blown out from the bottom through the wedge opening.

Carbonitriding Processes

The two principal casehardening processes which introduce both carbon and nitrogen into the surface of the steel are **cyaniding** (liquid carbonitriding) and **carbonitriding** (gas cyaniding).

Cyaniding

Cyaniding is a liquid carbonitriding process. The parts are cyanided to the proper case depth in a molten, cyaniding salt bath and quenched.

This produces a hardened case to depths which seldom exceed 0.5 mm (0.020"). The case depth depends upon the cyaniding temperature and the composition of the salt bath.

The salt bath usually is composed of cyanide-carbonate-chloride salts. The proportion of cyanide salts may vary from about 30% to 97%. An increase in cyanide concentration increases the proportion of carbon in the surface case.

The cyaniding temperature is above the lower critical temperature of the steel, usually in the range from 760° to 871° C (1400° to 1600° F). For greater surface depths, the range from 843° to 871° C (1550° to 1600° F) is used most frequently.

Cyaniding generally produces a surface case which is relatively shallow in comparison with carburizing processes. It is most efficient for short immersion periods, usually from 30 to 60 minutes. At a cyaniding temperature of 843° C (1550° F), the following approximate case depths can be obtained: 0.076 to 0.13 mm (0.003" to 0.005") in 30 minutes, 0.13 to 0.25 mm (0.005" to 0.010") in 1 hour, and about 0.38 mm (0.015") in 2 hours.

Cyanided parts may be quenched directly from the salt bath, into water, brine, or mineral oil, depending on the type of steel. Although most of the salt will be removed during the quench, the parts should be thoroughly rinsed or washed off with hot water or steam after quenching. The parts may be tempered if desired, usually at temperatures from 121 to 149° C (250° to 300° F).

CAUTION

Again, cyanide salts are highly poisonous. They are fatal if taken internally, if exposed to scratches or open wounds, or if fumes are inhaled. The safety precautions outlined on page 562 should be strictly followed when cyaniding or working with salt baths containing cyanides.

Carbonitriding

Carbonitriding frequently is termed **dry cyaniding** or **gas cyaniding.** It is a process which affects steel in a manner similar to cyaniding (liquid carbonitriding) by similarly introducing carbon and nitrogen into the surface of steel parts to form a hard, wear-resistant surface.

The process involves soaking steel parts at an elevated temperature in a gaseous atmosphere. The surface of the parts absorbs both carbon and nitrogen, simultaneously, from the gas atmosphere (composed of a mixture of carburizing gas and ammonia). The ammonia supplies the nitrogen.

A special furnace, somewhat similar to a gas carburizing furnace, is used for carbonitriding. The parts do not come in contact with air, thus preventing oxidation or the buildup of scale on the parts.

The depth of the surface case depends upon the temperature and time. Temperatures ranging from about 732° to 899° C (1350° to 1650° F) may be used. The case depth generally is shallower than with gas carburizing. Case depths up to about 0.75 mm (0.030") can be produced during a period of 4 or 5 hours at 871° C (1600° F). However, lower temperatures often are used in order to introduce a larger proportion of nitrogen into the surface. A large percentage of carbonitrided parts is produced with case depths from 0.13 to 0.25 mm (0.005" to 0.010") during a period of 1½ hours, at a temperature of about 788° C (1450° F).

Carbonitrided surfaces possess greater hardenability than carburized surfaces. Generally they are quenched to maximum hardness in oil. The oil quench results in less distortion of the parts.

Nitriding Processes

Nitriding is a casehardening process used to produce an exceptionally hard surface on certain alloy steel parts which have been previously machined and heat-treated. The process adds nitrogen to the surface of special alloy steels which are alloyed with nitride-forming elements. Because nitrides are exceptionally hard, they produce a harder surface than the previously heat-treated surface.

The alloy steels which are nitrided include various proportions of the following kinds of nitride-forming elements: aluminum, chromium, vanadium, and molybdenum. Nickel also is frequently alloyed with these steels to produce toughness.

Parts are nitrided by heating them at temperatures below the lower critical temperature, in a nitrogeneous atmosphere — either gas or a salt bath.

Gas Nitriding

For gas nitriding, ammonia gas is used most frequently. A protected atmosphere furnace is used. The nitriding temperature range usually is from 482° to 566° C (900° to 1050° F). Consequently, steels which are nitrided must have a tempering temperature which is at least 28° C (50° F) above the nitriding temperature.

The nitriding process is slow in comparison with other casehardening processes. The time required may take from 1 to 3 days, or longer, depending upon the case depth desired. Three days usually are required to produce a nitrided case about 0.38 mm (0.015") deep. Shorter periods are utilized for shallower case depths. Nitrided surfaces from 0.025 to 0.076 mm (0.001" to 0.003") commonly are used on many high-speed steel cutting tools. No quenching is required. The low operating temperatures produce very little distortion.

Nitriding produces a harder surface than any other form of heat treatment. The surface hardness may range from Rockwell C-70 to C-75. The process does not affect the core properties of the steel.

Salt Bath

Nitriding also can be performed in a molten salt bath. The nitriding salt introduces nitrogen only into the surface of the steel. Several kinds of salt mixtures, which include cyanide salts, are available for this purpose. The working temperatures used are about the same as those used for gas nitriding. Nitriding salts, like some carburizing salts and all cyaniding salts, are highly poisonous. They are dangerous if not handled properly and safely. The safety precautions which should be followed when using nitriding salts are listed on page 562.

Nitriding improves surface hardness, wear resistance, resistance to galling, and fatigue resistance. It also improves corrosive resistance, except on stainless steels. Items made from high-speed steel, tool and die steel, and stainless steel often are nitrided. Common items include taps, dies, reamers, milling cutters and other cutting tools. The tool life of cutting tools often is doubled or tripled by nitriding. .

Unit
142
Special Hardening Processes

This unit explains three heat treating processes used in special situations. **Flame hardening** and **induction hardening** require no furnace and the two are similar except for their source of heat. **Austempering** is a furnace-hardening process using a special quench to produce a grain structure (bainite) which is especially tough.

Flame Hardening

Flame hardening is a process used to harden selected surface areas of ferrous metals to depths from about 0.81 to 6.35 mm (1/32″ to 1/4″), as desired. The selected area is heated rapidly to above the upper critical temperature with an intensely hot flame. Quenching follows immediately. The part is spray-quenched with water from water jets located conveniently near the flame burners. Air-hardened steels usually are quenched with compressed air or circulating air. A cross-sectional view of a gear tooth which has been flame-hardened is shown in Fig. 142-1.

The flame is produced by mixtures of fuel gas and air or by an oxyacetylene flame. Burners or torches of special design are used to concentrate the flame in the area where hardening is desired. Small parts often are located in a stationary position under the flame. With very large parts, the flame may be guided steadily over the area to be hardened. In other setups, the part may be guided or revolved over the flame. A setup used to flame-harden the ways on a lathe bed is shown in Fig. 142-2. In this case, the flame is guided automatically along the lathe bed. The surface is heated and quenched in a single pass.

(Union Carbide Corporation — Linde Division)

Fig. 142-1.   Photograph of etched section of a flame-hardened gear tooth.

(Clausing Div. — Atlas Press)

Fig. 142-2.   Flame hardening the ways on a lathe bed.

Flame-hardened parts generally should be tempered immediately following hardening. Usually they are tempered in the same manner in which furnace-hardened parts are tempered. Very large parts may be tempered by the **flame method.** This involves the use of a special low-temperature flame head designed to follow immediately behind the quench, thus heating the surface to the desired tempering temperature. On some very large parts, there is considerable residual heat below the hardened surfaces after quenching. This heat often is sufficient to temper the hardened surface.

Flame hardening may be applied to ferrous metals, including steels and certain kinds of cast iron, which are capable of being hardened. Large cast-iron castings are very often flame-hardened, as in Fig. 142-2. Flame hardening also may be used to through-harden steel up to about 76.2 mm (3″) in diameter, depending on the hardenability of the steel.

Applications

Flame hardening has definite advantages for the following applications:

1. It permits making parts with the less-costly medium-carbon steels, rather than the more costly high-carbon steels.
2. It may be used where only selected areas require hardening.
3. It may be used for applications where through-hardening would cause excess distortion.
4. It may be used for very large objects which cannot be furnace-hardened efficiently.
5. The following items often are flame-hardened: ways and flat bearing surfaces of machine tools; gear teeth, bearing surfaces on shafts, splines on shafts, ends of push rods, cams, large rolls, forming dies, crankshafts, and camshafts.

Induction Hardening

Induction hardening, like flame hardening, is a process used to harden selected areas of ferrous metals to depths ranging up to 6.35 mm (¼″). The heat for induction hardening is generated electromagnetically by a high-frequency induction coil. Parts which are cylindrical in shape are heated by inserting them through an induction

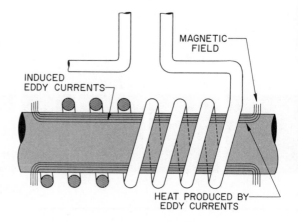

Fig. 142-3.   Heating surface of steel shaft with high-frequency induction coil.

(TOCCO Division — Ohio Crankshaft Company)

Fig. 142-4.   Cross section of induction-hardened gear achieved with high-frequency heating.

coil as shown in Fig. 142-3. The selected surface to be hardened is heated during several seconds to above the upper critical temperature. Hardening by quenching follows immediately. Water-hardening steels are quenched with a spray of water from a water jacket located conveniently near the coil. The parts shown in Figs. 142-4 and 142-5 were heated and quenched in this manner. The hardened parts should be tempered as soon as possible after hardening.

The principles involved in induction heating can be understood quite readily by studying Fig. 142-3. The figure shows an induction coil of the

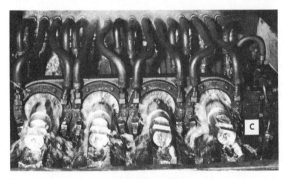

Fig. 142-5. High-frequency induction hardening of four track rollers simultaneously.
A. Assembly setup
B. Parts being heated
C. Parts being water-quenched

type commonly used to heat parts which are cylindrical in shape. When high-frequency alternating current passes through the coil, a magnetic field is established around the coil and through the core. In this case, the core is the workpiece to be hardened. The magnetic field causes eddy currents to be developed in the surface of the workpiece. These generate the heat required for hardening the surface layer of the workpiece.

Frequencies ranging from 1 kHz to 2 MHz per second may be used. The lower frequencies are used for heating to the deeper hardening depths, which may range from 1.5 to 6.35 mm (1/16″ to 1/4″). The higher frequencies generally are used for heating to shallow hardening depths, ranging from 0.25 to 1.5 mm (0.010″ to 1/16″). The very high frequencies tend to heat only a skin layer on the surface.

Induction coils of special design, and in various arrangements, are used to heat parts with different shapes. The pattern of the heated area is determined by the shape and design of the coil, the operating frequency, and the input current. Induction heating equipment is available for either manual or automatic operation. With the latter, the complete heating, quenching, and tempering cycle may be controlled automatically.

Quenching media used for induction hardening include water, oil, or compressed air, depending on the type of metal being hardened. Either spray-quenching or bath-quenching methods may be used, according to the production procedure.

Induction-hardened parts may be tempered in a tempering furnace in the same manner in which regular quench-hardened parts are tempered. They also may be heated for tempering with an induction coil at low temperatures. This is done in a manner similar to the method employed in heating for induction hardening.

Many different steels and certain grades of cast iron may be induction-hardened. Medium-carbon steels and many alloy steels are well suited for induction hardening.

Induction heating also may be applied to through-hardening applications. Shafts or rods may be through-hardened by using continuous line heating and quenching equipment. Selected areas such as gear teeth also may be through-hardened by using induction coils of standard design. Through heating is achieved by using lower frequencies, usually in the range from 180 to 3000 Hz.

Applications

Induction hardening may be used for many of the applications for which flame hardening is used. The following are typical advantages and applications of induction hardening:

1. It permits making parts with less-costly, medium-carbon steels, rather than the more-costly, high-carbon steels.
2. It may be used for selective hardening of surfaces or areas.
3. It may be used for long bars which cannot be furnace-hardened efficiently.
4. Generally, it results in less distortion, less oxidation or scaling, and less warpage than conventional quench-hardening.
5. The following items often are induction-hardened: gear teeth, splines on shafts, bearing surfaces on shafts, rolls, tubing, crankshafts, camshafts, and connecting rods.

Austempering

Austempering is a process which involves heat-treating steel in a manner which will produce a **bainite** grain structure, which is softer than martensite, but harder than pearlite. (Bainite will be explained shortly.) Austempering generally involves the following basic procedure:

1. The steel parts are heated uniformly to the hardening temperature, thus producing an austenitic grain structure.

2. The parts are quenched in a molten salt bath at a temperature which produces a bainite grain structure. The quenching temperatures for bainite, as shown in Fig. 142-8 may range from about 204° to 482° C (400° to 900° F); however, those from about 260° to 399° C (500° to 750° F) generally are used. The salt bath is maintained at the temperature which produces the desired kind of bainite.
3. The part is soaked in the salt bath for sufficient time to permit complete transformation from austenite to bainite.
4. The part is then removed from the salt bath, and it is rinsed in hot water or steam to remove the salt.

Bainite Formation

If steel is quenched from the critical temperature to a temperature in the range from about 204° to 482° C (400° to 900° F), and if it is soaked at this temperature long enough, the austenite will transform to a grain structure called **bainite,** as shown in Figs. 142-6 and 142-7. Bainite is a finer mixture of ferrite and cementite than is pearlite. It appears more acicular or needlelike than the pearlite grain, which has

(United States Steel Corporation)

(United States Steel Corporation)

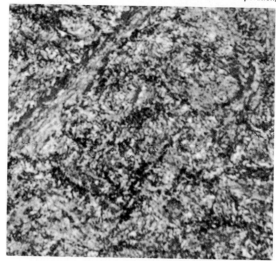

Fig. 142-6. Microstructure called bainite (partial) magnified 2500 times.

Fig. 142-7. Microstructure called bainite (complete) magnified 2500 times.

TRANSFORMATION TEMPERATURE	Degrees		Austenite Transforms to:	Hardness of Transformation Product in 0.80%-Carbon Steel Brinell
	C	F		
	704	1300	**PEARLITE** Coarse	170
	649	1200	Medium	250
	538	1000	Fine	400
	427	800	**BAINITE** Upper	400
	316	600	Lower	575
	204	400	**MARTENSITE**	600
	93	200		650
	21	70		

Fig. 142-8. Transformation products from austenite, and their hardness.

alternate layers of ferrite and cementite. Bainite is formed in the transformation temperature range below the pearlite range, but above the martensite range, as shown in Fig. 142-8. Its hardness also lies in the range between the softer pearlite and the harder martensite.

In order for bainite to form, the austenite must be quenched and cooled rapidly to the desired austempering temperature, about 316° C (600° F), in order to avoid pearlite formation. Since pearlite begins to form at 538° C (1000° F), the steel must be cooled to a point below this temperature in less than one second. However, more time may be allowed for further cooling to the 316° C (600° F) temperature selected. The molten salt bath should be maintained at (or slightly below) the desired austempering temperature.

The time required for complete transformation to bainite depends on the kind of steel and the austempering temperature selected. For plain carbon steel, the transformation period varies from about one minute at a temperature of 427° C (800° F), four minutes at 371° C (700° F), fifteen minutes at 316° C (600° F), to one hour at 260° C (500° F).

Applications

Austempering has several advantages over regular quench-hardening and tempering for certain applications. First, greater toughness is obtainable at equal hardness values. This is particularly true for parts which would suffer a loss of notch-toughness when tempered in the brittle temperature range. Austempered parts in the hardness range from Rockwell C-45 to C-55 generally possess much greater notch-toughness than quench-hardened and tempered parts of equal hardness.

A second major advantage of austempering is that it produces less distortion. There is also less possibility of quench cracks because of the higher quenching temperatures.

Austempering, however, has several limitations. Its use is limited to certain types of steel. Plain high-carbon steels and certain alloy steels may be austempered with good results. Plain carbon steels with less than 0.50% carbon usually cannot be austempered efficiently.

Austempering applications generally are limited to thicknesses of less than 4.76 mm (3/16") where a complete bainite structure is required. Plain carbon steels with less than 0.80% carbon generally are limited to less thickness. With modern production methods, austempering is much less expensive than the quench-harden and temper method. Typical examples of austempered parts include items such as lock washers, stamped nuts, spring nuts, spring-type toolholders, steel shanks in shoes, and many other thin steel parts which require hardness, toughness, and resilience. See Fig. 142-9.

(United States Steel Corporation)

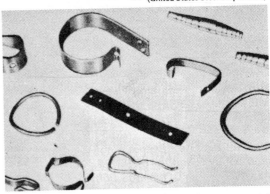

Fig. 142-9. Variety of industrial articles austempered.

Unit
143

Hardness Testing

The hardness of metals or metal parts normally is designated by a hardness number. Several common hardness-number systems may be used, depending on the type of hardness-testing instrument employed. The more common systems include:

Rockwell C Scale (RC or Rc)
Rockwell B Scale (RB or Rb)
Rockwell Superficial hardness scales
Brinell hardness number (BHN)
Shore scleroscope hardness number
Knoop hardness scale
Diamond Pyramid hardness number (DPH),
  also know as the Vickers Scale

The hardness value for a part to be made usually is specified on the drawing or blueprint for the part. A hardness number designated according to one system, such as a Rockwell number, may be converted to an approximate equivalent hardness number of another system through the use of special tables. For example, equivalent hardness values for steel may be determined by reading horizontally across Table A-6, Appendix.

Hardness values for thin sheet metal, thin metal parts, tubing, or for the thin case-hardened surfaces, usually are inaccurate when determined with conventional Brinell or Rockwell testers. Instead, these products may be measured by the less-common Rockwell Superficial hardness scales, including both the Rockwell N and Rockwell T scales. The hardness of these kinds of products also may be determined more accurately with a microhardness tester, as explained later in this unit.

The general principles involved in hardness testing according to the Rockwell, Brinell, Shore scleroscope, and Knoop hardness number scales are explained in this unit. However, there are numerous variations in the mechanical design of instruments developed by various manufacturers for testing hardness according to these scales. Therefore, the specific procedures recommended by the instrument manufacturer should be reviewed and followed before using a specific hardness tester.

Estimating Hardness with a File

The approximate hardness of a piece of metal may be estimated **roughly** simply by using the corner of a sharp file and noting the ease or difficulty encountered in making the file cut. The corner of the file should be used, since the teeth on the face of the file could be ruined on very hard steel. The data in Table 143-1 is a guide for estimating the hardness of metal parts.

Table 143-1
Estimating Hardness with a File

Rockwell C Hardness No.	Action of File on Steel
20	File removes metal easily with slight pressure
30	File starts to resist cutting metal
40	File cuts metal with difficulty
50	File barely cuts metal with great difficulty
57	File glides over metal without cutting

(American Chain and Cable — Wilson Mech. Instr. Div.)

(Ames Precision Machine Works)

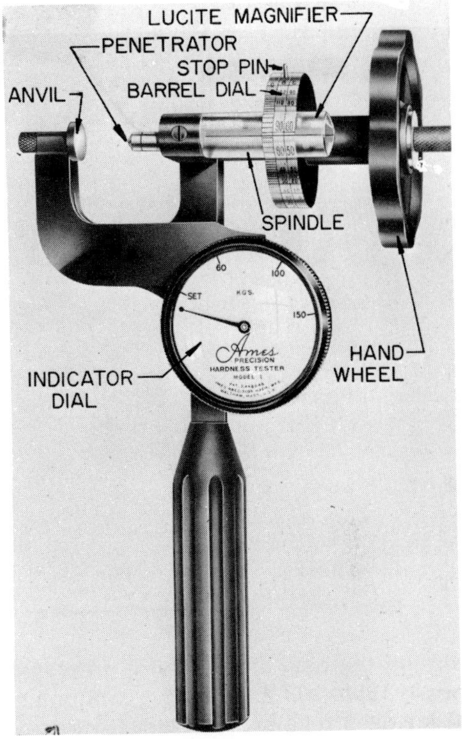

Fig. 143-1.   Motorized combination tester for Rockwell hardness testing and for Rockwell Superficial hardness testing.

Fig. 143-2.   Portable tester for Rockwell hardness testing.

Rockwell Hardness Test

The Rockwell hardness test is based on the depth of penetration made by a specific type of penetrator, into the surface of a metal specimen, while under a specific load. The hardness number is based on the difference in depth of penetration caused by a minor load and a major load on the penetrator. Deep penetration indicates a low hardness number; shallower penetration, a higher hardness number. The hardness number generally is read directly in Rockwell numbers from the dial or digital display unit of the tester. The Rockwell hardness test is simple to perform, requires only a few seconds, and is widely used.

Several types of hardness testers are available for measuring the hardness of metals according to standard Rockwell scales. Some are portable, while others are stationary, usually being mounted on a bench or stand, Fig. 143-1. The load on Rockwell testers generally is applied through a system involving weights, levers, screws, or a combination of these mechanisms. Some testers are hand-operated, while others are motor-driven. A hand-operated portable tester is shown in Fig. 143-2.

Testers are available for testing according to the **standard Rockwell hardness scales** only or the **Rockwell Superficial hardness scales** only. Combination models, which may be used for testing according to either scale, also are available. A motor-driven combination tester of this type is shown in Fig. 143-1.

Several different Rockwell hardness scales may be used for determining the hardness of various metals and other materials. Each scale is based on the use of a specified size and type of penetrator, which is used with a specified load. The various Rockwell hardness scales, the loads ap-

Table 143-2
Rockwell Hardness Scales

Scale or Prefix	Type of Penetrator	Major Loads in kgs	Dial Hardness Numbers
	Standard Scales		
B	1.5 mm (1/16″) Ball	100	Red
C	Diamond	150	Black
	Special Scales		
A	Diamond	60	Black
D	Diamond	100	Black
E	3.2 mm (1/8″) Ball	100	Red
F	1.5 mm (1/16″) Ball	60	Red
G	1.5 mm (1/16″) Ball	150	Red
H	3.2 mm (1/8″) Ball	60	Red
K	3.2 mm (1/8″) Ball	150	Red
	Superficial Hardness Scales		
15N	Diamond	15	N (Green)
30N	Diamond	30	N (Green)
45N	Diamond	45	N (Green)
15T	1.5 mm (1/16″) Ball	15	T (Green)
30T	1.5 mm (1/16″) Ball	30	T (Green)
45T	1.5 mm (1/16″) Ball	45	T (Green)

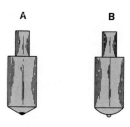

Fig. 143-3.   Penetrator points used on Rockwell hardness tester.
A. 120° diamond point
B. 1.5 mm (1/16″) diameter ball point

plied, and the type of penetrator employed are shown in Table 143-2. The most common scales used for determining the hardness of steel and other metals, however, include the Rockwell C and the Rockwell B scales. The hardness numbers for these scales are included in Table A-6, Appendix.

Standard Rockwell Hardness Scales

The Rockwell C scale is used to determine the hardness of hardened steel and other metals which are harder than RB 100. In testing hardness according to the RC scale, a standard 120°, sphero-conical, diamond penetrator (called a **Brale,** Fig. 143-3) is used with a **major load** of 150 kg (330.7 lbs.). A minor load of 10 kg (22 lbs.) is used.

The Rockwell B scale is used to determine the hardness of unhardened steel, cast iron, and nonferrous metals which have a hardness of less than RC 20. In testing hardness according to the RB scale, a 1.5 mm (1/16″) diameter hardened steel ball penetrator (Fig. 143-3) is used, with a minor load of 10 kg and a major load of 100 kg (220.5 lbs.). The RB hardness numbers are read

according to the red figures on dial-equipped machines.

The penetrator always should be in good condition, and the tester should be checked for accuracy before use. It is checked with test blocks which have known hardness values. These generally are supplied with the tester.

Rockwell hardness test values are more accurate when three hardness tests are made, each at a different spot on the specimen. An average of the three readings then is the correct hardness number value.

Rockwell Superficial Hardness Scales

These hardness number scales are used to determine the hardness of materials or products which are limited to very shallow penetration or to the application of lighter loads. With a lighter load, the indentation left on the surface of the part is less severe. Thus, less damage is caused on thin tubing, thin strips, and highly-finished surfaces.

The following kinds of products often are tested according to the Rockwell Superficial hardness scales: thin strip metal, safety razor blades, cutters on electric shavers, casehardened steel parts, plug and ring gages, wire and small rounds, laminated metal.

The Rockwell Superficial hardness scales, as shown in Table 143-2 include both the N scales and the T scales. With these, major load weights of 15, 30, or 45 kg are applied with a minor load weight of 3 kg. A special N diamond (Brale) penetrator is used when measuring according to the N scales. A 1.5 mm (1/16″) diameter ball penetrator is used with the T scales. The Rockwell Superficial hardness numbers for the N scales are shown in Table A-6, Appendix.

Procedure

Become familiar with the particular Rockwell hardness tester available. Read and follow any special recommendations made by the manufacturer of the tester.

The specimen or part to be tested should be free from rust, scale, or deep scratches. Hardened steel should be tested with a diamond penetrator. Soft steel, cast iron, and nonferrous metals generally should be tested with a 1.5 mm (1/16″) diameter hardened steel ball penetrator.

The following are the general steps of procedure to be followed in Rockwell hardness testing:

1. Select the proper type of penetrator. For the RC scale, use a diamond penetrator. For the RB scale, use a 1.5 mm (1/16″) diameter steel ball. See Table 143-2 for other Rockwell scales.
2. Check that the indicator hand is in the proper position for starting a test. On an instrument of the type shown in Fig. 143-2, the indicator hand must be over the **dot**. The **dial** may be turned to align the hand over the dot.
3. Place the specimen to be tested on the anvil of the tester. Then bring the penetrator into contact with the specimen. This is done by raising the anvil or lowering the penetrator, depending upon the design of the tester. On the tester in Fig. 143-2, the penetrator is brought into contact with the specimen by turning the handwheel.
4. Apply a minor load of 10 kg, as indicated by the appropriate mark on the tester dial. (On the tester in Fig. 143-2, the minor load is applied when the indicator hand is over the **set** mark.) Then set the hardness reading dial to zero. (On the tester in Fig. 143-2, this is done by turning the **barrel dial** on the spindle until the stop pin touches the lucite magnifier.)
5. Apply the major load. For the RC scale, a major load of 150 kg is applied. For the RB scale, 100 kg is applied. The major load for other Rockwell scales is shown in Table 143-2.
6. Reduce the major load to the minor load setting; then read the hardness directly, accord-

ing to the Rockwell hardness number scale being used. Generally, the RC numbers are black, while the RB numbers are red. On the tester in Fig. 143-2, the hardness number is on the barrel dial and is read through the lucite magnifier.

7. Release the minor load and remove the specimen from the tester. Several tests should be made. The correct hardness number is an average of these. The accuracy of the tester should be checked periodically by making tests on the test blocks provided with the tester.

Brinell Hardness Test

The Brinell hardness of metal is determined with a testing machine which forces a hardened ball into the smooth surface of a metal specimen or part, as in Fig. 143-4. The hardened ball is a specified size, and it is forced into the surface under a special load for a definite period of time. The ball is 10 mm (0.3937″) in diameter and may be a hardened steel ball, a hultgren ball, or a carbide ball.

For the standard Brinell hardness test on steel, the load is 3000 kg (6600 lbs.). The load must be applied steadily for a period of at least 15 sec-

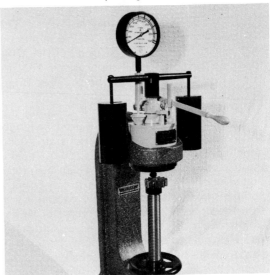

Fig. 143-4.   Brinell hardness tester.

(Bausch and Lomb)

Fig. 143-5.   Brinell microscope for measuring diameter of impression made by Brinell hardness tester.

onds for iron or steel and at least 30 seconds for nonferrous metals.

A standard load of 500 kg (1100 lbs.) generally is applied for nonferrous metals in the hardness range from 26-100 BHN. A load of 1500 kg (3300 lbs.) sometimes is used for testing nonferrous metals in the range from 80 to 300 BHN. Other loads are used for special applications. This unit is concerned primarily with the standard Brinell hardness for steel.

The Brinell hardness number is determined by measuring the width of the dent produced in the surface of the metal specimen. Generally, this is done through the use of a microscope (similar to the one shown in Fig. 143-5), which has a calibrated lens for measuring the width of the dent. A comparison chart supplied with the tester is used to determine the hardness for dents of various widths. Wide dents are produced in soft metals which have a low hardness number. The dents in harder metals have less width, thus resulting in a higher hardness number. The Brinell hardness numbers for steel are shown in Table A-6, Appendix.

On a Brinell hardness tester, the load generally is applied through a system of weights, levers, and screws. Brinell hardness testers are available in hand-operated models, Fig. 143-4, or in motorized models.

A Brinell hardness tester generally works best on metals which do not rank exceedingly high in hardness. This includes nonferrous metals, soft steels, and medium hardened steels. It is not recommended for testing materials harder than BHN 630. On highly hardened steels, the dent is so small that its size is difficult to measure accurately. The Brinell hardness of steel may range from 150 for low-carbon annealed steel to 739 for hardened high-carbon tool steel.

A carbide ball may be used for testing the hardness of hardened steel and other materials with hardness up to BHN 630; a hultgren ball, for materials with hardness up to BHN 500; and a steel ball, up to BHN 450.

The Shore Scleroscope Test

The Shore scleroscope system of hardness testing utilizes a tester that measures the height to which a diamond-tipped hammer rebounds after being dropped on the surface of a specimen being tested. See Figs. 143-6 and 143-7. The

(Shore Instrument and Mfg. Co., Inc.)

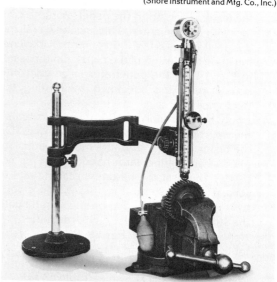

Fig. 143-6.   Shore scleroscope hardness tester mounted on swinging arm and post.

rebounding is caused by elasticity. Thus, the scleroscope actually measures elasticity. However, since elasticity is closely associated with hardness, the hardness is measured indirectly. Harder metals cause the hammer to rebound to greater heights, thus resulting in higher hardness numbers. See Table A-6, Appendix.

The scleroscope hardness test is essentially a nonmarring test, particularly on the harder metals. A minute dent may result on softer metals. For accurate readings, the scleroscope should be held in a vertical position.

Scleroscope hardness testers are available in several models. The hardness numbers may be located on the vertical barrel column, as shown in Fig. 143-6. On another model, the hardness numbers are located on a dial which also shows equivalent Rockwell C numbers and Brinell hardness numbers, Fig. 143-7.

The scleroscope shown in Fig. 143-6 may be removed from the stand for portable use on large workpieces. However, care must be taken to hold the instrument steadily, firmly against the test specimen, and in a vertical position.

The scleroscope shown in Fig. 143-7 is mounted in a clamping stand which should be placed on a sturdy bench to minimize vibration. The instrument is equipped with an adjustment screw and a spirit level for use in leveling the stand. Small parts of many sizes and shapes may be held firmly in the clamping stand for testing; small round parts are placed on a vee block. The scleroscope tester shown in Fig. 143-6 also may be equipped with a clamping stand.

Steel of the following minimum thicknesses may be tested with a scleroscope tester: hardened steel, 0.13 mm (0.005"); cold, rolled, unannealed brass and steel, about 0.25 mm (0.010"); and annealed sheets, about 0.38 mm (0.015").

Procedure

The test specimen should be clean, smooth, and free from dust or scale. The tester should be placed on a sturdy bench or stand to eliminate vibration.

1. Mount the tester on the swing arm and post (Fig. 143-6) or on the clamping stand (Fig. 143-7), depending on the shape and size of the specimen to be tested. Small parts up to 76 mm (3") in height may be held in the

(Shore Instrument and Mfg. Co., Inc.)

Fig. 143-7.   Dial-recording Shore scleroscope hardness tester.

clamping stand. Adjust the leveling screws and spirit level so that the barrel is as nearly vertical as possible.

2. Mount the specimen in the clamping stand (Fig. 143-7) or in a vise (Fig. 143-6), depending on the nature of the test specimen. With the use of the hand knob, lower the vertical barrel, holding the barrel cap firmly in contact with the test specimen until the test is complete.

3. Draw the hammer to the **up** position by rapidly squeezing the rubber bulb, Fig. 143-6. While holding the barrel firmly against the specimen, again squeeze and release the rubber bulb. At the same time, observe the extreme height to which the hammer rebounds on the first bounce. Immediately again squeeze the rubber bulb to raise the hammer to the up position, preparatory for the next test. This will prevent the hammer from rebounding a second time. The height of the first rebound indicates the hardness number. Record the hardness reading. On the tester shown in Fig. 143-7, the hammer is released with a control knob instead of a rubber bulb.

4. Repeat step 3 until at least three tests have been made. Each test should be made at a different spot, since the surface is partially work hardened by the impact of the hammer. Thus, additional tests on the same spot will read high. The average hardness is the average of several tests made on the test specimen. The accuracy of the tester should be checked periodically by testing the hardness of the **reference** test bar supplied with the tester.

Microhardness Tester

A microhardness tester is used to determine the microhardness of metals and other hard materials according to the Knoop hardness scale. Some microhardness testers also may be used to determine hardness according to the Diamond Pyramid hardness number (Vickers Scale), provided that a 136° Diamond Pyramid indenter (Vickers Type) is used. With the use of the Knoop diamond and the Diamond Pyramid indenter, a microhardness tester of the type shown in Fig. 143-8 may be used to determine hardness on both hardness scales. Microhardness testers are used widely to determine the hardness of individual grains or constituents in the microstructure of metals and other materials.

In a microhardness test, a diamond penetrator is pressed into the surface of the specimen being tested. The surface should be very smooth and free from nicks, scale, or other surface irregularities. A given force is applied to the penetrator for a given period of time. The depth of penetration will depend upon the amount of load and the hardness of the test specimen. Since the pressure applied is low, in the range from 25 to 3 600 grams (about 8 lbs. maximum), the impression in the surface is minute. It is so small that its size must be determined with the microscope provided on the tester. The hardness is established by the size of the dent in the surface.

Several advantages accrue through the use of the microhardness tester. The dent produced is so small that it does not damage the surface significantly. The use of the microhardness tester is by no means limited to the testing of microstructures. It may be used to determine the hardness of very small or thin parts, such as parts for watches and precision instruments and thinly hardened surfaces, thin tubing, thin sheet metals, or thin parts like razor blades.

The microhardness tester may be used with the Knoop diamond penetrator to determine the hardness of minerals and other materials which are too hard to test by other methods. The Knoop hardness scale, as used to determine the hardness of abrasives under a 100-gram load, is explained in Unit 98. The Knoop hardness number is based on a relationship between the load and the long dimension of the indentation in the surface. A conversion chart which includes these factors usually is supplied with the tester.

Test Your Knowledge of Section 18

Unit 135: Introduction to Heat Treatment

1. List several properties of metals which can be changed by heat treatment.
2. Give several examples of tools which must be hardened and tempered.
3. Why must a hardened tool be tempered before use?
4. List the three steps involved in the time-temperature cycle of all heat-treatment operations.
5. List several types of toolroom and production furnaces which may be used for hardening.

(E. Leitz, Inc.)

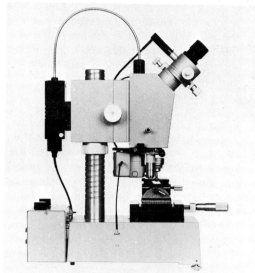

Fig. 143-8.   Microhardness tester.

6. Name two ways of heating parts for hardening which prevent scaling and decarburization.
7. List several kinds of molten baths which are used in pot-type furnaces for heat-treatment operations.
8. List three different methods by which temperatures of steel can be determined for heat-treatment purposes.

Unit 136: Grain Structure of Steel

1. What is a metallograph, and for what purposes is it used?
2. Describe briefly how a metal specimen is prepared for investigation with a metallurgical microscope.
3. What features of an etched metal specimen are investigated with a microscope?
4. How is the study of the grain structure of metals related to the study of heat treatment?
5. List four common types of space lattice units or cells.
6. Describe a body-centered cubic space lattice unit.
7. Describe a face-centered cubic space lattice unit.
8. How does the form and arrangement of carbon change in steel at the hardening temperature?
9. What is meant by a transformation temperature range?
10. Why should metals be allowed to soak at certain temperatures during heat-treatment operations? What is the general rule for determining soaking time?
11. Describe the nature of the grain structure of steel at the hardening temperature.
12. How does the speed of cooling generally affect grain fineness?
13. Describe one method used to determine the grain size of metals.

Unit 137: How Carbon Content Affects Hardening

1. List the basic steps involved in hardening steel.
2. What largely determines the degree of hardness obtainable in alloy steels through heat treatment?
3. In comparison with alloy steels, how do carbon steels generally rank in hardenability?

4. Describe pearlite and the form in which carbon exists in it.
5. How does an increase in pearlite content affect the strength of unhardened steel?
6. Describe the hardening temperature of steel in relation to the upper critical temperature on the iron-carbon phase diagram.
7. Explain how a magnet may be used as an aid in determining the hardening temperature of steel.
8. Describe the properties of martensite and tell how it is formed.
9. How do alloying elements in steel generally affect the speed of cooling required for hardening?
10. How does excessively rapid cooling affect steel which is quenched?

Unit 138: Quenching Media and Their Applications

1. List four quenching media which are used to quench steel in heat-treatment operations.
2. Compare the rate of cooling for oil, water, air, and brine.
3. How can the rate of cooling be increased for a given quenching medium?
4. At what temperatures should water, brine, and oil be maintained for best quenching results?
5. What may be expected if the wrong quenching medium is used to quench steel?
6. What quenching medium generally is used for most plain carbon steels?
7. What quenching medium generally is used for most alloy steels?
8. Explain why agitation is important when quenching.
9. Describe the method generally used to quench high-speed steels for hardening.
10. Why should hardened steel be tempered soon after hardening?

Unit 139: Tempering, Annealing and Normalizing

1. Explain how tempering affects hardened steel.
2. Describe briefly how steel is tempered.
3. What changes in properties take place in hardened steel as tempering temperatures are increased?

4. What factors determine the tempering temperature of steel?
5. What is the range of tempering temperatures which may be used for various tempering applications?
6. What range of tempering temperatures is used for tempering most cutting tools?
7. What is meant by selective heat treatment?
8. Explain how notched-bar toughness is determined.
9. Explain the meaning of the brittle temperature range.
10. Why are notch-bar toughness ratings sometimes misleading?
11. What is the purpose of annealing?
12. Explain the full annealing process.
13. Define process annealing and its purpose.
14. Define the nature and purpose of spheroidizing.
15. What is the purpose of normalizing?
16. Compare the grain structure of normalized steel with that of fully annealed steel.
17. Explain the procedure for normalizing a steel part.

Unit 141: Casehardening Processes

1. What is casehardening?
2. What types of steel most frequently are casehardened?
3. List several kinds of tools or parts which commonly are casehardened.
4. What is the range in depth for casehardened surfaces?
5. What is the range in Rockwell hardness for most casehardened surfaces?
6. Upon what does the core hardness of casehardened parts depend?
7. What substances can be added to the surface of steel to cause it to harden when quenched?
8. Explain the basic procedures which are involved in most casehardening processes.
9. At what temperatures are casehardened parts usually tempered?
10. List several common casehardening processes.
11. In general terms, explain the carburizing process.
12. Upon what factors does the time required for carburizing depend?
13. What kind of carbonaceous substances are used for pack carburizing?

14. What temperatures are used most often for pack carburizing?
15. Briefly describe several different hardening procedures which may follow pack carburizing.
16. Briefly describe two methods of casehardening with Kasenit.
17. Briefly describe the cyaniding process.
18. How do the depths of the hardened case compare for carburized and cyanided surfaces?
19. Briefly describe the carbonitriding process.
20. Briefly describe the nitriding process.
21. How does a nitrided surface affect the life of cutting tools?

Unit 142: Special Hardening Processes

1. Describe the flame-hardening process.
2. To what depths may surfaces be hardened by flame hardening?
3. What types of metal may be flame-hardened?
4. List several advantages of flame hardening in comparison with the standard furnace heating and quenching method for hardening.
5. List several products which may be flame hardened.
6. Describe the induction-hardening process.
7. List several advantages of induction hardening in comparison with the standard furnace heating and quenching method for hardening.
8. List several typical items which may be induction-hardened.
9. How is a bainite grain structure produced in steel?
10. Describe a bainite grain structure and its approximate hardness.
11. Explain the basic procedures used for austempering steel parts.
12. List several advantages of austempering in comparison with regular quench hardening and tempering.
13. List several limitations of the austempering process.
14. List several kinds of items which may be austempered.

Unit 143: Hardness Testing

1. List four hardness number systems.
2. Describe how the hardness of steel may be determined with a file.
3. Explain the Rockwell hardness testing system.

4. What are the most commonly used Rockwell hardness scales?
5. For what range of hardness is the Rockwell C scale recommended?
6. For what range of hardness is the Rockwell B scale used?
7. Describe the type of penetrator used for the Rockwell C and B scales.
8. Explain how a Brinell hardness tester operates.
9. What types of penetrators are used on Brinell hardness testers?
10. Explain how a scleroscope hardness tester operates.
11. For what hardness scale is a microhardness tester generally calibrated?
12. Explain how a microhardness tester functions.
13. List several advantages of a microhardness tester in comparison with other standard hardness testers.

Section
19

Career Opportunities in Machine Tool Processing

Unit
144

Looking at Careers in Metalworking

Fig. 144-1. A skilled machinist operating large vertical boring and turning machine.

(G.A. Gray Company)

Nearly every product of modern industry either contains metal parts or is produced on a machine that is made of metal parts. Many of these parts require machining and must be made by workers skilled in machine tool processing. A major portion of the labor force in America is employed directly or indirectly in metalworking occupations. Of all of the occupational groups classified under metalworking, the machining occupations employ the largest number of workers. In 1974, more than one million workers were employed in machining occupations. An estimated 350,000 were employed as machinists and instrument makers; 170,000 as tool and die makers; 50,000 as setup workers and layout artists; and 600,000 as machine tool operators.

Machine tool processing workers manufacture useful metal parts by machining them from metal castings, forgings, bar stock, thick plate, and stampings. Many high-volume parts are made on specialized automatic production machines. Once set up by a skilled setup worker, these machines require only an unskilled operator to load and unload parts. Most parts, however, are made in lot sizes of less than 50. This requires the services of highly skilled machine tool operators, machinists, and tool and die makers. See Fig. 144-1.

Skilled machining workers must be able to read working drawings and use measuring instruments skillfully. Accuracy in workmanship is a requirement since many metal parts must be machined to tolerances of plus or minus 0.002 5 mm (0.000 1″). Some highly skilled toolmakers make special precision gage blocks and other specialized measuring tools which are accurate to plus or minus 0.000 058 mm [0.000 002″ (two microinches)].

Learning Through Apprenticeship

A common method of entering a skilled metal machining career is through serving an apprenticeship. Apprenticeship is a period of formal, on-the-job training during which the worker learns all of the aspects of a chosen trade. The apprentice machinist learns blueprint reading, shop mathematics, machine tool operation, and science related to the trade. Apprentices also learn other related or technical information essential for becoming an efficient worker.

Apprenticeship involves training on the job under the supervision of a skilled worker. It also includes classroom instruction in related trade information by a skilled worker or qualified instructor. It is possible to become an all-around machinist or a tool and die maker, or to prepare to become a setup worker or an instrument maker by serving an apprenticeship. The typical apprenticeship period for becoming a machinist is four years (although some companies have training programs through which they qualify machinists in less than four years). It usually requires a total minimum of 8,000 hours (2,000 hours per year) of supervised training. This usually includes a specified minimum number of hours on each of the basic machine tools and also experience on several specialized machine tools. See Fig. 144-2. In addition, the apprentice usually must

receive a minimum of 144 hours per year of instruction in theory related to the trade. The apprenticeship period for a tool and die maker varies from four to five years.

Apprenticeship training programs may be sponsored independently by large industrial companies. They also may be sponsored by employers in joint cooperation with local, state, and federal **joint apprenticeship committees.** The latter committees cooperate with the Federal Bureau of Apprenticeship and Training, under the Department of Labor. The Federal Bureau of Apprenticeship makes recommendations concerning the training requirements of apprentices in various occupations, including several skilled machining occupations.

Upon completion of a recognized apprenticeship, the apprentice becomes a journeyman and has met minimum qualifications for entrance into the trade. The apprentice usually receives a written document which indicates satisfactory completion of the apprenticeship. This document is recognized by many employers and labor unions as satisfactory qualification for employment in the trade.

To qualify for an apprenticeship in a skilled metal machining occupation, the worker should have better than average mechanical ability. The worker also should have the ability to learn blueprint reading, shop mathematics, basic principles

Fig. 144-2.   Apprentice die maker receiving instructions on safe die-tryout procedures.

(General Motors Corporation)

of machines, and scientific principles related to the machining of metals. Many sponsors of apprenticeship programs require that applicants be high school graduates or possess equivalent education. Graduates of a high school or vocational school, who have a good background of mathematics, science, English, drafting, and machine shop or metalworking, frequently are sought as apprentices to learn the machinist or tool and die making trades.

The apprentice earns while learning the trade. Usually the apprentice is placed on a graduated pay scale with periodic increases throughout the training period. Upon completion of the apprenticeship, journeyman's pay as a skilled worker is received.

The Pick-Up Method of Learning

Most machine tool operators and many machinists, tool and die makers, and instrument makers have entered the machining occupations by the "pick-up" method. They acquired their skills and knowledge by working on different machine tools on several different jobs. Many of these workers also attended vocational and/or technical schools and studied blueprint reading, shop mathematics, and technical theory related to their work in order to improve their qualifications. Nevertheless, it is becoming increasingly difficult to enter the skilled metal machining occupations through this method.

Machinists

Machinists are skilled workers who are able to make all types of machined metal parts with machine tools. They must be able to set up and operate all of the basic machine tools and many different specialized production machine tools. They must know how to read blueprints and how to use all of the common metalworking hand tools and precision measuring instruments, Fig. 144-3. There is much variety in the work, since the machinist frequently is called upon to perform many different types of machining jobs on many different machines.

The machinist must know of the properties and characteristics of the different metals to be machined. The heat-treating properties of many metals must also be known, since these will be

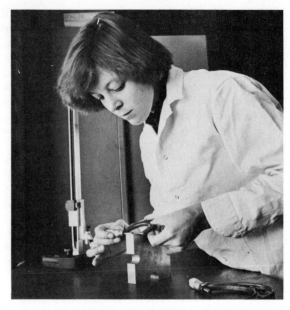

Fig. 144-3.   A machine tool worker uses a precision tool to measure the size of a die.

used in making metal parts and cutting tools which must be heat-treated after machining. Machinists also must be able to select correct cutting speeds and feeds intelligently, and compute them if necessary.

Types of Shops

The machinist may be employed in any one of three types of machine shops — a production shop, a job shop, or a maintenance shop.

A **production shop** usually is a large plant or factory which makes a great number of identical machined parts. Examples of items machined in a production shop are tractor and automotive engine blocks, crankshafts, and plumbing fixtures. See Fig. 144-4.

A **job shop** usually is a smaller machine shop which makes limited numbers of machined parts for manufacturers. The job shop also may build special machines or equipment ranging from small experimental items to very large machines.

The machinist employed in a **maintenance machine shop** must perform all kinds of machine work with many types of machine tools. The work

Fig. 144-4.   Machine operator milling engine blocks in large production machine shop.

involves the repair and maintenance of all types of machines, both machine tools and other manufacturing machinery. In all large factories (both metalworking and nonmetalworking), machinery must be repaired and adjusted by the maintenance machinist. These workers sometimes build new or specialized production machinery needed in a factory. Examples of plants or factories which employ maintenance machinists are farm machine factories, large foundries, electrical appliance factories, chemical plants, paper mills, textile plants, and large food making and packaging plants.

The pay for all-around machinists compares favorably with other highly skilled factory workers. Most companies employing machinists provide paid holidays, paid vacations, and other benefits such as paid life insurance, medical insurance, and retirement pensions.

Machine Tool Operators

The machine tool operator is a semiskilled worker. In 1974 more than 600,000 machine tool operators were employed in metal machining work in the United States. The machine tool operator usually is skilled in the operation of a single machine tool, such as a drill press, lathe, or milling machine, Fig. 144-4. Often the work is repetitive, making many identical parts.

The training period for machine tool operators may vary from several days to more than a year. For example, it may take only a few days to learn the operation of a small drill press. On the other hand, it may take more than a year to become highly skilled on a large radial drill press or universal milling machine. There are no specific educational requirements for the machine tool operator. However, a background of study in drafting, blueprint reading, industrial arts or vocational classes in metalwork is valuable in securing employment and advancing more rapidly. Machine tool operators frequently are paid according to their rating, which depends on training and experience. They often are rated as class A, class B, or class C, with class A being the highest rating and receiving the highest pay.

With continued development of automatic and more versatile machine tools, production per worker will increase. Continued development of automated machine tools also may affect the number employed. However, the machine tool operators with the previously mentioned educational qualifications can easily adjust to technical changes and secure new jobs.

Tool and Die Makers

The tool and die maker is a highly skilled worker who makes many types of cutting tools, dies, punches, and holding devices. These products are used on production machine tools and are the basis for mass production of many products, both metal and nonmetal. The tool and die maker produces dies for stamping sheet metal parts in punch presses, dies for die casting machines which make cast metal parts, and dies for use in plastic molding machines. An example of a die used for stamping automobile radiator mounting brackets from sheet metal is shown in Fig. 144-5. The die is mounted for use in the large press in Fig. 144-2. The tool and die maker makes **jigs** and **fixtures** (holding or guiding devices) for use on production machines and for certain types of rapid assembly work.

The tool and die maker operates all of the basic machine tools. This requires ability to work ac-

(General Motors Corporation)

Fig. 144-5.   Die used to cam pierce holes in automobile radiator mounting bracket.

curately with all metalworking hand tools, and to use all of the basic measuring tools and precision measuring instruments. Also required is a broad knowledge of shop mathematics, blueprint reading, machining operations, and the various metals and their heat-treatment properties.

Approximately 170,000 tool and die makers were employed in the United States in 1974. They are employed in large industrial metalworking plants, smaller tool and die making shops, and in job shops. Many also are employed in industrial maintenance shops where they repair or make dies and fixtures for production machine tools and other machinery.

The best method for learning to become a tool and die maker is by serving an apprenticeship. The training program is sometimes broader and requires more related theory than that required for the machinist.

Frequently, machine tool operators become tool and die makers after many years of experience and study. They acquire extensive training in vocational and/or technical schools, through correspondence study, and through independent study. With the knowledge and ability to perform all of the machining operations necessary for tool and die making, they are employed as tool and die makers. Similarly, some machinists also become tool and die makers.

Because of their extensive skills and the shortage of workers in this field, there is little prob-

ability of unemployment. Tool and die makers are needed in many industries, and their experience usually enables them to transfer for employment as machinists, instrument makers, and other occupations in industry.

Instrument Makers

Instrument makers are highly skilled machine tool workers. They make precision instruments, gages, experimental equipment, nonstandard instruments, and experimental models. They work very closely with scientists and engineers, translating new ideas into working models. Their work is very important because of the increasing use of instruments, gages, and special controls needed with modern production methods.

The skills and knowledge required of the instrument maker are similar to those required of the machinist and tool and die maker, except that more experience and broader training are usually required. Generally, the instrument maker must work to closer tolerances than the average machinist, Fig. 144-6. Instrument makers work with many different materials and precision measuring devices which usually are not required for the machinist or tool maker. They must be more creative, often working from very rough sketches

(United States Steel Corporation)

Fig. 144-6.   Skilled instrument maker machining a typical job to very close tolerances.

rather than from finished blueprints. Usually they perform all of the operations necessary to make the complete instrument.

Instrument workers are employed by industrial research centers, the federal government, and university scientific laboratories. They also are employed in conjunction with many other experimental and developmental laboratories.

Inspectors

The inspector in the production machine shop checks machined parts to determine whether they meet the specifications indicated on the drawing or blueprint. An inspector must be able to read drawings and blueprints and must be able to use many types of precision measuring tools, gages, and instruments. See Fig. 144-7. If the parts do not meet specifications, they will not fit together in the final assembly, or they will not be interchangeable with other similar parts.

The work of the inspector may require varying degrees of skill, depending on the size, type, cost, and tolerances required on the machined workpieces. Some simple inspection jobs require little training, while others require considerable technical knowledge and skill. The inspector's pay depends largely on the training and skill acquired. The range in pay may vary from the pay

Fig. 144-8.   An estimator conducts a cost accounting study in which all costs involved in making a product are determined.

of a semiskilled machine operator to that of a skilled machinist.

There are no specific educational requirements for becoming an inspector. The training may vary from several weeks to several years on the job, depending on the particular job requirements. However, a good knowledge of blueprint reading and basic shop mathematics plus experience in industrial arts, vocational and/or technical metalwork courses are valuable in securing employment and seeking advancement in this field.

Setup Workers

Setup workers are skilled workers employed in large production plants where they "set up" machine tools. The machines then are operated by semiskilled machine tool operators or, in some cases, by unskilled operators. Setup workers frequently specialize in setting up one type of production machine, such as a turret lathe. They may maintain and install the proper cutting tools, fixtures, and gages necessary for machining a particular part. They set the proper cutting speeds and feeds on the machine and make the first several pieces in order to check the tool settings. Then they often must instruct the machine tool operator in the proper methods of checking the parts being machined.

(Caterpillar Tractor Company)

Fig. 144-7.   Inspector using precision measuring tools to check accuracy of machined work.

Some setup workers are highly skilled machinists who can set up many types of production machines or any of the basic machine tools.

The setup worker must work from blueprints and must have a broad knowledge of different metals and their machining properties. The worker must know cutting lubricants, coolants, and cutting speeds and feeds. To qualify for employment, a setup worker must be an all-around machinist or a highly qualified specialist on one of the basic production-type machine tools. Setup positions frequently are filled from within the shop by promotion or reassignment.

Layout Artists

The layout artist makes marks on metal castings, forgings, and other metal workpieces, indicating where and how they are to be machined. The layout artist must be able to read blueprints and analyze the sequence of machine operations. The operation of all the basic and many of the specialized machine tools must be understood. A broad knowledge of the properties of many metals is important. The layout artist is a specialist who is particularly skilled in using all of the various hand layout tools, measuring instruments, and gages.

In order to qualify as a layout artist, a worker must serve an apprenticeship as a machinist or must have equivalent training. In addition, the worker must have considerable broad machining experience in order to understand the various possible sequences of metal machining operations.

There is little probability of unemployment for the layout artist, since the skills enable transfer to other machining work if the demand for layout artists decreases.

N/C Machine Tool Programmers

Numerical control machine tool programmers must be as thoroughly familiar with machining methods and procedures as skilled machine operators. They must be able to read working drawings accurately, select the best tool for each machining operation, and know acceptable cutting speeds and feeds for cutting different metals with different tool materials. This necessitates an educational background which includes some formal training in machine tool processing and N/C programming.

Fig. 144-9. Engineers constantly explore better designs for products. Here, the supervising engineers discuss redesign of a machined part.

Many vocational schools, technical schools, and universities offer programs which provide entry level or better skills for N/C programming positions. Some offer instruction in at least one N/C computer-assisted programming language. With the rapidly growing number of N/C machines being placed in service, the outlook for N/C programmers appears good for the forseeable future. This is particularly true for those with a knowledge of computer-assisted programming.

Related Occupations

Some knowledge and experience with the basic methods, hand tools, and machine tools used in the machining of metals is valuable in several occupations which do not directly involve the machining of metals.

One group of such occupations includes those in which workers manage or are responsible for supervising plant production. Workers such as managers, superintendents, foremen, production or quality control supervisors, estimators, salespeople, teachers, training directors, and technologists of several specialties are professionals or semiprofessionals. They must have extensive knowledge of machines, materials, processes, the jobs of other workers, and marketing.

Another group includes the machine drafting artists, machine designers, tool designers, and mechanical engineers who design and make drawings of machines, machine tools, and other mechanical devices to be constructed. They can understand their work and be more creative when they have a background knowledge of the basic machine operations. See Fig. 144-9.

In a machine shop where basic metalworking hand tools, machine tools, and precision measuring tools are used, the worker learns about the need for accuracy of measurement.

The student who takes a basic course in machine tool processing develops work habits and understandings which are valuable in many mechanical occupations involving the repair and maintenance of various types of mechanical equipment.

Examples of some workers who can profit from such knowledge and experience in basic machine shop work are auto mechanics, aircraft mechanics, diesel engine mechanics and electrical appliance, office machine, and factory maintenance repair personnel.

Test Your Knowledge of Section 19

Unit 144: Looking at Careers in Metalworking

1. What was the approximate number of workers employed in the metal machining occupations in 1974?
2. Most metal parts are made in lot sizes of less than what quantity?
3. What skilled machining occupations may one prepare for by serving an apprenticeship?
4. How long is the typical apprenticeship period for becoming a machinist?
5. Who sponsors or provides apprenticeship programs for training skilled metal machining workers?
6. What is meant by a **journeyman** in a trade?
7. What qualifications must one usually have in order to be employed as an apprentice in a skilled metal machining occupation?
8. Explain what it means to learn a trade through the pick-up method.
9. What types of work must the all-around machinist be able to do?
10. Explain the type of machine work done in job shops, production shops, and maintenance shops.
11. What types of industries employ maintenance machinists?
12. What type of work is performed by a machine tool operator?
13. What are the educational requirements for a machine tool operator?
14. How long is the training period for a machine tool operator?
15. What type of work does a tool and die maker do?
16. What basic knowledge should a tool and die maker possess?
17. How can one learn the tool and die making trade?
18. What type of work is performed by instrument makers?
19. What are the training requirements for becoming an instrument maker?
20. What type of work is performed by a setup worker?
21. What are the educational or training requirements for becoming a setup worker?
22. What type of work is performed by a layout artist?
23. What are the educational and training requirements for becoming a layout artist?
24. What knowledges and skills must N/C machine tool programmers have?
25. What is the employment outlook for a career in N/C programming?
26. What knowledges and skills are required in order to become an inspector in a machine shop?
27. Why is a basic knowledge of metal machining valuable to drafting artists, tool and machine designers, and mechanical engineers?
28. List five nonmachining occupations where some knowledge of the machining of metals would be valuable for success.

Appendix

Table A-1
Job Selector for Friction Cutting

Steels - SAE	Saw Velocity		
	Material Thickness		
	1.6-3mm (1/16"-1/8")	3-6.35 mm (1/8"-1/4")	6.35-12.7 mm (1/4"-1/2")
Carbon steel, 1010-1095	914 (3000)*	1524 (5000)	3658 (12,000)
Manganese steel, T1330-1350	914 (3000)	1524 (5000)	3658 (12,000)
Free machining	914 (3000)	1524 (5000)	3658 (12,000)
Nickel steels	914 (3000)	1828 (6000)	3962 (13,000)
Nickel chromium	914 (3000)	1828 (6000)	3962 (13,000)
Molybdenum steel	914 (3000)	1828 (6000)	3962 (13,000)
Chromium steels	914 (3000)	1828 (6000)	3658 (12,000)
Chromium steels	1524 (5000)	3048 (10,000)	4267 (14,000)
Chromium vanadium, 6115-6195	1524 (5000)	3658 (12,000)	4572 (15,000)
Tungsten steel, 7260-71360	1524 (5000)	3658 (12,000)	4572 (15,000)
N.E. steels, 8024-8949	1524 (5000)	3658 (12,000)	4572 (15,000)
Silicon manganese, 9255-9260	1524 (5000)	3658 (12,000)	4572 (15,000)
Other Steels			
Armor plate	914 (3000)	2743 (9000)	3962 (13,000)
Stainless steel, 18-8	914 (3000)	2743 (9000)	4267 (14,000)
Illium	1219 (4000)	3658 (12,000)	4572 (15,000)
Cast steel	914 (3000)	2743 (9000)	3658 (12,000)
Cast Irons			
Gray cast iron	914 (3000)	1524 (5000)	2134 (7000)
Malleable cast iron	914 (3000)	1524 (5000)	2134 (7000)
Meehanite castings	914 (3000)	1524 (5000)	2134 (7000)
	Saw Pitch		
	1.4 (18) mm (TPI)	1.8 (14) mm (TPI)	2.5 (10) mm (TPI)

*Meters per minute (feet per minute).

Table A-2
Copper and Copper Alloys

Name and Number	Composition	Machinability	Typical Applications
102 Oxygen-free Copper	99.95 Cu	20	Busbars, waveguides. Can be formed by coining, forging, stamping, swaging, spinning.
110 Electrolytic tough pitch copper	99.90 Cu 0.04 O	20	Downspouts, gutters, roofing, nails, gaskets, auto radiators, busbars, roofing, printing rolls, rivets, radio parts.
113, 114, 116 Silver-bearing tough pitch copper	99.90 Cu 0.04 O Ag	20	Gaskets, radiators, busbars, windings, switches, chemical process equipment, clad metals, printed circuit foil.
112 Phosphorous deoxidized copper	99.90 Cu 0.02 P	20	Gas and heater lines, oil burner, plumbing, condenser, dairy, distiller, steam, water, gas tubing.
172 Beryllium copper	99.5 Cu 1.9 Be 0.20 Co	20	Bellows, bourdon tubing, diaphragms, fuse clips, fasteners, lock washers, springs, switch parts, roll pins.
210 Gilding	95.0 Cu 5.0 Zn	20	Coins, medals, bullet jackets, fuse caps, primers, jewelry, base for gold plate.
220 Commercial bronze	90.0 Cu 10.0 Zn	20	Etching bronze, grillwork, screen cloth, weather-stripping, lipstick cases, compacts, marine hardware, screws, rivets.
240 Low brass	80.0 Cu 20.0 Zn	30	Battery caps, bellows, musical instruments, clock dials, pump lines.
260 Cartridge brass	70.0 Cu 30.0 Zn	30	Radiator cores and tanks, flashlight shells, lamp fixtures, fasteners, locks, hinges, ammunition parts, plumbing parts, pins, rivets.
268, 270 Yellow brass	65.0 Cu 35.0 Zn	30	Uses are the same as for 260, except for ammunition.
314 Leaded commercial bronze	89.0 Cu 1.75 Pb 9.25 Zn	80	Screws, machine parts, pickling crates.
360 Free-cutting brass	61.5 Cu 3.0 Pb 35.5 Zn	100	Gears, pinions, automatic high-speed screw machine parts.
464-467 Naval brass	60.0 Cu 39.25 Zn 0.75 Sn	30	Aircraft turnbuckle barrels, bolts, marine hardware, nuts, rivets, propeller shafts, valve stems, condenser plates, welding rods.
554 Free-cutting phosphorous bronze	88.0 Cu 4.0 Pb 4.0 Zn 4.0 Sn	80	Bearings, bushings, gears, shafts, thrust washers, valve parts.

Table A-3
Aluminum and Aluminum Alloys

Alloy	Chief Characteristics	Temper	Machinability	Typical Uses
1100	Excellent forming qualities, resistance to corrosion, weldability, electrical conductivity.	O H14 F	D C C	Dials, name plates, cooking utensils, giftware, reflectors, chemical equipment, tank cars, heat exchangers, pressure vessels, storage tanks.
3003	Similar to 1100 but with slightly higher strength.	O H14 F	D C C	Ductwork, cooking utensils, ice cube trays, garage doors, awning slats, trailer and truck panels, refrigerator panels, gas lines, gasoline tanks, heat exchangers, pressure vessels, storage tanks.
5052	Excellent resistance to corrosion, especially marine environments; good workability, higher strength than 1100 or 3003.	O H32 H34 F	D C C C	Kitchen cabinets, home freezers milk crates, bus and truck bodies, refrigerator trays, aircraft tubing, fencing, fan blades, shoe eyelets.
2011	Good machinability, free-cutting, good mechanical properties	T3	A	Screw-machine products, machine parts, atomizer and hose parts, pipe stems, cigarette holders, tube fittings.
2017	Relatively high strength, fair workability, good machinability.	T4	B	Pulleys, crochet and knitting needles, screw-machine products, coat hangers, gages, tube fittings.
2024	A high-strength material which has largely superseded 2017 for structural applications. Use of 2024-0 is not recommended unless subsequently heat-treated.	O T3 T4	C B B	Aircraft parts, piano hinges, luggage, scientific instruments, ski poles, fastening devices, orthopedic equipment.
Alclad 2024	Combines high strength of 2024 with excellent resistance to corrosion in T3 and T4 temper.	O T3 T4	C B B	Aircraft frames and skins, venetian blind slats, railroad car roofs and sides.
6061	Combines relatively high strength, good workability and high resistance to corrosion.	O T4 T6	D C C	Sailboats, canoes, truck and bus bodies, scaffolding, transmission towers, mine skips, furniture, chemical equipment, awnings, marine equipment, fire ladders.
6063	High resistance to corrosion. Pleasing natural finish greatly enhanced by the Alumilite® process.	T42 T5 T832	C C C	Irrigation pipe, awning support windows, architectural trim, stormsashes, thresholds, stair rails, general utility pipe.
6262	Relatively high strength, good machinability, good corrosion resistance.	T9	B	Marine hardware, hydraulic pistons, valve parts, pipe adapters, hand-rail fittings, candlesticks.
Alclad 7075	Very high strength, excellent resistance to corrosion in the T6 temper.	O T6	C B	Aircraft and highly stressed structural parts.

Table A-4
Recommended Heat Treatments for Various Grades of Steel
(No tempering treatments are given, as these temperatures depend upon the desired hardness.)

AISI or SAE Number	Normalizing Temperature °C	Normalizing Temperature °F	Annealing Temperature °C	Annealing Temperature °F	Hardening Temperature °C	Hardening Temperature °F	Quenching Medium
1030	885/941	1625/1725	829/857	1525/1575	843/871	1550/1600	water or brine
1040	871/927	1600/1700	802/829	1475/1525	816/843	1500/1550	water or brine
1050	843/899	1550/1650	788/816	1450/1500	802/829	1475/1525	water or brine
1060	816/871	1500/1600	774/802	1425/1475	788/816	1450/1500	water or brine
1070	816/871	1500/1600	744/802	1425/1475	788/816	1450/1500	water or brine
1080	802/857	1475/1575	746/774	1375/1425	760/788	1400/1450	water or brine
1090	802/857	1475/1575	746/774	1375/1425	760/788	1400/1450	water or brine
1132	885/941	1625/1725	829/857	1525/1575	843/871	1550/1600	water or brine
1140	871/927	1600/1700	802/829	1475/1525	816/843	1500/1550	water or brine
1151	843/899	1550/1650	788/816	1450/1500	802/829	1475/1525	water or brine
1330	871/927	1600/1700	816/843	1500/1550	829/857	1525/1575	oil or water
1340	843/899	1550/1650	802/829	1475/1525	816/843	1500/1550	oil or water
3140	843/899	1550/1650	802/829	1475/1525	802/829	1475/1525	oil
4028	871/927	1600/1700	829/857	1525/1575	843/871	1550/1600	oil or water
4042	843/899	1550/1650	802/829	1475/1525	816/843	1500/1550	oil
4063	843/899	1550/1650	788/816	1450/1500	802/829	1475/1525	oil
4130	871/927	1600/1700	829/857	1525/1575	843/871	1550/1600	oil or water
4140	871/927	1600/1700	816/843	1500/1550	829/857	1525/1575	oil
4150	871/927	1600/1700	802/829	1475/1525	816/843	1500/1550	oil
4340	871/927	1600/1700	816/843	1500/1550	816/843	1500/1550	oil
5046	871/927	1600/1700	802/829	1475/1525	816/843	1500/1550	oil
5130	871/927	1600/1700	816/843	1500/1550	843/871	1550/1600	oil or water
5145	871/927	1600/1700	802/829	1475/1525	829/857	1525/1575	oil
5160	871/927	1600/1700	788/816	1450/1500	816/843	1500/1550	oil
50100	871/927	1600/1700	760/788	1400/1450	788/816	1450/1500	oil
51100	871/927	1600/1700	760/788	1400/1450	802/829	1475/1525	oil
52100	871/927	1600/1700	760/788	1400/1450	816/843	1500/1550	oil
6150	871/927	1600/1700	816/843	1500/1550	843/871	1550/1600	oil
8630	871/927	1600/1700	816/843	1500/1550	829/857	1525/1575	oil
8645	871/927	1660/1700	802/829	1475/1525	816/843	1500/1550	oil
8660	871/927	1600/1700	788/816	1450/1500	802/829	1475/1525	oil
8735	871/927	1600/1700	816/843	1500/1550	829/857	1525/1575	oil
8742	871/927	1600/1700	816/843	1500/1550	829/857	1525/1575	oil
9260	871/927	1600/1700	816/843	1500/1550	843/871	1550/1600	oil
9840	871/927	1600/1700	802/829	1475/1525	816/843	1500/1550	oil
9850	871/927	1600/1700	802/829	1475/1525	816/843	1500/1550	oil

Table A-5

Recommendations for Carburizing Various Grades of Steel

AISI or SAE Number	Carburizing Temperature °C (°F)	Cooling Method	Reheat °C (°F)	Cooling Medium	Second Reheat °C (°F)	Cooling Medium	Tempering Temperature °C (°F)
1008 to 1024; 1108 to 1119; 1211 to 1215	899/927 (1650/1700)	brine or water					
		water or oil					
		cool slowly	760/788 (1400/1450)	brine or water	760/788 (1400/1450)	brine or water	121/149 (250/300)
		cool slowly	760/788 (1400/1450)	brine or water			
			899/927 (1650/1700)	water or oil			
3310	899/927 (1650/1700)	oil					
		oil					
		cool slowly	746/774 (1375/1425)	oil	746/774 (1375/1425)	oil	121/149 (250/300)
		cool slowly	746/744 (1375/1425)	oil			
			829/857 (1525/1575)	oil			
4012 to 4024; 4118	899/927 (1650/1700)	oil					
		oil					
		cool slowly	774/802 (1425/1475)	oil	774/802 (1425/1475)	oil	121/149 (250/300)
		cool slowly	774/802 (1425/1475)	oil			
			871/899 (1600/1650)	oil			
4419, 4422 and 4427; 8822	899/927 (1650/1700)	oil					
		oil					
		cool slowly	788/816 (1450/1500)	oil	788/816 (1450/1500)	oil	121/149 (250/300)
		cool slowly	788/816 (1450/1500)	oil			
			885/913 (1625/1675)	oil			
4520	899/927 (1650/1700)	oil					
		oil					
		cool slowly	802/829 (1475/1525)	oil	802/829 (1475/1525)	oil	121/149 (250/300)
		cool slowly	802/829 (1475/1525)	oil			
			899/927 (1650/1700)	oil			

AISI or SAE Number	Carburizing Temperature °C (°F)	Cooling Method	Reheat °C (°F)	Cooling Medium	Second Reheat °C (°F)	Cooling Medium	Tempering Temperature °C (°F)
4320; 4615 to 4626; 4718 and 4720	899/927 (1650/1700)	oil oil cool slowly cool slowly	774/802 (1425/1475) 746/774 (1375/1425) 829/857 (1525/1575)	oil oil oil	746/774 (1375/1425)	oil	121/149 (250/300)
4815 to 4820	899/927 (1650/1700)	oil oil cool slowly cool slowly	732/760 (1350/1400) 732/760 (1350/1400) 843/871 (1550/1600)	oil oil oil	732/760 (1350/1400)	oil	121/149 (250/300)
5015; 5115 and 5120	899/927 (1650/1700)	oil oil cool slowly cool slowly	774/802 (1425/1475) 774/802 (1425/1475) 871/899 (1600/1650)	oil oil oil or water	774/802 (1425/1475)	oil	121/149 (250/300)
6118 and 6120	899/927 (1650/1700)	oil oil cool slowly cool slowly	788/816 (1450/1500) 788/816 (1450/1500) 871/899 (1600/1650)	oil oil oil	788/816 (1450/1500)	oil or water	149/204 (300/400)
8115; 8615 to 8622; 8720; 8822; 9310 and 94B17	899/927 (1650/1700)	oil oil cool slowly cool slowly	788/816 (1450/1500) 788/816 (1450/1500) 843/871 (1550/1600)	oil oil oil	788/816 (1450/1500)	oil	121/149 (250/300)

Table A-6

Hardness Numbers for Steel Approximately Equivalent to Rockwell C Scale

Rockwell C-Scale Hardness No.	Diamond Pyramid Hardness No.	Brinell Hardness No. (10-mm Ball, 3000-kg load)			Rockwell Hardness No.			Rockwell Superficial Hardness No. (Superficial Brale Penetrator)			Shore Scleroscope Hardness No.	Tensile Strength (Approx.) 1000 psi
		Standard Ball	Hultgren Ball	Carbide Ball	A Scale (60-kg Brale)	B Scale (100-kg 1/16" Ball)	D Scale (100-kg Brale)	15-N Scale (15 kg)	30-N Scale (30 kg)	45-N Scale (45 kg)		
68	940	...	...	...	85.6	...	76.9	93.2	84.4	75.4	97	...
67	900	...	...	...	85.0	...	76.1	92.9	83.6	74.2	95	...
66	865	...	...	...	84.5	...	75.4	92.5	82.8	73.3	92	...
65	832	...	...	(739)	83.9	...	74.5	92.2	81.9	72.0	91	...
64	800	...	...	(722)	83.4	...	73.8	91.8	81.1	71.0	88	...
63	772	...	...	(705)	82.8	...	73.0	91.4	80.1	69.9	87	...
62	746	...	...	(688)	82.3	...	72.2	91.1	79.3	68.8	85	...
61	720	...	(613)	(670)	81.8	...	71.5	90.7	78.4	67.7	83	...
60	697	...	(599)	(654)	81.2	...	70.7	90.2	77.5	66.6	81	...
59	674	...	...	(634)	80.7	...	69.9	89.8	76.6	65.5	80	326
58	653	...	(587)	615	80.1	...	69.2	89.3	75.7	64.3	78	315
57	633	...	(575)	595	79.6	...	68.5	88.9	74.8	63.2	76	305
56	613	...	(561)	577	79.0	...	67.7	88.3	73.9	62.0	75	295
55	595	...	(546)	560	78.5	...	66.9	87.9	73.0	60.9	74	287
54	577	...	(534)	543	78.0	...	66.1	87.4	72.0	59.8	72	278
53	560	...	(519)	525	77.4	...	65.4	86.9	71.2	58.6	71	269
52	544	(500)	(508)	512	76.8	...	64.6	86.4	70.2	57.4	69	262
51	528	(487)	494	496	76.3	...	63.8	85.9	69.4	56.1	68	253
50	513	(475)	481	481	75.9	...	63.1	85.5	68.5	55.0	67	245
49	498	(464)	469	469	75.2	...	62.1	85.0	67.6	53.8	66	239
48	484	451	455	455	74.7	...	61.4	84.5	66.7	52.5	64	232
47	471	442	443	443	74.1	...	60.8	83.9	65.8	51.4	63	225
46	458	432	432	432	73.6	...	60.0	83.5	64.8	50.3	62	219
45	446	421	421	421	73.1	...	59.2	83.0	64.0	49.0	60	212
44	434	409	409	409	72.5	...	58.5	82.5	63.1	47.8	58	206
43	423	400	400	400	72.0	...	57.7	82.0	62.2	46.7	57	201
42	412	390	390	390	71.5	...	56.9	81.5	61.3	45.5	56	196
41	402	381	381	381	70.9	...	56.2	80.9	60.4	44.3	55	191
40	392	371	371	371	70.4	...	55.4	80.4	59.5	43.1	54	186
39	382	362	362	362	69.9	...	54.6	79.9	58.6	41.9	52	181
38	372	353	353	353	69.4	...	53.8	79.4	57.7	40.8	51	176
37	363	344	344	344	68.9	(109.0)	53.1	78.8	56.8	39.6	50	172
36	354	336	336	336	68.4	(108.5)	52.3	78.3	55.9	38.4	49	168
35	345	327	327	327	67.9	(108.0)	51.5	77.7	55.0	37.2	48	163
34	336	319	319	319	67.4	...	50.8	77.2	54.2	36.1	47	159

(Continued)

Rockwell C-Scale Hardness No.	Diamond Pyramid Hardness No.	Brinell Hardness No. (10-mm Ball, 3000-kg load)			Rockwell Hardness No.			Rockwell Superficial Hardness No. (Superficial Brale Penetrator)			Shore Scleroscope Hardness No.	Tensile Strength (Approx.) 1000 psi
		Standard Ball	Hultgren Ball	Carbide Ball	A Scale (60-kg Brale)	B Scale (100-kg 1/16" Ball)	D Scale (100-kg Brale)	15-N Scale (15 kg)	30-N Scale (30 kg)	45-N Scale (45 kg)		
33	327	311	311	311	66.8	(107.5)	50.0	76.6	53.3	34.9	46	154
32	318	301	301	301	66.3	(107.0)	49.2	76.1	52.1	33.7	44	150
31	310	294	294	294	65.8	(106.0)	48.4	75.6	51.3	32.5	43	146
30	302	286	286	286	65.3	(105.5)	47.7	75.0	50.4	31.3	42	142
29	294	279	279	279	64.7	(104.5)	47.0	74.5	49.5	30.1	41	138
28	286	271	271	271	64.3	(104.0)	46.1	73.9	48.6	28.9	41	134
27	279	264	264	264	63.8	(103.0)	45.2	73.3	47.7	27.8	40	131
26	272	258	258	258	63.3	(102.5)	44.6	72.8	46.8	26.7	38	127
25	266	253	253	253	62.8	(101.5)	43.8	72.2	45.9	25.5	38	124
24	260	247	247	247	62.4	(101.0)	43.1	71.6	45.0	24.3	37	121
23	254	243	243	243	62.0	100.0	42.1	71.0	44.0	23.1	36	118
22	248	237	237	237	61.5	99.0	41.6	70.5	43.2	22.0	35	115
21	243	231	231	231	61.0	98.5	40.9	69.9	42.3	20.7	35	113
20	238	226	226	226	60.5	97.8	40.1	69.4	41.5	19.6	34	110
(18)	230	219	219	219	...	96.7	...	...	...	...	33	106
(16)	222	212	212	212	...	95.5	...	...	...	...	32	102
(14)	213	203	203	203	...	93.9	...	...	...	...	31	98
(12)	204	194	194	194	...	92.3	...	...	...	...	29	94
(10)	196	187	187	187	...	90.7	...	...	...	...	28	90
(8)	188	179	179	179	...	89.5	...	...	...	...	27	87
(6)	180	171	171	171	...	87.1	...	...	...	...	26	84
(4)	173	165	165	165	...	85.5	...	...	...	...	25	80
(2)	166	158	158	158	...	83.5	...	...	...	...	24	77
(0)	160	152	152	152	...	81.7	...	...	...	...	24	75

The values in boldface type correspond to the values in the joint SAE-ASM-ASTM hardness conversions as printed in ASTM E140-65, Table 2. Values in parentheses are beyond normal range and are given for information only. Data from Metals Handbook, 8th Edition, American Society for Metals. (Reprinted with permission.)

Table A-7
ISO Metric Threads

| Nominal Size mm | COARSE SERIES | | | FINE SERIES | |
| | Approximately 75% Thread | | | | |
	Pitch mm	Tap Drill mm	Clearance Drill mm	Pitch mm	Tap Drill mm
1.4	0.3	1.1	1.55	—	—
1.6	0.35	1.25	1.8	—	—
2	0.4	1.6	2.2	—	—
2.5	0.45	2.05	2.6	—	—
3	0.5	2.5	3.2	—	—
4	0.7	3.3	4.2	—	—
5	0.8	4.2	5.2	—	—
6	1.0	5.0	6.2	—	—
8	1.25	6.75	8.2	1	7.0
10	1.5	8.5	10.2	1.25	8.75
12	1.75	10.25	12.2	1.25	10.50
14	2	12.00	14.2	1.5	12.50
16	2	14.00	16.45	1.5	14.50
18	2.5	15.50	18.20	1.5	16.50
20	2.5	17.50	20.50	1.5	18.50
22	2.5	19.50	22.80	1.5	20.50
24	3	21.00	24.60	2	22.00
27	3	24.00	27.95	2	25.00

Table A-8
Coarse Thread Dimensions — Unified and American National Coarse Series, UNC, NC

Size of Thread and Threads Per Inch	Major Diameter in Inches	Pitch Diameter in Inches	Minor Diameter of External Threads in Inches*	Commercial Tap Drill for About 75% Thread	Decimal Equivalent of Tap Drill in Inches
1 × 64	0.0730	0.0629	0.0538	No. 53	0.0595
2 × 56	0.0860	0.0744	0.0641	No. 50	0.0700
3 × 48	0.0990	0.0855	0.0734	No. 47	0.0785
4 × 40	0.1120	0.0958	0.0813	No. 43	0.0890
5 × 40	0.1250	0.1088	0.0943	No. 38	0.1015
6 × 32	0.1380	0.1177	0.0997	No. 36	0.1065
8 × 32	0.1640	0.1437	0.1257	No. 29	0.1360
10 × 24	0.1900	0.1629	0.1389	No. 25	0.1495
12 × 24	0.2160	0.1889	0.1649	No. 16	0.1770
1/4 × 20	0.2500	0.2175	0.1887	No. 7	0.2010
5/16 × 18	0.3125	0.2764	0.2443	F	0.2570
3/8 × 16	0.3750	0.3344	0.2983	5/16	0.3125
7/16 × 14	0.4375	0.3911	0.3499	U	0.3680
1/2 × 13	0.5000	0.4500	0.4056	27/64	0.4219
9/16 × 12	0.5625	0.5084	0.4603	31/64	0.4844
5/8 × 11	0.6250	0.5660	0.5135	17/32	0.5312
3/4 × 10	0.7500	0.6850	0.6273	21/32	0.6562
7/8 × 9	0.8750	0.8028	0.7387	49/64	0.7656
1 × 8	1.0000	0.9188	0.8466	7/8	0.8750
1-1/8 × 7	1.1250	1.0322	0.9497	63/64	0.9844
1-1/4 × 7	1.2500	1.1572	1.0747	1-7/64	1.1094
1-3/8 × 6	1.3750	1.2667	1.1705	1-7/32	1.2187
1-1/2 × 6	1.5000	1.3917	1.2955	1-21/64	1.3281
1-3/4 × 5	1.7500	1.6201	1.5046	1-35/64	1.5469
2 × 4-1/2	2.0000	1.8557	1.7274	1-25/32	1.7812
2-1/4 × 4-1/2	2.2500	2.1057	1.9774	2-1/32	2.0312
2-1/2 × 4	2.5000	2.3376	2.1933	2-1/4	2.2500
2-3/4 × 4	2.7500	2.5876	2.4433	2-1/2	2.5000
3 × 4	3.0000	2.8376	2.6933	2-3/4	2.7500
3-1/4 × 4	3.2500	3.0876	2.9433	3	3.0000
3-1/2 × 4	3.5000	3.3376	3.1933	3-1/4	3.2500
3-3/4 × 4	3.7500	3.5876	3.4433	3-1/2	3.5000
4 × 4	4.0000	3.8376	3.6933	3-3/4	3.7500

*Design form, maximum metal condition. Based on external thread height = 0.61343 Pitch.

Table A-9
Fine Thread Dimensions — Unified and American National Fine Series, UNF, NF

Size of Thread and Threads Per Inch	Major Diameter in Inches	Pitch Diameter in Inches	Minor Diameter of External Threads in Inches*	Commercial Tap Drill for About 75% Thread	Decimal Equivalent of Tap Drill in Inches
0 × 80	0.0600	0.0519	0.0447	3/64	0.0469
1 × 72	0.0730	0.0640	0.0560	No. 53	0.0595
2 × 64	0.0860	0.0759	0.0668	No. 50	0.0700
3 × 56	0.0990	0.0874	0.0771	No. 45	0.0820
4 × 48	0.1120	0.0985	0.0864	No. 42	0.0935
5 × 44	0.1250	0.1102	0.0971	No. 37	0.1040
6 × 40	0.1380	0.1218	0.1073	No. 33	0.1130
8 × 36	0.1640	0.1460	0.1299	No. 29	0.1360
10 × 32	0.1900	0.1697	0.1517	No. 21	0.1590
12 × 28	0.2160	0.1928	0.1722	No. 14	0.1820
1/4 × 28	0.2500	0.2268	0.2062	No. 3	0.2130
5/16 × 24	0.3125	0.2854	0.2614	I	0.2720
3/8 × 24	0.3750	0.3479	0.3239	Q	0.3320
7/16 × 20	0.4375	0.4050	0.3762	25/64	0.3906
1/2 × 20	0.5000	0.4675	0.4387	29/64	0.4531
9/16 × 18	0.5625	0.5264	0.4943	33/64	0.5156
5/8 × 18	0.6250	0.5889	0.5568	37/64	0.5781
3/4 × 16	0.7500	0.7094	0.6733	11/16	0.6875
7/8 × 14	0.8750	0.8286	0.7874	13/16	0.8125
1 × 14	1.0000	0.9536	0.8978	15/16	0.9375
1-1/8 × 12	1.1250	1.0709	1.0228	1-3/64	1.0469
1-1/4 × 12	1.2500	1.1959	1.1478	1-11/64	1.1719
1-3/8 × 12	1.3750	1.3209	1.2728	1-19/64	1.2969
1-1/2 × 12	1.5000	1.4459	1.3978	1-27/64	1.4219

*Design form, maximum metal condition.
Based on external thread height = 0.61343 Pitch

Table A-10
Cutting Speeds for Various Diameters

Meters Per Minute		15.2	18.3	21.3	24.4	27.4	30.5	33.5	36.6	39.6	42.7	45.7	61	94.1
Feet Per Minute		50'	60'	70'	80'	90'	100'	110'	120'	130'	140'	150'	200'	300'
Diameter		Revolutions Per Minute												
mm	inches													
1.6	1/16	3056	3667	4278	4889	5500	6111	6722	7334	7945	8556	9167	12229	18344
3.2	1/8	1528	1833	2139	2445	2750	3056	3361	3667	3973	4278	4584	6115	9172
4.8	3/16	1019	1222	1426	1630	1833	2037	2241	2445	2648	2852	3056	4076	6115
6.4	1/4	764	917	1070	1222	1375	1528	1681	1833	1986	2139	2292	3057	4586
7.9	5/16	611	733	856	978	1100	1222	1345	1467	1589	1711	1833	2446	3669
9.5	3/8	509	611	713	815	917	1019	1120	1222	1324	1426	1528	2038	3057
11.1	7/16	437	524	611	698	786	873	960	1048	1135	1222	1310	1747	2621
12.7	1/2	382	458	535	611	688	764	840	917	993	1070	1146	1529	2293
15.9	5/8	306	367	428	489	550	611	672	733	794	856	917	1223	1834
19.1	3/4	255	306	357	407	458	509	560	611	662	713	764	1019	1529
22.2	7/8	218	262	306	349	393	436	480	524	568	611	655	874	1310
25.4	1	191	229	267	306	344	382	420	458	497	535	573	764	1146
28.6	1-1/8	170	204	238	272	306	340	373	407	441	475	509	679	1019
31.8	1-1/4	153	183	214	244	275	306	336	367	397	428	458	612	918
34.9	1-3/8	139	167	194	222	250	278	306	333	361	389	417	556	834
38.1	1-1/2	127	153	178	204	229	255	280	306	331	357	382	510	765
41.3	1-5/8	117	141	165	188	212	235	259	282	306	329	353	470	705
44.5	1-3/4	109	131	153	175	196	218	240	262	284	306	327	436	654
47.6	1-7/8	102	122	143	163	183	204	224	244	265	285	306	408	612
50.8	2	95	115	134	153	172	191	210	229	248	267	287	382	573
57.2	2-1/4	85	102	119	136	153	170	187	204	221	238	255	340	510
63.5	2-1/2	76	92	107	122	137	153	168	183	199	214	229	306	459
69.9	2-3/4	69	83	97	111	125	139	153	167	181	194	208	278	417
76.2	3	64	76	89	102	115	127	140	153	166	178	191	254	381

Table A-11
Selection Chart for Cutting Fluids

		FERROUS METALS				NON-FERROUS METALS	
	Group:	I	II	III	IV	V	VI
	Machinability:*	Above 70%	50-70%	40-50%	Below 40%	Above 100%	Below 100%
	Materials:	Low-carbon Steels, High-carbon Steels, Malleable Iron, Cast Steel, Stainless Iron	Cast Iron	Ingot Iron, Wrought Iron	Stainless Steels, Tool Steels, High-speed Steels	Aluminum and Alloys, Brasses and Bronzes, Magnesium and Alloys, Zinc	Copper, Nickel, Inconel, Monel
Severity	Type of Machining Operation						
1. (Greatest)	Broaching; internal	Em. Sul.	Sul. Em.	Sul Em.	Sul. Em.	MO. Em.	Sul. ML.
2.	Broaching; surface	Em. Sul.	Em. Sul.	Sul. Em.	Sul. Em.	MO. Em.	Sul. ML.
2.	Threading; pipe	Sul.	Sul. ML.	Sul.	Sul.		Sul. †
3.	Tapping; plain	Sul.	Sul.	Sul.	Sul.	Em. Dry	Sul. ML.
3.	Threading; plain	Sul.	Sul.	Sul.	Sul.	Em. Sul.	Sul. †
4.	Gear shaving	Sul. L.	Sul. L.	Sul. L.	Sul. L.		
4.	Reaming; plain	ML. Sul.	ML. Sul.	ML. Sul.	ML. Sul.	ML. MO. Em.	ML. MO. Sul.
4.	Gear cutting	Sul. ML. Em.	Sul.	Sul.	Sul. ML.		Sul. ML.
5.	Drilling; deep	Em. ML.	Em. Sul.	Em.	Sul.	MO. ML. Em.	Sul. ML.
6.	Milling; plain	Em. ML. Sul.	Em.	Em.	Sul.	Em. MO. Dry	Sul. Em.
6.	Milling; multiple cutter	ML.	Sul.	Sul.	Sul. ML.	Em. MO. Dry	Sul. Em.
7.	Boring; multiple head	Sul. Em.	Sul. HDS	Sul. HDS	Sul. ML.	K. Dry Em.	Sul. Em.
7.	Multiple-spindle automatic screw machines and turret lathes: drilling, forming, turning, reaming, cutting-off, tapping threading	Sul. Em. ML.	Sul. Em. ML.	Sul. Em. ML. HDS	Sul. ML. Em. HDS	Em. Dry ML.	Sul.
8.	High speed, light feed automatic screw machines: drilling, forming, tapping, threading turning, reaming, box milling, cutting off	Sul. Em. ML.	Sul. Em. ML.	Sul Em. ML.	Sul. ML. Em.	Em. Dry ML.	Sul.
9.	Drilling	Em.	Em.	Em.	Em. Sul.	Em. Dry	Em.
9.	Planing, shaping	Em. Sul. ML.	Em. Sul. ML.	Em. Sul. ML.	Em. Sul.	Em. Dry	Em. Sul.
9.	Turning; single point tool, form tools	Sul. ML. ML.	Sul. Em. ML.	Em. Sul. ML.	Em. Sul. ML.	Dry MO. Em.	Em. Dry ML.
10.	Sawing; circular, hack	Sul. ML. Em.	Sul. ML. Em.	Sul. Em. ML.	Sul. ML. Em.	Em.	Em. Sul. ML.
9.	Grinding: 1. plain	Em.	Em.	Em.	Em.	MO. Sul.	Em.
10. (Least)	2. form (thread, etc.)	Sul.	Sul.	Sul.	Sul.		Sul.

*Machinability rating based on 100% for cold drawn Bessemer screw stock (specification B 1112).

†Palm oil is frequently used to thread copper.

Compiled from **Metals Handbook, Machinery's Handbook,** and **AISI Steel Products Manual.**

KEY
K.—Kerosene
L.—Lard Oil
MO.—Mineral oils
ML.—Mineral-lard oils
Sul.—Sulphurized oils, with or without chlorine
Em.—Soluble or emulsifiable oils and compounds
Dry—No cutting fluid needed
HDS—Heavy duty soluble oil

Table A-12
Millimeters to Decimal Inches

mm	Inches	mm	Inches	mm	Inches	mm	Inches
0.01 =	0.00039	0.34 =	0.01339	0.67 =	0.02638	1.0 =	0.03937
0.02 =	0.00079	0.35 =	0.01378	0.68 =	0.02677	2.0 =	0.07874
0.03 =	0.00118	0.36 =	0.01417	0.69 =	0.02717	3.0 =	0.11811
0.04 =	0.00157	0.37 =	0.01457	0.70 =	0.02756	4.0 =	0.15748
0.05 =	0.00197	0.38 =	0.01496	0.71 =	0.02795	5.0 =	0.19685
0.06 =	0.00236	0.39 =	0.01535	0.72 =	0.02835	6.0 =	0.23622
0.07 =	0.00276	0.40 =	0.01575	0.73 =	0.02874	7.0 =	0.27559
0.08 =	0.00315	0.41 =	0.01614	0.74 =	0.02913	8.0 =	0.31496
0.09 =	0.00354	0.42 =	0.01654	0.75 =	0.02953	9.0 =	0.35433
0.10 =	0.00394	0.43 =	0.01693	0.76 =	0.02992	10.0 =	0.39370
0.11 =	0.00433	0.44 =	0.01732	0.77 =	0.03032	11.0 =	0.43307
0.12 =	0.00472	0.45 =	0.01772	0.78 =	0.03071	12.0 =	0.47244
0.13 =	0.00512	0.46 =	0.01811	0.79 =	0.03110	13.0 =	0.51181
0.14 =	0.00551	0.47 =	0.01850	0.80 =	0.03150	14.0 =	0.55118
0.15 =	0.00591	0.48 =	0.01890	0.81 =	0.03189	15.0 =	0.59055
0.16 =	0.00630	0.49 =	0.01929	0.82 =	0.03228	16.0 =	0.62992
0.17 =	0.00669	0.50 =	0.01969	0.83 =	0.03268	17.0 =	0.66929
0.18 =	0.00709	0.51 =	0.02008	0.84 =	0.03307	18.0 =	0.70866
0.19 =	0.00748	0.52 =	0.02047	0.85 =	0.03346	19.0 =	0.74803
0.20 =	0.00787	0.53 =	0.02087	0.86 =	0.03386	20.0 =	0.78740
0.21 =	0.00827	0.54 =	0.02126	0.87 =	0.03425	21.0 =	0.82677
0.22 =	0.00866	0.55 =	0.02165	0.88 =	0.03465	22.0 =	0.86614
0.23 =	0.00906	0.56 =	0.02205	0.89 =	0.03504	23.0 =	0.90551
0.24 =	0.00945	0.57 =	0.02244	0.90 =	0.03543	24.0 =	0.94488
0.25 =	0.00984	0.58 =	0.02283	0.91 =	0.03583	25.0 =	0.98425
0.26 =	0.01024	0.59 =	0.02323	0.92 =	0.03622	26.0 =	1.02362
0.27 =	0.01063	0.60 =	0.02362	0.93 =	0.03661	27.0 =	1.06299
0.28 =	0.01102	0.61 =	0.02402	0.94 =	0.03701	28.0 =	1.10236
0.29 =	0.01142	0.62 =	0.02441	0.95 =	0.03740	29.0 =	1.14173
0.30 =	0.01181	0.63 =	0.02480	0.96 =	0.03780	30.0 =	1.18110
0.31 =	0.01220	0.64 =	0.02520	0.97 =	0.03819	31.0 =	1.22047
0.32 =	0.01260	0.65 =	0.02559	0.98 =	0.03858	32.0 =	1.25984
0.33 =	0.01299	0.66 =	0.02598	0.99 =	0.03898	33.0 =	1.29921

Fractional Inches to Millimeters

Inches	mm	Inches	mm	Inches	mm	Inches	mm
1/64 =	0.397	17/64 =	6.747	33/64 =	13.097	49/64 =	19.447
1/32 =	0.794	9/32 =	7.144	17/32 =	13.494	25/32 =	19.844
3/64 =	1.191	19/64 =	7.541	35/64 =	13.890	51/64 =	20.240
1/16 =	1.587	5/16 =	7.937	9/16 =	14.287	13/16 =	20.637
5/64 =	1.984	21/64 =	8.334	37/64 =	14.684	53/64 =	21.034
3/32 =	2.381	11/32 =	8.731	19/32 =	15.081	27/32 =	21.431
7/64 =	2.778	23/64 =	9.128	39/64 =	15.478	55/64 =	21.828
1/8 =	3.175	3/8 =	9.525	5/8 =	15.875	7/8 =	22.225
9/64 =	3.572	25/64 =	9.922	41/64 =	16.272	57/64 =	22.622
5/32 =	3.969	13/32 =	10.319	21/32 =	16.669	29/32 =	23.019
11/64 =	4.366	27/64 =	10.716	43/64 =	17.065	59/64 =	23.415
3/16 =	4.762	7/16 =	11.113	11/16 =	17.462	15/16 =	23.812
13/64 =	5.159	29/64 =	11.509	45/64 =	17.859	61/64 =	24.209
7/32 =	5.556	15/32 =	11.906	23/32 =	18.256	31/32 =	24.606
15/64 =	5.953	31/64 =	12.303	47/64 =	18.653	63/64 =	25.003
1/4 =	6.350	1/2 =	12.700	3/4 =	19.050	1 =	25.400

Table A-13
Common Conversion
Metric to U.S. — U.S. to Metric

Length

Metric to U.S.	U.S. to Metric
1 millimeter = 0.03937 inch 1 centimeter = 0.3937 inch 1 meter = 39.37 inches 1 meter = 3.2808 feet 1 meter = 1.0936 yards 1 kilometer = 0.62137 mile	1 inch = 25.40 millimeters 1 inch = 2.540 centimeters 1 foot = 30.480 centimeters 1 foot = 0.3048 meter 1 yard = 91.440 centimeters 1 yard = 0.9144 meter 1 mile = 1.609 kilometers

Area

Metric to U.S.	U.S. to Metric
1 sq. millimeter = 0.00155 sq. inch 1 sq. centimeter = 0.1550 sq. inch 1 sq. meter = 10.7640 sq. feet 1 sq. meter = 1.196 sq. yards 1 sq. hectometer = 2.471 acres 1 hectare = 2.471 acres 1 sq. kilometer = 0.386 sq. mile	1 sq. inch = 645.16 sq. millimeters 1 sq. inch = 6.4516 sq. centimeters 1 sq. foot = 929.03 sq. centimeters 1 sq. foot = 0.0929 sq. meter 1 sq. yard = 0.836 sq. meter 1 acre = 0.4047 sq. hectometer 1 acre = 0.4047 hectare 1 sq. mile = 2.59 sq. kilometers

Mass (Weight)

Metric to U.S.	U.S. to Metric
1 gram = 0.03527 ounce 1 kilogram = 2.2046 pounds 1 metric ton = 2 204.6 pounds 1 metric ton = 1.102 tons (short)	1 ounce (dry) = 28.35 grams 1 pound = 0.4536 kilogram 1 short ton (2000 lb.) = 907.2 kilograms 1 short ton (2000 lb.) = 0.9072 metric ton

Volume (Capacity)

Metric to U.S.	U.S. to Metric
1 centiliter = 10 cm^3 = 0.338 fluid ounce 1 deciliter = 100 cm^3 = 0.0528 pint (liq.) 1 liter = 1 dm^3 = 1.0567 quarts (liq.) 1 liter = 1 dm^3 = 0.26417 gallon (liq.)	1 fluid ounce = 2.957 centiliters = 29.57 cm^3 1 pint (liq.) = 4.732 deciliters = 473.2 cm^3 1 quart (liq.) = 0.9463 liter = 0.9463 dm^3 1 gallon (liq.) = 3.7853 liters = 3.7853 dm^3

Table A-14
Decimal Equivalents of Common Fractions

1/64		0.015625	33/64	0.515625
	1/32	0.03125	17/32	0.53125
3/64		0.046875	35/64	0.546875
	1/16	0.0625	9/16	0.5625
5/64		0.078125	37/64	0.578125
	3/32	0.09375	19/32	0.59375
7/64		0.109375	39/64	0.609375
	1/8	0.125	5/8	0.625
9/64		0.140625	41/64	0.640625
	5/32	0.15625	21/32	0.65625
11/64		0.171875	43/64	0.671875
	3/16	0.1875	11/16	0.6875
13/64		0.203125	45/64	0.703125
	7/32	0.21875	23/32	0.71875
15/64		0.234375	47/64	0.734375
	1/4	0.250	3/4	0.750
17/64		0.265625	49/64	0.765625
	9/32	0.28125	25/32	0.78125
19/64		0.296875	51/64	0.796875
	5/16	0.3125	13/16	0.8125
21/64		0.328125	53/64	0.828125
	11/32	0.34375	27/32	0.84375
23/64		0.359375	55/64	0.859375
	3/8	0.375	7/8	0.875
25/64		0.390625	57/64	0.890625
	13/32	0.40625	29/32	0.90625
27/64		0.421875	59/64	0.921875
	7/16	0.4375	15/16	0.9375
29/64		0.453125	61/64	0.953125
	15/32	0.46875	31/32	0.96875
31/64		0.484375	63/64	0.984375
	1/2	0.500	1	1.0000

Table A-15
Four-Place Trigonometric Functions

Angles	Sines Nat.	Log.	Cosines Nat.	Log.	Tangents Nat.	Log.	Cotangents Nat.	Log.	Angles
6°00'	0.1045	9.0192	0.9945	9.9976	0.1051	9.0216	9.5144	0.9784	84°00'
10	0.1074	0311	0.9942	9975	0.1080	0336	9.2553	9664	50
20	0.1103	0426	0.9939	9973	0.1110	0453	9.0098	9547	40
30	0.1132	0539	0.9936	9972	0.1139	0567	8.7769	9433	30
40	0.1161	0648	0.9932	9971	0.1169	0678	8.5555	9322	20
50	0.1190	0755	0.9929	9969	0.1198	0786	8.3450	9214	10
7°00'	0.1219	9.0859	0.9925	9.9968	0.1228	9.0891	8.1443	0.9109	83°00'
10	0.1248	0961	0.9922	9966	0.1257	0995	7.9530	9005	50
20	0.1276	1060	0.9918	9964	0.1287	1096	7.7704	8904	40
30	0.1305	1157	0.9914	9963	0.1317	1194	7.5958	8806	30
40	0.1334	1252	0.9911	9961	0.1346	1291	7.4287	8709	20
50	0.1363	1345	0.9907	9959	0.1376	1385	7.2687	8615	10
8°00'	0.1392	9.1436	0.9903	9.9958	0.1405	9.1478	7.1154	0.8522	82°00'
10	0.1421	1525	0.9899	9956	0.1435	1569	6.9682	8431	50
20	0.1449	1612	0.9894	9954	0.1465	1658	6.8269	8342	40
30	0.1478	1697	0.9890	9952	0.1495	1745	6.6912	8255	30
40	0.1507	1781	0.9886	9950	0.1524	1831	6.5606	8169	20
50	0.1536	1863	0.9881	9948	0.1554	1915	6.4348	8085	10
9°00'	0.1564	9.1943	0.9877	9.9946	0.1584	9.1997	6.3138	0.8003	81°00'
10	0.1593	2022	0.9872	9944	0.1614	2078	6.1970	7922	50
20	0.1622	2100	0.9868	9942	0.1644	2158	6.0844	7842	40
30	0.1650	2176	0.9863	9940	0.1673	2236	5.9758	7764	30
40	0.1679	2251	0.9858	9938	0.1703	2313	5.8708	7687	20
50	0.1708	2324	0.9853	9936	0.1733	2389	5.7694	7611	10
10°00'	0.1736	9.2397	0.9848	9.9934	0.1763	9.2463	5.6713	0.7537	80°00'
10	0.1765	2468	0.9843	9931	0.1793	2536	5.5764	7464	50
20	0.1794	2538	0.9838	9929	0.1823	2609	5.4845	7391	40
30	0.1822	2606	0.9833	9927	0.1853	2680	5.3955	7320	30
40	0.1851	2674	0.9827	9924	0.1883	2750	5.3093	7250	20
50	0.1880	2740	0.9822	9922	0.1914	2819	5.2257	7181	10
11°00'	0.1908	9.2806	0.9816	9.9919	0.1944	9.2887	5.1446	0.7113	79°00'
10	0.1937	2870	0.9811	9917	0.1974	2953	5.0658	7047	50
20	0.1965	2934	0.9805	9914	0.2004	3020	4.9894	6980	40
30	0.1994	2997	0.9799	9912	0.2035	3085	4.9152	6915	30
40	0.2022	3058	0.9793	9909	0.2065	3149	4.8430	6851	20
50	0.2051	3119	0.9787	9907	0.2095	3212	4.7729	6788	10
Angles	Nat. Log. Cosines		Nat. Log. Sines		Nat. Log. Cotangents		Nat. Log. Tangents		Angles

Angles	Sines Nat.	Log.	Cosines Nat.	Log.	Tangents Nat.	Log.	Cotangents Nat.	Log.	Angles
0°00'	0.0000	∞	1.0000	0.0000	0.0000	∞	∞	∞	90°00'
10	0.0029	7.4637	1.0000	0.0000	0.0029	7.4637	343.77	2.5363	50
20	0.0058	7648	1.0000	0.0000	0.0058	7648	171.89	2352	40
30	0.0087	9408	1.0000	0000	0.0087	9409	114.59	0591	30
40	0.0116	8.0658	0.9999	0000	0.0116	8.0658	85.940	1.9342	20
50	0.0145	1627	0.9999	0000	0.0145	1627	68.750	8373	10
1°00'	0.0175	8.2419	0.9998	9.9999	0.0175	8.2419	57.290	1.7581	89°00'
10	0.0204	3088	0.9998	9999	0.0204	3089	49.104	6911	50
20	0.0233	3668	0.9997	9999	0.0233	3669	42.964	6331	40
30	0.0262	4179	0.9997	9999	0.0262	4181	38.188	5819	30
40	0.0291	4637	0.9996	9998	0.0291	4638	34.368	5362	20
50	0.0320	5050	0.9995	9998	0.0320	5053	31.242	4947	10
2°00'	0.0349	8.5428	0.9994	9.9997	0.0349	8.5431	28.636	1.4569	88°00'
10	0.0378	5776	0.9993	9997	0.0378	5779	26.432	4221	50
20	0.0407	6097	0.9992	9996	0.0407	6101	24.542	3899	40
30	0.0436	6397	0.9990	9996	0.0437	6401	22.904	3599	30
40	0.0465	6677	0.9989	9995	0.0466	6682	21.470	3318	20
50	0.0494	6940	0.9988	9995	0.0495	6945	20.206	3055	10
3°00'	0.0523	8.7188	0.9986	9.9994	0.0524	8.7194	19.081	1.2806	87°00'
10	0.0552	7423	0.9985	9993	0.0553	7429	18.075	2571	50
20	0.0581	7645	0.9983	9993	0.0582	7652	17.169	2348	40
30	0.0610	7857	0.9981	9992	0.0612	7685	16.350	2135	30
40	0.0640	8059	0.9980	9991	0.0641	8067	15.605	1933	20
50	0.0669	8251	0.9978	9990	0.0670	8261	14.924	1739	10
4°00'	0.0698	8.8436	0.9976	9.9989	0.0699	8.8446	14.301	1.1554	86°00'
10	0.0727	8613	0.9974	9989	0.0729	8624	13.727	1376	50
20	0.0756	8783	0.9971	9988	0.0758	8795	13.197	1205	40
30	0.0785	8946	0.9969	9987	0.0787	8960	12.706	1040	30
40	0.0814	9104	0.9967	9986	0.0816	9118	12.251	0882	20
50	0.0843	9256	0.9964	9985	0.0846	9272	11.826	0728	10
5°00'	0.0872	8.9403	0.9962	9.9983	0.0875	8.9420	11.430	1.0580	85°00'
10	0.0901	9545	0.9959	9982	0.0904	9563	11.059	0437	50
20	0.0929	9682	0.9957	9981	0.0934	9701	10.712	0299	40
30	0.0958	9816	0.9954	9980	0.0963	9836	10.385	0164	30
40	0.0987	9945	0.9951	9979	0.0992	9966	10.078	0034	20
50	0.1016	9.0070	0.9948	9977	0.1022	9.0093	9.7882	0.9907	10
Angles	Nat. Log. Cosines		Nat. Log. Sines		Nat. Log. Cotangents		Nat. Log. Tangents		Angles

Continued

Table A-15 (Cont.)
Four-Place Trigonometric Functions

Angles	Sines Nat.	Log.	Tangents Nat.	Log.	Cosines Nat.	Log.	Cotangents Nat.	Log.	Angles
18°00'	0.3090	9.4900	0.3249	9.5118	0.9511	9.9782	3.0777	0.4882	72°00'
10	0.3118	4939	0.3281	5161	0.9502	9778	3.0475	4839	50
20	0.3145	4977	0.3314	5203	0.9492	9774	3.0178	4797	40
30	0.3173	5015	0.3346	5245	0.9483	9770	2.9887	4755	30
40	0.3201	5052	0.3378	5287	0.9474	9765	2.9600	4713	20
50	0.3228	5090	0.3411	5329	0.9465	9761	2.9319	4671	10
19°00'	0.3256	9.5126	0.3443	9.5370	0.9455	9.9757	2.9042	0.4630	71°00'
10	0.3283	5163	0.3476	5411	0.9446	9752	2.8770	4589	50
20	0.3311	5199	0.3508	5451	0.9436	9748	2.8502	4549	40
30	0.3338	5235	0.3541	5491	0.9426	9743	2.8239	4509	30
40	0.3365	5270	0.3574	5531	0.9417	9739	2.7980	4469	20
50	0.3393	5306	0.3607	5571	0.9407	9734	2.7725	4429	10
20°00'	0.3420	9.5341	0.3640	9.5611	0.9397	9.9730	2.7475	0.4389	70°00'
10	0.3448	5375	0.3673	5650	0.9387	9725	2.7228	4350	50
20	0.3475	5409	0.3706	5689	0.9377	9721	2.6985	4311	40
30	0.3502	5443	0.3739	5727	0.9367	9716	2.6746	4273	30
40	0.3529	5477	0.3772	5766	0.9356	9711	2.6511	4234	20
50	0.3557	5510	0.3805	5804	0.9346	9706	2.6279	4196	10
21°00'	0.3584	9.5543	0.3839	9.5842	0.9336	9.9702	2.6051	0.4158	69°00'
10	0.3611	5576	0.3872	5879	0.9325	9697	2.5826	4121	50
20	0.3638	5609	0.3906	5917	0.9315	9692	2.5605	4083	40
30	0.3665	5641	0.3939	5954	0.9304	9687	2.5386	4046	30
40	0.3692	5673	0.3973	5991	0.9293	9682	2.5172	4009	20
50	0.3719	5704	0.4006	6028	0.9283	9677	2.4960	3972	10
22°00'	0.3746	9.5736	0.4040	9.6064	0.9272	9.9672	2.4751	0.3936	68°00'
10	0.3773	5767	0.4074	6100	0.9261	9667	2.4545	3900	50
20	0.3800	5798	0.4108	6136	0.9250	9661	2.4342	3864	40
30	0.3827	5828	0.4142	6172	0.9239	9656	2.4142	3828	30
40	0.3854	5859	0.4176	6208	0.9228	9651	2.3945	3792	20
50	0.3881	5889	0.4210	6243	0.9216	9646	2.3750	3757	10
23°00'	0.3907	9.5919	0.4245	9.6279	0.9205	9.9640	2.3559	0.3721	67°00'
10	0.3934	5948	0.4279	6314	0.9194	9635	2.3369	3686	50
20	0.3961	5978	0.4314	6348	0.9182	9629	2.3183	3652	40
30	0.3987	6007	0.4348	6383	0.9171	9624	2.2998	3617	30
40	0.4014	6036	0.4383	6417	0.9159	9618	2.2817	3583	20
50	0.4041	6065	0.4417	6452	0.9147	9613	2.2637	3548	10
Angles	Nat.	Log.	Nat.	Log.	Nat.	Log.	Nat.	Log.	Angles
	Cosines		Cotangents		Sines		Tangents		

Angles	Sines Nat.	Log.	Cosines Nat.	Log.	Tangents Nat.	Log.	Cotangents Nat.	Log.	Angles
12°00'	0.2079	9.3179	0.9781	9.9904	0.2126	9.3275	4.7046	0.6725	78°00'
10	0.2108	3238	0.9775	9901	0.2156	3336	4.6382	6664	50
20	0.2136	3296	0.9769	9899	0.2186	3397	4.5736	6603	40
30	0.2164	3353	0.9763	9896	0.2217	3458	4.5107	6542	30
40	0.2193	3410	0.9757	9893	0.2247	3517	4.4494	6483	20
50	0.2221	3466	0.9750	9890	0.2278	3576	4.3897	6424	10
13°00'	0.2250	9.3521	0.9744	9.9887	0.2309	9.3634	4.3315	0.6366	77°00'
10	0.2278	3575	0.9737	9884	0.2339	3691	4.2747	6309	50
20	0.2306	3629	0.9730	9881	0.2370	3748	4.2193	6252	40
30	0.2334	3682	0.9724	9878	0.2401	3804	4.1653	6196	30
40	0.2363	3734	0.9717	9875	0.2432	3859	4.1126	6141	20
50	0.2391	3786	0.9710	9872	0.2462	3914	4.0611	6086	10
14°00'	0.2419	9.3837	0.9703	9.9869	0.2493	9.3968	4.0108	0.6032	76°00'
10	0.2447	3887	0.9696	9866	0.2524	4021	3.9617	5979	50
20	0.2476	3937	0.9689	9863	0.2555	4074	3.9136	5926	40
30	0.2504	3986	0.9681	9859	0.2586	4127	3.8667	5873	30
40	0.2532	4035	0.9674	9856	0.2617	4178	3.8208	5822	20
50	0.2560	4083	0.9667	9853	0.2648	4230	3.7760	5770	10
15°00'	0.2588	9.4130	0.9659	9.9849	0.2679	9.4281	3.7321	0.5719	75°00'
10	0.2616	4177	0.9652	9846	0.2711	4331	3.6891	5669	50
20	0.2644	4223	0.9644	9843	0.2742	4381	3.6470	5619	40
30	0.2672	4269	0.9636	9839	0.2773	4430	3.6059	5570	30
40	0.2700	4314	0.9628	9836	0.2805	4479	3.5656	5521	20
50	0.2728	4359	0.9621	9832	0.2836	4527	3.5261	5473	10
16°00'	0.2756	9.4403	0.9613	9.9828	0.2867	9.4575	3.4874	0.5425	74°00'
10	0.2784	4447	0.9605	9825	0.2899	4622	3.4495	5378	50
20	0.2812	4491	0.9596	9821	0.2931	4669	3.4124	5331	40
30	0.2840	4533	0.9588	9817	0.2962	4716	3.3759	5284	30
40	0.2868	4576	0.9580	9814	0.2994	4762	3.3402	5238	20
50	0.2896	4618	0.9572	9810	0.3026	4808	3.3052	5192	10
17°00'	0.2924	9.4659	0.9563	9.9806	0.3057	9.4853	3.2709	0.5147	73°00'
10	0.2952	4700	0.9555	9802	0.3089	4898	3.2371	5102	50
20	0.2979	4741	0.9546	9798	0.3121	4943	3.2041	5057	40
30	0.3007	4781	0.9537	9794	0.3153	4987	3.1716	5013	30
40	0.3035	4821	0.9528	9790	0.3185	5031	3.1397	4969	20
50	0.3062	4861	0.9520	9786	0.3217	5075	3.1084	4925	10
Angles	Nat.	Log.	Nat.	Log.	Nat.	Log.	Nat.	Log.	Angles
	Cosines		Sines		Cotangents		Tangents		

Continued

Table A-15 (Cont.)
Four-Place Trigonometric Functions

Angles	Sines Nat.	Sines Log.	Cosines Nat.	Cosines Log.	Tangents Nat.	Tangents Log.	Cotangents Nat.	Cotangents Log.	Angles
30°00'	0.5000	9.6990	0.8660	9.9375	0.5574	9.7614	1.7321	0.2386	60°00'
10	0.5025	7012	0.8646	9368	0.5812	7644	1.7205	2356	50
20	0.5050	7033	0.8631	9361	0.5851	7673	1.7090	2327	40
30	0.5075	7055	0.8616	9353	0.5890	7701	1.6977	2299	30
40	0.5100	7076	0.8601	9346	0.5930	7730	1.6864	2270	20
50	0.5125	7097	0.8587	9338	0.5969	7759	1.6753	2241	10
31°00'	0.5150	9.7118	0.8572	9.9331	0.6009	9.7788	1.6643	0.2212	59°00'
10	0.5175	7139	0.8557	9323	0.6048	7816	1.6534	2184	50
20	0.5200	7160	0.8542	9315	0.6088	7845	1.6426	2155	40
30	0.5225	7181	0.8526	9308	0.6128	7873	1.6319	2127	30
40	0.5250	7201	0.8511	9300	0.6168	7902	1.6212	2098	20
50	0.5275	7222	0.8496	9292	0.6208	7930	1.6107	2070	10
32°00'	0.5299	9.7242	0.8480	9.9284	0.6249	9.7958	1.6003	0.2042	58°00'
10	0.5324	7262	0.8465	9276	0.6289	7986	1.5900	2014	50
20	0.5348	7282	0.8450	9268	0.6330	8014	1.5798	1986	40
30	0.5373	7302	0.8434	9260	0.6371	8042	1.5697	1958	30
40	0.5398	7322	0.8418	9252	0.6412	8070	1.5597	1930	20
50	0.5422	7342	0.8403	9244	0.6453	8097	1.5497	1903	10
33°00'	0.5446	9.7361	0.8387	9.9236	0.6494	9.8125	1.5399	0.1875	57°00'
10	0.5471	7380	0.8371	9228	0.6536	8153	1.5301	1847	50
20	0.5495	7400	0.8355	9219	0.6577	8180	1.5204	1820	40
30	0.5519	7419	0.8339	9211	0.6619	8208	1.5108	1792	30
40	0.5544	7438	0.8323	9203	0.6661	8235	1.5013	1765	20
50	0.5568	7457	0.8307	9194	0.6703	8263	1.4919	1737	10
34°00'	0.5592	9.7476	0.8290	9.9186	0.6745	9.8290	1.4826	0.1710	56°00'
10	0.5616	7494	0.8274	9177	0.6787	8317	1.4733	1683	50
20	0.5640	7513	0.8258	9169	0.6830	8344	1.4641	1656	40
30	0.5664	7531	0.8241	9160	0.6873	8371	1.4550	1629	30
40	0.5688	7550	0.8225	9151	0.6916	8398	1.4460	1602	20
50	0.5712	7568	0.8208	9142	0.6959	8425	1.4370	1575	10
35°00'	0.5736	9.7586	0.8192	9.9134	0.7002	9.8452	1.4281	0.1548	55°00'
10	0.5760	7604	0.8175	9125	0.7046	8479	1.4193	1521	50
20	0.5783	7622	0.8158	9116	0.7089	8506	1.4106	1494	40
30	0.5807	7640	0.8141	9107	0.7133	8533	1.4019	1467	30
40	0.5831	7657	0.8124	9098	0.7177	8559	1.3934	1441	20
50	0.5854	7675	0.8107	9089	0.7221	8586	1.3848	1414	10
Angles	Nat.	Log. Cosines	Nat.	Log. Sines	Nat.	Log. Cotangents	Nat.	Log. Tangents	Angles

Angles	Sines Nat.	Sines Log.	Cosines Nat.	Cosines Log.	Tangents Nat.	Tangents Log.	Cotangents Nat.	Cotangents Log.	Angles
24°00'	0.4067	9.6093	0.9135	9.9607	0.4452	9.6486	2.2460	0.3514	66°00'
10	0.4094	6121	0.9124	9602	0.4487	6520	2.2286	3480	50
20	0.4120	6149	0.9112	9596	0.4522	6553	2.2113	3447	40
30	0.4147	6177	0.9100	9590	0.4557	6587	2.1943	3413	30
40	0.4173	6205	0.9088	9584	0.4592	6620	2.1775	3380	20
50	0.4200	6232	0.9075	9579	0.4628	6654	2.1609	3346	10
25°00'	0.4226	9.6259	0.9063	9.9573	0.4663	9.6687	2.1445	0.3313	65°00'
10	0.4253	6286	0.9051	9567	0.4699	6720	2.1283	3280	50
20	0.4279	6313	0.9038	9561	0.4734	6752	2.1123	3248	40
30	0.4305	6340	0.9026	9555	0.4770	6785	2.0965	3215	30
40	0.4331	6366	0.9013	9549	0.4806	6817	2.0809	3183	20
50	0.4358	6392	0.9001	9543	0.4841	6850	2.0655	3150	10
26°00'	0.4384	9.6418	0.8988	9.9537	0.4877	9.6882	2.0503	0.3118	64°00'
10	0.4410	6444	0.8975	9530	0.4913	6914	2.0353	3086	50
20	0.4436	6470	0.8962	9524	0.4950	6946	2.0204	3054	40
30	0.4462	6495	0.8949	9518	0.4986	6977	2.0057	3023	30
40	0.4488	6521	0.8936	9512	0.5022	7009	1.9912	2991	20
50	0.4514	6546	0.8923	9505	0.5059	7040	1.9768	2960	10
27°00'	0.4540	9.6570	0.8910	9.9499	0.5095	9.7072	1.9626	0.2928	63°00'
10	0.4566	6595	0.8897	9492	0.5132	7103	1.9486	2897	50
20	0.4592	6620	0.8884	9486	0.5169	7134	1.9347	2866	40
30	0.4617	6644	0.8870	9479	0.5206	7165	1.9210	2835	30
40	0.4643	6668	0.8857	9473	0.5243	7196	1.9074	2804	20
50	0.4670	6692	0.8843	9466	0.5280	7226	1.8940	2774	10
28°00'	0.4695	9.6716	0.8829	9.9459	0.5317	9.7257	1.8807	0.2743	62°00'
10	0.4720	6740	0.8816	9453	0.5354	7287	1.8676	2713	50
20	0.4746	6763	0.8802	9446	0.5392	7317	1.8546	2683	40
30	0.4772	6787	0.8788	9439	0.5430	7348	1.8418	2652	30
40	0.4797	6810	0.8774	9432	0.5467	7378	1.8291	2622	20
50	0.4823	6833	0.8760	9425	0.5505	7408	1.8165	2592	10
29°00'	0.4848	9.6856	0.8746	9.9418	0.5543	9.7438	1.8040	0.2562	61°00'
10	0.4874	6878	0.8732	9411	0.5581	7467	1.7917	2533	50
20	0.4899	6901	0.8718	9404	0.5619	7497	1.7796	2503	40
30	0.4924	6923	0.8704	9397	0.5658	7526	1.7675	2474	30
40	0.4950	6946	0.8689	9390	0.5696	7556	1.7556	2444	20
50	0.4975	6968	0.8675	9383	0.5735	7585	1.7437	2415	10
Angles	Nat.	Log. Cosines	Nat.	Log. Sines	Nat.	Log. Cotangents	Nat.	Log. Tangents	Angles

Continued

Table A-15 (Cont.)
Four-Place Trigonometric Functions

Angles	Sines Nat.	Sines Log.	Cosines Nat.	Cosines Log.	Tangents Nat.	Tangents Log.	Cotangents Nat.	Cotangents Log.	Angles
41°00'	0.6561	9.8169	0.7547	9.8778	0.8693	9.9392	1.1504	0.0608	49°00'
10	0.6583	8184	0.7528	8767	0.8744	9417	1.1436	0583	50
20	0.6604	8198	0.7509	8756	0.8796	9443	1.1369	0557	40
30	0.6626	8213	0.7490	8745	0.8847	9468	1.1303	0532	30
40	0.6648	8227	0.7470	8733	0.8899	9494	1.1237	0506	20
50	0.6670	8241	0.7451	8722	0.8952	9519	1.1171	0481	10
42°00'	0.6691	9.8255	0.7431	9.8711	0.9004	9.9544	1.1106	0.0456	48°00'
10	0.6713	8269	0.7412	8699	0.9057	9570	1.1041	0430	50
20	0.6734	8283	0.7392	8688	0.9110	9595	1.0977	0405	40
30	0.6756	8297	0.7373	8676	0.9163	9621	1.0913	0379	30
40	0.6777	8311	0.7353	8665	0.9217	9646	1.0850	0354	20
50	0.6799	8324	0.7333	8653	0.9271	9671	1.0786	0329	10
43°00'	0.6820	9.8338	0.7314	9.8641	0.9325	9.9697	1.0724	0.0303	47°00'
10	0.6841	8351	0.7294	8629	0.9380	9722	1.0661	0278	50
20	0.6862	8365	0.7274	8618	0.9435	9747	1.0599	0253	40
30	0.6884	8378	0.7254	8606	0.9490	9772	1.0538	0228	30
40	0.6905	8391	0.7234	8594	0.9545	9798	1.0477	0202	20
50	0.6926	8405	0.7214	8582	0.9601	9823	1.0416	0177	10
44°00'	0.6947	9.8418	0.7193	9.8569	0.9657	9.9848	1.0355	0.0152	46°00'
10	0.6967	8431	0.7173	8557	0.9713	9874	1.0295	0126	50
20	0.6988	8444	0.7153	8545	0.9770	9899	1.0235	0101	40
30	0.7009	8457	0.7133	8532	0.9827	9924	1.0176	0076	30
40	0.7030	8469	0.7112	8520	0.9884	9949	1.0117	0051	20
50	0.7050	8482	0.7092	8507	0.9942	9975	1.0058	0025	10
45°00'	0.7071	9.8495	0.7071	9.8495	1.0000	0.000	1.0000	0.0000	45°00'
Angles	Nat.	Log. Cosines	Nat.	Log. Sines	Nat.	Log. Cotangents	Nat.	Log. Tangents	Angles

Angles	Sines Nat.	Sines Log.	Cosines Nat.	Cosines Log.	Tangents Nat.	Tangents Log.	Cotangents Nat.	Cotangents Log.	Angles
36°00'	0.5878	9.7692	0.8090	9.9080	0.7265	9.8613	1.3764	0.1387	54°00'
10	0.5901	7710	0.8073	9070	0.7310	8639	1.3680	1361	50
20	0.5925	7727	0.8056	9061	0.7355	8666	1.3597	1334	40
30	0.5948	7744	0.8039	9052	0.7400	8692	1.3514	1308	30
40	0.5972	7761	0.8021	9042	0.7445	8718	1.3432	1282	20
50	0.5995	7778	0.8004	9033	0.7490	8745	1.3351	1255	10
37°00'	0.6018	9.7795	0.7986	9.9023	0.7536	9.8771	1.3270	0.1229	53°00'
10	0.6041	7811	0.7969	9014	0.7581	8797	1.3190	1203	50
20	0.6065	7828	0.7951	9004	0.7627	8824	1.3111	1176	40
30	0.6088	7844	0.7934	8995	0.7673	8850	1.3032	1150	30
40	0.6111	7861	0.7916	8985	0.7720	8876	1.2954	1124	20
50	0.6134	7877	0.7898	8975	0.7766	8902	1.2876	1098	10
38°00'	0.6157	9.7893	0.7880	9.8965	0.7813	9.8928	1.2799	0.1072	52°00'
10	0.6180	7910	0.7862	8955	0.7860	8954	1.2723	1046	50
20	0.6202	7926	0.7844	8945	0.7907	8980	1.2647	1020	40
30	0.6225	7941	0.7826	8935	0.7954	9006	1.2572	0994	30
40	0.6248	7957	0.7808	8925	0.8002	9032	1.2497	0968	20
50	0.6271	7973	0.7790	8915	0.8050	9058	1.2423	0942	10
39°00'	0.6293	9.7989	0.7771	9.8905	0.8098	9.9084	1.2349	0.0916	51°00'
10	0.6316	8004	0.7753	8895	0.8146	9110	1.2276	0890	50
20	0.6338	8020	0.7735	8884	0.8195	9135	1.2203	0865	40
30	0.6361	8035	0.7716	8874	0.8243	9161	1.2131	0839	30
40	0.6383	8050	0.7698	8864	0.8292	9187	1.2059	0813	20
50	0.6406	8066	0.7679	8853	0.8342	9212	1.1988	0788	10
40°00'	0.6428	9.8081	0.7660	9.8843	0.8391	9.9238	1.1918	0.0762	50°00'
10	0.6450	8096	0.7642	8832	0.8441	9264	1.1847	0736	50
20	0.6472	8111	0.7623	8821	0.8491	9289	1.1778	0711	40
30	0.6494	8125	0.7604	8810	0.8541	9315	1.1708	0685	30
40	0.6517	8140	0.7585	8800	0.8591	9341	1.1640	0659	20
50	0.6539	8155	0.7566	8789	0.8642	9366	1.1571	0634	10
Angles	Nat.	Log. Cosines	Nat.	Log. Sines	Nat.	Log. Cotangents	Nat.	Log. Tangents	Angles

Table A-16
Metric Drill Sizes

(See Table A-17 for American drill sizes.)

Dia. mm	Dec. Equiv. Inch	Dia. mm	Dec. Equiv. Inch	Dia. mm	Dec. Equiv. Inch	Dia. mm	Dec. Equiv. Inch
0.15	0.0059	1.20	0.0472	3.75	0.1476	7.20	0.2835
0.16	0.0063	1.25	0.0492	3.80	0.1496	7.25	0.2854
0.17	0.0067	1.30	0.0512	3.90	0.1535	7.30	0.2874
0.18	0.0071	1.35	0.0531	4.00	0.1575	7.40	0.2913
0.19	0.0075	1.40	0.0551	4.10	0.1614	7.50	0.2953
0.20	0.0079	1.45	0.0571	4.20	0.1654	7.60	0.2992
0.21	0.0083	1.50	0.0591	4.25	0.1673	7.70	0.3031
0.22	0.0087	1.55	0.0610	4.30	0.1693	7.75	0.3051
0.23	0.0091	1.60	0.0630	4.40	0.1732	7.80	0.3071
0.24	0.0094	1.65	0.0650	4.50	0.1772	7.90	0.3110
0.25	0.0098	1.70	0.0669	4.60	0.1811	8.00	0.3150
0.26	0.0102	1.75	0.0689	4.70	0.1850	8.10	0.3189
0.27	0.0106	1.80	0.0709	4.75	0.1870	8.20	0.3228
0.28	0.0110	1.85	0.0728	4.80	0.1890	8.25	0.3248
0.29	0.0114	1.90	0.0748	4.90	0.1929	8.30	0.3268
0.30	0.0118	1.95	0.0768	5.00	0.1969	8.40	0.3307
0.32	0.0126	2.00	0.0787	5.10	0.2008	8.50	0.3346
0.34	0.0134	2.05	0.0807	5.20	0.2047	8.60	0.3386
0.35	0.0138	2.10	0.0827	5.25	0.2067	8.70	0.3425
0.36	0.0142	2.15	0.0846	5.30	0.2087	8.75	0.3445
0.38	0.0150	2.20	0.0866	5.40	0.2126	8.80	0.3465
0.40	0.0157	2.25	0.0886	5.50	0.2165	8.90	0.3504
0.42	0.0165	2.30	0.0906	5.60	0.2205	9.00	0.3543
0.44	0.0173	2.35	0.0925	5.70	0.2244	9.10	0.3583
0.45	0.0178	2.40	0.0945	5.75	0.2264	9.20	0.3622
0.46	0.0181	2.45	0.0965	5.80	0.2283	9.25	0.3642
0.48	0.0189	2.50	0.0984	5.90	0.2323	9.30	0.3661
0.50	0.0197	2.60	0.1024	6.00	0.2362	9.40	0.3701
0.55	0.0217	2.70	0.1063	6.10	0.2402	9.50	0.3740
0.60	0.0236	2.75	0.1083	6.20	0.2441	9.60	0.3780
0.65	0.0256	2.80	0.1102	6.25	0.2461	9.70	0.3819
0.70	0.0276	2.90	0.1142	6.30	0.2480	9.75	0.3839
0.75	0.0295	3.00	0.1181	6.40	0.2520	9.80	0.3858
0.80	0.0315	3.10	0.1220	6.50	0.2559	9.90	0.3898
0.85	0.0350	3.20	0.1260	6.60	0.2598	10.00	0.3937
0.90	0.0354	3.25	0.1280	6.70	0.2638	10.50	0.4134
0.95	0.0374	3.30	0.1299	6.75	0.2657	11.00	0.4331
1.00	0.0394	3.40	0.1339	6.80	0.2677	11.50	0.4528
1.05	0.0413	3.50	0.1378	6.90	0.2717	12.00	0.4724
1.10	0.0433	3.60	0.1417	7.00	0.2756	12.50	0.4921
1.15	0.0453	3.70	0.1457	7.10	0.2795		

American Drill Sizes ■ Number, Letter and Fractional

Note that letter drills are larger than number drills, and they begin where number drills end.
(See Table A-16 for Metric drill sizes.)

Number and Letter Drills	Fractional Drills	Decimal Equivalents
80		0.0135
79		0.0145
	1/64	0.0156
78		0.0160
77		0.0180
76		0.0200
75		0.0210
74		0.0225
73		0.0240
72		0.0250
71		0.0260
70		0.0280
69		0.0292
68		0.0310
	1/32	0.0312
67		0.0320
66		0.0330
65		0.0350
64		0.0360
63		0.0370
62		0.0380
61		0.0390
60		0.0400
59		0.0410
58		0.0420
57		0.0430
56		0.0465
	3/64	0.0469
55		0.0520
54		0.0550
53		0.0595
	1/16	0.0625
52		0.0635
51		0.0670
50		0.0700
49		0.0730
48		0.0760
	5/64	0.0781
47		0.0785
46		0.0810
45		0.0820
44		0.0860
43		0.0890
42		0.0935
	3/32	0.0937
41		0.0960
40		0.0980
39		0.0995
38		0.1015
37		0.1040
36		0.1065
	7/64	0.1094
35		0.1100
34		0.1110
33		0.1130
32		0.1160
31		0.1200
	1/8	0.1250
30		0.1285
29		0.1360
28		0.1405
	9/64	0.1406
27		0.1440
26		0.1470
25		0.1495
24		0.1520
23		0.1540
	5/32	0.1562
22		0.1570
21		0.1590
20		0.1610
19		0.1660
18		0.1695
	11/64	0.1719
17		0.1720
16		0.1770
15		0.1800
14		0.1820
13		0.1850
	3/16	0.1875
12		0.1890
11		0.1910
10		0.1935
9		0.1960
8		0.1990
7		0.2010
	13/64	0.2031
6		0.2040
5		0.2055
4		0.2090
3		0.2130
	7/32	0.2187
2		0.2210
1		0.2280
A		0.2340
	15/64	0.2344
B		0.2380
C		0.2420
D		0.2460
E	1/4	0.2500
F		0.2570
G		0.2610
	17/64	0.2656
H		0.2660
I		0.2720
J		0.2770
K		0.2810
	9/32	0.2812
L		0.2900
M		0.2950
	19/64	0.2969
N		0.3020
	5/16	0.3125
O		0.3160
P		0.3230
	21/64	0.3281
Q		0.3320
R		0.3390
	11/32	0.3437
S		0.3480
T		0.3580
	23/64	0.3594
U		0.3680
	3/8	0.3750
V		0.3770
W		0.3860
	25/64	0.3906
X		0.3970
Y		0.4040
	13/32	0.4062
Z		0.4130
	27/64	0.4219
	7/16	0.4375
	29/64	0.4531
	15/32	0.4687
	31/64	0.4844
	1/2	0.5000
	33/64	0.5156
	17/32	0.5312
	35/64	0.5469
	9/16	0.5625
	37/64	0.5781
	19/32	0.5937
	39/64	0.6094
	5/8	0.6250
	41/64	0.6406
	21/32	0.6562
	43/64	0.6719
	11/16	0.6875
	45/64	0.7031
	23/32	0.7187
	47/64	0.7344
	3/4	0.7500
	49/64	0.7656
	25/32	0.7812
	51/64	0.7969
	13/16	0.8125
	53/64	0.8281
	27/32	0.8437
	55/64	0.8594
	7/8	0.8750
	57/64	0.8906
	29/32	0.9062
	59/64	0.9219
	15/16	0.9375
	61/64	0.9531
	31/32	0.9687
	63/64	0.9844
	1	1.0000

Table A-18
Melting Points of Metals and Alloys of Practical Importance

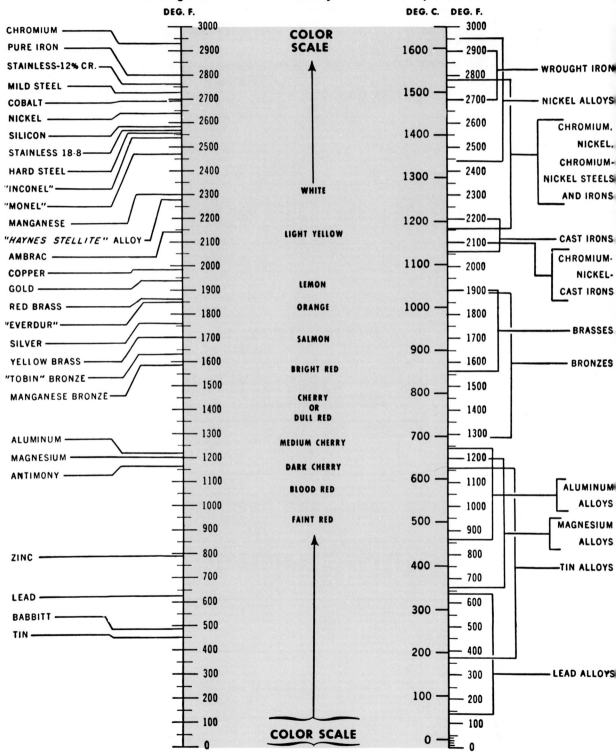

(Linde Co., Div. of Union Carbide Corp.)

Index